Avril Robarts LRC

Liverpool John Moores University

is to be returned

SCIENCE AND TECHNOLOGY OF WOOD

Structure, Properties, Utilization

SCIENCE AND TECHNOLOGY OF WOOD

Structure, Properties, Utilization

George Tsoumis

Emeritus Professor
Aristotelian University
Thessaloniki, Greece

VAN NOSTRAND REINHOLD
New York

Copyright © 1991 by Van Nostrand Reinhold

Library of Congress Catalog Card Number 90-45340
ISBN 0-442-23985-8

All rights reserved. No part of this work covered by the copyright hereon may be reproduced or used in any form or by any means—graphic, electronic, or mechanical, including photocopying, recording, taping, or information storage and retrieval systems—without written permission of the publisher.

Manufactured in the United States of America

Published by Van Nostrand Reinhold
115 Fifth Avenue
New York, New York 10003

Chapman and Hall
2-6 Boundary Row
London, SE1 8HN, England

Thomas Nelson Australia
102 Dodds Street
South Melbourne 3205
Victoria, Australia

Nelson Canada
1120 Birchmount Road
Scarborough, Ontario M1K 5G4, Canada

16 15 14 13 12 11 10 9 8 7 6 5 4 3 2 1

Library of Congress Cataloging-in-Publication Data

Tsoumis, George T.
   Science and technology of wood: structure, properties, utilization / George T. Tsoumis.
     p.   cm.
   Includes index.
   ISBN 0-442-23985-8
    1. Wood.   I. Title.
TA419.T77   1991   90-45340
620.1′2—dc20   CIP

Contents

Prologue vii

Introduction ix

I STRUCTURE
1. Macroscopic Characteristics of Wood 3
2. Physical Characteristics of Wood 9
3. Wood Under the Microscope 14
4. Chemical Composition and Ultrastructure of Wood 34
5. The Mechanism of Wood Formation 57
6. Variation of Wood Structure 66
7. Abnormalities in Wood 84

II PROPERTIES
8. Density and Specific Gravity 111
9. Hygroscopicity 128
10. Shrinkage and Swelling 145
11. Mechanical Properties 160
12. Thermal Properties 194
13. Acoustical Properties 204
14. Electrical Properties 208
15. Degradation of Wood 213

III UTILIZATION
16. Roundwood Products 237
17. Lumber 239
18. Drying 264
19. Preservative Treatment 293
20. Veneer 309
21. Adhesion and Adhesives 327
22. Plywood 339
23. Laminated Wood 351
24. Particleboard 361
25. Fiberboard 388
26. Paper 399

APPENDIXES
I. Other Wood and Forest Products 421
II. Temperate and Tropical Woods 431
III. Bark as a Material 468

Subject Index 483

Species Index 491

Prologue

The field of Science and Technology of Wood is very broad. Separate books exist that cover subjects dealt with in only chapters or parts of this book. For example, there are books dealing with anatomical structure, wood–water relationships, mechanical properties, various industrial products (lumber, plywood, particleboard, etc.), and other aspects of wood (and bark). This book, however, is a concise, comprehensive presentation of the *total* subject.

The contents is presented in four parts, and developed into 29 chapters. *Part I* is a discussion of *wood structure*, dealing with macroscopic, physical and microscopic characteristics, chemical composition and ultrastructure, the mechanism of wood formation by trees, variation of structure, and abnormalities in wood.

Part II deals with *properties:* density (specific gravity), hygroscopicity, shrinkage and swelling (dimensional changes), mechanical, thermal, acoustical, and electrical properties, and degradation of wood by bacteria, fungi, insects, and other destructive agents.

Part III is devoted to *utilization* (i.e., to products made by primary processing of wood)—namely, roundwood products, lumber, veneer, plywood, laminated wood, particleboard, fiberboard, and pulp and paper. Additional chapters refer to drying, preservation, adhesion and adhesives.

An *Appendix* includes: (i) a brief discussion of secondary products made by mechanical processing (furniture, etc.), products of chemical utilization, wood as a source of energy, and other forest products (foliage, pine resin, etc.); (ii) a detailed treatment of specific woods (North American, European, and some important tropical species) with regard to identification, geographical source, properties and uses; and (iii) a discussion of bark (including cork) with regard to structure, properties, and utilization.

Part of this book, mainly on anatomical structure and identification, is a revised excerpt of my earlier book *Wood as Raw Material* (Pergamon Press, 1968/1969), and is included here with permission. The total contents of this book represent the expansion of an article entitled "Wood and Wood Products" ("Wood Production" after the 1985 printing), which I originally wrote for the *Encyclopaedia Britannica*.

The contents of this book are universal in application, but some aspects (e.g., grading specifications and lumber dimensions) present differences in various countries (or regions), and it would not be practical to present such detail in this book. The problem is circumvented by demonstrating, for example, the importance of defects in grading (Tables 11-4 and 17-4, Figure 17-25), or by reference to proposals for international standardization of dimensions (Table 17-1). Literature references are cited in the text, and extensive lists are included at the end of each chapter, allowing the reader to seek further specialized information. Measuring units are international (SI) and English.

This book was originally published in Greek (1983), and this is a version of it. My sincerest thanks are due to colleagues for reviewing the re-worked manuscript and for their kind assistance. I am particularly thankful to Dr. F. F. Wangaard (emeritus head, Department of Forest and Wood Sciences, Colorado State University), who encouraged this work, previewed the total manuscript, and offered many constructive suggestions, especially on properties. Other colleagues who previewed and commented on specific chapters are Drs. J. Bodig and H. Schroeder (Colorado State University), E. Biblis (Auburn University, Alabama), T. Amburgey, D. Nicholas, T. Sellers, F. Taylor, and F. Wagner (Mississippi State University), I. Goldstein (North Carolina State University), B. Thunell (Royal Institute of Technology, Stockholm, Sweden), and G. Stegmann (Fraunhofer Institut für Holzforschung, Braunschweig, Germany). I am indebted to all but, of course, the responsibility for the contents belongs to the author.

Fellow scientists, laboratories, companies making equipment or products, and others have contributed artwork (photographs, drawings), and their contributions are acknowedged where they appear in the book. Acknowledgment is also made to publishers for permission to reproduce copyright material. Most reproduced drawings, especially the graphs, have been redrawn. It should also be noted that any mention of specific industrial companies or trademarks is only an acknowledgment of source and not an endorsement.

Parts of the manuscript have been re-worked during a sabbatical at the Wood Science Laboratory, Colorado State University, and the Forest Products Laboratory, Auburn University, Alabama, and I acknowledge this cooperation on the part of the respective forestry schools.

GEORGE TSOUMIS
Thessaloniki

Introduction

Wood has served man since he appeared on Earth, and has decisively contributed to his survival and to the development of civilization.[1] Moveover, wood continues to be the raw material for a large number of products even in modern times, although other competitive materials (metals, cement, plastics) are available. The value of wood is preserved in many traditional uses, and grows steadily with its use in new products to meet the increasing needs of man (Figures 1 and 2).

After harvesting in the forest, the wood is converted into a great number of products by sawing, slicing, gluing, chipping, pulping, modification by impregnation with chemicals, or chemical processing. In chemical products, the change is so drastic that their wood origin cannot be recognized. Products of primary industrial processing include poles, posts, lumber, laminated wood, veneer, plywood, particleboard, fiberboard, pulp and paper—and, in turn, these are made into products for final use (furniture, etc.). Products of chemical processing are synthetic fibers, photographic films, explosives, chemicals, and many others.

Wood is also an important fuel material for cooking, heating, and production of steam, which may be utilized as a source of energy. About half of the world's production of wood is used as fuel. With the existing energy problems, wood, as a renewable product of nature, is acquiring a greater importance as fuel.

These multiple services are due to certain advantages: wood is aesthetically unrivaled as a material (3), because it is available in a great variety of colors, textures, and figures; it gives a feeling of "warmth" to touch and sight, which is not possessed by competitive materials; it is very strong mechanically in relation to its weight; it is insulating to heat and electricity, exhibits little thermal contraction and expansion, and has good acoustical properties (utilized in making musical instruments); it does not oxidize (rust) and shows considerable resistance to mild concentrations of acids; it may be easily machined with small consumption of energy; nailing or bonding with metal connectors, as well as gluing, is easily achieved; wood is the main source of cellulose, which is the basis of numerous products; it is found in most parts of the world, and is a renewable resource—in contrast to petroleum, metal ores, and coal, which are gradually but steadily exhausted (4, 8, 11) (Figure 3); it is biodegradable; it is a source of energy (i.e., gives heat by direct burning or produces combustible gases).

Wood has disadvantages as well: it is hygroscopic—holds moisture in contact with liquid water or water vapor; the gain or loss of moisture, within certain limits, results in dimensional changes; it is an anisotropic material—presents differential mechanical strength and differential dimensional changes in different structural directions; it may burn and decay; it has variable structure and properties, because it is a product of biological processes—it is produced by many tree species, and its production is influenced by environmental factors and heredity.

As with any other material, sound knowledge of its advantages and disadvantages is prerequisite to rational utilization of wood.[2] Such knowledge allows for improvement of the

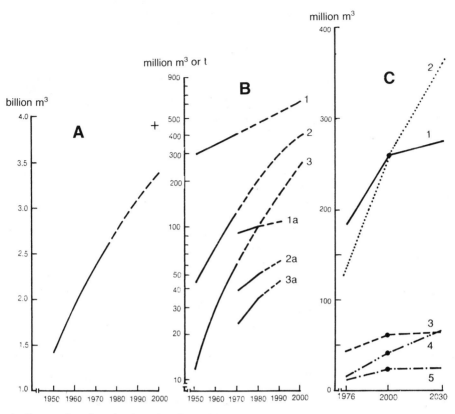

Figure 1. Consumption of wood and wood products with projected demands. (A) World (roundwood). (B) World (1, 2, 3) and Europe (1a, 2a, 3a); 1 and 1a, lumber (m^3); 2 and 2a, paper and paperboard (tons); 3 and 3a, panel products (m^3). (C) U.S.A. (roundwood): 1, sawlogs; 2, pulpwood; 3, veneer logs; 4, fuelwood; 5, other. [Based on data from the following sources: (A) FAO Yearbooks of Forest Products and Ref. 2; + according to Ref. 1; (B) Refs. 6 and 9; (C) Ref. 14.] (1 m^3 = 35.3 ft^3).

quality of wood produced in the forest, better use of the numerous available species, limitation of disadvantages, making products of the best possible quality, and reduction of waste.

The possibility of improved wood quality in the forest is of basic importance and may be realized, within limits, with silvicultural and other measures, such as pruning and spacing of trees, protection from strong winds, microorganisms, and other adverse factors, selection and propagation of genetically superior individuals, and careful harvesting.

With regard to the disadvantages, the following remarks may be made: hygroscopicity and the related dimensional changes may be controlled for practical purposes by proper drying or modification of wood, and thus undesirable effects (checking, warping, etc.) may be avoided; anisotropy is not always a disadvantage (mechanical anisotropy is advantageous in certain types of loading) and, in practice, it may be mitigated in such products as plywood, particleboard, fiberboard, and paper; wood may be protected from fire by impregnation with fire-retardant chemicals and, in the same manner, the durability of wood (i.e., its resistance to insects, fungi, and other destructive agents) may be greatly increased; the variability of wood with regard to anatomical structure and properties, within and between trees and species, may not cause problems if such variability is known and taken into consideration when various products are made. It may be concluded, therefore, that although

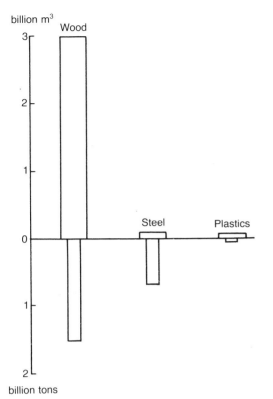

Figure 2. World production of wood, steel, and plastics (1985). (Based on data from FAO and professional societies: schematic presentation after Ref. 10.)

wood has disadvantages, there are possibilities of control in order to ensure the best possible utilization of this precious natural material.

A matter of great concern, both scientific and practical, is that large amounts of wood are usually wasted. It is rather optimistic to say that one third of the volume produced by the trees is finally utilized. Considerable quantities are left in the forest, in the form of logging residues, or form residues during various manufacturing processes. In converting logs to lumber, the residues are on the order of 30–50%. In the pulp and paper industry, 50% or more of the original volume of wood may be wasted. Lignin, a major component of wood, is virtually wasted due to incomplete knowledge of its chemical nature. Problems arise not only from the waste of material, but also from the concurrent pollution of the environment (e.g., water bodies, where residues and chemicals of the pulp and paper industry, if not treated properly, are discharged).

In this area of waste, great efforts are being made and some outstanding successes have been achieved. The best example is the particleboard industry, where residues of sawmills, veneer factories, and logging can be utilized. Current plans include utilizing all the biomass

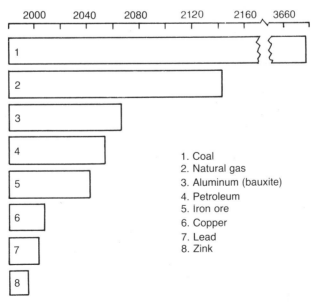

1. Coal
2. Natural gas
3. Aluminum (bauxite)
4. Petroleum
5. Iron ore
6. Copper
7. Lead
8. Zink

Figure 3. Maximum "life expectancy" of world's resources of raw materials; these estimates are being challenged. (Based on data from Ref. 15.)

produced by trees—not only stems, but also stumps, roots, branches, and foliage. Bark is finding various uses. Foliage is used to produce animal food, pharmaceuticals, and other products of mechanical and chemical processing (see Appendix I).[3]

The progress made and the prospects for the future are based on the fact that wood (tree biomass, in general) is the object of great scientific and technological interest, the purpose of which is to acquire a better knowledge of the structure and properties of this material, and to improve its processing methods in industry. These topics are now taught in universities and occupy researchers in most countries. Although the field is relatively new, very important progress has been made, mainly with research conducted in forestry schools and forest-products laboratories.

It may be said that the study of wood started in ancient Greece with Theophrastus (12), a student of Aristotle; however, for a very long period of time only macroscopic and empirical observations could be made. Microscopic structure was studied after the invention of the microscope (Hooke first observed cork cells in 1605), and the first studies of properties were made in the eighteenth century (7). Today, all means of modern research are employed in the study of wood, but progress in the field of utilization is also related to the general advancement of technology.

In order to understand the reason for scientific interest in wood, one must take into consideration that this material is produced and consumed in very large quantities, which are expected to increase in the future[4] (Figures 1 and 2). To meet such demand, it will be necessary to increase wood production from the forests[5] and to improve its utilization.

REFERENCES

1. Commision of the European Communities. 1986. Community action in the forestry sector, Brussels (mimeo.)
2. Costea, C. and D. Vacaroiu. 1979. Model matematic pentru prognoza consumului del lemn. *Bull. Univ. Brasov*, Romania **21**:95-98.
3. Dietz, A. G. H. 1972. Wood in the competition of materials. In *Design and Aesthetics in Wood*, ed, E. A. Anderson and G. F. Earle, pp. 166-174. Syracuse, New York, Syracuse University Press.
4. Goeller, H. E., and A. Zucker. 1984. Infinite resources: The Ultimate strategy. *Science* **223**:456-462.
5. Keays, J. L., and J. V. Hatton. 1976. The implication of full-forest utilization on worldwide supplies of wood by year 2000. *World Wood* **17**(1):12-15.
6. Madas, A. 1974. *World Consumption of Wood*. Budapest: Akademiae Kiadó.
7. Mantel, K. 1964. History of the international science of forestry with special consideration of Central Europe. In *International Review of Forestry Research*, ed. J. Romberger and P. Mikola, pp. 1-37. New York: Academic Press.
8. Meadows, D. L. 1972. *The Limits of Growth*. New York: Universal Books.
9. Peck, T. J. 1981. Trends and prospects for forest products in Europe. *J. Inst. Wood Sci.* **9**(1):2-17.
10. Schulz, H. 1974. Unsere Enkel und ihr Wald. *Holz Roh- Werkstoff* **32**:205-211.
11. Simon, J. L., and H. Kahn. 1984. *The Resourceful Earth: A Response to Global 2000*. Oxford; Blackwell.
12. Theophrastus. *Enquiry into Plants*. English translation by A. Hort. Cambridge, Massachusetts: Harvard University Press, 1916.
13. Tsoumis, G. 1985. The depletion of forests in the Mediterranean region: A historical review from ancient times to the present. *Annals Dept. Forestry and Natural Environ*. Aristotelian University, KH:265-301 (Greek, English).
14. U.S. Department of Agriculture, Forest Service. 1982. An analysis of the timber situation in the United States 1952-2030. Forest Resource Report No. 23.
15. U.S. Government Printing Office. 1980. *The Global Report 2000*. Report to the President. Washington D.C.

FOOTNOTES

[1]In ancient Greek literature and the Bible (Greek text), the word for *material* (hyle) is also used to mean *wood* (tree and forest).

[2]The main competitive materials possess the following advantages and disadvantages (3): *Concrete*: does not burn; is not attacked by insects and decay fungi; has low strength—needs reinforcement; is heavy in comparison to wood (sp. gr. 2.5); hygroscopic; may be easily shaped. *Steel*: high strength; does not burn but structures may succumb under high temperatures of fire; heavy (sp. gr. 8); may oxidize (rust); stiff (difficult to bend); noninsulating to heat and electricity. *Aluminum*: heavy (sp. gr. 2.8); stiff; noninsulating; expensive. *Plastics*: low strength (high if reinforced); light in comparison to steel and aluminum but heavier than most woods (sp. gr. 0.9-1.4); expensive; uncertain durability (new materials); may be destroyed by fire or high temperatures.

[3] The following remarks should be noted: (a) As proved by the example of utilization of residues in the particleboard industry, such residues are not "waste." However, prerequisites to residue utilization are such factors as quantity available, cost of collection and transportation, and "quality" (dimensions, presence of bark, etc.). For example, small quantities of residues from small sawmills or workshops cannot be utilized if far from a particleboard plant and the cost of transportation is high. In that case, burning for production of steam is a form of residue utilization. (b) Utilization of the whole biomass of a tree has adverse silvicultural implications (site deterioration due to removal of nutrients), and this constitutes a limitation to such utilization.

[4] The current world production (and consumption) of wood exceeds 3 billion cubic meters (about 100 billion ft^3). It has been estimated that by the year 2000, the needs of consumption will be about 4 billion m^3 (see Figure 1), and there will be a deficit of 200 million m^3 if the present "waste" of wood continues, in the sense that only part of the biomass produced in the forests is finally utilized. Full utilization may triple the production of wood fiber (for pulp, fiberboard, paper, etc.) from the same trees (5).

[5] Although wood consumption is increasing, the forests are being destroyed at a fast pace (13), especially in the tropics. It has been estimated that by the year 2000 40% of the forests of the developing countries will disappear.

SCIENCE AND TECHNOLOGY OF WOOD

Structure, Properties, Utilization

I

STRUCTURE

Structure is the architectural organization of wood—that is, the nature and arrangement of its physical (macroscopic, microscopic, ultramicroscopic) and chemical building components. In addition to biological interest, the knowledge of structure is of multiple, practical importance: it explains the behavior of wood as a material, due to the close relationship of structure, properties, and utilization; it is helpful in identification and selection of various woods—a service of interest in commerce, industry, engineering, architecture, etc.; it is useful in understanding the mechanism of tree growth and, therefore, it can contribute to improvement in the quality of wood produced in a forest.

1

Macroscopic Characteristics of Wood

Macroscopic characteristics are those features visible with the naked eye, or with a hand lens capable of magnifying 2–3 times. The macroscopic appearance of wood varies according to the plane at which it is sectioned and viewed—in relation to the longitudinal axis of a tree.

CHARACTERISTICS OF A TRANSVERSE SURFACE

A transverse or cross section of a stem is normally circular (Figure 1-1). Three parts may be distinguished: *pith*, *wood*, and *bark*. Between wood and bark there is a tissue visible only with a microscope. This is the *cambium*, which produces wood and bark.

Pith is normally at the center of the stem. It may vary in size from very small and barely visible with the naked eye in certain species to large and conspicuous in others, such as elder (*Sambucus*) and tree-of-heaven (*Ailanthus*). In softwood, pith is fairly uniform, but in hardwoods[1] its shape, color, and structure vary. It is star-shaped in oak; triangular in beech, birch, and alder; ellipsoid in basswood, ash, and maple; circular in walnut, elm, willow, and dogwood (*Cornus*); and squarish in teak (*Tectona grandis*) (8). Pith may vary in color from black to whitish, and its structure (best seen in longitudinal sections) may be continuous (solid), spongy (porous), chambered, or hollow (7).

Wood is characterized by the presence of more or less conspicuous concentric layers, known as *growth rings* or *annual rings*. This pattern is due to the mechanism of tree growth which takes place by superposition of structurally different conoid layers (Figure 1-2). In the temperate zones there is, as a rule, one such wood layer (and one bark layer) added during each season of growth. It is preferred, however, to call these layers *growth rings* rather than annual rings, because there are abnormal cases in which more than one such layer may be formed during a year, whereas in other cases certain rings may be locally discontinuous.

In tropical species, growth rings are not always distinct. When distinct, they correspond to alternating wet and dry periods (i.e., seasonability in rainfall or flooding), but in areas with fairly uniform rainfall ("everwet" rain forests), the reasons for delineation of growth rings are not well understood (1, 3).[2]

In most species, growth rings can be easily distinguished from one another because of differences between *earlywood* and *latewood*. These tissues, known also as springwood and summerwood, respectively, may differ in density, color, and other structural features which reflect their cellular (microscopic) structure. In softwoods, earlywood and latewood differ in density and color; latewood is darker in color and of higher density (Figure 1-3). In hardwoods, structural features are more helpful. Characteristic of this category is the presence of pores—small roundish openings within the growth ring. These pores may sometimes be seen with the naked eye, and are always visible with a hand lens.

On the basis of the distinction of pores within a growth ring, hardwoods are classified into two large groups: *ring-porous*, such as oak and

Figure 1-1. A transverse section of a stem of oak showing *pith* (center), *heartwood* (dark-colored inner portion), *sapwood* (light-colored outer zone), *growth rings* (with light-colored *earlywood* and dark-colored *latewood*), pronounced *rays* radiating from the pith, *inner bark* (forming a complete, light-colored circle outside sapwood), and *outer bark* (dark-colored and broken in parts.)

chestnut, in which the pores of earlywood are conspicuously larger and arranged in a ring (around the pith); and *diffuse-porous*, such as beech and poplar, in which the pores are fairly uniform in size and are scattered (Figure 1-4). Growth rings are generally more distinct in ring-porous than in diffuse-porous woods. In diffuse-porous woods, distinction is often possible due to some reduction of pore size by the end of the growing season; presence of visible marks that reflect cellular differences at the ring boundaries are also helpful. Some species, such as walnut, are not clearly ring-porous or diffuse-porous. These, called *semi-ring-porous* or *semi-diffuse-porous*, exhibit a gradual transition of pore size from earlywood to latewood, resembling more *diffuse-porous* than ring-porous woods.

The number of growth rings, as counted on a cross section near the ground, may be used to find the age of a tree. A number of years should be added, however, to compensate for the time taken by the young plant to reach that height; this depends on the rate of height growth of the particular species and of the individual tree. Correct determination should also account for the possible presence of false or discontinuous rings.

Growth rings may be narrow or relatively wide. Differences exist within a tree, between trees, and between species. Some species may be fast-growing (due to genetic constitution),

Figure 1-2. Diagrammatic representation of the growth of a tree by deposition of successive growth layers in the trunk and a branch. The layers are shown in axial (longitudinal) and transverse sections (----------, pith).

while others will grow more slowly under the same conditions. In very old trees, the outer rings tend to become very narrow. In general, the width as well as the pattern of variation of successive rings are largely influenced by growth conditions; the availability of space both above and below ground is important.[3]

In many species, the cross section of a stem does not have a uniform color, but the inner portion is darker than the peripheral; these portions are called *heartwood* and *sapwood*, respectively. Heartwood of darker color occurs in pine, Douglas-fir, redwood, hemlock, cypress (*Cupressus*), oak, walnut, chestnut, black locust, elm, hickory, and others. Several species exhibit no macroscopically apparent difference between heartwood and sapwood,[4] although heartwood is present in all trees after a certain age, irrespective of the existence or absence of a macroscopic color differentiation. The distinction of heartwood and sapwood is a functional one. As the diameter of the main stem (and of the branches and roots) increases with growth, the older growth rings gradually stop participating in the life processes of the tree. They no longer take part in translocation and storage of food, but provide only mechanical support. This functional change is associated with physiological, structural, and chemical changes. Because of the latter, it is possible by use of chemical methods to produce color differences between heartwood and sapwood and, therefore, bring about their distinction in certain species in which color differences do not occur naturally (9).[5]

Heartwood starts forming in older growth rings near the pith; therefore, its diameter normally decreases from the bottom of the tree upward. The relative amounts of heartwood and sapwood within a tree differ according to species, age, and environment of growth.

An additional macroscopic feature of certain softwoods is the presence of *resin canals*. These normally occur in wood of pine, larch, spruce and Douglas-fir. They are generally more numerous and relatively larger in pine species. Resin canals appear to the naked eye or under a hand lens as small, dark or whitish dots (Figure 1-4A). In hardwoods, *gum canals* are respective features, but are seldom found in species of the temperate zone.

Finally, all woods possess *rays* which, on a cross section, appear as lines extending in the general direction from pith to bark. In some hardwoods, such as oak, beech, and sycamore, the rays are wide and very conspicuous. In others (and in all softwoods), they are more or less fine and sometimes difficult to distinguish—even with a hand lens. Careful observation reveals that all rays do not start from the pith. They may start within any one growth ring but, once started, they usually continue toward the bark, which they also enter.

Surrounding the central cylinder of wood is the *bark*. The macroscopic appearance of this tissue differs according to species and age. In

6   I / STRUCTURE

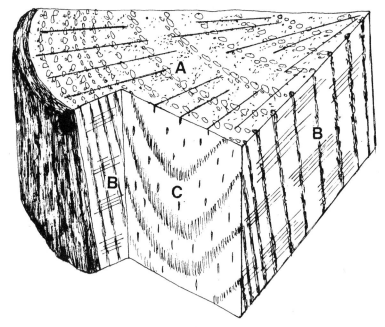

Figure 1-3. Schematic representation of a sector of wood (ring-porous hardwood) showing (A) transverse, (B) radial, and (C) tangential surfaces. The different configuration of growth rings, earlywood and latewood, and rays is shown on all surfaces.

some very old trees, such as redwood and Douglas-fir of the U.S. west coast, bark may exceed 30 cm in thickness. Growth layers in the bark are not macroscopically demarcated, as are the growth rings in wood (see Appendix). However, in older trees, two portions of bark may be recognized; *inner bark* (relatively light in color, narrow, and moist) and *outer bark* (dark, dry, and corky). The outer layers of inner bark are gradually changing into outer

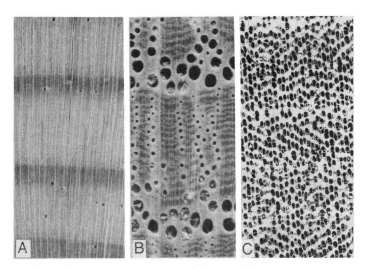

Figure 1-4. Growth rings in transverse section: A, softwood with resin canals (pine); B, ring-porous hardwood (red oak); C, diffuse-porous hardwood (a tropical wood, Afrormosia, without distinct growth rings).

bark in a process that may be considered analogous to heartwood formation. On the other hand, the outer layers of outer bark are gradually falling off. The external appearance of outer bark, as seen on standing trees or logs, is usually characteristic of the species and, therefore, of diagnostic value in dendrology (7).

CHARACTERISTICS OF RADIAL AND TANGENTIAL SURFACES

Radial and tangential sections produce surfaces that are characteristically different from each other, and from cross-sectional surfaces (Figure 1-3). A radial surface is produced by sectioning a stem through its pith. The various features described above—pith, growth rings, earlywood and latewood, heartwood and sapwood, inner and outer bark—appear as longitudinal strips, but some (pith, sapwood or heartwood, bark) may not be represented in a given sample, depending on its location. Resin canals (and gum canals when present) and the larger pores of hardwoods show as fine longitudinal lines or indentations of different color. The rays run crosswise. In woods possessing wide rays, these appear as large, conspicuous flecks; in the case of oak, a characteristic decorative figure, known as silver grain, is thus produced (see Figure 2-1A).

A distinctly different picture is presented by sectioning wood in a tangent to the growth rings. The tangential surface has a more or less pronounced wavy appearance, depending on the contrast between earlywood and latewood. The pith is not exposed, but all other macroscopic features may be represented according to the level of sectioning in relation to the pith. The rays, cut transversely, appear as longitudinal, often spindle-shaped lines of varying length and width; accordingly, they may be conspicuous with the naked eye or difficult to see even with a hand lens.

The macroscopic characteristics of wood, as described above, presume normal tree growth. Presence of various defects may greatly modify the appearance, particularly in the radial and tangential surfaces. Such surfaces may also be modified intentionally by sectioning not at truly radial or tangential planes. Practical use of this possibility is made in the manufacture of decorative veneer (see Chapter 20).

REFERENCES

1. Alvim De, T. P. 1964. Tree growth periodicity in tropical climates. In *The Formation of Wood in Forest Trees*, ed. M. H. Zimmermann, pp. 479-495. New York: Academic Press.
2. Baas, P., and R. E. Vetter, eds. 1989. Growth Rings in Tropical Trees. *IAWA Bull*. n.s. 10(2):95-174.
3. Bornmann, F. H., and G. Berlyn, eds. 1981. *Age and Growth Rate in Tropical Trees: New Directions for Research*, New Haven: Yale University, Bull. No. 94.
4. Douglass, A. E. 1937. Tree rings and climate. *Phys. Sci. Bull*. No. 1, Univ. Arizona.
5. Fletcher, J. ed. 1978. *Dendrochronology in Europe*. Oxford: BAR Intern. Series 51.
6. Fritts, H. C. 1978. *Tree Rings and Climate*. New York: Academic Press.
7. Harlow, W. M., and E. S. Harrar. 1958. *Textbook of Dendrology*. New York: McGraw-Hill.
8. Jane, F. W. 1970. *The Structure of Wood*, London: A&C Black.
9. Kutscha, N. P., and I. B. Sachs. 1962. Color tests for differentiating heartwood and sapwood in certain softwood tree species. U.S. For. Prod. Lab. Report No. 2246.
10. Martin de P. 1974. *Analyse des Cernes (Dendrochronologie et Dendroclimatologie)*. Paris: Masson et Cie.
11. Yang, K. C. 1987. Growth ring enhancement and the differentiation of sapwood and heartwood zones. *Wood Fiber Sci*. 19(4):339-342.

FOOTNOTES

[1] Softwoods are produced by conifers and hardwoods by broad-leaved species. Botanically, conifers belong to *Gymnosperms* and broad-leaved species to *Angiosperms* (*Dicotyledons*). The structure of wood is different between these two classes of plants, as explained mainly in Chapter 3. Note that the terms "softwood" and "hardwood" do not always correspond to respective wood hardness; some hardwoods (willow, poplar, basswood, Balsa) are softer than some softwoods (hard pines, Douglas-fir, Mediterranean cypress, yew).

[2] Delineation of growth rings is necessary in order to determine age and growth rate of trees. Microscopic observation may reveal boundaries of successive growth, although more sophisticated techniques have been used, such as artificial induction of growth boundaries, radiocarbon dating, and x-ray densitometry (2).

[3] In certain regions, the variability of ring width (and ring structure) is closely related to environmental conditions,

such as height of rainfall and air temperature. Thus, it is possible to study the climate of ancient times and to date wooden archaeological findings. These relationships work better in places where one factor plays a decisive role in the growth of trees (rain in dry regions, warm weather in cold regions); however, modern developments, including the use of computers, have extended their application. The fields of science that deal with these subjects are called *Dendrochronology* and *Dendroclimatology*. Dating has been applied to about 6000 B.C., and is considered so accurate as to be used to check radiocarbon dating (4, 5, 6, 10).

[4] Additional information regarding species with and without colored heartwood is given in the Appendix (II).

[5] The growth rings and boundary between sapwood and heartwood may be enhanced, in certain species, by smoothing and wetting the cross-sectional surface of a dried sample, then taking its image with a paper-copying machine (the effect has been attributed to differential permeability and capillarity of cell elements) (11).

2

Physical Characteristics of Wood

In addition to the macroscopic characteristics discussed in Chapter 1, there are a number of other physical characteristics, such as color, luster, odor, taste, texture, grain, and figure, which are likewise useful in describing a piece of wood in macroscopic terms. Weight and hardness may also be included here, as far as they can be judged by simple means, such as lifting by hand in the case of weight, or pressing with the thumbnail in the case of hardness.

COLOR

Wood comes in a variety of natural colors which may range from almost white, as in the sapwood of many species, to the jet black of the heartwood of black ebony (*Diospyros ebenum*). Color differences may exist, however, in a single sample of wood, as between sapwood and heartwood, earlywood and latewood, or between ray tissue and the surrounding wood. Heartwood presents a wide variation of colors, predominantly browns of various shades. Sapwood is always lighter.

Color characterizes various species, but is a feature difficult to describe with words. Also, it may vary within a species, and is subject to change due to exposure or treatment. Usually, color is estimated visually, but it may also be measured by technical means (5, 9, 14).

The color of wood exposed to the atmosphere frequently darkens; sapwood usually darkens more than heartwood. Such changes are generally chemical in nature: they result from oxidation of organic compounds contained in wood. Change of color may take place soon after felling trees in the forest, or after sawing green logs to lumber. The wood of alder quickly changes from whitish to reddish, then fades to a pale brown. The heartwood of black locust turns from light green to dark brown. The heartwood of mahogany (*Swietenia macrophylla*) changes, with time and the action of light, from reddish-white to reddish-brown. Douglas-fir becomes reddish. Long exposure of light-colored woods to the sun, especially in high elevations, changes their color to brown; whereas long exposure to rain or high humidity changes them to dark gray.

Color is imparted to wood by extraneous materials (tannins, etc.) called *extractives* (see Chapter 4). In addition to previously mentioned factors, extractives are influenced by growth conditions. For example, color differences in walnut wood from different geographical locations are attributed to soil characteristics (11). Chemical differences of extractives make it possible to distinguish between wood species, or bring coloration to noncolored heartwood by application of chemicals (7, 10). Some woods, such as black locust, honeylocust, and several tropical species (1, 4, 12) are fluorescent[1] due to their extractives.

Natural coloration in many woods is very attractive and may be preserved with transparent finishes, or it may be artificially changed by dyeing or bleaching. Color changes may be also produced by the action of water or steam. Oak becomes almost black after prolonged storage under water. European beech is often steamed to darken its color and make it more desirable for the furniture market. Walnut and sweetgum are also treated with steam to darken their sapwood so that it is better matched with heartwood (3). In inexpensive furniture, the natural color of various woods is often imitated by dyeing less expensive species.

Irregular deposition of coloring materials may cause local variation in color. Color changes may also result from fungal and bacterial attack, or from other reasons (see Chapter 15).

LUSTER

Some woods possess a natural luster, which may be distinguished from artificial luster (applied by polishing) in that the former has depth while the latter is superficial. As a rule, woods exhibit more luster on radial surfaces due to exposure of rays. Germans have named the radial surface *Spiegelschnitt* (i.e., mirror-section). Luster is also affected by the angle of light reflection.

Among lustrous woods are spruce, ash, sycamore, basswood, and poplar. On the contrary, the surfaces of other woods feel greasy, as for example baldcypress, olive (*Olea europaea*), teak (*Tectona grandis*), and lignum vitae (*Guaiacum officinale*).

ODOR

Odor in woods is due to volatile extraneous materials. Such materials, when present, are mostly deposited in heartwood, where odor is therefore more pronounced. Due to the volatility of these materials, odor gradually fades upon exposure. For this reason, it is more prevalent in freshly exposed surfaces.

Like color, odor is not an easy feature to describe. Cedars and cypress (*Cupressus*) possess an aromatic odor, and the odor of pine is resinous. Among other species, odor is characteristic of sassafras, baldcypress has sometimes an unpleasant odor, and catalpa heartwood is said to suggest kerosene (13). Maple (*Acer pseudoplatanus*), when hot and wet, has been described to smell like strawberry jam, teak like burnt leather, and the tropical coachwood (*Ceratopetalum apelatum*) like newmown hay (6) or caramel (8)! Other tropical species possess various odors; some have been given suggestive names [camphor wood, garlic wood, raspberry-jam wood, stinkwood, etc. (8, 13)].

Odor may be an advantageous characteristic, as in the case of Spanish cedar (*Cedrela odorata*, a hardwood), which is used for cigar boxes, or some cedars and cypress (*Cupressus*), which are preferred for clothes chests. On the other hand, odor is undesirable in wood used for baskets, boxes, or crates for packing food.

Aside from occurring normally, odors may be produced during the decomposition of wood by microorganisms.

TASTE

Taste is due to volatile deposits. Thus, it is more pronounced in fresh material; it is also more distinctive in heartwood than in sapwood. Woods like oak and chestnut, which contain appreciable amounts of tannin, possess a bitter taste. Taste is not a very important diagnostic feature, but it may help in some cases in separating similar woods. For example, the American cedars—incense cedar (*Libocedrus decurrens*) and western red cedar (*Thuja plicata*)—are woods similar in structure and appearance; however, the former has a spicy taste, whereas the latter is faintly bitter.

TEXTURE, GRAIN, AND FIGURE

Texture and *grain* are terms often misused in practice. Expressions, such as *coarse*, *fine*, or *medium* texture, and *even* or *uneven* texture, are frequently employed as synonymous to analogous qualifications of grain. The connotations—coarse, fine, and medium—refer to the relative size and proportion of wood elements (cells) as seen with the naked eye or with a hand lens. Large diameter cells produce a coarse (open) texture or grain (i.e., a greater macroscopic porosity in wood), whereas small diameter cells form a fine texture or grain. Fast-growing trees often produce coarse wood.

Even or uneven texture or grain refer to the

degree of uniformity of appearance; they indicate differences in structure within a growth ring. Ring-porous woods are generally uneven in comparison to diffuse-porous woods. Likewise, softwoods with an abrupt transition from earlywood to latewood have an uneven texture or grain.

The term *grain* has an additional meaning; it denotes direction of wood elements. The connotations *straight grain*, *spiral grain*, *interlocked grain*, *diagonal grain*, *cross grain*, and *wavy grain* or *curly grain* are employed for this purpose.

Figure, a less disputed term, is used to describe the natural design or pattern on wood surfaces. As explained in Chapter 1, the various macroscopic characteristics of wood (heartwood, sapwood, growth rings, earlywood, latewood, rays, resin canals, hardwood pores) create certain designs, depending on the plane of sectioning (Figure 2-1). While the figure of wood of normal structure may be pleasing, more attractive figure may, in some cases, be produced by structural abnormalities. Grain deviations, burls (rounded outgrowths on stems) and crotches (forked portions of stems), eccentric growth, uneven deposition of color, and other irregularities may produce beautiful figure (2) (Figure 2-2). Natural designs are sometimes accentuated, for example, by sandblasting tangential surfaces; this treatment results in superficial removal of earlywood, whereas latewood, being harder, is little affected.

Figure in wood is sought particularly by makers of veneer for furniture and interior paneling. Woods with natural figure are, however, scarce and expensive, so imitations are used extensively. These are manufactured by reproducing, by printing or other means, the figure of choice species on paper, plastic, or other overlay material, then bonding to panels of common wood, plywood, particleboard, or fiberboard.

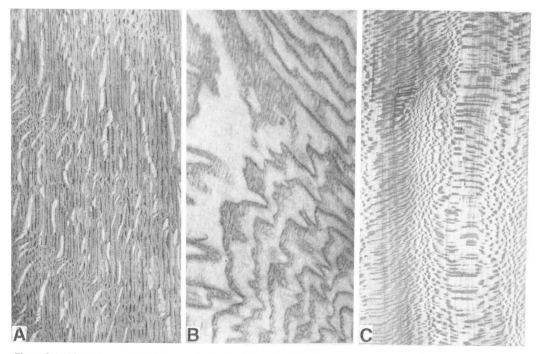

Figure 2-1. Figure in wood. A, Silver grain in oak; light-colored flecks are rays; radial surface. B, Figure in ash; the pattern is produced by alternating bands of earlywood (light-colored) and latewood (dark-colored); tangential surface. C, Figure in plane (*Platanus*); dark-colored flecks are rays on radial surface. (Reproduced by permission from Pergamon Press.)

Figure 2-2. A symmetric pattern of figure in walnut produced by matching and gluing four consecutive veneer sheets. (Courtesy of The Times Veneer Co. Ltd.)

WEIGHT AND HARDNESS

Weight, estimated simply by lifting a sample of wood, is an added helpful physical characteristic for purposes of description and identification, among others. However, it should be remembered that weight is influenced by various factors, such as sapwood and heartwood, proportion of earlywood and latewood, and especially by moisture content. To be valid, any comparison should be made under approximately identical conditions. Moisture content is very important, because in light woods the weight of included moisture may exceed many times the weight of the wood substance itself. Moisture is also subject to continuous variation when wood is exposed under varying atmospheric conditions (see Chapters 8 and 9).

Hardness is also considered here as a feature that may be judged simply by pressing the wood with a thumbnail. The ensuing ease or difficulty with which the wood can be indented is taken as a measure of its hardness. (Hardness may be accurately measured in the laboratory; see Chapter 11.) Hardness is closely related to weight—heavier woods are harder. However, the effect of moisture is opposite; with increasing moisture content hardness decreases. Hardness is also different on transverse, radial, and tangential surfaces.

REFERENCES

1. Avella, T., R. Dechamps, and M. Bastin. 1988. Fluorescence study of 10,610 woody species from the Tervuren (Tw) collection, Belgium. *IAWA Bull.* n.s. 9(4):346–352.
2. Beals, H. O., and T. C. Davis. 1977. Figure in Wood. *Ag. Expt. Station Bull.* 486, Auburn, Alabama.
3. Brauner, A., and E. M. Conway. 1964. Steaming walnut for color. *For. Prod. J.* 14:525–527.
4. Dyer, S. T. 1988. Wood fluorescence of indigenous African trees. *IAWA Bull.* n.s. 9(1):75–87.
5. Gray, V. R. 1961. The colour of wood and its changes. *J. Inst. Wood Sci.* 8:35–37.
6. Jane, F. W. 1970. *The Structure of Wood.* London: A.&C. Black.
7. Kutscha, N. P., and T. B. Sachs. 1962. Color tests for differentiating heartwood and sapwood in certain softwood tree species. U.S. For. Prod. Lab. Report No. 2246.
8. Latham, B. 1964. *Wood: From Forest to Man.* London: Harrap & Co.
9. Loos, W. E., and W. A. Coppock, 1964. Measuring wood color with precision. *For. Prod. J.* 14:85–86.
10. Miller, R. B., Quirk, J. T., and D. J. Christensen. 1985. Identifying white oak logs with sodium nitrite. *For. Prod. J.* 35(2):33–38.
11. Nelson, N. D., R. R. Maeglin, and H. E. Wahlgren. 1969. Relationship of black walnut wood color to soil properties and site. *Wood and Fiber* 1(1):29–37.

12. Panshin, A., and C. De Zeew. 1980. *Textbook of Wood Technology*, 4th ed. New York: McGraw-Hill.
13. Record, S. J. 1934. *Identification of the Timbers of Temperate North America*. New York: J. Wiley & Sons.
14. Vetter, R. E., Coradin, V. R., Martino, E. C., and J. A. A. Camargos. 1990. Wood color—A comparison between determination methods. *IAWA Bull. n.s.* 11(4):429–439.

FOOTNOTE

[1]Fluorescence is the property of producing color by action of ultraviolet light. Heartwood usually appears as bright yellow, while sapwood takes some shade of blue (12); the property may be examined with wood or extractives (water, ethanol). Fluorescence is a useful tool in wood identification (to be included in computerized characteristics or otherwise), because it was found to be strongly linked to taxonomic delimitation at the genus or family level. An extensive study has shown that fluorescence is not an uncommon occurrence; out of 10,610 species examined (mostly African and South American, and mostly in the form of heartwood), 1237 (about 12%) were clearly fluorescent, and another 2272 species (21%) showed a weak fluorescence. Yellow color dominated (94%), violet was represented by about 6% of total fluorescent species, and a few species showed green or orange fluorescence (1).

3

Wood Under the Microscope

Wood is composed of a multitude of minute units, called *cells*.[1] Their recognition became possible only after the invention of the microscope. Hooke (16) first observed the cells of cork in 1665.

GENERAL APPEARANCE OF INDIVIDUAL WOOD CELLS

Cells are connected together in various ways to form the mass of wood. By chemical means, it is possible to dissolve the substance which cements one to another. Mechanical separation is also possible, but this results in a large number of broken cells. The process of separating the cells is called *maceration*. Microscopic observation of macerated material,[2] under low magnification, reveals the following gross characteristics of cellular morphology.

Cells of softwood species differ in appearance from cells of hardwood species (Figures 3-1 and 3-2). Softwoods are mainly composed of long and narrow tube-like cells, with closed and pointed or blunt ends. These cells, called *tracheids*, have thick or thin walls, and are equipped with discontinuities called *pits*. Among the tracheids are a few small, rectangular (brick-like) *parenchyma cells*. These also have pits. In some softwood species, macerated material includes a small number of ray tracheids; under low magnification, these may resemble parenchyma cells, but are irregular in shape and have pits that are similar to those of long tracheids.

Hardwood cells show greater variation in size and shape. The majority are long and narrow with closed and pointed ends; they have a general resemblance to, but are much shorter than, softwood tracheids. These cells, called *fibers*, may have thin or thick walls, depending on the species. In macerated material, brick-like parenchyma cells are also present. Also included are a relatively small number of cells with open ends. These cells are called *vessel members* or *vessel segments*, and are generally shorter than fibers but vary in size and shape. Some are long and narrow, some are short and wide, and others may be more wide than long. In some hardwood species *tracheids* are also present; however, these are morphologically different from the tracheids of softwoods, as is further explained below. Pits are present throughout, being especially abundant in vessel members and tracheids.

WOOD TISSUE; CELL LUMINA, WALLS, AND CEMENTING SUBSTANCE

The above cell types are combined to form wood tissue. Most cells (axial tracheids, fibers, vessel members, certain parenchyma cells) are placed longitudinally (i.e., parallel to the length of the trunk, branches, and roots). Other cells (ray tracheids and parenchyma cells that compose rays) are elongated in the radial direction (i.e., from pith to bark). The arrangement varies with different species of wood and in combination with varying cell morphology presents a different image in transverse, tangential, and radial surfaces. These surfaces may be viewed separately (on thin, transparent wood sections), but the scanning electron microscope (SEM, see Chapter 4) allows a simultaneous (three-

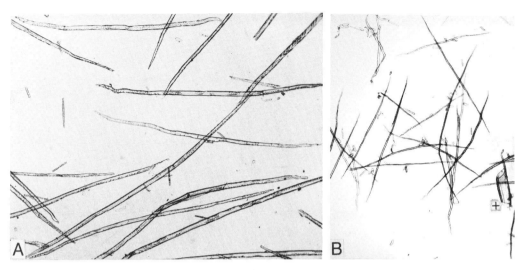

Figure 3-1. Macerated cells (30×). (A) Axial tracheids of pine, (B) fibers and a vessel member (+) of mixed hardwoods. Parenchyma cells are also present but cannot be clearly seen at this small magnification. (Reproduced by permission from Pergamon Press.)

dimensional) observation and photography (Figure 3-3).

Mature wood cells are dead (i.e., devoid of protoplasm and nucleus), even in the living tree. Thus, most cell lumina are empty. Protoplasm and nucleus are present in the early stages of the short life of a wood cell, but disappear in the process of cell development. The exceptions are few rows of young cells produced during current growth by the cambium, and parenchyma cells located in sapwood.[3]

The cell wall of the mature cell is composed of two layers—a thin outer layer called the *primary wall*, and the thicker inner layer toward the cell cavity called the *secondary wall*. The substance that cements the cells together, like mortar cements bricks in a wall, is called the *intercellular layer* or *middle lamella*. In common microscopic preparations, the primary wall cannot be distinguished from the middle lamella. Both stain similarly, due to related chemical composition. Thus, in sections of wood tissue, the primary walls of two adjoining cells and the middle lamella, which cements these cells together, appear as one layer. This three-ply layer is called *compound middle lamella* (Figure 3-4).

STRUCTURE AND ARRANGEMENT OF PITS

A careful study of pits shows that they are discontinuities in the secondary wall. Pits serve as passages of communication between neighboring cells. They are visible in all views (transverse, radial, tangential), but in most cells their structure is best seen on radial and tangential sections. There are two main types of pits—*simple* and *bordered*. All pits have two essential components—the *pit cavity* and the *pit membrane*. In the simple pit, the cavity is nearly constant in width, perhaps only gradually widening or narrowing toward the cell lumen. In the bordered pit, the cavity narrows more or less abruptly toward the cell lumen; typically, the membrane is overarched by the secondary cell wall (20) (Figures 3-5 to 3-7).

The pit membrane consists of primary wall and middle lamella. As a rule, pits in the walls of adjoining cells appear in pairs (called *pit-pairs*), and the common membrane is therefore composed of two primary walls and middle lamella.

Two complementary simple pits form a simple pit-pair, and two bordered pits form a bor-

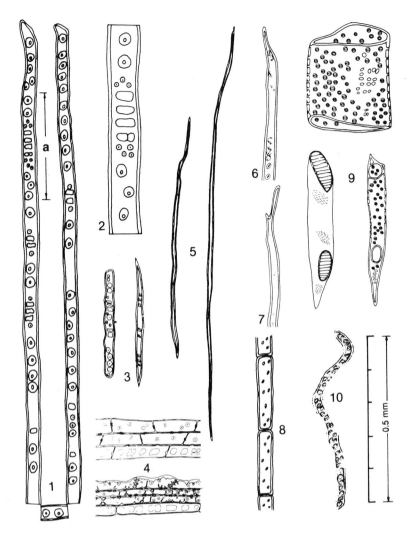

Figure 3-2. Cells of wood: 1, axial earlywood tracheid (Scots pine); 2, enlargement of part *a*; 3, axial tracheid of earlywood (left) and latewood (right) in indicative drawing (not in scale); 4, ray tracheids at the upper boundaries of rays (smooth and dentate inner walls, and parenchyma cells with window-like pits); 5, fibers (oak); 6 and 7, parts of fibers in enlargement (6, fiber tracheid—7, libriform fiber); 8, axial parenchyma (simple pits); 9, vessel members (simple and scalariform perforations); 10, vasicentric tracheid (oak). Note that the scale applies to numbers 1, 5, 9 (vessel members are very variable in size), and 10. The enlargement of 2, 4, and 8 is about double. (After Refs. 4, 6 and 19; 1, 2 and 5 drawn from photographs in Ref. 6.)

dered pit-pair. Some pit-pairs are *half-bordered* (*semi-bordered*); this is a pairing of a simple and a bordered pit. When a pit is not paired but solitary, it is called a *blind pit*.

The type of pitting—whether simple or bordered—is largely characteristic of various cell types. Thus, bordered pits occur in the walls of tracheids; mostly bordered, and sometimes simple (37), in vessel members; simple pits in parenchyma cells (Figure 3-6C and D); and simple or bordered pits in fibers.

The structure of pit membranes varies. In the bordered pits of most softwoods, the central portion of the membrane is thickened; this is called *torus* (Figure 3-6B). The portion that surrounds the torus is called *margo* (see Figure 4-9A), and the outer thickened rim of the membrane is called *annulus*. In cedar (*Cedrus*) the

WOOD UNDER THE MICROSCOPE 17

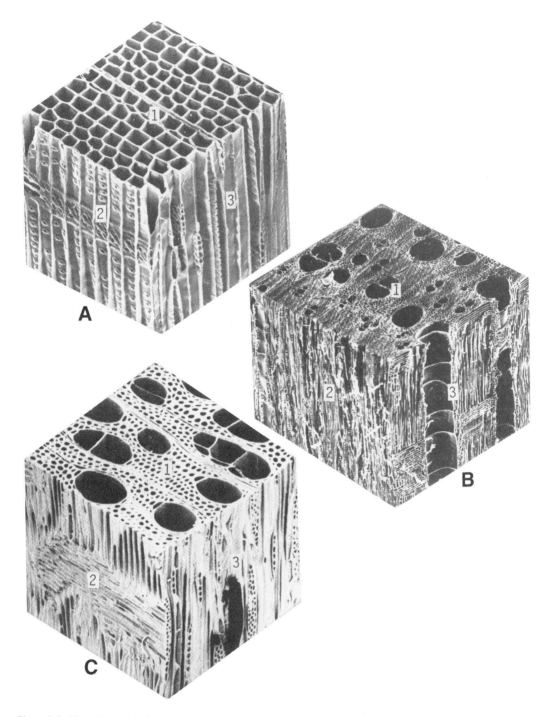

Figure 3-3. Three-dimensional appearance of small cubes of wood: (A) pine (sugar, 40×), (B) ash (50×), (C) birch (100×). 1, transvese; 2, radial; 3, tangential surface. (Photographs A and C adapted from N. C. Brown Center for Ultrastructure Studies, Syracuse, New York; B courtesy of W. C. McMillin)

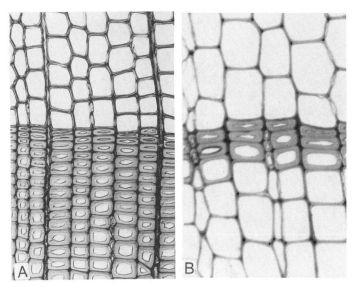

Figure 3-4. Transverse views showing earlywood and latewood cells (tracheids) in softwoods: (A) spruce, (B) redwood (*Sequoia*). All cells are empty, thin-walled in earlywood and thick-walled in latewood. In redwood, only 2–3 rows of cells constitute latewood. The compound middle lamella may be roughly distinguished from the secondary wall in latewood tracheids. Bordered pits are seen in cross-section. (A 150×, B 230×).

margins of tori are *scalloped* (i.e., indented). There is no torus in bordered pits of hardwoods, nor in simple or half-bordered pits (see Chapter 4).

The opening of a pit (toward the cell lumen) is the *pit aperture* or pit mouth. Pit cavity is the space from the aperture to the membrane. In bordered pits of thick-walled cells this space is separated in two parts, the *pit canal* and the *pit chamber*; the former is the passage from the lumen to the chamber, and the latter is the remaining space up to the membrane. The opening of the pit canal into the cell lumen is called *inner aperture*, that of the pit canal into the pit chamber *outer aperture* (Figure 3-6C).

Pits that occur in cross-fields[4] of softwoods are termed *window-like* (*fenestriform, fenestrate*), *pinoid, cupressoid, piceoid,* and *taxodioid*, according to appearance (39) (Figures 3-8 and 3-9). Distinction of these pits is best made in earlywood. All are half-bordered pits, but in window-like and pinoid pits the borders are often indistinct or very narrow.

In some hardwoods (e.g., black locust and eucalypts), *vestured* pits may be present. These are bordered pits with the pit cavity wholly or partially lined with projections (outgrowths) from the pit border (i.e., from the secondary wall). In front view, vestured pits have a characteristic zigzag appearance (see Figure 4-12).

A common modification of bordered pit-pairs is the lateral displacement of the membrane. This phenomenon, called *aspiration*, usually occurs when sapwood is transformed into heartwood or when wood dries. Apparently, it results from high tension forces set up by menisci formed in pit apertures and in pit membrane openings through which water (sap) is moving out. In softwoods, the torus seals one of the pit apertures and, therefore, blocks the passage through the pit (Figure 3-7B). Aspiration makes the wood of fir, spruce, and Douglas-fir difficult to impregnate with preservatives.

DESCRIPTION OF CELL TYPES

The various cell types that compose wood—namely, tracheids, vessel members, fibers, and parenchyma cells—will now be morphologically described in detail. These cells are often

Figure 3-5. Three-dimensional view of a transverse section of spruce wood (earlywood). Axial tracheids are seen in cross-section with bordered pits (x), and semi-bordered pits (xx) in the walls connecting tracheids and ray parenchyma cells. A ray is diagonal at the upper left side (gamma-irradiated wood, 1200×). (Courtesy of W.A. Côté.)

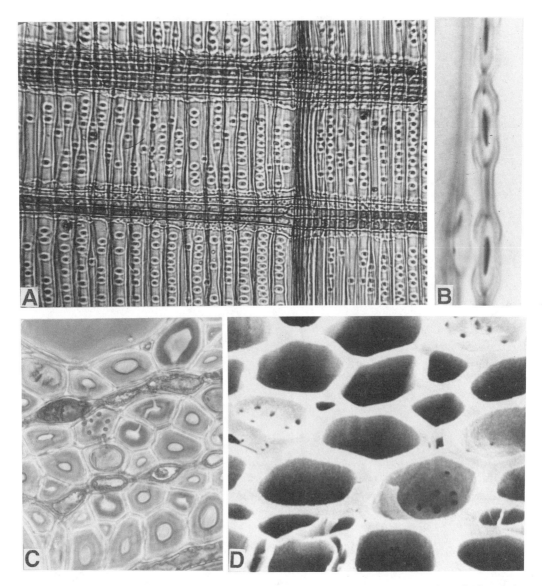

Figure 3-6. Pits. (A) Radial view (white pine, *Pinus strobus*): bordered pits and rays crossing horizontally through the boundary of earlywood-latewood with ray tracheids on their borders (above and below). (B) Tangential view (spruce): bordered pits with tori cut longitudinally. (C) Simple pits appear as small, rounded openings in the transverse wall of an axial parenchyma cell of beech; cells filled with dark contents are ray parenchyma, and thick-walled cells are fibers; part of a large vessel member is shown on the upper left. (D) A similar but three-dimensional view of oak wood; simple pits of parenchyma cells are seen as little holes. (A, 175×, B, 1400×; C, 750×; D, 1500×.). (Photograph A reproduced by permission from Pergamon Press; D courtesy of E. Voulgaridis.)

grouped into *tracheary*, *parenchymatous*, and *prosenchymatous* elements. Tracheids and vessel members are called tracheary elements; these are the principal water-conducting cells. Parenchyma or parenchymatous elements are cells primarily concerned with translocation and storage of food. The fibers are mainly strengthening, but some are involved in conduction and even in storage. Finally, prosenchyma (or prosenchymatous elements) is a general term for elongated cells with tapering ends—including the fibers and tracheids, and sometimes the

WOOD UNDER THE MICROSCOPE 21

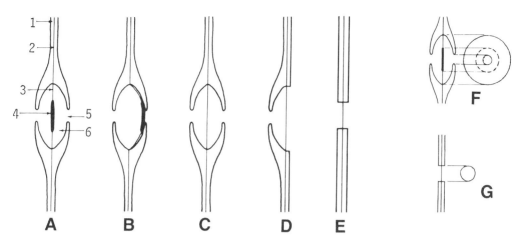

Figure 3-7. Pit-pairs. (A, B, C) Bordered pits (1, secondary wall; 2, compound middle lamella; 3, membrane; 4, torus; 5, pit aperture; 6, pit chamber (B) As in A, but the pit is aspirated (one pit aperture is sealed by the torus). (C) Bordered pit in hardwood (membrane without torus). (D). Semi-bordered pit (bordered in one cell, simple in the other). (E) Simple pit. Figures A–E show pits in tangential (or transverse) view, and F and G front (radial) views of a bordered (F) and a simple pit (G). (A and B from J. Dinwoodie.)

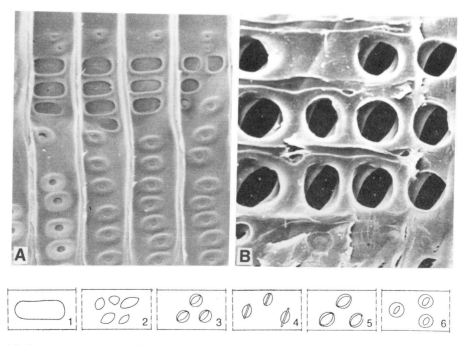

Figure 3-8. Types of pits in cross-fields (ray crossings) of softwoods. (A) Window-like (fenestriform) in Scots pine (265×). (B) Pinoid with border in Swiss stone pine (*Pinus cembra*, 725×). 1–6, Schematic representation of pit types according to Phillips (39): 1, window-like; 2, pinoid (without a distinct border); 3, pinoid with border; 4, piceoid; 5, taxodioid; 6, cupressoid. (Photographs by Alice Hirzel, ETH Zurich, courtesy of L.J. Kucera.)

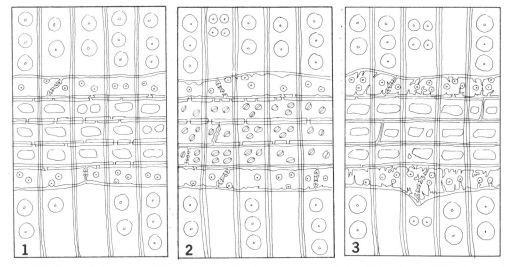

Figure 3-9. Types of pits in cross-fields and inner walls of ray tracheids as distinguishing characteristics of pines: 1, window-like and smooth inner walls (white pine); 2, pinoid (with border) and light dentation (Aleppo, umbrella pine); 3, window-like and pronounced (articulate) dentation (black, Scots pine).

vessel members; this term is used in contrast to parenchyma.

Tracheids

The tracheids of softwoods are mostly or exclusively vertical. In some species a few tracheids may be placed horizontally in association with rays. The former are called vertical or axial tracheids, and the latter horizontal or ray tracheids.

Axial tracheids comprise 90% or more of the volume of softwoods. They are long, narrow cells of a length about 75–200 times (mostly 100 times) their diameter. In most softwoods, average lengths of mature tracheids vary from about 3 to 5 mm (Table 3-1).[5] Lengths smaller than 2 mm and greater than 5 mm are exceptional. Diameters are on the average 0.02–0.04 mm, but may range from 0.015 to 0.080 mm (15–80 μm) (21, 22, 37, 44).

The morphology of tracheids is different in the earlywood and latewood of the same growth ring (Figure 3-4). The tracheids of earlywood are relatively thin-walled, polygonic to squarish in cross-section, and have large lumina. Bordered pits are typical, located (as a rule) on their radial walls. On the contrary, the tracheids of latewood possess thick walls, smaller lumina, and they tend to be rectangular in shape and elongated in the tangential direction. The bordered pits are smaller and fewer.[6] Latewood tracheids are on the average about 10% longer than earlywood tracheids (9, 42).

The transition from thin-walled earlywood tracheids to the thick-walled latewood tracheids may be abrupt or gradual. This is largely a species characteristic. For example, in Douglas-fir, larch, redwood, baldcypress, and the hard pines (see keys), the transition is abrupt. However, this feature may be affected by growth conditions, since in adjacent growth rings the manner of transition may sometimes change. The proportion of thick-walled latewood tracheids within a growth ring is also affected by species and growth conditions.

In certain species, as in Douglas-fir and yew, the tracheids possess *spiral thickenings* (i.e., helical ridges on the inner faces of their walls) (Figure 3-10). In Douglas-fir, these are well-developed in earlywood tracheids, but are fewer or absent in latewood tracheids. In yew, thickenings are present throughout. Sporadically, spiral thickenings occur also in tracheids of spruce (23, 37).

The ends of all tracheids, as already mentioned, are closed. In typical axial tracheids the ends are blunt (rounded) in earlywood and ta-

Table 3-1. Cell Length (mm)[a]

Softwoods (axial tracheids)		Hardwoods (fibers)	
North American		**1. Temperate**	
Baldcypress	6.2	**North American**	
Douglas-fir	3.9	Alder, red	1.2
Fir, balsam	3.5	Ash, white	1.2
Hemlock, western	4.2	Aspen, quaking	1.0
Pine, eastern white	3.0	Cottonwood, eastern	1.0
Pine, loblolly	3.6	Beech	1.2
Pine, ponderosa	3.6	Birch	1.5–1.8
Pine, parana[b]	7.2[c]	Hickory	1.3
Redcedar, western	3.5	Maple	0.8
Redwood	7.0	Sweetgum	1.7
Spruce, Engelmann	3.0	Yellow poplar	1.9
European		**European**	
Cypress, Mediterranean	2.6	Alder	0.9
Fir, Grecian	3.6	Ash, flowering	0.8
Fir, white	4.3[c]	Beech	1.1
Larch	3.4[c]	Birch, Swedish	1.4, 1.0[c]
Pine, Aleppo	4.4	Chestnut	1.2
Pine, black (Austrian)	3.3	Elm, mountain	1.3
Pine, Scots	2.9, 3.1[c]	Maple, field	0.8
Pine, umbrella	3.4	Oak, red	1.1
Spruce	3.6, 2.9[c]	Oak, white (English)	1.2, 0.8[c]
Yew	1.6	Plane, oriental	1.4
		Poplar, hybrid (I-214)	1.1
		Willow	1.0
		2. Tropical	
		Balsa	2.2[c]
		Bete	1.0[c]
		Dibetou, Iroko, Mahogany	1.3[c]
		Makoré, Palissander	1.2[c]
		Okoumé, Sipo	1.1[c]
		Sapele	1.4[c]
		Teak	0.7–1.4[c]
		Tiama	1.9[c]

[a] Mean values of "mature" wood (see Chapter 6) of some North American, European, and tropical woods. (Latin names are found in Appendix II). Values for North American species from Ref. 22; hickory from Ref. 37.
[b] South American species (*Araucaria angustifolia*), botanically not a pine.
[c] Values from Ref. 48.

pering in latewood. In examining sections of wood tissue, the ends of earlywood tracheids, in particular, appear different in radial and tangential views. In the former they are blunt, and in the latter they are tapered; this is due to their wedged form.

In radial or tangential sections of wood tissue, some axial tracheids may occasionally be mistaken as parenchyma cells. Such tracheids, called *septate* or *strand tracheids*, are short cells which usually have both end walls horizontal and are placed endwise in series (strands). Strand tracheids may be easily recognized from parenchyma, however, in that they are empty and have bordered pits in their longitudinal and end walls. On occasion, parenchyma cells may be mixed in with such a series of tracheids.

Axial tracheids may sometimes possess special features—namely, *trabeculae* or *crassulae*. Trabeculae are rod-like cell-wall projections, running radially across the lumen. Their ends at the points contacting the tangential walls are enlarged, and they may run through several tracheids [7] (Figure 3-11). On the other hand, crassulae are localized thickenings of the inter-

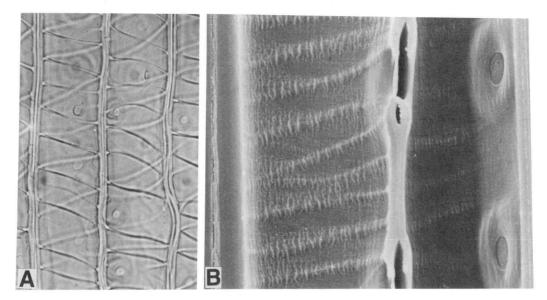

Figure 3-10. Spiral (helical) thickenings. (A) Yew (600×), (B) Douglas-fir (SEM, 1200×). (Photograph A courtesy of R. W. Hess, reproduced by permission from Pergamon Press; B by A. Hirzel, ETH Zurich, courtesy of L.J. Kucera.)

cellular layer and the primary walls. On radial sections, such thickenings may be observed above and below bordered pits.

Horizontal or *ray tracheids* are a regular feature in pine, spruce, larch, hemlock, and Douglas-fir. In certain cedars (*Cedrus* spp. and *Chamaecyparis nootkatensis*) they are always present; whereas in others (*Chamaecyparis*, *Juniperus*, *Libocedrus*, *Thuja*), and in fir and redwood, they occur only occasionally (23, 24, 37). Ray tracheids form the margins of rays (Figures 3-6A and 3-9) or they may be both

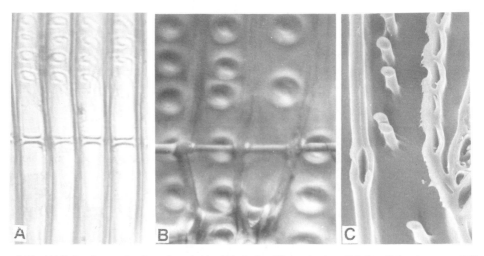

Figure 3-11. (A) Trabeculae passing through tracheids of black pine (*Pinus nigra*), radial view, light microscope (175×); (B) western white pine (*P. monticola*), radial view, incident light (420×); (C) Sitka spruce (*Picea sitchensis*), cross-cut trabeculae projecting from the lumen wall of a tracheid, tangential view (650×). (Photograph B courtesy of R. E. Pentoney; C, after Ref. 36 and courtesy of J. Ohtani.)

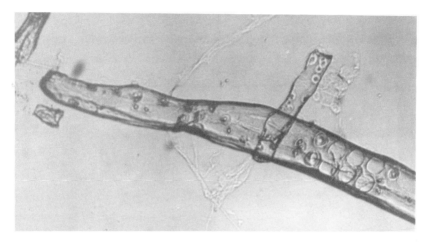

Figure 3-12. A ray tracheid attached to an axial earlywood tracheid of white pine (macerated material, 430×). (Reproduced by permission from Pergamon Press.)

marginal and interspersed between rows of ray parenchyma. In exceptional cases they constitute entire rays.

Ray tracheids are very short cells in comparison to axial tracheids (Figure 3-12), with an average length of about 0.1–0.2 mm. They are only about 5–10 times more long than wide. They have a general resemblance to parenchyma cells, but are different in that they are empty, tend (especially the marginal ones) to attain more irregular shapes, and possess small bordered pits. The inner walls of ray tracheids may be smooth, dentate, or reticulate (net-like; Figure 3-9). The degree of dentation is helpful in identifiying pines (4, 18, 24, 37, 45). Shallow dentations may also be seen in spruce and larch. In Douglas-fir the ray tracheids, like axial tracheids, have spiral thickenings.

Tracheids, in hardwoods, are not the constant feature they are in softwoods. Present only in a few hardwood species, they are considered to be transitional elements related to vessel members or fibers. There are two types in hardwoods—*vascular and vasicentric*—and they are discussed briefly later in this chapter.

Vessel Members

Vessel members occur only in hardwoods.[8] In wood tissue an indefinite number of such cells are connected endwise to form a pipe-like structure of indeterminate length, which is called a *vessel*. The end walls of the component vessel members disappear wholly or partially in the process of development after being formed by the cambium. The area of the adjacent end walls involved in endwise connection of two vessel members is called *perforation plate*. Perforations may be *simple* or *multiple*. Simple perforation is a single, usually large, more or less rounded opening; multiple perforation plates consist of several openings, and when these openings are elongated, parallel, and separated by bars (remnants of cell walls), the plates are called *scalariform* (i.e., ladder-like) (Figures 3-2, 9, 3-13). The number of bars per perforation plate has diagnostic value (2).

The end walls of vessel members may be either horizontal or oblique, as may be seen in macerated material. Plates in the oblique position may be seen in longitudinal sections of wood tissue. In such sections, remnants of end walls may also be observed; these remnants indicate the length of vessel members that compose a vessel.[9] In cross-sections, vessels appear as *solitary* pores, or in *multiples*, *chains*, or *clusters*, and their shape may be circular, ellipsoid, or angular. A chain is made of adjacent solitary pores arranged in line. A multiple is also made of adjacent pores, but these tend to be flattened along the lines of contact, thus

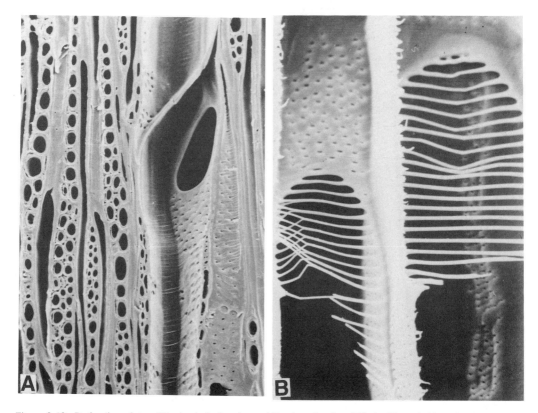

Figure 3-13. Perforation plates: (A) simple in hornbeam (*Carpinus betulus*, 250×); (B) scalariform in alder (*Alnus glutinosa*, 580×). In A, at left, an aggregate ray. (Photographs by A. Hirzel, ETH Zurich, courtesy of L. J. Kucera.)

appearing as subdivisions of a single large pore. In a cluster, several pores form an irregular group.

The size of vessel members varies widely; in ring-porous hardwoods, differences within a growth ring are much greater than differences between species. Vessel members (particularly in earlywood of ring-porous hardwoods) are the most massive wood cells. Some are more short than wide. Compared to tracheids, vessel members are short, ranging in average length from 0.2 to 1.3 mm (22, 37, 44). Diameters may vary from 0.005 to 0.5 mm (44).

Vessel members have, as a rule, bordered pits. The inner surface of their walls may possess *spiral thickenings*. The presence or absence of such thickenings may have diagnostic value. For example, the distinction of maple and birch under the microscope can be based, among others, on this feature; maple has spiral thickenings, but birch does not. An additional feature of the inner (lumen) surface of vessel members (as well as of tracheids and fibers in many species) is the presence of a *warty* layer. As a rule, an electron microscope is needed to see this feature; however, in some species, such as beech and cedar (*Cedrus*), warts may be seen with the light microscope (29).

Vessels which no longer participate in sap conduction, mainly when sapwood is transformed into heartwood, may be plugged with *tyloses*[10] (Figure 3-14). Tylosis is an outgrowth from an adjoining ray or vertical parenchyma cell through a pit-pair into the lumen of a vessel. This is apparently a result of differential pressure in the lumina of adjoining vessels and parenchyma cells. Tylosis consists of protoplasm of the parenchyma cell and of storage materials in the form of starch, crystals, resins, gums, and others.

Formation of tyloses has been related to the size of pit apertures. It was observed that they

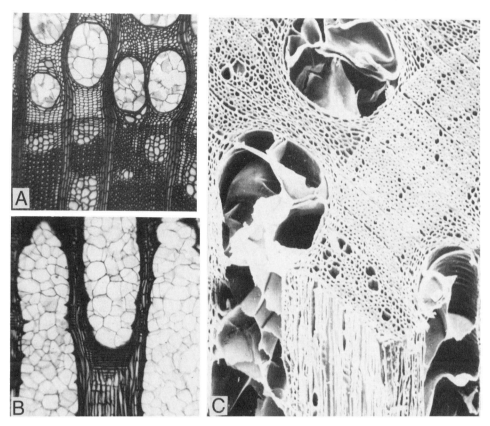

Figure 3-14. Tyloses in vessel members. (A, B) Black locust (80×) (A, transverse; B, radial section), (C) white oak, SEM 110×). (Photographs A and B reproduced by permission from Pergamon Press; C courtesy of N. C. Brown Center for Ultrastructure Studies, Syracuse, New York.)

can be formed only when the width of the aperture toward the vessel is about 10 μm or greater. No tyloses are formed when the aperture is 8 μm or smaller, but extraneous materials are extruded from the dying parenchyma cell into the lumen of the vessel (7). Tyloses are common in white oak, black locust, mulberry, osage orange, hickory, walnut, chestnut, teak, and other woods.[11]

In addition to their normal occurrence, tyloses may form pathologically, as a result of mechanical injury, fungus, or virus infection (27).

Vascular and Vasicentric Tracheids

Vascular and vasicentric tracheids are both rare cell elements occurring only in a few hardwoods in association with vessels. *Vascular tracheids* resemble small vessel members but have no perforated ends. Sometimes, they are called "imperfect" (20) or "degenerate" vessel members. They have many bordered pits and may possess spiral thickenings. Vascular tracheids are found in the wavy bands of latewood in elm and hackberry (6, 37), for example, together with small vessels.

Vasicentric tracheids are mainly found near the large earlywood vessels in some ring-porous woods, such as certain oaks, ash, and chestnut (6, 23, 37). They also have closed ends and many bordered pits, but differ from vascular tracheids in that they are mostly longer and irregular in shape. In radial sections of wood tissue, vasicentric tracheids appear to have characteristic overlapping. Their twisted shape and heavy pitting make them easily detectable in macerated material (Figure 3-2, 10).

Parenchyma

Parenchyma cells are typically prismatic (brick-like) cells and have simple pits. As a rule, they appear full of material, which may be living cell contents (in sapwood) or various inorganic inclusions, sometimes in the form of crystals. According to orientation in the tree, parenchyma is classified as vertical or axial and radial or ray parenchyma.

Axial parenchyma is not present in all woods. In softwoods, it may be entirely absent (pine, spruce, yew); absent or very sparce (hemlock, fir, Douglas-fir); or present in variable proportions (redwood, baldcypress, cypress, cedars). When present, parenchyma cells may be *diffuse* among the tracheids; *zonate* or *banded* (i.e., in tangential lines or bands); and *boundary* (*initial* or *terminal*), if placed at the boundaries of growth rings. All these terms relate to the appearance of such parenchyma on transverse sections of wood (Figure 3-15).

In hardwoods, axial parenchyma is seldom absent [e.g., sometimes in cherry (37)]. In gen-

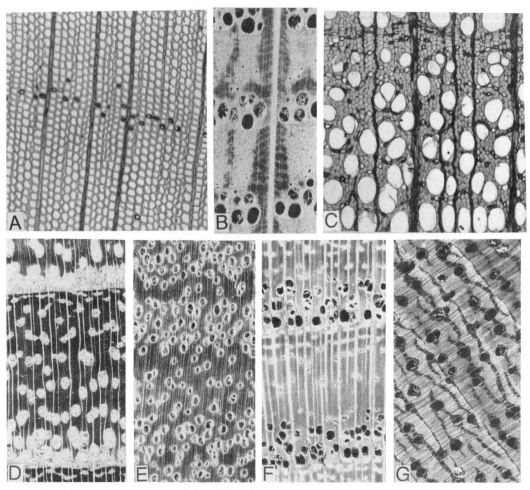

Figure 3-15. Parenchyma: (A) diffuse (Mediterranean cypress); (B) paratracheal and apotracheal in white oak (parenchyma occupies most of the light-colored, flame-like tissue in latewood with small pores embedded, and also banded, in tangential lines on both sides of the ray—see also red oak in Figures 1-4B and 6-6B); (C) apotracheal (diffuse, banded, terminal) in beech; (D) paratracheal (vasicentric, confluent, and terminal) in black locust; (E) paratracheal (vasicentric) in eucalyptus; (F) paratracheal (vasicentric, confluent, and terminal in tree-of-heaven, *Ailanthus*); (G) paratracheal (aliform, confluent) in Iroko. (A, C, 80×; B, D–G, 10×.) Note that various configurations of parenchyma are also shown in Appendix II, especially in tropical woods.

eral, it is more abundant than in softwoods; in some species, it is conspicuous to the naked eye or with a hand lens (see keys). According to its position relative to vessels, hardwood axial parenchyma may be *paratracheal* or *apotracheal*. Paratracheal (from the Greek *para* = near) is parenchyma located adjacent to vessels, and apotracheal (from the Greek *apo* = away) when not in contact with vessels.[12] Paratracheal or apotracheal parenchyma placed at the boundaries of growth rings is called *boundary* (*initial* or *terminal*).

In both softwoods and hardwoods, more than one of the above types of axial parenchyma may be present in a growth ring.

Ray parenchyma cells constitute the rays wholly or in part. Rays in hardwoods are composed entirely of ray parenchyma cells; however, in softwoods ray tracheids may also be present, as may resin canals. Rays consisting of radially elongated parenchyma cells of practically equal height, or of cells all squarish or upright, are called *homocellular* (*homogeneous*). If certain rows (usually the marginal) are made of squarish or upright cells, or if ray tracheids are present, the rays are termed *heterocellular* (*heterogeneous*) (20). Rays that include resin canals are *fusiform*.

Furthermore, rays may be *uniseriate* (one cell wide as seen in tangential view), *biseriate*, *multiseriate*, and *aggregate* (Figure 3-16). An aggregate ray appears to the naked eye or through a hand lens to be a single, relatively broad ray. Under the microscope, in tangential sections or cross-sections, aggregate rays are revealed to be composed of several uniseriate and wider rays separated by rows of fibers and sometimes by rows of small vessels. Aggregate rays are present, for example, in hornbeam, alder, hazelnut, and evergreen (live) oaks. In tangential sections, all rays appear to terminate their margins in a single cell and are, therefore, spindle-shaped. In their middle portion, rays may vary in width from one cell to several (30 or more); and their height may vary from one to many hundreds of cells.

In some species, the rays are arranged in horizontal series, as seen on tangential sections. Such stratified or "storied" arrangement of rays (and sometimes of axial cells) gives rise to *ripple marks* (i.e., fine horizontal striations macroscopically visible on tangential surfaces of certain woods—such as buckeye, persimmon, and some tropical species) (20, 37).

As a rule, rays of softwoods are uniseriate. In hardwoods, rays of varying width may be present in the same species. Accordingly, the volume of ray tissue varies in different species; and there is a much greater variation in hardwoods than in softwoods. In hardwoods, average ray volumes range from about 5% (e.g., basswood) to about 30% (oak). In softwoods, the range is from about 5 to 10% of the total volume of wood (34, 37).

In general, parenchyma cells are smaller than other cells, similar in size only to ray tracheids. Their length is about 0.1–0.22 mm, with widths ranging from about 0.01–0.05 mm (35, 37, 44). Recognition of axial and ray parenchyma cells is practically impossible in macerated material.

In addition to the parenchyma cells described above, specialized parenchyma cells, associated with intercellular canals (see below), may be present in certain woods. The pith

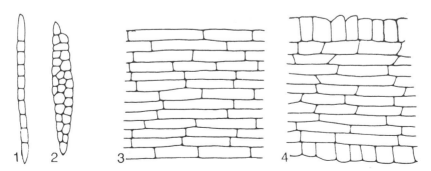

Figure 3-16. Rays: 1, uniseriate; 2, triseriate; 3, homocellular; 4, heterocellular (1, 2 tangential and 3, 4 radial view).

is also chiefly parenchymatous. Furthermore, abnormal parenchyma cells may be produced as a result of injuries to the cambium.

Fibers

Fibers are present only in hardwoods. They are long narrow cells with a general resemblance to latewood tracheids. Length varies between species, ranging on the average between 1 and 2 mm (22, 37)—exceptionally 0.5–2.5 mm.[13] Diameters range from about 0.01 to 0.05 mm (44). Fibers have closed ends, mostly pointed, and are sometimes forked or equipped with dentations. The walls may be thick or thin, and the lumina narrow or large. This varies mainly with species, but fibers produced near the end of the growing period are, as a rule, thick-walled and tangentially flattened. The latter characteristics are sometimes helpful in defining the boundaries of growth rings in some diffuse-porous species.

Fibers are classified into *fiber tracheids* and *libriform fibers*. The basis of distinction is the nature of pitting; fiber tracheids have bordered pits, while pits are simple (sometimes minutely bordered) (1) in libriform fibers. Libriform fibers are also usually smaller than fiber tracheids in length and diameter, and have narrow lumina that are often difficult to see under low magnifications. The primary function of fibers is to provide mechanical support to the living tree; however, they (fiber tracheids, especially) may also participate in conduction. Some fibers are *septate* (i.e., with thin transverse walls across the lumen).

The proportion of fibers is variable in different hardwoods. In many species, they may contribute 50% or more of the total wood volume.[14]

INTERCELLULAR CANALS AND SPACES

Intercellular canals are spaces in wood tissue. They are not cells. Tubular in structure and indeterminate in length,[15] they are lined with specialized parenchyma cells called *epithelial cells* or *epithelium*. Intercellular canals may occur in both softwoods and hardwoods. In softwoods they are called *resin canals* or *resin ducts*, and in hardwoods *gum canals* or *gum ducts*. These terms are sometimes used interchangeably.

Resin canals are always found in pine, spruce, larch, and Douglas-fir. They extend axially among vertical tracheids and radially within rays. In general, axial canals have larger diameters than radial, but both are interconnected and form a network within the tree.[16] Radial resin canals are contained in fusiform rays (Figure 3-17).

The number and size of resin canals vary. Size (diameter), number, and arrangement of axial canals are valuable clues in macroscopic and microscopic identification of pine, spruce, larch, and Douglas-fir. Resin canals are larger and more numerous in pine, with differences existing between species (see keys in Appendix II). However, the size, number, and arrangement of canals may vary (within limits) between adjacent growth rings and in different heights of the tree (38, 47).

Wall thickness of epithelial cells is also of diagnostic value. Epithelical cells in pine are thin-walled, while in spruce, larch, and Douglas-fir they are thick-walled. The number of epithelial cells surrounding a radial resin canal (in fusiform rays) is sometimes considered distinctive in larch and spruce (more than nine cells in larch—less than nine in spruce) (23). However, systematic study has shown this to be rather unreliable (17). Epithelial cells are considered the source of resin, which is an important forest product[17] (see Appendix I).

Wounding of the cambium leads to formation of *traumatic* (wound) canals, which differ from normal resin canals in structure and arrangement (see Chapter 7).

When sapwood transforms into heartwood, resin canals may become plugged with *tylosoids*. This phenomenon is analogous to the formation of tyloses in hardwoods, but differs in that a tylosoid derives from an epithelial cell and does not pass through a pit cavity.

Gum canals are, as mentioned above, structural features of hardwoods respective to the resin canals of softwoods. They may be axial or radial, but both types seldom occur in the same wood. Gum canals may be normal or

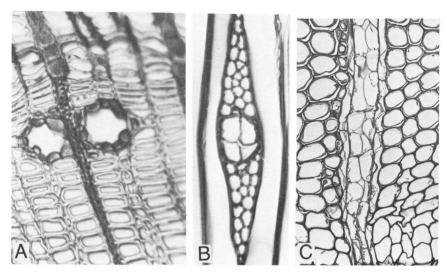

Figure 3-17. Resin canals: (A) transverse section; (B) tangential view (a resin canal in a fusiform ray). (A) Spruce (thick-walled epithelial cells, canals near dormant cambium); (B) pine (thin-walled epithelial cells). (C) Longitudinal section of a radial canal in Aleppo pine, *P. halepensis* (this is the path of resin flow when harvesting pine resin by removing strips of bark from living trees). (A, 180×; B, 250×; C, 150×.) (Photograph A reproduced by permission from Pergamon Press; C courtesy of J. Dahl Møller.)

traumatic. Normal gum canals do not appear in temperate woods of commercial importance.

Intercellular spaces, not canals, sometimes exist when adjacent cell contact is not tight. Such spaces are characteristic of compression wood in softwoods (see Chapter 7), but they have also been observed in normal wood (e.g., in eastern redcedar, *Juniperus virginiana*) between tracheids or parenchyma cells (axial and radial) (32).[18]

REFERENCES

1. Baas, P. 1985. A new multilingual glossary of terms used in wood anatomy. *IAWA Bull.* n.s. 6(2):83.
2. Bhat, K. M. 1983. A review of wood anatomy and selected properties of stems, branches and roots of birch trees (abstract). *IAWA Bull.* n.s. 4(2, 3):70.
3. Berlyn, G. P. and J. P. Miksche. 1976. *Botanical Microtechnique and Cytochemistry*. Ames, Iowa: The Iowa State University Press.
4. Bosshard, H. H. 1983. *Holzkunde*, 2nd ed. Basel: Birkhäuser Verlag.
5. Bosshard, H. H. and U. E. Hug. 1980. The anastomoses of the resin canal system in *Picea abies* (L.) Karst., *Larix decidua* Mill. and *Pinus silvestris* L. *Holz Roh- Werkstoff* 38:325–328.
6. Carpenter, C. H., and L. Leney. 1952. *Photomicrographs of Paper Making Fibers*. Syracuse, New York: College of Forestry, Syracuse University.
7. Chattaway, M. M. 1949. The development of tyloses and secretion of gum in heartwood formation. *Austral. J. Sci. Res. B.* 2:227–240.
8. Core, H. A., W. A. Côté, and A. C. Day. 1976. *Wood Structure and Identification*. Syracuse, New York: Syracuse University Press.
9. Dinwoodie, J. M. 1961. Tracheid and fiber length in timber: A review of literature. *Forestry* 34:125–144.
10. Esau, K. 1965. *Plant Anatomy*, 2nd ed. New York: John Wiley & Sons.
11. Fahn, A., and B. Leshem. 1963. Wood fibers with living protoplasts. *New Phytol.* 62:91–98.
12. Fahn, A. 1979. *Secretory Tissues in Plants*. Academic Press, London/N.Y.
13. Good, H. M., and C. D. Nelson. 1951. A histological study of sugar maple decayed by *Polyporus glomeratus* Peck. *Can. J. Bot.* 29:215–223.
14. Gottwald, H. P. J. 1972. Tyloses in fibre tracheids. *Wood Sci.* 6(2):121–127.
15. Grosser, D. 1986. On the occurrence of trabeculae with special consideration of diseased trees. *IAWA Bull.* n.s. 7(4):319–341.
16. Hooke, R. 1665. *Micrographia*. London: J. Martyn & J. Allestry.
17. Hudson, R. H. 1958. The value of the fusiform ray in separating the genera *Picea* and *Larix*. *J. Inst. Wood Sci.* 2:22–30.
18. Hudson, R. H. 1959. The anatomy of the genus *Pinus* in relation to its classification. *J. Inst. Wood Sci.* 6:26–46.
19. Ingenieurschule f. Holztechnik, Dresden. 1965. *Taschenbuch der Holztechnologie*. Leipzig: VEB Fachbuch-Verlag.

20. International Association of Wood Anatomists. 1964. *Multilingual Glossary of Terms Used in Wood Anatomy*. Winterthur, Switzerland: Konkordia.
21. Isenberg, I. H. 1963. The structure of wood. In *The Chemistry of Wood*, ed. B. L. Browning, pp. 7–55. New York: Interscience.
22. Isenberg, I. H. 1980. *Pulpwoods of the United States and Canada. I. Conifers, II. Hardwoods*, 3rd ed., revised by M. L. Harder and L. Louden. Appleton, Wisconsin: The Institute of Paper Chemistry.
23. Jane, F. W. 1970. *The Structure of Wood*. London: A. & C. Black.
24. Jaquiot, C. 1955. *Atlas d'Anatomie des Bois des Conifères*. Paris: Centre Technique du Bois.
25. Johansen, D. A. 1940. *Plant Microtechnique*. New York: McGraw-Hill.
26. Koran, Z. 1974. Intertracheid pitting in the radial walls of black spruce tracheids. *Wood Sci*. 7(2):111–115.
27. Koran, Z., and W. A. Côté. 1965. The ultrastructure of tyloses. In *Cellular Ultrastructure of Woody Plants*, ed. W. A. Côté, pp. 319–333. Syracuse, New York: Syracuse University Press.
28. Kucera, L. J. 1985. Zur Morphologie der Interzellularen in den Markstrahlen. *Viertelj. Natur. Gesell. Zurich* 130(2):157–198.
29. Liese, W. 1965. The warty layer. In *Cellular Ultrastructure of Wood Plants*, ed. W. A. Côté, pp. 251–269. Syracuse, New York: Syracuse University Press.
30. Maeglin, R. R., and J. T. Quirk. 1984. Tissue proportions and cell dimensions for red and white oak groups. *Can. J. For. Res*. 14:101–106.
31. McDougal, D. T., and G. M. Smith. 1927. Long-lived cells of the redwood. *Science* 66(1715):456–457.
32. McGinnes, E. A., and J. E. Phelps. 1972. Intercellular spaces in eastern redcedar (*Juniperus virginiana* L.). *Wood Sci*. 4(4):225–229.
33. Murmanis L. 1975. Formation of tyloses in felled *Quercus rubra* L. *Wood Sci*. 9(1):3–14.
34. Myer, J. E. 1922. Ray volumes of the commercial woods of the United States and their significance. *J. For*. 20:337–351.
35. Nyren, V., and E. Back. 1960. The dimensions of tracheidal and parenchymatous ray cells of *Picea abies* (Karts.) pulpwood. *Sv. Papperstidn*. 63:619–627.
36. Ohtani, J., K. Fukazawa, and T. Fukumorita. 1987. SEM observations on indented rings. *IAWA Bull*. n.s. 8(2):113–124.
37. Panshin, A. J., and C. De Zeew. 1980. *Textbook of Wood Technology*, 4th ed. New York: McGraw-Hill.
38. Pejoski, B. 1956. Recherches concernant les canaux résinifères, le gemmage et la gemme de *Pinus peuce* Grieseb. et comparaison avec les canaux résinifères des autres pins indigènes. *Ann. Fac. Agr. Silv*. 9:5–106, Univ. Skopje, Yugoslavia.
39. Phillips, E. W. J. 1959. Identification of softwoods by their microscopic structure. *For. Prod. Res. Bull*. No. 22, Princes Risborough, England.
40. Sass, J. E. 1958. *Botanical Microtechnique*. Ames, Iowa: The Iowa State College Press.
41. Schweingruber, F. H. 1978. *Mikroskopische Holzanatomie*. Birmensdorf, Switzerland: Zürcher AG.
42. Spurr, S. H., and M. J. Hyvärinen. 1954. Wood fiber length as related to position in tree and growth. *Bot. Review* 20:561–575.
43. Thomas, R. J. 1977. Wood: structure and chemical composition. In *Wood Technology: Chemical Aspects*, ed. I. S. Goldstein, pp. 1–23. Washington, D.C.: American Chemical Society.
44. Trendelenburg, R., and H. Mayer-Wegelin. 1956. *Das Holz als Rohstoff*. München: Hanser Verlag.
45. Tsoumis, G. 1968. *Wood as Raw Material*. New York: Pergamon Press (reprinted 1969).
46. Tsoumis, G. 1991. *Harvesting Forest Products*. (In press).
47. Vasiljevic, S. 1959. Periodicity of number of resin canals of *Pinus heldreichii* Christ. *Bull. College For*. 16:431–439. Univ. Belgrade, Yugoslavia.
48. Wagenführ, R. 1980. *Anatomie des Holzes*. Leipzig: VEB Fachbuchverlag.
49. Werker, E., and A. Fahn. 1969. Resin ducts of *Pinus halepensis* Mill: Their structure, development and pattern of arrangement. *Bot. J. Lin. Soc*. 62:379–411.
50. Werker, E., and P. Baas. 1981. Trabeculae of Sanio in secondary tissues of *Inula viscosa* (L.) Desf. and *Salvia fruticosa* Mill. *IAWA Bull*. n.s. 2(2/3):69–76.
51. Zimmermann, M. 1980. Annual report of the Harvard Forest, 1979/80, p. 8.

FOOTNOTES

[1] One cubic centimeter (1/16 in.^{3}) contains about 350,000–500,000 softwood cells and about 2–3 million hardwoods cells.

[2] Wood to be macerated is first reduced to slivers about half a millimeter thick and 1–2 cm (about 0.5–1 in.) in length. The slivers are placed in a test tube or other small glass container, and chemicals are added. A simple method employs a mixture of equal parts of glacial acetic acid and hydrogen peroxide (20 vol). The preparation (slivers and chemicals) is placed in an oven at 60°C (140°F) for 48 h. A little higher temperature (70°C, 160°F) and stronger peroxide will produce satisfactory results in a much shorter time (4–8 h), depending on wood species. Techniques for microscopic examination of wood are described in Tsoumis' *Wood as Raw Material* (45) and other books (3, 25, 40).

[3] The exceptional presence of parenchyma with live, nucleate protoplasm has been observed in the heartwood of redwood (*Sequoia*) (31) and sugar maple (13). In addition to parenchyma, fibers were also found to retain their living protoplasm in sapwood of many woody plants (mainly shrubs) in the Mediterranean region (11).

[4] Cross-field or ray-crossing is the rectangle formed (as observed on radial sections) by the horizontal walls of a ray

parenchyma and the walls of a vertical (axial) tracheid. The term is applied to softwoods only (20).

[5]The size of cells, in general, varies within a tree: cells near the pith are much shorter than "mature" cells. Therefore, the source of material is a very important factor in cell-size considerations (see Chapter 6).

[6]Earlywood tracheids of spruce were found to have, on the average, 50 (29–77) pits per radial wall, whereas latewood tracheids had only 15. Pit diameters were, on the average, 16.4 μm in earlywood and 6.1 μm (smaller diameter) in latewood. The average diameter of earlywood tracheids was 36 μm, and of latewood tracheids 12 μm (26).

[7]Trabeculae are not limited to softwood tracheids, but are also found in hardwoods; they may be present in all axial cell types and in rays, and are suggested to be produced as a result of injury (50). Trabeculae were also observed in diseased trees, in spruce wood with indented growth rings, and in compression wood (15, 36).

[8]Species of the order *Gnetales* (*Gnetum, Ephedra*) have also vessel members, although they are softwoods (*Gymnosperms*). However, in general, these are small plants, and their wood is not considered in this book. On the other hand, the wood of an angiosperm (*Drimys*) is without vessel members, and its fibers are quite similar to softwood tracheids (45).

[9]The number of vessel members that are connected to form a vessel is large. In ring-porous hardwoods, the length of a vessel is sometimes equal to the length of the tree trunk. In diffuse-porous hardwoods, the longest measured vessels were about 1 m (3 ft) long; however, in the majority (80%), they were shorter than 20 cm (8 in.) (51).

[10]*Tylosis* is a Greek word; it is characteristic that in certain regions of Greece, *tylos* is the stopper that is used to plug barrels.

[11]In red oak, which has few or no tyloses, their formation was observed in sapwood after felling and during storage of the wood (33). Besides vessels, the presence of tyloses was reported in fiber tracheids (14).

[12]According to arrangement, further subdivisions of these two main types are recognized. Paratracheal parenchyma is subdivided into *scanty* (occasionally around vessels), *unilateral* (limited to one side of a vessel), *vasicentric* (forming a complete sheath around a vessel), *aliform* (with wing-like lateral extensions), *confluent* (coalesced aliform), and *banded* (in tangential bands). Apotracheal parenchyma may be *diffuse*, *diffuse-in-aggregates*, and *banded* (20). Examples of such arrangements, often conspicuous to the naked eye, are offered especially by tropical woods (see Figure 3-15 and Appendix II).

[13]In the tropical species Apanit (*Mastixia philippinensis*), fibers up to 4.2 mm long have been measured (F. N. Tamolang, personal communication; see also Figure 6-3).

[14]Differences exist depending on rate of growth, and apply to all cell elements. For example, in 11 American red and white oaks, the proportion by weight for fibers varied from 74% for fast-grown trees to 38% for slow-grown trees; for axial and ray parenchyma from 24% for fast-grown to 54% for slow-grown; and for vessels from 2% for fast-grown to 8% for slow-grown (30).

[15]The length of axial canals was found to vary in pines from about 1 cm to 1 m (0.4 in.–3 ft), depending on species, age, and position in a tree (Ref. 12 and LaPasha, C.A., and E.A. Wheeler, *IAWA Bull. n.s.* 11/3:227–238, 1990).

[16]Resin canals do not form a three-dimensional, but many two-dimensional networks in radial planes. Each radial canal starts from an axial canal. Radial canals continue inside the bark, but the cavities in wood and bark do not communicate; communication is interrupted in the cambial region (12, 49).

[17]Resin is exuded from resin canals when these are exposed (e.g., in tapping pine trees to harvest the resin). When tapping is practiced by cutting into sapwood, both axial and radial canals are severed. In tapping by debarking (and application of chemical stimulants, 46), only radial canals are exposed, but resin still flows out of both, because of their interconnections.

[18]The morphology of intercellular spaces was studied in detail in rays of oak and beech; they are triangular, four-cornered or circular, radially, tangentially, or axially oriented, and differ with regard to lining, connections, sculpturing, and content, as well as between species. Intercellular spaces are usually empty, but sometimes contain extraneous materials (28).

4

Chemical Composition and Ultrastructure of Wood

ELEMENTARY CHEMICAL COMPOSITION

With regard to elementary chemical composition, there are no important differences among woods. The principal chemical elements of wood are carbon (C), hydrogen (H), and oxygen (O); small amounts of nitrogen (N) are also present. Chemical analysis of a number of species, including softwoods and hardwoods, shows the proportion of elements, in percent of the oven-dry weight of wood, to be approximately as follows: carbon, 49–50%; hydrogen, 6%; oxygen, 44–45%; and nitrogen, only 0.1–1%. In addition to the above, small amounts of mineral elements—principally calcium (Ca), potassium (K), and magnesium (Mg)—are found in wood ash. Usually, ash content is seldom lower than 0.2% or higher than 1%[1] of the oven-dry weight of wood (13, 32, 65, 127).

ORGANIC COMPONENTS OF WOOD

Carbon, hydrogen, and oxygen combine to form the principal organic components of wood substance, namely *cellulose*, *hemicelluloses*, and *lignin*; small amounts of *pectic substances* are also present.

These organic components of wood are not chemical entities that can be easily identified. The terms (cellulose, hemicelluloses, etc.) are generic, and each includes a number of chemically related compounds. Separation and quantitative determination of each component is accomplished in the laboratory through the use of solvents and specific techniques (2, 108). Differences in analytical procedure will show variation of chemical composition in the same sample of wood. As a result, literature reports are not always in agreement,[2] and doubts exist whether laboratory preparations are quantitatively and qualitatively representative of components as they exist in wood in its natural state (97, 127).

The proportions of cellulose, hemicelluloses, and lignin, based on standardized methods of preparation, are approximately as follows (in percent of the oven-dry weight of wood): cellulose, 40–45% (about the same in softwoods and hardwoods); lignin, 25–35% in softwoods and 17–25% in hardwoods; hemicelluloses, 20% in softwoods and 15–35% in hardwoods (Table 4-1). The proportion of pectic substances is small (15, 32, 111).

Cellulose is composed of molecules of glucose ($C_6H_{12}O_6$), a monosaccharide formed through photosynthesis from atmospheric carbon dioxide (CO_2). Glucose molecules are linked together to form long cellulose chain molecules. Linkage of any two glucose molecules is accompanied by the elimination of one molecule of water, and each glucose molecule added to the chain is rotated 180°. This process of repeated, indefinite addition of glucose units (monomers) is called *polymerization*. The empirical formula for cellulose is $(C_6H_{10}O_5)_n$; where n is the *degree of polymerization* or number of glucose monomers per cellulose chain. This number in "native" cellulose (i.e., cellulose in its natural state) lies on the average between 8000 and 10,000 (111).[3] The struc-

CHEMICAL COMPOSITION AND ULTRASTRUCTURE OF WOOD

Table 4-1. Chemical Composition of Woods (%)[a]

Components	Softwoods	Hardwoods
Holocellulose	59.8–80.9	71.0–89.1
Cellulose	30.1–60.7	31.1–64.4
Polyoses[b]	12.5–29.1	18.0–41.2
Pentosans[b]	4.5–17.5	12.6–32.3
Lignin	21.7–37.0	14.0–34.6
Extractives, hot water	0.2–14.4	0.3–11.0
Extractives, cold water	0.5–10.6	0.2–8.9
Extractives, ether	0.2–8.5	0.1–7.7
Ash	0.02–1.1	0.1–5.4

[a] Range of values from 153 botanical species of the temperate zone. As a rule, extractives are based on oven-dry weight of wood, and other components on oven-dry weight without extractives (31). In a study of tropical woods, total extractives were found to range from 0.62 to 19.8% (129).
[b] Hemicelluloses.

tural formula for a portion of a cellulose chain molecule is shown in Figure 4-1.

Hemicelluloses are chemically related to cellulose in that both are carbohydrates. Carbohydrates are chemical substances composed of carbon, hydrogen, and oxygen, in which the last two elements are present in the same proportions as they are in water. Separation of cellulose and hemicelluloses is based on their respective solubility in alkali; cellulose is not soluble in a 17.5% solution of caustic soda (NaOH), whereas hemicelluloses are soluble. The molecules of hemicelluloses are also chain-like, as in cellulose, but the degree of polymerization is much smaller (on the average about 150). Unlike cellulose, which is exclusively composed of glucose, hemicelluloses include a variety of monosaccharides. In softwoods these units are mostly mannose (a 6-carbon sugar, like glucose) and some xylose, whereas in hardwoods they are mostly xylose (a 5-carbon sugar) and little mannose (15, 65, 127).

Pectic substances are also carbohydrates or related compounds. They are prominent in cambial tissues, where they form the membrane that separates the young daughter cells produced by the cambium. According to some reports, pectic substances are absent from older wood (having changed into lignin-like compounds); however, the prevailing opinion is that they are present, although in small proportions—they are mainly located in the middle lamella and the primary wall (15, 79, 127).

All carbohydrates (cellulose, hemicelluloses, and pectic substances) are sometimes called, summarily, *holocellulose*. On the basis of solubility in 17.5% caustic soda, holocellulose is subdivided into insoluble *α-cellulose*, which is synonymous to cellulose as used in this text, and the soluble *β-* and *γ-celluloses*.

Lignin is the cell-wall component that differentiates wood from other cellulosic materials produced by nature. Lignification, namely deposition of lignin, constitutes the last stage of cell-wall development. Lignin is produced only by living cells. Completion of lignification practically coincides with consumption of the protoplasm—and cell death.[4] It is interesting to note that lignin always occurs in association with cellulose, whereas cellulose may be found almost pure in nature (e.g., in cotton).

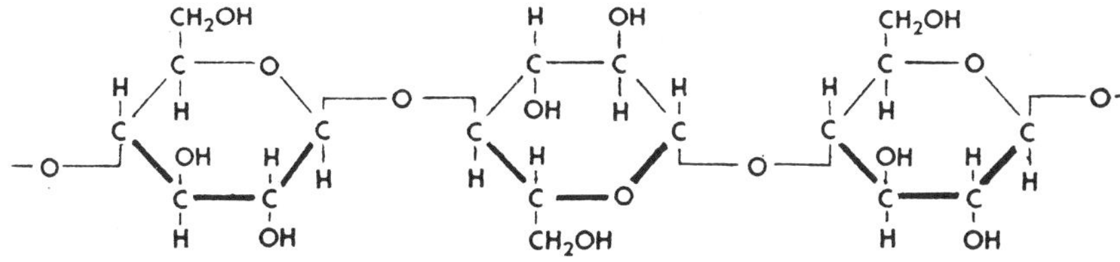

Figure 4-1. Structural formula of a portion of a cellulose chain molecule.

Lignin is not a carbohydrate; it is predominantly aromatic in nature.[5] However, the type of chemical structure and the reactivity of lignin are not completely known, and its isolation is still a problem. The composition of lignin differs between softwoods ("guaiacyl" lignin) and hardwoods ("syringyl" lignin), and also varies especially among different hardwood species (13, 15, 32, 127).

EXTRACTIVES

Wood may contain various inclusions (mainly organic) that are collectively called *extraneous materials* or *extractives*. They are not part of the wood substance, but are deposited in cell lumina and cell walls.

Extractives are compounds of varying chemical composition, such as gums, fats, resins, sugars, oils, starches, alkaloids, and tannins. The term is based on their possible (at least partial) extraction from wood with cold or hot water or neutral organic solvents, such as alcohol, benzene, acetone, or ether. The proportion of extractives varies from less than 1% (e.g., poplar) to more than 10% (e.g., redwood) of the oven-dry weight of wood (31). However, in some tropical species extractives may amount to about 20% (see Table 4-1). Variations exist not only between species, but also within a single tree, mainly between sapwood and heartwood.

Certain inorganic materials, such as calcium salts and inclusions of silica, are not soluble in the solvents mentioned above, but are sometimes considered as extractives since they are not cell-wall components. Under this concept, all inorganic (ash) materials can be considered extractives. Those that reside in cell walls are not structural components of the walls, even though they are dispersed in them (16).

ULTRASTRUCTURAL ARCHITECTURE OF WOOD

Microfibrils and Their Structure

The manner in which the organic components of wood are arranged in order to build its cell walls has been investigated through a variety of research methods, including polarization microscopy, x-rays, and electron microscopy. Nägeli (86), an early investigator, suggested in 1858 that the cell wall is built with microscopically invisible rod-like crystals oriented parallel to themselves and to the cell axis. He hypothesized that these crystals, which he named *micellae* (from *mica*), are embedded in a medium of lignin, water, and other materials, much the same as bricks arranged in a parallel manner in a brick wall are separated by mortar. He based his hypothesis on observations with the polarizing microscope and on the differential shrinkage and swelling of wood, which is greater across than it is along the grain (see Chapter 10).

Despite the distance of time, Nägeli's hypothesis has basic similarities with the present concept of cell-wall ultrastructure. The crystalline nature of cell walls is now established by polarization microscopy, as well as x-ray and electron diffraction studies (94). These techniques do not provide the means to observe individual crystals, but they do provide proof that cell walls are of crystalline structure. They reveal the general orientation of cell-wall crystals, and also provide clues as to their dimensions. The removal of lignin and other noncellulosic components does not change the patterns obtained by the above techniques; this implies that cellulose is the substance with the crystalline properties.

Based on findings through such studies as polarized light and diffraction analyses, observations with electron microscopes,[6] and additional physical and chemical evidence, the now-accepted picture of cell-wall ultrastructure is as follows (34, 85, 95).

The skeleton of wood is built with cellulose chain molecules. Assuming a degree of polymerization of 10,000, the length of such a molecule would be 5 μm, which would make it visible with a light microscope; cellulose molecules are not visible, however, due to their limited width.[7]

The smallest structural units of cell walls, which may be seen readily with an electron microscope, are called *microfibrils*. In electron photomicrographs of replicas of wood surfaces (113), microfibrils appear to be roughly cylindrical[8] and about 10–30 μm in diameter.

It is impossible to measure the length of a microfibril, however, because they are always present in great numbers, they are mostly closely packed, and a single microfibril cannot be followed for a long distance (Figure 4-2). Microfibrils should not be confused with the fibrillar structures that are sometimes visible with the light microscope (5); these structures, called *macrofibrils*, are aggregates of microfibrils.

Each microfibril is a bundle of a number of cellulose chain molecules. The molecules are generally arranged lengthwise with regard to the microfibril axis, but are parallel to each other only in portions. In these portions, which are called *crystalline regions* or *crystallites*, the molecules are strongly connected to each other by hydrogen bridges. The crystallites correspond to the micelles proposed by Nägeli (in fact, crystallites are sometimes referred to as micelles), although they are separate structures.

Parallelism of the cellulose molecules is followed by portions in which these molecules are somewhat disorganized in arrangement—not parallel to one another and not strongly connected; these portions of low molecular order are called *paracrystalline* or *amorphous regions* (Figure 4-3). Cellulose molecules run, therefore, through several crystalline and amorphous regions. These molecules are over 5000 μm in length. Crystallites are at least 60 μm long, 10 μm wide, and 3 μm thick. The transition from crystalline to amorphous regions is gradual (47, 94, 95).[9]

The *degree of crystallinity* (i.e., the relative quantities of crystalline and amorphous regions) varies; however, in general, it is high. According to some investigators, about two thirds of the cellulosic substance is more or less crystalline and about one third amorphous; whereas others believe that only about 5–10% is truly amorphous. It should be noted that measurement of crystallinity is connected with experimental difficulties, and the results obtained by different methods are rarely in agreement.[10] Crystallinity varies from pith outward, and between earlywood and latewood (see Chapter 6); also, primary walls are reported to have smaller degrees of crystallinity and smaller sized crystallites in comparison to secondary walls (34, 49, 94, 95).

Microfibrils are pictured to be composed of a crystalline core surrounded by an amorphous or paracrystalline sheath (94, 95) (Figure 4-3). There is no general agreement on this concept, however. According to another theory, microfibrils are composed of smaller, thread-like units called *elementary fibrils* (34). These have an average diameter of 3.5 μm, and each contains about 40 cellulose chains. Two or more such elementary fibrils are fasciculated to form a microfibril (85). Others doubt the existence of elementary fibrils (45).

Between microfibrils there are interstices or *intermicrofibrillar* spaces that are about 10 μm in width, and within microfibrils there are *intramicrofibrillar* spaces about 1 μm wide (119). These spaces are partly or wholly occupied by noncellulosic materials[11]—hemicelluloses, lignin, pectic substances—which are collectively called *incrusting* materials (94), whereas cellulose is a framework or skeletal substance. Noncellulosic materials do not form microfibrils.[12] It is believed, however, that some lignin may be associated with the amorphous phase of cellulose (120), and also that hemi-

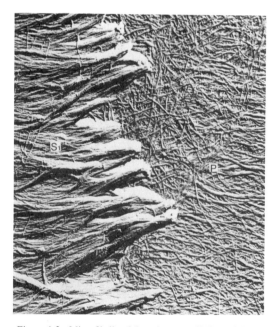

Figure 4-2. Microfibrils of the primary wall (P) and the S_1 layer. Pine (loblolly, 27,500×). (From Ref. 109; reproduced by permission from Materials Education Council, Pennsylvania State University, University Park, Pennsylvania.)

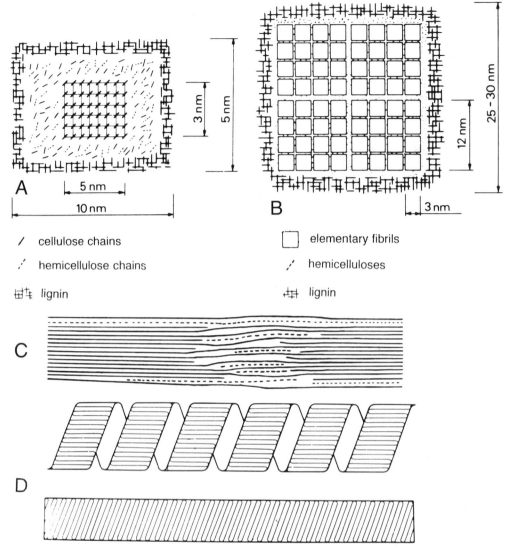

Figure 4-3. Models for ultrastructural organization of a microfibril. (A, B) Transverse section (A, a crystalline core is surrounded by an amorphous layer and lignin; B, the core is composed of "elementary fibrils"). (C) Longitudinal section showing crystalline and amorphous regions. (D) The "folded chain" theory. (A and C after Ref. 94; B after Ref. 30, and D after Ref. 72; reproduced by permission from Chapman and Hall, U.S. Technical Association of the Pulp and Paper Industry, and MacMillan Magazines Ltd; A and B as adapted by J. Dinwoodie.)

cellulose chains may replace some of the cellulose chains of a microfibril (94). Molecules of hemicelluloses are parallel to cellulose molecules and connect cellulose and lignin (87). Lignin is amorphous.

Orientation of Microfibrils: Cell-Wall Layers

The arrangement of microfibrils is not homogeneous within a cell wall. Differences in orientation help to distinguish the primary and the secondary walls of mature cells, and also to recognize component layers in the latter. Before the advent of the electron microscope, Bailey and Kerr (7) determined by iodine staining and polarization microscopy that the secondary wall is composed of three layers, which they designated as outer (S_1), middle (S_2) and inner (S_3); the S_3 layer is toward the cell lumen.

When cross-sections of wood are examined in a polarizing microscope, the outer (S_1) and

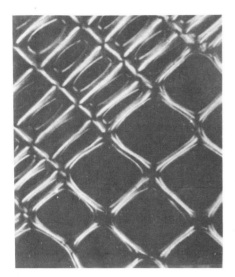

Figure 4-4. A transverse section of pine photographed with polarized light. The inner (S_3) layer appears bright (birefringent), and the middle layer (S_2) is dark (extinct). An outer layer is also bright but the component parts (S_1 and primary wall) cannot be separated (500×). (From Ref. 107; reproduced by permission from the Society of Wood Science and Technology, USA.)

inner (S_3) layers become bright and the middle (S_2) layer darkens (Figure 4-4). This is attributed to the different orientation of microfibrils (crystallites) in these three layers (7). Measurements of the extinction position with the polarizing microscope have shown that the microfibrils form flat spirals (about transverse to cell axis) in the S_1 and S_3 layers, and steep spirals (about parallel to cell axis) in the S_2. The primary wall is rarely visible because, as a rule, it is very thin and cannot be distinguished from the S_1 layer.

X-ray diffraction analysis supports the findings of polarization microscopy with respect to arrangement of the microfibrils in cell walls (94). This technique is limited, however, because a considerable number of cells must be placed in the x-ray beam to produce a distinct diagram. The results, therefore, are dominated by the arrangement of the microfibrils in the S_2 layer as this is the thickest.

The electron microscope provides a unique tool for investigating cell-wall ultrastructure. In addition to direct observation of microfibrils and other cellular characteristics, electron microscopes can be adapted for studies of cell-wall crystallinity; with crystalline materials, electron rays produce similar diffraction patterns as those obtained by the use of x-rays.

Electron photomicrographs reveal that the middle lamella has no special structure—at least as far as the microscope can resolve. The primary wall exhibits a loose weaving of microfibrils (Figure 4-5). Differences have been de-

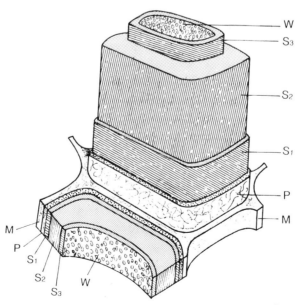

Figure 4-5. Schematic representation of cell-wall layers in a tracheid or fiber with respective orientation of microfibrils (M, middle lamella; P, primary wall; S_1, S_2, S_3 layers of the secondary wall; W, warty layer). (Adapted from Refs. 21, 81, and 91.)

tected between the inner and the outer surfaces (119, 120); the former show a preferred transverse orientation of microfibrils, while in the latter the microfibrils are dispersed. This wall is very thin and does not show the lamellation observed in the secondary wall.

In the secondary wall the microfibrils are closely packed. The S_1 layer is thin and consists of a few (4–6) lamellae. It has a crossed microfibrillar texture, with its lamellae exhibiting an alternating left-handed and right-handed helical arrangement. These helices are designated by S and Z, respectively—S denoting that the microfibrils spiral upward to the left of the observer, and Z to his right.[13] In each lamella, the helical angle is about 50–90°, as measured from the longitudinal axis of the cell (70, 119).

The S_2 layer is thick, especially in latewood tracheids and thick-walled fibers. It is composed of several (30–150) lamellae, each about 60–70 μm thick (70, 119). Adjoining lamellae were observed to exhibit a similar (not crossed) microfibrillar orientation, but deviations in orientation and even direction have also been reported. The microfibrils show a high degree of parallelism (Figure 4-5) in all lamellae (with only a small dispersion), and they run approximately parallel to cell axis, usually not exceeding an angle of about 30°. A Z helical orientation exists in the tracheids of most softwoods (82).

The S_3 layer is usually thinner than the S_1, and it is lamellate (up to 6 lamellae) (119). The angle of microfibrils likewise varies from about 50 to 90°,[14] and the orientation is also alternating S and Z. The microfibrils are relatively loosely textured when compared to those in the S_2, and are often characteristically joined in small bundles which "criss-cross" (Figure 4-6) each other at an angle—usually 20–30° and sometimes as much as 50° (83, 97). The S_3 layer may sometimes be missing (75). It should be noted, however, that absence of birefringence of the innermost layer in the polarizing microscope does not necessarily mean that the S_3 layer is missing. This may be due to a steeper helical arrangement of its microfibrils; such a case has been reported for spruce and birch (76).

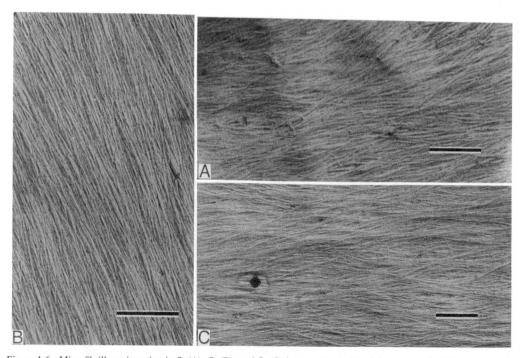

Figure 4-6. Microfibrillar orientation in S_1 (A), S_2 (B), and S_3 (C) layers of tracheids of Japanese cedar (*Chamaecyparis obtusa*) in relation to the longitudinal cell axis (A, 28,000×; B, 34,000×; C, 22,000×; bars = 0.5 μm). (Photographs by Y. Kataoka; courtesy of H. Saiki.)

The transition between cell-wall layers is gradual. There may be several lamellae of intermediate orientation between S_1 and S_2, as well as between the S_2 and S_3 layers (41, 119).

The cell-wall organization described above is largely based on observations of softwood tracheids and hardwood fibers. Vessel members and parenchyma cells present variations which are more difficult to investigate (42, 119). The extensive presence of pits in vessel members results in deviations. The typical three-layer (S_1, S_2, S_3) organization is often present in their walls, but an unlayered structure (undetectable layers) also exists, as well as a multilayered arrangement (with more than four layers and a progressively shifting microfibrillar orientation from 0–90°); differences have been observed between species and position of the vessels (earlywood, latewood) (43).

The three-layered secondary wall is also present in axial and ray parenchyma cells of several softwoods and hardwoods, but this is not the rule (38, 73). In softwoods, a secondary wall may be missing, or an inner "protective" layer may be present, or the structure may be lamellated, consisting of several lamellae. In hardwoods, the primary wall may be lamellated, and an inner amorphous (protective) layer may be present (rich in hemicelluloses, containing some pectic substances along with scattered cellulose microfibrils) (43). In the three-layered structure, the microfibrils are generally quite parallel to cell axis in the central (S_2) layer, and they are arranged in flat spirals (at an angle of about 30–80°) in the S_1 and S_3 layers. Microfibrillar orientation appears to be similar in axial and ray parenchyma cells (42, 43).

The thickness of cell-wall layers may be measured on electron photomicrographs of ultrathin cross-sections of wood tissue. Measurements of softwood tracheids show that most of the thickness of the cell wall (80% or more) is occupied by the S_2 layer[15] (Figure 4-7). This layer was also found to be the thickest in vessel members (proportional thickness was found to be about 50% in beech), but exceptions may occur in parenchyma cells; in ray parenchyma cells of cedar (*Cryptomeria*), the S_2 layer was observed to be thin, whereas S_1 and S_3 were comparatively thick (42).

The orientation of microfibrils in the various cell-wall layers may be directly observed in the electron microscope, although there is the limitation that only a small portion of a cell can be seen at a time, and observations should be care-

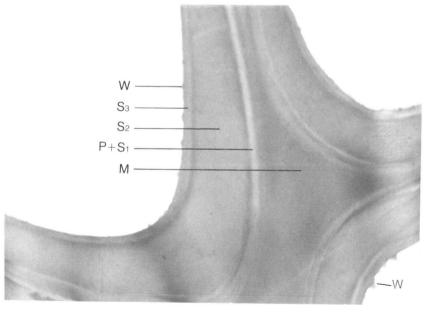

Figure 4-7. Cell-wall layers in transverse section. M, middle lamella; P + S_1, primary wall and S_1 layer, layers S_2 and S_3, and W, warty layer (Scots pine, 11,200×).

fully correlated to cell axis. Microfibril angles may also be measured with x-rays and polarization microscopy, but it is not always necessary to use these techniques. Most important (with respect to wood properties) is the magnitude of the angle of microfibrils in the S_2 layer; this dominating angle can be measured with a light microscope—from the direction of elongated pit mouths in relation to cell axis. This angle is normally small; as already mentioned, the microfibrils in the S_2 layer are nearly parallel to cell axis. However, angles are affected by cell size: shorter and wider cells have larger angles (94). The presence of many pits (as in radial walls of axial softwood tracheids, and in the walls of vessel members) causes deviations in orientation of microfibrils, resulting in larger angles.

Spiral Thickenings and Perforation Plates

In the electron microscope, spiral thickenings are shown to be composed of microfibrils, and to lie on the S_3 layer of the secondary wall. On the surface of such thickenings the microfibrils are interwoven as in the S_3 layer, but their orientation does not always coincide. In certain woods (yew), spiral thickenings are loosely attached, while in others (Douglas-fir, spruce, larch) they have broad bases blended tightly with the microfibrils of the S_3 layer (88).

The electron microscope has also been used to study perforation plates in vessel members; it was observed that the bars of scalariform perforations (see Figure 3-13) may be connected with microfibrils (80).

Ultrastructure of Pits

The structure of pits was adequately described long before the application of electron microscopy to wood research. Use of the instrument has revealed previously invisible structural characteristics of pit membranes, but even here much was learned from the light microscope. For example, it was observed that the membranes of bordered pits in most softwoods possess a torus. It was learned that the portion of the membrane surrounding the torus (i.e., the margo) is not homogeneous, but is equipped with radial filaments. Bailey (3), in 1913, was able to demonstrate that it has "minute openings" (Figure 4-8).

The electron microscope is a unique tool for the study of pit structure under high resolution. Most of this work has been concerned with bordered pits of softwoods. In such pits it was shown that the S_3 layer covers the pit border inside (pit cavity) and out (lumen side, Figure 4-6). Microfibrils in the walls of pit cavities exhibit a circular pattern around the aperture. Within the pit border there is a centripetal de-

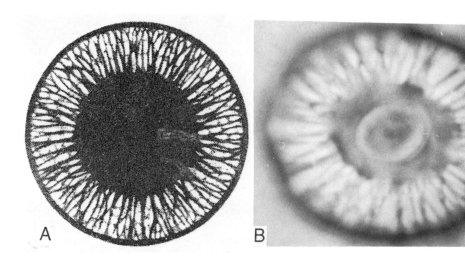

Figure 4-8. (A) The membrane of a softwood bordered pit, as drawn by Bailey in 1913. (B) Membranes (with torus and radial filaments) have also been photographed in an optical (light) microscope. (Photograph B courtesy of G.P. Berlyn.)

position of lamellae of the S_2 layer; such lamellae are deposited until the final size of the pit aperture is reached (41).

Electron microscopic studies of pit membranes reveal differences, especially between bordered pits of softwoods and hardwoods. Almost all important softwoods were found to possess a torus (Figures 4-9A, B, 4-10A). In its typical form, the torus is composed of three layers: the primary walls of the two adjoining cells and the middle lamella. No trace of the latter appears in the margo (i.e., the portion of the membrane surrounding the torus) (22, 44). Additional incrusting (non-microfibrillar) materials are deposited to form the torus disc, and an apparently secondary deposition of microfi-

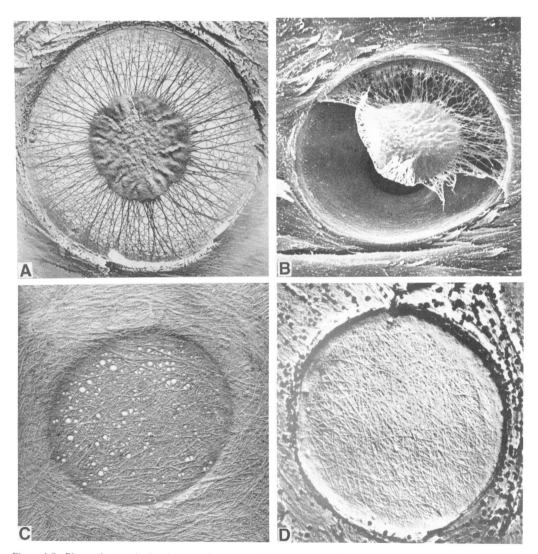

Figure 4-9. Pit membrances in the electron microscope. (A) Membrane of a bordered pit in fir (*Abies*). (B) Membrane of a bordered pit in pine; the pit chamber and part of the pit aperture may be seen behind the broken membrane. (C) The central part of a membrane of a bordered pit of spruce—after a delignification treatment that removed the incrusting material forming the torus and revealed a network of microfibrils over an imprint of the pit aperture. Note an apparently secondary deposition of microfibrils on the edge of the torus. (D) Membrane of a simple pit in oak (A, 4000×; B, 3000X; C, 7400; D, 9000×). (Photograph A courtesy of A. Petty; C reproduced by permission from Pergamon Press; B and D from Refs. 81 and 125, reproduced by permission from Chapman and Hall, and the Society of Wood Science and Technology, USA.)

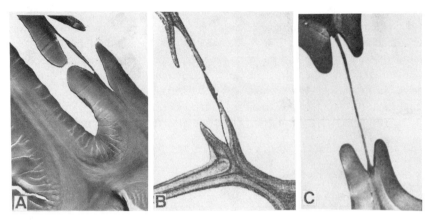

Figure 4-10. Pit membranes in section. (A, B) Softwood (with torus); (C) hardwood (without torus). (A) Larch (*Larix laricina*, 2400×, compression wood—see Chapter 7); (B) redwood (*Sequoia sempervirens*, aspirated, 1500×); (C) red oak (*Quercus rubra*, 7200×). (Photographs A and C courtesy of N.C. Brown Center for Ultrastructure Studies, Syracuse, New York; B by R. Krahmer and W.A. Côté, reproduced by permission from Springer Verlag.)

brils also takes place; these microfibrils may be observed on the torus—arranged in a preferentially circular and parallel manner. The torus is supported in a central position by microfibrils and microfibrillar bundles, among which there are openings ranging in size from a few nanometers (nm) to about 1 μm (43, 69, 102, 110).

Such a typical organization of softwood pit membranes exists in pine, spruce, larch, Douglas-fir, fir, hemlock, and cedar (*Cedrus*) (9); *scalloped* tori occur in *Cedrus* (96) (Figure 4-11), and *extended* tori (with irregular margins) in hemlock. In other softwoods—for example cedars (*Thuja, Cryptomeria*) and redwood— the pit membrane consists of densely packed and radially oriented microfibrils, which may be visible or slightly obstructed by incrustations at the center of the membrane; circularly arranged microfibrils, as in typical tori, also do not exist (69).[16]

Variations in details of pit membrane structure exist not only between genera, but also within a tree and between adjoining tracheids (6). In general, the porosity of membranes is reduced by incrustations with increasing age, especially as sapwood is transformed into heartwood. The appearance of the torus disc changes. In old heartwood, the membrane often changes so that torus and supporting microfibrils are hardly discernible (36, 61, 63, 102).

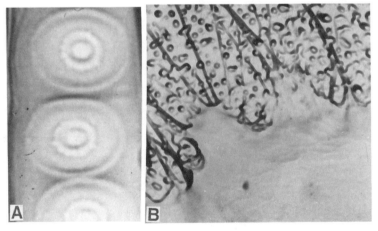

Figure 4-11. Scalloped tori in cedar (*Cedrus libani*). (A) appearance in a light microscope (600×); (B) portion of a membrane (11000×).

The observation of aspirated pit membranes with the electron microscope has shown that the torus does not always seal the pit aperture tightly; the presence of warts on the wall of the pit cavity may prevent such a tight seal (20). It was also observed that pit aspiration is not irreversible; pit membranes were experimentally aspirated and deaspirated by solvent exchange drying (110). The appearance of such deaspirated membranes provides additional support to the fact that the torus is a normal feature of membranes of softwood bordered pits—and not a result of aspiration, as has been suggested (52).

Chemical treatment (delignification) removes the incrusting materials from the torus, and reveals the underlying primary wall microfibrils. When present, the secondary deposition of circularly arranged microfibrils becomes distinct on the edge of the torus (96, 114) (Figure 4-9C).

Membranes of hardwood bordered pits do not possess a torus (Figures 4-9D, 4-10B). Tori also do not exist in half-bordered pits and in simple pits both of softwoods and hardwoods.[17] These membranes are made of randomly arranged microfibrils with a primary wall texture (19, 101, 123, 125). Three types of such membranes have been recognized: *intervascular*, *vessel-parenchyma*, and *interparenchymatous* (101). Membranes between vessel members (intervascular) were observed to consist of various microfibrillar layers alternating with layers of incrusting (amorphous) materials. Membranes of pits connecting vessel members and parenchyma cells (vessel-parenchyma membranes) show an additional "protection layer" on the side of the parenchyma cell. Finally, membranes of pits between parenchyma cells (interparenchymatous) possess characteristic small openings which are former passages of plasmodesmata.[18]

Electron microscope study of the structure of vestured pits has confirmed earlier observations by Bailey (4), who described vestures as outgrowths or deposits in pit cavities and not openings in pit membranes. Vestures are non-microfibrillar protrusions that occur at vessel pits (Figure 4-12) and sometimes on vessel walls (rarely on fibers and axial parenchyma

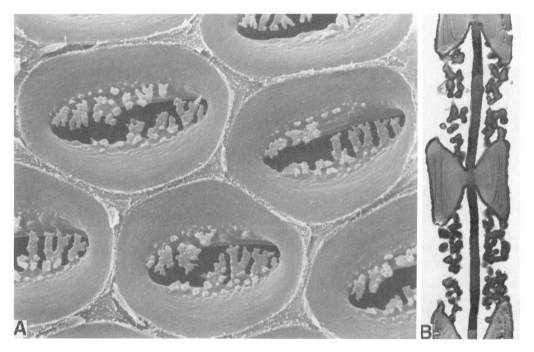

Figure 4-12. Vestured pits in (A) black locust (radial view, 4800×); (B) *Parashorea plicata*, tangential section (4300×). (Photograph A courtesy of J. Ohtani and S. Ishida; B courtesy of N. C. Brown Center for Ultrastructure Studies, Syracuse, New York.)

cells); they may be branched or unbranched, and their presence is of diagnostic value (23, 89).

The Warty Layer

The electron microscope has also revealed that a layer or thin membrane may line the lumina and pit cavities of tracheids, fibers, and vessels of many softwoods and hardwoods. When present, this layer is attached to the S_3 layer, and may occur on top of spiral thickenings. Examined from the inside of cell lumina or pit cavities, it appears to be granular—that is, covered with warts (Figure 4-13). The underlying microfibrils may be partly seen between these warts, or may be totally hidden by an amorphous layer associated with the warts. The presence, size, and distribution of warts may be of some taxonomic value; existence of a relationship between warts and dentation of ray tracheids has been proposed, but this does not seem to be absolute (35, 67). In certain species—beech and cedar (*Cedrus*), for exam-

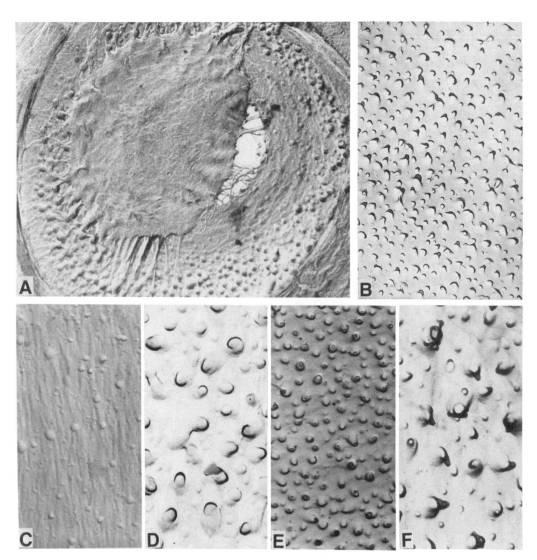

Figure 4-13. Warty layers. (A) On the wall of the pit chamber of a bordered pit of pine; the membrane is displaced revealing part of the pit aperture. Warty layers in (B) fir (*Abies*), (C) Scots pine, (D) beech, (E) redwood (*Sequoia*), (F) cedar (*Cedrus*). (A, 5300×; B, 13,600×; C, 10,700×; D, 20,400×; E, 13,200×; F, 19,100×.)

ple—warts may be seen with a light microscope.

The warty layer forms during the final stages of cell-wall development, and is regarded as a structure arising from the dying protoplasm (deposited by the living protoplasm prior to its degeneration). Warts appear morphologically similar to vestures—which have been observed on parenchyma cells, trabeculae, and spiral thickenings (88, 89). Warts and vestures often coexist, and are considered similar in origin and chemical composition (in fact, it has been proposed that the warty layer be named "vestured layer") (90). The major chemical component of both warts and vestures is lignin; hemicelluloses are also present, and some pectin and cellulose have been reported to occur in vestures (23, 68, 100). Some evidence suggests that the warty layer is related to extractives; extraction of wood surfaces with water and neutral organic solvents was observed to entirely remove the warty layer or to erode the warts (115).

Ultrastructure of Tyloses

Tyloses are initiated when membranes of pit-pairs connecting vessel members with adjoining parenchyma cells rupture or disintegrate—probably as a result of increased osmotic pressure in parenchyma cells.[19] The living protoplasms of such cells enter the lumina of adjoining vessels, and there become enlarged and develop walls that are microfibrillar in nature. Tyloses walls are composed of the same chemical constituents as the cell walls, and may consist of one, two, or three layers (a suberized layer may also be present) (93).[20] When two tyloses meet, intertylosic pit-pairs are formed, the membranes of which are similar to membranes of hardwood pit-pairs (Figure 4-14). Adjoining tyloses are bonded together with a cementing substance that appears to be similar in nature to the middle lamella (62).

Distribution of Chemical Components in Cell Walls

Chemical determination of the organic cell-wall components (cellulose, hemicelluloses, lignin) is made by various methods—usually by use of wood reduced to small particles (wood flour), and only after the extractives have been removed (2, 108). When desired to examine the distribution of these components in the cell wall, different techniques are required. In traditional botanical microtechnique, stain reagents are applied on microtome sections of wood. Several stain combinations may be used for this purpose. Stains are helpful because they stain cellulose and lignin differently. By use of stains, it is also possible to follow the progress of chemical differentiation of cell walls, beginning with separation from the cambium until lignification is complete.

Related techniques have been applied for the selective staining of cell-wall layers. For example, application of Victoria blue (in combination with swelling) stained the primary wall and the S_3 layer blue, and the S_2 layer red. This

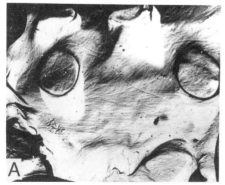

Figure 4-14. Ultrastructure of tyloses. (A) Tylosis wall showing microfibrils and intertylosic pits; (B) section of a pit membrane. (Black locust, 8600×). (Courtesy of Z. Koran.)

made it possible to show that the S_3 layer lines the cavities of bordered pits, and is therefore connected to the primary wall.

In addition to staining, the distribution of cell-wall components has been studied by several other techniques, such as chromatography, spectrography, chemical analysis of isolated cells at various stages of development, dissolution of nonlignin compounds, microdissection followed by chemical analysis, and ultraviolet and fluorescence microscopy (76). Biological attack of wood by fungi has also been used, as different cell-wall components are preferred by different fungi (25, 79). On the basis of such work, in addition to observations with the electron microscope, the following conclusions may be drawn.

There is no cellulose in the middle lamella; this is mostly lignin. Lignin concentration is higher at cell corners,[21] and it is also reported to be higher between radial rather than between tangential walls (120). The primary wall—originally mostly cellulose—acquires large deposits of lignin during the process of lignification. As a result, the compound middle lamella (middle lamella and primary walls) has a very high lignin content.[22] In the compound middle lamella there is also a high concentration of pectic materials. In the secondary wall the amount of lignin is low, about 10–20% (10, 97) (Figure 4-15). Lignin content varies in the three layers of the secondary wall. Reports concerning the mode of variation do not agree. According to some studies, there is a gradual reduction toward the lumen (the rate of reduction reported to be less pronounced in softwoods than in hardwoods) (25), but others report increased concentrations in the S_3 layer (99, 100). As already mentioned, lignin is located between microfibrils. In addition, it is apparently concentrated between lamellae in the secondary wall, as evidenced by concentric patterns obtained by differential solubility techniques. Radioconcentric patterns are also observed; however, these are attributed to fissures in the cellulosic framework, which are occupied by lignin (120).

The proportion of hemicelluloses is also high in the middle lamella and the primary walls, where cellulose content is lowest (66, 79, 100). Cellulose is mostly concentrated in the secondary wall. A detailed study of the relative distribution of cellulose and hemicelluloses across the cell walls of tracheids and fibers of Scots pine, Norway spruce, and European birch has shown that cellulose comprises about 40% (33.4–41.4%) and hemicelluloses about 60% in the compound middle lamella. In the secondary wall, cellulose content ranged from about 50% to over 60% (47.5–66.5%), however, with respective reduction of hemicelluloses. The proportion of cellulose in the different layers is shown in Table 4-2.

Differences were also observed between cell types: in birch, parenchyma cells had a very low proportion of cellulose (14%) in relation to hemicelluloses; whereas in other cell types (fibers and vessel members of birch—axial tracheids, and parenchyma cells together with ray tracheids in Scots pine), the proportion of cellulose was about equal to that of hemicelluloses (79). In addition, differences in chemical composition exist between earlywood and latewood (15, 49, 126), and between wood located at different distances from pith (see Chapter 6). Cellulose and lignin content and distribution differ between normal wood and reaction wood (see Chapter 7).

The nature of linkage between cellulose and lignin is not clearly understood. Some believe that cellulose is mechanically encased by lignin, whereas others suggest that the linkage is

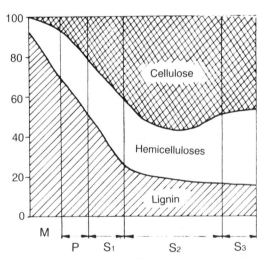

Figure 4-15. Distribution (%) of cellulose, hemicelluloses, and lignin in the cell wall of a hardwood. (After Ref. 13; reproduced by permission from VEB Fachbuchverlag.)

Table 4-2. Distribution of Cellulose in Cell-Wall Layers.

Species	ML + P	S_1	S_2 (outer part)	S_2 (inner part) + S_3
Pine	33.4[a]	55.2	64.3	63.6
Spruce	35.5	61.5	66.5	47.5
Birch	41.4	49.8	48.0	60.0

[a] In percent of the total amount of cellulose and hemicelluloses. Figures complementary to 100.0 designate in all cases the respective percentages of hemicelluloses (79).

chemical, at least in part. Cellulose is chemically linked to hemicelluloses, and there is little dispute that the linkage of hemicelluloses to lignin is also chemical (15, 28, 56, 66, 75, 79).

Extractives are not a component of cell-wall substance. These are deposited in cell cavities and in cell walls, and they are also present in the warty layer (115). Sometimes, they occupy intercellular spaces.

EFFECTS OF CHEMICAL COMPONENTS ON WOOD PROPERTIES AND USES

Important to the concept of wood as raw material is the contribution of the chemical components of wood to its properties and utilization. As already mentioned in the Introduction, cellulose is the main contributor to chemical wood utilization—making wood a valuable material for pulp, paper, and a multitude of other products.

Strength is imparted to wood by all its chemical components. Cellulose is primarily responsible for the very high strength of wood in axial tension, because of the axial (parallel to tree length) arrangement of the microfibrils and cellulose chain molecules in the S_2 (the thickest) layer of the secondary wall.[23] Hemicelluloses and lignin bind the cells together and support the cellulosic framework, thus contributing desirable elasticity and compressive strength (59, 98, 120).[24]

Removal of lignin and hemicelluloses drastically decreases the strength of wood in wet condition. Experiments with beech and poplar have shown that removal of lignin reduced wet strength (in axial tension, compression, and bending) to only about 10–20% of values pertaining to wood in its natural state (57). Similarly, removal of lignin and most of the hemicelluloses reduced the wet strength of birch and poplar up to zero (58, 59, 98). Dry strength was observed to increase on the basis of unit area of remaining tissue (57, 59). The difference in behavior of wood in wet and dry condition results from the fact that in dry condition pectic substances and hemicelluloses still bond cells to one another, whereas in wet condition they are not capable of performing such a function (57, 59, 74). Thus, the coherence between cells is destroyed and strength is drastically reduced or eliminated.

Individual fibers (cells) are subject to less reduction of strength, because the incrusting substances (lignin, in particular) are located in large amounts between cells. However, comparison of fibers is complicated because of difficulties in isolating them in natural state. Some preliminary work with mechanically macerated and "holocellulose" fibers from a tropical hardwood (*Pentacme contorta*) indicates only a small reduction of axial strength in the latter—about 15% (in air-dry condition) (106). Similar tests with delignified softwood tracheids, from which varying amounts of hemicelluloses were progressively removed, showed an increase of axial tensile strength at first, then a decrease (on the basis of cross-sectional area of remaining tissue); however, such a decrease was not as great as was reported above for wood tissue (74).

The hygroscopicity of wood (i.e., its affinity for water and other liquids) is due to its chemical composition—mainly to free –OH groups on cellulose chain molecules; it is also due to the presence of hydrophilic components, such as pectic substances and hemicelluloses. Wood expands upon gaining moisture. The anisotropic character of swelling (and inversely of shrinking)—namely, the negligible dimensional changes in the axial direction (as opposed to comparatively high changes in the transverse direction)—is fundamentally due to the orientation of microfibrils in the S_2 layer of the secondary wall. However, the almost transverse to cell axis orientation of microfibrils in the S_1 and S_3 layers exercises a restraining effect on transverse dimensional changes. If all microfibrils were about parallel to cell axis, as in the S_2 layer, swelling of wood would be unlimited.

Lignin also imparts dimensional stability to wood, because it occupies spaces in cell walls which would be taken up by water. Nonlignified cell walls shrink more than lignified ones (118).

The angle of microfibrils in the S_2 layer is a very important characteristic of wood with respect to its properties. Deviations of the microfibrils from parallelism to cell axis increase axial shrinkage and swelling, and decrease the strength of wood in the axial direction. In addition, the angle of microfibrils affects heat conductivity and other wood properties (94).

Degree of crystallinity is also very important. The proportion of crystalline cellulose is related to many properties of wood, fibers (individual cells), and paper. Among the properties affected are bending, extensibility, swelling and shrinking, bonding of fibers, staining, tear, and resistance to chemical attack (65).

Extractives have profound effects, and are greatly responsible for differences of properties among species. They influence color, odor, taste, fluorescence, durability, inflammability, wood–moisture relations, gluing, pulping, and other properties (51, 64). For example, toxic extractives impart durability. Woods with comparatively less extractives can hold more water in their cell walls than woods with a higher extractive content; removal of extractives increases the capacity of cell walls for water—therefore, the magnitude of shrinkage and swelling and other wood–moisture relations. Certain extractives cause wear and corrosion of cutting tools (due to acidic constitution), or affect the adhesion of coatings, and the setting of glues (e.g., in making particleboard, including cement-particleboard). The processes and economics of pulping are also greatly influenced by the amount and kind of extractives (extractives influence chemical reactions, increase the consumption of bleaches and water, and may create problems in the processing equipment, e.g., pine resin). The dust produced by machining certain woods, mainly tropical, due to the nature of their extractives, may be injurious to health (causing dermatitis, asthma, and other diseases), and are suspected of causing cancer. Some extractives, such as resins and tannins, obtained from wood (and bark) are valuable products in themselves.

BIOLOGICAL ASPECTS

Electron microscopy and other research techniques have contributed significantly to our comprehension of the structural organization of wood. However, biological aspects of cell-wall ultrastructure present questions not yet answered. Thus, it is largely unknown where and how cellulose molecules are synthesized, and how and where microfibrils arise from them (18, 40, 84). It is assumed that no preforming of cellulose occurs, and that both the molecules and the microfibrils grow in length at the same time and at the same point—namely, at their tips (94, 97). Growth is supposed by some to be performed by action of enzymes which carry monomers to the ends of molecules. The more or less uniform thickness of microfibrils (plus the fact that no tendency to transverse growth is detectable) (18) supports the assumption that growth in diameter does not occur. Synthesis of cellulose microfibrils has not been achieved in the laboratory (in vitro), and this creates difficulties of precisely understanding the mechanisms involved (71).

With regard to microfibrillar orientation, a widely accepted concept is the so-called *multinet growth* (97), which visualizes an apposition of successive layers of microfibrils, with the earlier layers becoming modified in orientation by wall extension during cell enlargement. Another hypothesis explains the orientation through forces generated in the living cell: the cell membrane (plasmalemma) is considered to sense the strains developed during extension of the cell wall and to convey this information to the protoplasm, which in turn directs the formation and orientation of microfibrils to satisfy strength requirements; this is called the *strain hypothesis* (14).

The biosynthesis of hemicelluloses and lignin is also not clearly understood. Hemicelluloses are believed to be synthesized by enzymes in a manner similar to the formation of entirely amorphous cellulose (50, 97).

The origin of lignin is likewise not clear. Enzymatic reactions are again involved, and the precursors (coniferyl alcohol, sinapyl alcohol) are believed to form in lignifying cells and not transported from the cambium or phloem. Softwood and hardwood lignin differ in biosynthesis (37, 39, 50, 122).

There is also lack of knowledge concerning the systems controlling the formation of tannins and other extractives. Oleoresin is regarded to result from biosynthetic activities of epithelial cells lining the resin canals; it is reportedly made in plastids of epithelial cells but the precursors are transported from the phloem (28, 50).

REFERENCES

1. Adler, E. 1977. Lignin chemistry: Past, present and future. *Wood Sci. Tech.* 11(3):169–218.
2. American Society for Testing and Materials. 1985. *Book of ASTM Standards*. Philadelphia.
3. Bailey, I. W. 1913. The preservative treatment of wood. II. The structure of the pit membranes in the tracheids of conifers and their relation to the penetration of gases, liquids and finely divided solids into green and seasoned wood. *For. Quart.* 11:12–20.
4. Bailey, I. W. 1933. The cambium and its derivative tissues. VIII. Structure, distribution and diagnostic significance of vestured pits in dicotyledons. *J. Arnold Arbor.* 14:259–273.
5. Bailey, I. W. 1957. Aggregations of microfibrils and their orientations in the secondary wall of coniferous tracheids. *Amer. J. Bot.* 44:415–418.
6. Bailey, I. W. 1957. Die Struktur der Tüpfelmembranen bei den Tracheiden der Koniferen. *Holz Roh-Werkstoff* 15:210–213.
7. Bailey, I. W., and T. Kerr. 1934. The cambium and its derivative tissues. X. Structure, optical properties and chemical composition of the so-called middle lamella. *J. Arnold Arbor.* 15:327–349.
8. Bamber, R. K. 1981. Structure-property relationships of timber. *IAWA Bull.* n.s. 2(2/3):56.
9. Bauch, J., W. Liese, and R. Schulze. 1972. The morphological variability of bordered pit membranes in gymnosperms. *Wood Sci. Tech.* 6(3):166–184.
10. Berlyn, G. P. 1964. Recent advances in wood anatomy: The cell walls in secondary xylem. *For. Prod. J.* 14:467–476.
11. Berlyn, G. P., and R. E. Mark. 1965. Lignin distribution in wood cell walls. *For. Prod. J.* 15:140–141.
12. Blackwell, J. 1982. The macromolecular organization of cellulose and hitin. In *Cellulose and Other Natural Polymer Systems*, ed. R. Malcolm Brown, pp. 403–428. New York: Plenum Press.
13. Blazej, A., L. Suty, and P. Krkoska. 1979. *Chemie des Holzes*. Leipzig: VEB Fachbuchverlag.
14. Boyd, J. D. 1985. *Biophysical Control of Microfibril Orientation in Plant Cell Walls*. Dordrecht, The Netherlands: Nijhoff/Junk.
15. Browning, B. L., ed. 1963. *The Chemistry of Wood*. New York: Interscience Publishing.
16. Chatters, R. M. 1963. Siliceous skeletons of wood fibers. *For. Prod. J.* 13:368–372.
17. Chnazy, H., B. Henrissat, and R. Vuong. 1986. Structural changes of cellulose crystals during the reversible transformation of cellulose I-III in *Valonia*. *Holzforschung* 40 (Suppl.):25–30.
18. Colvin, J. R. 1964. The biosynthesis of cellulose. In *The Formation of Wood in Forest Trees*, ed. M. H. Zimmermann, pp. 189–201. New York: Academic Press.
19. Côté, W. A. 1958. Electron microscope studies of pit membrane structure: Implications in seasoning and preservation of wood. *For. Prod. J.* 8:296–301.
20. Côté, W. A. 1963. Structural factors affecting the permeability of wood. *J. Polymer Sci. C* 2:231–242.
21. Côté, W. A. 1967. *Wood Ultrastructure: An Atlas of Electron Micrographs*. Seattle: University of Washington Press.
22. Côté, W. A. 1976. Wood ultrastructure in relation to chemical composition. In *The Structure, Biosynthesis and Degradation of Wood*, eds. F. A. Loewus and V. C. Runeckles, pp. 1–44. New York: Plenum Press.
23. Côté, W. A., and A. C. Day. 1970. Vestured pits and apparent relationship with warts. *Tappi* 45:906–909.
24. Collet, B. M. 1970. Scanning electron microscopy: A review and report of research in wood science. *Wood and Fiber* 2(2):113–133.
25. Cowling, E. B. 1965. Microorganisms and microbial enzyme systems as selective tools in wood anatomy. In *Cellular Ultrastructure of Wood Plants*, ed. W. A. Côté, pp. 341–368. Syracuse, New York: Syracuse University Press.
26. Cutter, B. E., E. A. McGinnes, and P. W. Schmidt. 1980. X-ray scattering and x-ray diffraction techniques in studies of gamma-irradiated wood. *Wood and Fiber* 11(4):228–232.
27. Ellis, E. T. 1965. Inorganic elements in wood. In *Cellular Ultrastructure of Wood Plants*, ed. W. A. Côté, pp. 181–189. Syracuse, New York: Syracuse University Press.
28. Eriksson, O., D. A. I. Goring, and B. O. Lindgren. 1980. Structural studies on the chemical bonds be-

tween lignins and carbohydrates in spruce wood. *Wood Sci. Tech.* 14(4):267-279.
29. Fahn, A. 1979. *Secretory Tissues in Plants.* New York: Academic Press.
30. Fengel, D. 1970. Ultrastructural behavior of cell wall polysaccharides. *Tappi* 53(3):497-503.
31. Fengel, D., and D. Grosser. 1975. Chemische Zusammensetzung von Nadel- und Laubhölzern. *Holz Roh- Werkstoff* 33(1):32-34.
32. Fengel, D., and G. Wegener. 1984. *Wood: Chemistry, Ultrastructure, Reactions.* New York: Walter de Gruyter.
33. Fergus, B. J., A. R. Procter, J. A. N. Scott, and D. A. Goring. 1969. The distribution of lignin in sprucewood as determined by ultraviolet microscopy. *Wood Sci. Tech.* 3:117-138.
34. Frey-Wyssling, A. 1957. *Macromolecules in Cell Structure.* Cambridge, Massachusetts: Harvard University Press.
35. Frey-Wyssling, A., K. Mühlethaler, and H. H. Bosshard. 1955. Das Elecktronmikroskop im Dienste der Bestimmung von Pinusarten. *Holz Roh- Werkstoff* 13:245-249.
36. Frey-Wyssling, A., H. H. Bosshard, and K. Mühlethaler. 1956. Die submikroskopische Entwicklung der Hoftüpfel. *Planta* 47:115-126.
37. Freudenberg, K. 1964. The formation of lignin in the tissue and in vitro. In *The Formation of Wood in Forest Trees*, ed. M. H. Zimmermann, pp. 203-218. New York: Academic Press.
38. Fujii, T., H. Harada, and H. Saiki. 1979. The layered structure of ray parenchyma secondary wall in the wood of 49 japanese angiosperm species. *Mokuzai Gakkaishi* 25:251-257.
39. Gross, G. G. 1976. Biosynthesis of lignin and related monomers. In *The Structure, Biosynthesis and Degradation of Wood*, ed. F. A. Loewus and V. C. Runeckles, pp. 141-184. New York: Plenum Press.
40. Hall, J. L., T. J. Flowers, and R. M. Roberts. 1982. *Plant Cell Structure and Metabolism.* New York: Longman.
41. Harada, H. 1965. Ultrastructure and organization of gymnosperm cell walls. In *Cellular Ultrastructure of Woody Plants*, ed. W. A. Côté, pp. 215-233. Syracuse, New York: Syracuse University Press.
42. Harada, H. 1965. Ultrastructure of angiosperm vessels and ray parenchyma. In *Cellular Ultrastructure of Woody Plants*, ed. W. A. Côté, pp. 235-249. Syracuse, New York: Syracuse University Press.
43. Harada, H., Y. Miyasaki, and T. Wakashima. 1958. Electron microscopic investigation on the cell wall structure of wood. *Bull. Gov. For. Exp. St.* 104, Meguro, Tokyo, Japan.
44. Harada, H., and W. A. Côté. 1967. Cell wall organization in the pit border region of softwood tracheids. *Holzforschung* 21(3):81-85.
45. Harada, H., and T. Goto. 1982. The structure of cellulose microfibrils in Valonia. In *Cellulose and Other Natural Polymer Systems*, ed. R. Malcolm Brown, pp. 383-401. New York: Plenum Press.
46. Hayat, M. A. 1986. *Basic Techniques for Transmission Electron Microscopy.* Orlando, Florida: Academic Press.
47. Hearle, J. W. S. 1963. The fine structure of fibres and crystalline polymers. I. Fringe fibril structure. *J. Appl. Polym. Sci.* 7:1175-1193.
48. Hearle, J. W. S., J. T. Sparrow, and P. M. Grosse. 1972. *The Use of Scanning Electron Microscope.* Oxford: Pergamon Press.
49. Hermans, P. H. 1949. *Physics and Chemistry of Cellulose Fibers.* New York: Elsevier.
50. Higuchi, T., ed. 1985. *Biosynthesis and Biodegradation of Wood Components.* New York: Academic Press.
51. Hillis, W. E. 1987. *Heartwood and Tree Exudates.* New York: Springer-Verlag.
52. Jayme, G., G. Hunger, and D. Fengel. 1960. Das elektronmikroskopische Bild des Cellulosefeinbaus verschlossener und unverschlossener Hoftüpel der Nadelhölzer. *Holzforschung* 14:97-105.
53. Jayme, G., and D. Fengel. 1961. Beitrag zur Kentniss des Feinbaus der Frühholztracheiden. Beobachtungen an Ultradünnschnitten von Fichtenholz. *Holz Roh- Werkstoff* 19:50-55.
54. Jutte, S. M., and B. J. Spit. 1963. The submicroscopic structure of bordered pits on the radial walls of tracheids in Parana pine, Kauri and European spruce. *Holzforschung* 17:168-175.
55. Illston, J. M., J. M. Dinwoodie, and A. A. Smith. 1979. *Concrete, Timber and Metals.* New York: Van Nostrand Reinhold.
56. Iversen, T. 1985. Lignin-carbohydrate bonds in a lignin-carbohydrate complex isolated from spruce. *Wood Sci. Tech.* 19:243-251.
57. Klauditz, W. 1952. Zur biologisch-mechanischen Wirkung des Lignins im Stammholz der Nadel- und Laubhölzer. *Holzforschung* 6:70-82.
58. Klauditz, W. 1957. Zur biologisch-mechanischen Wirkung der Acetylgruppen im Festigungsgewebe der Laubhölzer. *Holzforschung* 11:47-55.
59. Klauditz, W. 1957. Zur biologisch-mechanischen Wirkung der Cellulose und Hemicellulose im Festigungsgewebe der Laubhölzer. *Holzforschung* 11:110-116.
60. Kollmann, F. 1951. *Technologie des Holzes und der Holzwerkstoffe*, Band 1. Berlin: Springer, Verlag.
61. Koran, Z. 1964. Air permeability and creosote retention of Douglas-fir. *For. Prod. J.* 14:159-166.
62. Koran, Z., and W. A. Côté. 1965. The ultrastructure of tyloses. In *Cellular Ultrastructure of Woody Plants*, ed. W. A. Côté, pp. 319-333. Syracuse, New York: Syracuse University Press.
63. Krahmer, R. L., and W. A. Côté. 1963. Changes in coniferous wood cells associated with heartwood formation. *Tappi* 46:42-49.
64. Krilov, A. 1987. Corrosive properties of some eucalypts. *Wood Sci. Tech.* 21:211-217.
65. Kürschner, K. 1962. *Chemie des Holzes.* Berlin: VEB Deutscher Verlag der Wissenschaften.
66. Lange, P. W. 1958. The distribution of the chemical

constituents throughout the cell wall. In *Fundamentals of Papermaking Fibers*, ed. F. Bolam, pp. 147–185. Kenley, Surrey, England: British Paper and Board Maker's Association.
67. Liese, W. 1956. Zur systematischen Bedeutung der submikroskopischen Warzenstruktur bei der Gattung *Pinus*. *Holz Roh- Werkstoff* 14:417–424.
68. Liese, W. 1965. The warty layer. In *Cellular Ultrastructure of Woody Plants*, ed. W. A. Côté, pp. 251–269. Syracuse, New York: Syracuse University Press.
69. Liese, W. 1965. The fine structure of bordered pits in softwoods. In *Cellular Ultrastructure of Woody Plants*, ed. W. A. Côté, pp. 271–290. Syracuse, New York: Syracuse University Press.
70. Liese, W., and W. A. Côté. 1960. Electron microscopy of wood: Results of the first ten years of research. *Proc. Fifth World For. Congr.* 2:1288–1298.
71. Loews, F. A., and V. R. Runeckles, eds. 1977. *The Structure, Biosynthesis and Degradation of Wood*. New York: Plenum Press.
72. Manley St. John, R. 1964. Fine structure of native cellulose microfibrils. MacMillan Magazines Ltd. *Nature* 20(4964):1155–1157.
73. Mann, P. T. 1974. Ray parenchyma cell wall ultrastructure and formation in *Pinus banksiana*. *Wood and Fiber* 6(1):18–25.
74. Mark, R. 1967. *Cell Wall Mechanics of Tracheids*. New Haven, Connecticut: Yale University Press.
75. Meier, H. 1955. Über den Zellwandabbau durch Holzvermoschungspilze und die submikroskopische Struktur von Fichtentracheiden und Birkenholzfasern. *Holz Roh- Werkstoff* 13:323–338.
76. Meier, H. 1957. Discussion of the cell wall organization of tracheids and fibers. *Holzforschung* 11:41–46.
77. Meier, H. 1960. Über die Feinstrucktur der Markstrahltracheiden von *Pinus silvestris*. *Beiheft z.Z. Scheiz. Forstw. Bd. 30*, pp. 49–54. Festschrift Prof. Frey-Wyssling.
78. Meier, H. 1962. Chemical and morphological aspects of the fine structure of wood. *Proc. Wood Chemistry Symposium*, pp. 37–52. London: Butterworths.
79. Meier, H. 1964. General chemistry of cell walls and distribution of the chemical constituents across the wall. In *The Formation of Wood in Forest Trees*, ed. M. H. Zimmermann, pp. 137–151. New York: Academic Press.
80. Meyer, R. W., and A. F. Muhammad. 1971. Scalariform perforation plate fine structure. *Wood and Fiber* 3(3):139–145.
81. Meylan, B. A., and B. G. Butterfield. 1972. *Three-Dimensional Structure of Wood*. London: Chapman and Hall.
82. Meylan, B. A., and B. G. Butterfield. 1978. Helical orientation of the microfibrils in tracheids, fibres and vessels. *Wood Sci. Tech.* 12(3):219–222.
83. Morey, P. R. 1973. *How Trees Grow*. London: Arnold.
84. Mühlethaler, K. 1965. Growth theories and the development of the cell wall. In *Cellular Ultrastructure of Woody Plants*, ed. W. A. Côté, pp. 51–60. Syracuse, New York: Syracuse University Press.
85. Mühlethaler, K. 1965. The fine structure of the cellulose microfibril. In *Cellular Ultrastructure of Woody Plants*, ed. W. A. Côté, pp. 191–198. Syracuse, New York: Syracuse University Press.
86. Naegeli, C. 1928. Die Micellartheorie. In *Oswald's Klassiker*, Nr. 227, ed. A. Frey. Leipzig.
87. Nikitin, N. I. 1962. *The Chemistry of Cellulose and Wood*. Moscow-Leningrad: Academy of the USSR. (English translation by J. Schmorak, 1966, Jerusalem: Israel Program for Scientific Translations).
88. Ohtani, J. 1985. SEM observations on trabeculae in *Abies sachalinensis*. *IAWA Bull.* n.s. 6(1):43–51.
89. Ohtani, J. 1986. Vestures in axial parenchyma cells. *IAWA Bull.* n.s. 7(1):39–45.
90. Ohtani, J., B. A. Meylan, and B. G. Butterfield. 1984. Vestures or warts: Proposed terminology. *IAWA Bull.* n.s. 5(1):3–8.
91. Panshin, A., and C. De Zeew. 1980. *Textbook of Wood Technology*, 4th ed. New York: McGraw-Hill.
92. Parameswaran, N., and W. Liese. 1981. Torus-like structures in interfibre pits of *Prunus* and *Pyrus*. *IAWA Bull.* n.s. 2(2):89–93.
93. Parameswaran, N., H. Knigge, and W. Liese. 1985. Electron microscopic demonstration of a suberized layer in the tylosis wall of beech and oak. *IAWA Bull.* n.s. 6(3):269–271.
94. Preston, R. D. 1974. *The Physical Biology of Plant Cell Walls*. London: Chapman & Hall.
95. Ranby, B. G. 1958. Cellulosen und ähnliche Wandsubstanzen von Kohlenhydratnatur, Z. B. Hemicellulosen (Struktur, Eigenschaften und Verbreitung). In *Handbuch der Pflanzenphysiologie*, ed. W. Ruhland, pp. 268–304. Berlin: Springer-Verlag.
96. Robarts, A. W. 1965. Electron microscopic observations on the bordered pits of *Cedrus libani* Loud. *J. Inst. Wood Sci.* 15:25–35.
97. Roelofsen, P. A. 1959. The plant cell-wall. In *Handbuch der Pflanzenanatomie*, Bd. III, T. 4. Berlin: Gebrüder Borntraeger.
98. Runger, H. G., and W. Klauditz. 1953. Über Beziehungen zwischen der chemischen Zusammensetzung und der Festigkeitseigenschaften des Stammholzes von Pappeln. *Holzforschung* 7:43–58.
99. Sachs, I. B. 1965. Evidence of lignin in the tertiary wall of certain wood cells. In *Cellular Ultrastructure of Woody Plants*, ed. W. A. Côté, pp. 335–339. Syracuse New York: Syracuse University Press.
100. Sachs, I. B., I. T. Clark, and J. C. Pew. 1965. Investigations of lignin distribution in the cell wall of certain woods. *J. Polym. Sci. C* 2:203–212.
101. Schmid, R. 1965. The fine structure of pits of hardwoods. In *Cellular Ultrastructure of Woody Plants* ed. W. A. Côté, pp. 291–304. Syracuse New York: Syracuse University Press.
102. Sebastian, L. P., W. A. Côté, and C. Skaar. 1965.

Relationships of gas permeability to ultrastructure of white spruce wood. *For. Prod. J.* 15:394–404.
103. Spit, B. J., and S. M. Jutte. 1965. Gas discharge etching applied on sections of beech and ash. *Acta Botanica Neerlandica* 14:403–408.
104. Stoll, M., and D. Fengel. 1989. Electron microscopic visualization of individual cellulose molecules. *Holzforschung* 43(1):7–10.
105. Sugiyama, J., Y. Otsuka, H. Murase, and H. Harada. 1986. Toward direct imaging of cellulose microfibrils in wood. *Holzforschung* 40 (Suppl.):31–36.
106. Tamolang, F. N. 1966. Personal communication.
107. Tang, R. C. 1973. The microfibrillar orientation in cell-wall layers of Virginia pine tracheids. *Wood Sci.* 5(3):181–186.
108. Technical Association of the Pulp and Paper Industry. *TAPPI Standards*.
109. Thomas, R. J. 1981. Wood anatomy and ultrastructure. In *Wood: Its Structure and Properties*, ed. F. F. Wangaard, pp. 109–146. Materials Education Council, University Park, Pennsylvania: The Pennsylvania State University.
110. Thomas, R. J., and D. D. Nicholas. 1966. Pit membrane structure in loblolly pine as influenced by solvent exchange drying. *For. Prod. J.* 16(3):53–56.
111. Timell, T. E. 1965. Wood and bark polysaccharides. In *Cellular Ultrastructure of Woody Plants*, ed. W. A. Côté, pp. 127–156. Syracuse, New York: Syracuse University Press.
112. Tsoumis, G. 1956. Une étude sur la résistance des bois aux attaques des xylophages marins en rapport avec la teneur en silice. *Revue Bois Appl.* 10(12):23–25.
113. Tsoumis, G. 1964. Preparation of preshadowed carbon replicas for electron microscopic investigation of wood. *Holzforschung* 18:117–119.
114. Tsoumis, G. 1965. Light and electron microscopic evidence on the structure of the membrane of bordered pits in tracheids of conifers. In *Cellular Ultrastructure of Woody Plants*, ed. W. A. Côté, pp. 305–317. Syracuse, New York: Syracuse University Press.
115. Tsoumis. G. 1965. Electron microscopic observations relate the warty layer to extractives in wood. *Tappi* 48:451–455.
116. Tsoumis, G. 1968. *Wood as Raw Material*. Oxford: Pergamon Press.
117. Tsoumis, G. 1983. SEM observations on irradiated old wood. *IAWA Bull.* n.s. 4(1):41–45.
118. Wardrop, A. B. 1957. The phase of lignification in the differentiation of wood fibers. *Tappi* 40:225–243.
119. Wardrop, A. B. 1964. The structure and formation of the cell wall in xylem. In *The Formation of Wood in Forest Trees*, ed. M. H. Zimmermann, pp. 87–134. New York: Academic Press.
120. Wardrop, A. B. 1965. Cellular differentiation in xylem. In *Cellular Ultrastructure in Woody Plants*, ed. W. A. Côté, pp. 61–97. Syracuse, New York: Syracuse University Press.
121. Wazny, H., and J. Wazny. 1964. Über das Auftreten von Spurenelementen im Holz. *Holz Roh- Werkstoff* 22:299–304.
122. Wenzl, H. F. J. 1970. *The Chemical Technology of Wood*. New York: Academic Press.
123. Wheeler, E. A. 1981. Intervascular pitting in *Fraxinus americana* L. *IAWA Bull.* n.s. 2(4):169–174.
124. Wheeler, E. A. 1983. Intervascular pit membranes in *Ulmus* and *Celtis* native to the United States. *IAWA Bull.* n.s. 4(2/3):79–88.
125. Wheeler, E. A., and R. J. Thomas. 1981. Ultrastructural characteristics of mature wood of southern red oak (*Quercus falcata* Michx.) and white oak (*Quercus alba* L.). *Wood and Fiber* 13(3):169–181.
126. Wilson J. W., and R. W. Wellwood. 1965. Intra-increment chemical properties of certain western Canadian coniferous species. In *Cellular Ultrastructure of Woody Plants*, ed. W. A. Côté, pp. 551–559. Syracuse, New York: Syracuse University Press.
127. Wise, L. E., and E. C. Jahn, eds. 1952. *Wood Chemistry*, Vol. 1. New York: Reinhold.
128. Wise, L. E., R. C. Rittenhouse, E. E. Dickey, O. H. Olson, and C. Garcia. 1952. The chemical composition of tropical woods. *For. Prod. J.* 2:227–249.
129. Yatagai, M., and T. Takahashi. 1980. Tropical wood extractives: Effects on durability, paint curing time and pulp sheet resin spotting. *Wood Sci.* 12(3):176–182.

FOOTNOTES

[1]Unusually high amounts of ash (sometimes about 3% or higher) are found in certain tropical woods. Aside from the above-mentioned elements, ash may contain minute quantities of a large number of other elements, such as phosphorus (P), sulfur (S), sodium (N), aluminum (Al), titanium (Ti), manganese (Mn), iron (Fe), cobalt (Co), nickel (Ni), silver (Ag), barium (Ba), lead (Pb), and gold (Au). In some tropical species, most of the ash is composed of silica (SiO_2) (16, 27, 112, 121, 127, 128).

[2]Discrepancies are sometimes large. For example, reports of cellulose content of wood of Norway spruce range from 37.2 to 64.6% (60), according to the procedure used for its separation. The high figure applies to the so-called "Cross and Bevan" cellulose, which still contains hemicelluloses (15). Other sources assign a 38.1–57.8% cellulose content to spruce wood (different species); pine is reported to contain 32.0–60.7% cellulose, oak 31.1–49.9%, and similar large variations were also observed in other species (31).

[3]Various treatments—such as grinding, heat, light, and chemical action (delignification, bleaching, hydrolysis), as well as prolonged aging of wood—result in a drastic decrease of the degree of polymerization (65). The average degree of polymerization of cellulose from other plants is similar to that of cellulose from wood (111).

[4] Exception is made by the parenchyma cells of sapwood, which keep their protoplasm and nucleus after lignification.

[5] Aromatic compounds are derivatives of a series of hydrocarbons, whose simplest (and perhaps more familiar) member is benzene (C_6H_6).

[6] There are two types of electron microscopes—*transmission* (TEM) and *scanning* (SEM). Transmission electron microscopes were invented in 1932 (marketed after 1940) and used with wood in the 1950s. The scanning electron microscope was developed later (1965). TEM microscopes transmit electrons through ultrathin sections of wood (less than 0.15 μm), while SEM microscopes may work with sizeable samples and give three-dimensional images (24, 48, 113).

Electron microscopes can magnify many thousands of times, but the power of magnification is not a valid basis for evaluating their capacity. A more realistic basis for comparison is the *resolving power*, which is the ability to separate two closely spaced points. The resolving power of the human eye is 25-100 μm, whereas light (optical) microscopes may resolve down to 0.1 μm, SEM microscopes to 0.5 nm (5 Å), and TEM microscopes to 0.2 nm (2 Å) (48).

1 μm (micrometer) = 0.001 mm (= 1 micron, μ)
1 nm (nanometer) = 0.000.001 mm (= 10 Å)

Note that *micrometer* is the new name for micron (μ) and *nanometer* has replaced Ångstrom (Å).

[7] Cellulose molecules have been visualized by electron microscopy (in highly diluted solutions of cellulose) as rod-like elements, which in some parts show a parallel arrangement. Individual rodlets were about 3-4 nm in diameter with a variable length (100-200 nm, some longer or shorter); longer, coiled structures were also observed (104).

[8] Electron photomicrographs of individual cellulose microfibrils of *Valonia* (green algae) show their cross-sectional shape to be nearly square (17, 45). Related efforts with poplar tension wood fibers and primary walls of pine were not so successful (105). Microfibrils in cross-section were also thought to have been seen after etching ultrathin sections of wood tissue by discharging pure oxygen (103, 116).

[9] This concept of cell-wall ultrastructure is known as the *fringe micellar theory* (47). (The term is apparently descriptive of "fringes," which are visualized to be formed by cellulose chain molecules as they enter from crystalline to amorphous regions). There is another, so-called *folded chain theory*, which advocates that cellulose chain molecules are folded and form a flat ribbon which is wound to a tight helix (Figure 4-3D). The existence of folded chains is considered unlikely (12, 74, 85). The parallel arrangement of chains is more in agreement with the assumed manner of cellulose biosynthesis (tip growth; see further in this chapter) and the high axial tensile strength of wood.

[10] The crystallinity of cellulose decreases under the action of atomic radiation (gamma rays) and is destroyed by high doses (26, 117); the wood becomes fragile.

[11] Water also enters these spaces between and within microfibrils. Initially, it is attracted by free $-$OH groups between crystallites and in the amorphous regions, where it forms at first a monomolecular layer. The additional water enlarges these spaces and increases the distance between microfibrils, causing wood to swell (see Chapter 10). The proportion of void spaces in dry cell walls is considered to be about 1-5% (see Chapter 8).

[12] It has been suggested, that some hemicelluloses form their own microfibrils, which may or may not have crystalline regions (78).

[13] Note that the letters S and Z, even if placed inversely, retain their identity, and designate (in their middle portion) left-handed and right-handed orientation, respectively.

[14] These angle sizes (30°, 50-90°) are average and approximate. Similar general data are reported in other bibliographical sources (S_1 layer 50-70°, S_2 layer 10-30°, S_3 layer 60-90°) (55). Detailed measurements of pine and Japanese cedar tracheids are as follows: S_1 layer 68-83°, S_2 layer 3-26°, S_3 layer 79-85°; differences were observed between species, earlywood and latewood, and radial and tangential walls (43).

[15] In earlywood tracheids of spruce, the primary wall was measured to be 0.23-0.34 μm thick, the S_1 layer 0.12-0.35 μm, the S_2 layer 1.77-3.68 μm, and the S_3 layer 0.10-0.15 μm (53). These figures are in general agreement with measurements of earlywood tracheids of pine (primary wall 0.06 μm, S_1 layer 0.31 μm, S_2 layer 1.93 μm, S_3 layer 0.17 μm) (41); a thickness of 6.94 μm is reported for the S_2 layer of latewood tracheids of pine (43).

[16] A similar organization exists in species of *Araucaria*, such as Parana pine (*Araucaria angustifolia*); in this species, microfibrils were observed to cross between adjoining pit membranes (54).

[17] Bordered pits between ray tracheids may sometimes lack a torus (e.g., as was reported for Scots pine) (77). As a rule, however, such pits have tori, as can be readily seen with a light microscope.

[18] Intervascular pit membranes of certain elm and hackberry (*Celtis*) species have a central thickening resembling a torus (124). A torus-like structure was also observed in interfiber pits of other species (*Prunus*, *Pyrus*) (92). It is also interesting to note that newly formed intervascular pit membranes of ash sapwood showed a seasonal variation in appearance, in the sense that there is a progressing infilling with extraneous materials; this coating is removed in the next spring (124).

[19] It has been demonstrated that tyloses develop from the protective layer of parenchyma cells (93).

[20] The presence of suberin may impede the penetration of liquids and gases, and therefore increase the difficulties of protective treatment of hardwoods with a large number of tyloses [e.g., white oak and discolored heartwood (red heartwood) of beech] (93).

[21] Earlywood of spruce was found to contain lignin as follows (in percent of the total in each location): secondary

wall 22.5%, middle lamella 49.7%, cell corners 84.8% (1).

[22]It is usually reported in the literature that about 60–90% of the total lignin is localized in the compound middle lamella (10, 79). Calculations based on the relative proportions of volume contributed by the compound middle lamella and the secondary wall suggest, however, that although most of the compound middle lamella is lignin, most of the lignin (> 60% of the total) is rather contained in the secondary wall (11). Specifically, in earlywood of spruce 72% of the lignin was found in the secondary wall, and in latewood 82% respectively, leaving only 28% and 18% in the compound middle lamella (33).

[23]The microscopic structure of wood also imparts strength through the morphology (length, diameter, wall thickness) and arrangement of cells (8). Wood is a composite material—in two levels. In the microscopic level, most cell elements (tracheids, fibers, etc.) are placed in an axial direction within a matrix of middle lamella. In the submicroscopic level, cellulose chains are also mostly axially directed and embedded in a matrix of hemicelluloses and lignin. This double synthesis imparts a very high axial tensile strength—about 40 times higher than transverse tensile strength (on a unit weight basis, wood is stronger than steel in axial tension, see Chapter 11). Some man-made products imitate natural synthesis (e.g., plastics incorporating glass or carbon fibers).

[24]Lignin increases the strength of wood to such an extent that huge trees (100 m, 300 ft high or more) can remain upright (32).

5

The Mechanism of Wood Formation

The study of a biological product, like wood, is not complete without a discussion of the mechanism of its production by trees—and this chapter deals briefly with that subject. Bark is a product of the same biological mechanism, but its production is discussed in Appendix III in the context of a general consideration of bark as a material.

Introducing the subject, it may be said that trees grow in two directions—in length and in diameter. In either case, growth is achieved by the multiplying of specialized cells capable of division; these cells compose the so-called generative or *meristematic* tissues (from the Greek *meristos* = divisible). Growth in length, called *primary* or *apical growth*, takes place by activity of *apical meristems* at the tips of the main stem, branches, and roots.[1] Growth in diameter, or *secondary growth*, is accomplished by *lateral meristems*—mainly by the *cambium*.[2] Tissues produced by apical meristems are called *primary tissues*; those produced by lateral meristems are called *secondary tissues*.

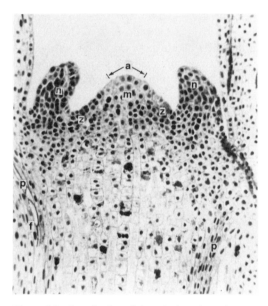

Figure 5-1. Organization of the apical meristem in sugar pine: apical initials (a), mother cells (m), and peripheral zone (z); procambium (p), first needles (n), and first primary xylem tracheids (f) (70×). (Courtesy of G. P. Berlyn; reproduced by permission from Pergamon Press.)

PRIMARY GROWTH

In forest trees (higher plants, in general), apical growth is initiated by a group of cells, called *apical initials*. The immediate derivatives of these initials (called *mother cells*) usually maintain a meristematic capability. Apical initials and mother cells, collectively designated as *protomeristem*, constitute the growing front of the apex (Figure 5-1).

As growth proceeds, the older tissues, which are left behind the protomeristem, gradually differentiate somewhat in size and shape, forming three more or less distinct regions: *protoderm* (the outermost layer), *procambium*, and *ground meristem*. These are primary meristems; by further differentiation they produce primary tissues.

Protoderm gives rise to epidermis; this is usually a one-cell thick protective layer, having the outer cell walls impregnated with waterproof substances (i.e., *cutin* and waxes). Cutin is also deposited as a continuous layer on the exterior. Thus, the epidermis is able to protect the underlying tissues from excessive loss of moisture. Except for *stomata* (apertures of special structure serving metabolism), epidermis forms a continuous protective cover of young stems. Procambium is differentiated into *primary xylem*, *primary phloem*, and *cambium*; cells located near the inner portion of cambium differentiate into primary xylem, while those located outward differentiate into primary phloem. These tissues expand radially, until finally a one-cell thick layer remains between them; this is cambium. Cambium is formed early, during the first season of growth, at a very short distance behind the apex. The ground meristem gives rise to *cortex* and *pith*, and the *rays* that start at the pith. Few to several cells in thickness, cortex is primary tissue that remains between the epidermis and the primary phloem; like the pith, it is parenchymatous in nature.

SECONDARY GROWTH

Secondary growth, or growth in diameter, begins with the formation of the cambium. The cambium produces *secondary xylem* and *secondary phloem*. Both primary and secondary tissues may be seen in cross-sections of stems cut beneath the apical meristem. The tissues appear, from the center outward, as *pith*, *primary xylem*, *secondary xylem*, *cambium*, *secondary phloem*, *primary phloem*, *cortex*, and *epidermis* (Figures 5-2 and 5-3).

Cambium is composed of a one-cell wide (as seen in cross-sections) layer of initials and of a small but varying number of layers of undifferentiated derivatives. Recognition of the initials is practically impossible, and for this reason the complex is usually called *cambial region* or *cambial zone* (12, 31). During dormancy this zone is relatively compact; its cells have nar-

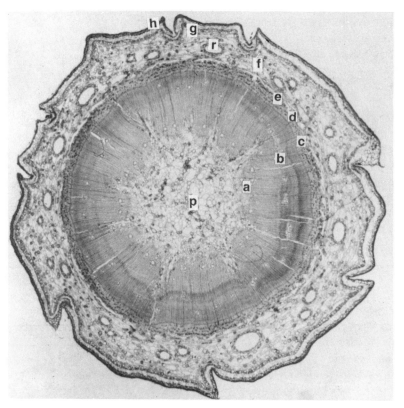

Figure 5-2. Young stem of white pine: pith (p), primary xylem (a), secondary xylem (b) with atypical development of earlywood and latewood, cambium (c), secondary phloem (d), primary phloem (e), cortex (f) with cortical resin canals (r), periderm (g), and epidermis (h) (10×). (Reproduced by permission from Pergamon Press.)

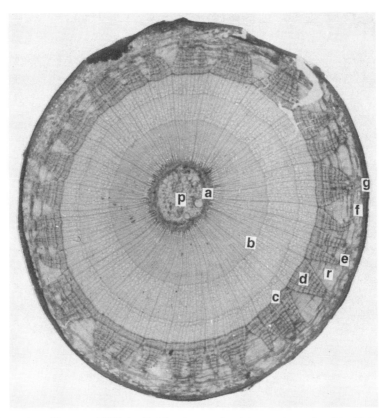

Figure 5-3. Young stem of basswood (*Tilia*): pith (p), primary xylem (a), secondary xylem (b) with three growth rings, cambium (c), secondary phloem (d) with characteristic ray flares (r), primary phloem (e), cortex (f), and periderm (g) (10×). (Reproduced by permission from Pergamon Press.)

row radial diameters and contain dense protoplasms. In the spring, before division starts, the cambial zone expands[3] (Figure 5-4A and B). The cells swell and increase in radial diameter, and the cell lumina are no longer entirely filled by the protoplasms. It is at this time that bark is most easily peeled off. Moreover, during dormancy the transition from cambial zone to xylem is abrupt, while during active growth it is gradual.

Cambium consists of *fusiform* and *ray initials*. The majority are fusiform—elongated and spindle-shaped (fusiform) as seen in tangential sections. Ray initials are narrow and short, slightly elongated to nearly isodiametric. Fusiform initials produce all the axial cells of xylem (tracheids, fibers, vessel members, axial parenchyma) and the respective axial cells of phloem, whereas rays originate from ray initials. In most species, the ends of fusiform initials do not lie in the same plane (they overlap); exceptions are found in *stratified cambia* (2) of certain hardwoods, such as black locust and persimmon. In these, the fusiform initials are short, nearly uniform in length, and arranged in parallel, horizontal series as seen in tangential sections.

Cambium initials divide periclinally, namely with walls parallel to the cambium layer (in a tangential-longitudinal plane), and produce *xylem* and *phloem mother cells*. These usually divide again, and their derivatives develop gradually into xylem (wood) and phloem (bark) cells. The interval between successive divisions varies with time (early or late in the growing season), species, and environmental conditions (7). Normally, divisions of cambium initials (or mother cells) occur more slowly than those of apical meristems. Alternation of xylem and phloem production is not

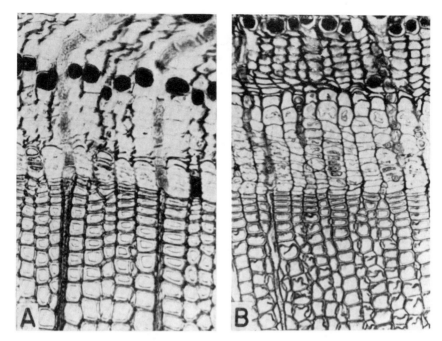

Figure 5-4. The cambial zone in spruce. (A) In dormant state. (B) Beginning activity; the cambial zone is expanded and division has started; narrowest radial diameters and exposure (to plane of sectioning) of nuclei in some cells is indicative of the approximate position of dividing cells (initials and mother cells); division is periclinal (i.e., with walls parallel to the cambium layer) (135×). (Photograph B Courtesy of P. R. White; reproduced by permission from Pergamon Press.)

regular. Within a growing season there is greater activity toward the xylem side, so that more xylem than phloem cells are produced.[4]

The mechanism that stimulates activity in the cambial zone is not clearly known. It is widely accepted that the stimulus comes from growth hormones—mainly auxins. Growth stimulants are believed to arise in the apical regions of growing stems; young leaves are rich sources of auxins (41). Activity starts below the swelling buds and spreads downward through the phloem; this downward spread moves more quickly in ring-porous than in diffuse-porous hardwoods and conifers (40, 44). However, there is evidence that initiation of the change of the cambium from the dormant to "active" state may take place locally and independently within individual cells or groups of cells of the cambial zone (8, 42).

In contrast to growth in height, division of the cambium, and therefore growth in diameter, depends primarily on current photosynthesis. Synthesized food stored from the previous season may be used for diameter growth early in the season (19). It is observed, for example, that in oaks the foliage appears only after the pores of earlywood are formed (16).

New cells produced by the cambium increase the girth of the stem, branches, and roots; therefore, it is necessary that the cambium ring increases, too. This is accomplished to some degree by increase of the tangential diameter of the initials. Sanio (32) observed that the diameter of the initials of a 2-year-old stem of Scots pine was 0.012 mm, while that of a 100-year-old trunk of the same species was 0.026 mm, or about double. Bailey (2) measured the diameter of the initials of a 1-year-old stem and a 60-year-old trunk of white pine (*Pinus strobus*); the average values for fusiform initials were 0.016 mm and 0.042 mm, and for ray initials 0.014 mm and 0.017 mm, respectively.

This increase in diameter alone of the initials is, obviously, not sufficient. The increase in girth of the cambium ring is accomplished mainly through the addition of new initials. In softwoods and most hardwoods, new fusiform initials are produced from existing ones by formation of *anticlinal pseudotransverse* (semi-transverse) walls; these walls are of various degrees of inclination with regard to the long axis of initials. The derivatives of these divisions elongate by *apical intrusive growth* (Figure 5-5), usually at both ends until they reach the size of the originating initial. In a few hardwoods with stratified cambia, the short fusiform initials divide by *radial-longitudinal* walls (2); the derivatives do not grow in length but only in diameter.

New initials are mainly produced toward the end of maximum cambial activity during each season of growth. Development of initials (and of their derivatives) apparently depends on the extent of their contact with rays through which synthesized food material is translocated. Insufficient contact will result in elimination of stunted existence of some, while others will give rise to ray initials that will start new rays (6, 20, 35).

The beginning, end, and length of time of production of xylem and phloem cells by the cambium may vary from year to year, as well as among species and individual trees. Within the growing season, cambial activity may be affected by various environmental factors, such as soil and air temperatures, light intensity, day length, and moisture and nutrient availability (36, 40). Day length in particular is believed to influence cambial activity; the time that growth ceases in the fall seems to be correlated with steadily decreasing day lengths of that season (40).[5]

In the temperate zones, cambial activity is restricted to the warmer months. The increments of annual growth become distinctive due to cellular differences between earlywood and latewood. Large diameter earlywood cells are produced during the beginning of each season, when growth in height is active; this is associated with high auxin synthesis. Narrow-diameter latewood cells are produced following cessation of height growth and reduction in auxin synthesis (22, 40, 41, 44). In the north temperate zone, 90% or more of the annual growth increment may be laid down by the end of July and sometimes earlier; however, cessation of cell division in the cambial zone takes place sometime between the end of August and the end of September (7)—depending on geographical location and altitude, although deposition of secondary walls may continue in October (43). In warm regions, such as certain Mediterranean countries, cambial division may, in some species, continue throughout the year (13).[6]

Cambium is a long-living tissue with a capability for renewal that may extend to hundreds and even thousands of years, and a power of healing local injuries. However, adverse environmental conditions (and sometimes unknown factors) cause abnormal divisions in the cambium, which lead to structural abnormalities in the wood produced (see Chapter 7).

CELL DEVELOPMENT

The cells produced by the cambium develop rapidly into mature wood and bark cells, capable of performing the tasks of conduction, food storage, and mechanical support of the

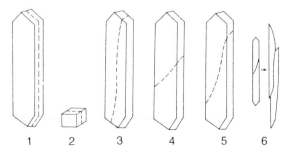

Figure 5-5. Diagrammatic representation of periclinal and anticlinal division of cambial initials. (1, 2) periclinal (1, fusiform; 2, ray initial); (3, 4, 5) various types of anticlinal, pseudotransverse division; the derivatives elongate as shown in 6. (Adapted from Ref. 12.)

living tree. Four stages of cell development may be recognized:

1. *Stage of origin.* This involves production of cells, and has been discussed above.

2. *Stage of enlargement (surface growth).* The new cells gradually enlarge toward their final size and shape by increasing in diameter and length. In most cases they become many times longer than the initials. Parenchyma cells do not exceed their initials in length, and also vessel members of some species, but vessel members may grow many times in diameter—especially in the earlywood of ring-porous hardwoods. Elongation of the new cells serves the function of conduction and support; in vessel members, this objective is fulfilled by end-connection to form long vessels.

Enlargement takes place by stretching of the primary wall. Stretching has the effect of loosening the microfibrillar network and reorientating the microfibrils in the direction of elongation (27). Reorientation is greatest in the first-produced (outer) microfibrils of the primary wall, while those produced later more or less maintain their posture at deposition (i.e., an orientation approximately transverse to cell axis) (see Chapter 4).

The increase in cell volume leads to *vacuolation* of cells—that is, restriction of the protoplasm to an inner lining of the cell lumen, and the formation of sap cavities (vacuoles).

3. *Stages of cell-wall thickening.* At the end of the enlargement stage, the boundaries of the new cells consist solely of primary wall. Deposition of the secondary wall by the protoplasm follows. The thickness of the secondary wall varies with species, cell types, and other factors; and also between earlywood and latewood cells, especially in the case of softwood tracheids.

Thickening of the cell wall by deposition of the secondary wall is closely associated with the forming of pits, although the origin of pits may be traced back to the deposition of the primary wall. The primary wall is indeed not deposited evenly over the middle lamella; certain areas remain very thin, and these constitute the *primary pit fields*. Pit fields are not covered during deposition of the secondary wall. The noncovered areas then become pits, and their structure (simple, bordered, etc.) depends upon the manner in which the secondary wall is deposited. Pit formation has been closely studied in bordered pits of softwoods, where pits appear to be delineated in the primary wall by a ring of microfibrils (9, 14, 15, 17) (Figure 5-6).

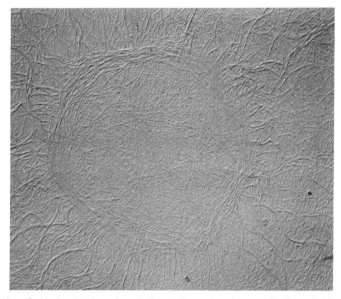

Figure 5-6. Delineation of a bordered pit by a ring of microfibrils on the primary wall of a tracheid of Scots pine (Replica from delignified sample, 8500×). (After Ref. 15; courtesy of A. Frey-Wyssling, and reproduced by permission from Pergamon Press.)

The last (S_3) layer of the secondary wall, some specialists believe, apparently extends over the inner side of the overhanging border, thus lining the cavities of these pits (11, 23). Deposition of the secondary wall is thought to occur progressively in both directions from the center of the cell to its ends (37). Completion of the secondary wall may be topped by formation of spiral thickenings.

4. *Stage of lignification*. Lignification is the final stage in the development of the cell. It involves chemical alteration of the wall (i.e., deposition of lignin). This was observed to occur first at cell corners, then to extend simultaneously to the middle lamella and the secondary wall (37). When lignification is completed, the protoplasm, as a rule, is consumed. Thus, with the exception of a few rows of young cells under development and parenchyma cells in sapwood (and rarely in heartwood), all mature cells are dead, because they are devoid of nucleus and protoplasm.

The *warty layer* is formed during this final stage, when lignification is complete or nearly so. According to some investigators, this layer is composed of the remainder of the dying protoplasm (37, 39), but others have shown that it develops prior to its degeneration.

It should be noted, that the above stages of cell development are not sharply demarcated; deposition of the secondary wall may start before the cell has reached its final size and shape, and lignification may, likewise, commence before deposition of the secondary wall is ended (27, 38). However, these stages proceed in immediate succession and are rapidly concluded—at a distance of a few rows of cells from the cambium[7]. In terms of required time, it was observed in softwood tracheids that about 20 days were necessary for all development stages to be completed (1 day was generally required to produce a new cell) (18).

The various processes of cell development undoubtedly serve essential needs of the living tree. Tracheids and fibers become channels for food translocation from roots to crown by their elongation in that direction. This same task is facilitated by the disappearance or reduction of end walls of vessel members. Among fibers, translocation of nutrients is best served by fiber tracheids, while libriform fibers in many instances are able to perform a storage function; such fibers may retain their living contents, in spite of secondary thickening and lignification of their walls (3). However, fibers are primarily strengthening elements in hardwoods; in softwoods, tracheids (especially those of latewood) perform a strengthening function—besides that of food translocation. The function of cell-wall thickenings is also mechanical. Parenchyma cells, while remaining alive, serve translocation and storage of synthesized food. Communication between all cell elements is provided by pit-pairs, and by perforation plates between vessel members. Such communication is highly efficient due to the perfect coincidence of these passages between adjoining cells.

Certain morphological variations of cell elements are attributed to evolution. For example, vessel members are considered to originate from hardwood tracheids. The most primitive vessels are composed of members that are very similar to tracheids in size, shape, wall thickness, and pitting; they differ from the latter in that the membranes have been dissolved in a number of the bordered pits in their overlapping ends. With increasing specialization, vessel members become shorter and wider with less extensively overlapping ends; their perforations change from multiple or scalariform to simple. The simple perforations are characteristic of the most highly specialized vessels (3, 4). Evolutionary changes of fibers include decrease in length, reduction in size of pit borders, and change in size and shape of pit apertures (5).

Ecological trends also exist in the sense that climatic conditions affect cellular changes. Specifically, it has been observed that in the sequence from boreal to temperate to mediterranean climates, there is a decreasing incidence of scalariform perforations, almost exclusively solitary vessels, and fiber tracheids; the incidence of different vessel size classes and vascular tracheids increases, while ring-porous tendencies and spiral vessel thickenings have their peak in the temperate zone (1).

In association with the processes of production of new cells, intercellular canals (resin canals in softwoods, and gum canals in hardwoods) are formed in certain species. Canals are postcambial in origin. Resin canals are said to be *schizogenous* (from the Greek *schizo* = split). They are formed within groups of cambial derivatives that fail to differentiate into tracheids. The process involves splitting of the middle lamella of centrally placed derivatives, and the formation of the epithelium by subsequent division of derivatives adjoining the resulting cavity (12, 29). It has been proposed (on the basis of work with Canadian white spruce) that resin canals originate from wounding (34), but the regularity of their distribution and repetition in growth rings, year after year, does not support this point of view. Gum canals are not normal features of temperate hardwoods of commercial importance. The formation of traumatic (wound) canals is discussed in Chapter 7.

REFERENCES

1. Baas, P., and F. H. Schweingruber. 1987. Ecological trends in the wood anatomy of trees, shrubs and climbers from Europe. *IAWA Bull.* n.s. 8(3):245–274.
2. Bailey, I. W. 1923. The cambium and its derivative tissues. IV. The increase in girth of the cambium. *Amer. J. Bot.* 10:499–509.
3. Bailey, I. W. 1953. Evolution of the tracheary tissue of land plants. *Amer. J. Bot.* 40:1–8.
4. Bailey, I. W. 1957. The potentialities and limitations of wood anatomy in the study of the phylogeny and classification of angiosperms. *J. Arnold Arb.* 38:243–254.
5. Bailey, I. W. and W. W. Tupper. 1918. Size variation in tracheary cells. I. A comparison between the secondary xylems of vascular cryptogams, gymnosperms and angiosperms. *Proc. Amer. Acad. Arts Sci.* 54:149–204.
6. Bannan, M. W. 1961. Anticlinal division and cell length in the cambium of conifers. *Intern. Ass. Wood Anatom. Bull.* 1:2–6.
7. Bannan, M. W. 1962. The vascular cambium and tree-ring development. In *Tree Growth*, ed. T. T. Kozlowski, pp. 3–21. New York: Ronald Press.
8. Barnard, J. E. 1965. A study of the phenology of cambial activity in sugar maple sprouts during early spring using tetrazolium chloride as an indicator. School of Forestry, Pennsylvania State University (unpublished).
9. Bauch, J., W. Liese, and F. Scholz. 1968. Über die Entwicklung und stoffliche Zusammensetzung der Hoftüpfelmembranen von Längstracheiden in Coniferen. 22(5):144–153.
10. Bauch, J., W. Schweers, and H. Bernt. 1974. Lignification during heartwood formation: Comparative study of rays and bordered pit membranes in coniferous woods. *Holzforschung* 28(3):86–91.
11. Bucher, H. 1957. Die Tertiärwand von Holzfasern und ihre Erscheinungsformen bei Coniferen. *Holzforschung* 11:2–16.
12. Esau, K. 1965. *Plant Anatomy*, 2nd ed. New York: John Wiley & Sons.
13. Fahn, A. 1962. Xylem structure and the annual rhythm of cambial activity in woody species of the east Mediterranean regions. *Intern. Ass. Wood Anatom. Bull.* 1:2–6.
14. Fengel, D. 1972. Structure and function of the membrane in softwood bordered pits. *Holzforschung* 26(1):1–9.
15. Frey-Wyssling, A., H. H. Bosshard, and K. Mühlethaler. 1956. Die submikroskopische Entwicklung der Hoftüpfel. *Planta* 47:115–126.
16. Huber, B. 1963. New discoveries in tree-ring research. *Intern. Ass. Wood Anatom. Bull.* 2:2–4.
17. Imamura, Y., and H. Harada. 1973. Electron microscopic study on the development of the bordered pit in coniferous tracheids. *Wood Sci. Tech.* 7(3):189–205.
18. Kennedy, R. W., and J. L. Farrar. 1965. Tracheid development in tilted seedings. In *Cellular Ultrastructure of Woody Plants*, ed. W. A. Côté, pp. 419–453. Syracuse, New York. Syracuse University Press.
19. Kozlowski, T. T. 1962. Photosynthesis, climate and tree growth. In *Tree Growth*, ed. T. T. Kozlowski, pp. 149–164. New York: Ronald Press.
20. Kremers, R. F. 1963. The chemistry of developing wood. In *The Chemistry of Wood*, ed. B. L. Browning, pp. 369–404. New York: Interscience.
21. Kutscha, N. P., F. Hyland, and J. M. Schwarzmann. 1975. Certain seasonal changes in balsam fir cambium and its derivatives. *Wood Sci. Tech.* 9(3):175–188.
22. Larson, P. R. 1960. A physiological consideration of springwood-summerwood transition in red pine. *For. Sci.* 6:110–122.
23. Liese, W. 1963. Tertiary wall and warty layer in wood cells. *J. Polym. Sci. C*, 2:213–219.
24. Liphschitz, N., S. Lev-Yadun, E. Rosen, and Y. Waisel. 1984. The annual rhythm of activity of the lateral meristems (cambium and phellogen) in *Pinus halepensis* Mill, and *Pinus pinea* L. *IAWA Bull.* n.s. 5(4):263–274.
25. Liphschitz, N., S. Lev-Yadun, and Y. Waisel. 1985. The annual rhythm of activity in the lateral meristems (cambium and phellogen) in *Pistacia lentiscus* L. *IAWA Bull.* n.s. 6(3):239–244.
26. Lipschitz, N., and W. Lev-Yadun. 1986. Cambial activity of evergreen and seasonal dimorphics around the Mediterranean. *IAWA Bull.* n.s. 7(2):145–153.
27. Mühlethaler, K. 1961. Plant cell walls. In *The Cell*, ed. J. Brachet and A. E. Mirsky, Vol. II, pp. 85–134. New York: Academic Press.
28. Necesany, V. 1971. Effect of some environmental factors on the cell wall structure. *Holzforschung* 25(1):4–8.

29. Panshin, A., and C. De Zeew. 1980. *Textbook of Wood Technology*, 4th ed. New York: McGraw-Hill.
30. Pauly, G. 1963. Les canaux sécréteurs du pin maritime. Travaux du Lab., Toulouse 6(1):1-32.
31. Philipson, W. R., J. M. Ward, and B. C. Butterfield. 1971. *The Vascular Cambium*. London: Chapman & Hall.
32. Sanio, K. 1873. Anatomie der gemeinen Kiefer (*Pinus sylvestris*). *Jahr.wiss.Bot.* 9:50-126.
33. Srivastava, L. M. 1966. Histochemical studies on lignin. *Tappi* 49:173-183.
34. Thomson, R. B., and H. B. Sifton. 1925. Resin canals in the Canadian spruce, *Picea canadensis* (Mill.) *B.S.P. Phil. Trans. Roy. Soc. B*, 214:63-111.
35. Vesilevic, S. 1959. Die Rolle der Markstrahlen bei der Verlangerung der Tracheiden bei *Abies pectinata* L. *Bull. College of Forestry* 16:101-121 (University of Belgrade).
36. Waisel, Y., and A. Fahn. 1965. The effects of environment on wood formation and cambial activity in *Robinia pseudoacacia* L. *New Phytologist* 64:436-442.
37. Wardrop, A. B. 1964. The structure and formation of the cell wall in xylem. In *The Formation of Wood in Forest Trees*, ed. M. H. Zimmermann, pp. 87-134. New York: Academic Press.
38. Wardrop, A. B. 1965. Cellular differentiation in xylem. In *Cellular Ultrastructure of Woody Plants*, ed. W. A. Côté, pp. 61-97. New York: Syracuse University Press.
39. Wardrop, A. B., W. Liese, and G. W. Davies. 1959. The nature of the wart structure in conifer tracheids. *Holzforschung* 13:115-120.
40. Wareing, P. F. 1958. The physiology of cambial activity. *J. Inst. Wood Sci.* 1:34-42.
41. Wareing, P. F., C. E. A. Hanney, and J. Digby. 1964. The role of endogeneous hormones in cambial activity and xylem differentiation. In *The Formation of Wood in Forest Trees*, ed. M. H. Zimmermann, p. 323. New York: Academic Press.
42. Westing, A. H. 1965. Formation and function of compression wood in gymnosperms. *Bot. Review* 31:381-480.
43. White, P. R., G. Tsoumis, and F. Hyland. 1966. Some seasonal aspects of the growth of tumors on the white spruce, *Picea glauca. Can. J. Bot.* 45:2229-2232.
44. Wort, D. J. 1962. Physiology of cambial activity. In *Tree Growth*, ed. T. T. Kozlowski, pp. 89-95. New York: Ronald Press.
45. Zimmermann, M. H., and C. L. Brown. 1971. *Trees—Structure and Function*. Berlin: Springer-Verlag.

FOOTNOTES

[1] The discussion that follows in this chapter (and in Appendix III, about bark) does not concern roots. The reader is also reminded that consideration is limited to forest trees.

[2] A second lateral meristem, called *phellogen* or *cork cambium*, operates in the bark. This also produces secondary tissues (see Appendix III).

[3] The cambial zone in balsam fir was 6 cells wide during dormancy (fall) and 15 cells wide in the season of growth (June) (21). In white pine, the numbers were 4-8 and 10-15, respectively (45).

[4] Xylem cells may outnumber phloem cells by 10 times or more, except in slow-growing trees. In balsam fir, it was observed that for every 14 xylem cells only one phloem cell was produced (21). In (northern white) cedar, the numbers were 30-60 and 2-4, respectively (45).

[5] Cell characteristics—such as length, diameter, wall thickness, and arrangement of microfibrils—are also affected by environmental conditions (28).

[6] Several variations exist in this region, such as: a condition similar to that in the temperate zone (growth in spring and summer); growth in two separate seasons (spring and fall); a second rest period in summer; start in the fall and last until early summer; growth throughout the year in irrigated plants (24-26).

[7] It is sometimes suggested that additional cell-wall synthesis (cell-wall thickening, lignification) takes place during transformation of sapwood to heartwood. Formation of heartwood does not involve such processes, but rather physiological and structural changes mentioned elsewhere (see Chapter 6)—that is, death of parenchyma cells, pit aspiration, formation of tyloses, and deposition of extractives. Exceptions may occur in parenchyma cells, including epithelial cells of resin canals (30), and pith (33), where (on the basis of stain reactions) delayed lignification has been reported; delayed lignification was also observed in ray parenchyma cells and bordered pit membranes of softwoods during formation of heartwood (9, 10, 14).

6

Variation of Wood Structure

The structure of wood varies, within limits, as trees grow from seedlings to old age. Cross-sections of stems show, in most cases clearly, that the width of growth rings is not uniform from pith to bark but varies. In addition to *ring width*, variations are exhibited by *growth ring structure*, *cell morphology*, *ultrastructure*, and *chemical composition*. These characteristics vary within a tree, and between trees of the same species.

VARIATION WITHIN A TREE

Within a tree, there is a *horizontal variation* of wood structure—from pith to bark, and a *vertical variation*—from base to top.

Horizontal Variation

In the direction from pith to bark, at any height level, there is a general pattern of variation of structural characteristics. Under this general pattern, there is variation within and between growth rings.

General Pattern of Horizontal Variation. Existing data indicate that growth ring structure (development of typical structure), cell morphology (length, diameter, wall thickness), ultrastructure (angle of microfibrils, degree of crystallinity), and cellulose and lignin content all vary progressively and rapidly until, after a number of growth rings, they attain a more or less "typical" level; in the outer rings of very old trees they differ again. Superimposed are changes associated with progressive transformation of sapwood to heartwood.

This general pattern suggests an influence of age on wood structure. In accordance with stages recognized in other living organisms (humans, animals), three periods of development have also been recognized in trees: (i) *juvenile* or *immature*, (ii) *adult* or *mature*, and (iii) *senescent* or *overmature* (25). The wood produced during these periods is called *juvenile* or *core wood*, *adult* or *mature wood*, and *overmature wood*, respectively. The duration of these three periods is difficult to define. For indicative purposes only, it may be said that the first period lasts up to 20+ years, and that the last starts after 200+ years. Further discussion below will qualify this generalization.

In the horizontal direction, from pith to bark, structural characteristics of wood vary as follows.

Growth ring structure. In softwoods, growth rings located near the pith have less pronounced latewood in comparison to the typical level attained later. In species that normally exhibit a distinct contrast and more or less abrupt transition from earlywood to latewood (e.g., hard pines, larch, Douglas-fir, etc.), the first growth rings show a less abrupt transition: latewood is not so dense and there tends to be less of it, and typical resin canal distribution develops gradually (Figure 6-1A). In ring-porous hardwoods, the ring-porous character develops gradually; the first rings appear diffuse-porous (Figure 6-1B). The outer growth rings of very old trees are also atypical; they are very narrow with little latewood.

Cell morphology. Investigations of variation in cell size from pith to bark have been primarily concerned with *cell length*. Possible changes of *cell diameter*, *wall thickness*, and *cavity* (lumen) *diameter* were given little atten-

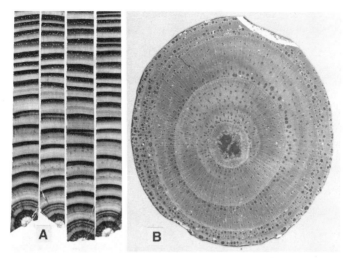

Figure 6-1. Development of typical structure. (A) Gradual formation of typical latewood and resin canals in black (Austrian) pine. (B) Gradual formation of ring-porous structure in black locust. (Photograph A courtesy of V. Mandaltsi.)

tion. The existing limited information based on studies of pines (11, 57, 110), redwood (9), and poplars (*Populus*) (87) indicates that the diameter of tracheids or fibers may (9, 11, 57) or may not (110) change from pitch outward, and that differences may exist between tracheids of earlywood and latewood (57), or between radial and tangential diameters (9, 11). Wall thickness appears to increase from pith outward (12), but differences were likewise observed between earlywood and latewood tracheids, and between radial and tangential walls. Wall thickness of earlywood tracheids was found to increase gradually with age, whereas in latewood thickness increased more rapidly but exhibited greater fluctuations (57). Cavity diameters were greater in tracheids near the pith, but in fast-grown poplars, fiber diameter, cavity diameter, and wall thickness, showed an opposite trend (12, 49) (Figure 6-2). The cells that comprise the outer growth rings of very old trees are thin-walled in comparison to wood formed during the adult period.

A horizontal variation in *cell length* was first

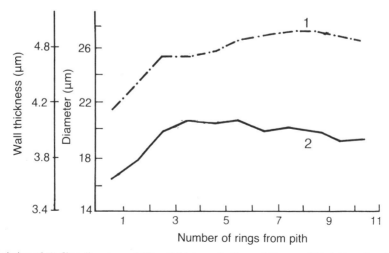

Figure 6-2. Variation of (1) fiber diameter and (2) wall thickness in fibers of 11-year-old hybrid poplar. (Adapted from Ref. 49.)

observed by Sanio (1872) (85). He conducted measurements on Scots pine (*Pinus sylvestris*) and found that "tracheids in the stem and branches generally increase in size from the inside toward the outside through a number of annual rings, until a definite size is reached, which then remains constant for the following annual rings." These observations were verified, in their substance, by many investigators. Bailey (3) (1920) traced these changes to the size of cambial initials. He concluded that the variation in length of tracheary cells of the secondary xylem parallel closely similar fluctuations of the fusiform initials. Such initials are relatively short in young stems of softwoods, but increase in length during subsequent growth for a number of years, until a certain maximum length is attained. Afterward, length fluctuates about the maximum. According to measurements made in white pine (*Pinus strobus*), the average length of fusiform initials in a 1-year-old stem was 0.87 mm, while those of a 60-year-old stem averaged 4 mm. In hardwoods, the increase is much smaller and maximum size is attained earlier. In hardwoods with stratified initials, the latter do not change in length with increasing age (4).

Horizontal variation in cell length has been investigated in a great number of studies. The results of such work may be summarized as follows: At any given height, the tracheids or fibers of the first growth ring are very short; softwood tracheids range in length from 0.5 to 1.5 mm, and hardwood fibers from 0.1 to 1.0 mm (27). Length increases rapidly in the second growth ring and in few subsequent ones; then the rate of increase declines, but continues until a maximum length is reached. In softwoods this is about 3–5 times greater than the initial length (24, 27). In hardwoods the increase is not nearly as great, but it may reach two times as much in most species (24). Species with stratified cambia, such as black locust, exhibit only a slight increase.[1] There is no general rule as to the number of years necessary for maximum length to be reached; this varies with species, and possibly also with the rate of growth. Maximum length may be reached in as little as 6–8 years, and may not be reached until a tree is 200–300 years of age; a continuous increase has been observed throughout the growth rings of a 455-year-old Douglas-fir tree (36), and over 2200 years in bristlecone pine (*Pinus longaeva*) (2). In young, plantation-grown material there is an apparent maximum length occurring between 10 and 20 years. After maximum is reached, there follows a period during which length may vary considerably about a mean maximum value. Finally, a noticeable decrease of length may take place in very old trees (27, 92). Patterns differ, however, and they have been summarized as follows: (a) constant length after a juvenile period; (b) continuous increase; and (c) increase to a maximum followed by decrease in a parabolic curve pattern (69) (Figure 6-3).

Data concerning other cell types are rare. With regard to vessel members, a study in ash (*Fraxinus excelsior*) has shown that earlywood members show a pattern of length variation similar to fibers, but latewood members did not change (17). Variation similar to fibers was observed in vessel members of a eucalypt (78), while only minor changes were reported in vessel members of black locust (39). In fast-grown hardwoods, the diameter of vessel members was observed to increase fast from the pith outward, while their quantitative representation in wood tissue decreased (12).

Ultrastructure and chemical composition. Variation in ultrastructure refers especially to the angle of microfibrils in the S_2 layer of the secondary wall. The angle is large in the first growth rings near the pith, but decreases outward for some years until it attains a more or less constant magnitude. This change is in agreement with the reported inverse relationship between angle of microfibrils and cell length (29, 75). Because the change in cell length is greater in softwoods than in hardwoods, the change in angle is also greater in softwoods; measurements showed such change to range from 55 to 20° in softwoods and from 28 to 10° in hardwoods (24).

Crystallinity was found to increase significantly through successive growth rings of hemlock (*Tsuga heterophylla*) from pith to about 15 years, after which it reached a more or less constant value (58).

Finally, there is a change in chemical com-

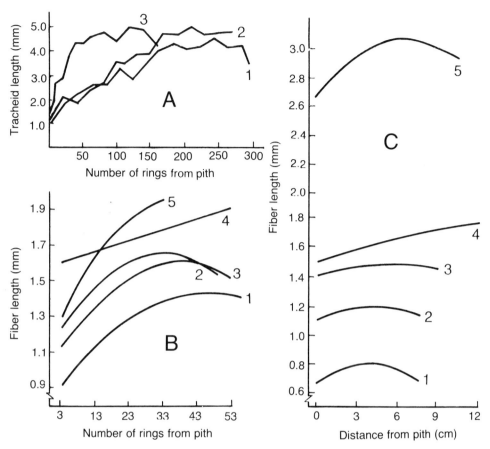

Figure 6-3. Horizontal variation of cell length. (A) Patterns of variation in Japanese spruce (*Picea glehni*) at different heights in the trunk: 1. 0.5 m (1.5 ft), 2. 4.5 m (13.5 ft), and 8.3 m (25 ft). (B) North American hardwoods (at 1.5 m, 5 ft height): 1, black gum (*Nyssa sylvatica*); 2, sweetgum (*Liquidambar styraciflua*); 3; southern red oak (*Quercus falcata*); 4, shagbark hickory (*Carya ovata*); 5, mockernut hickory (*Carya tomentosa*). (C) Tropical hardwoods (at stump height): 1, dillenia (*Dillenia pentagyna*); 2, erythrina (*Erythrina stricta*); 3, kindal (*Terminalia paniculata*); 4, teak (*Tectona grandis*); 5, cashew (*Anacardium occidentale*). (Adapted from Refs. 4, 34, and 95.)

position. Cellulose content was observed in Douglas-fir (38, 102) and species of pine (38, 57, 102, 110) to increase from pith outward, following the general trend of variation of tracheid length. Lignin content gradually declines in the same direction (57, 100). In the outer rings of old trees, cellulose content was found to be lower and lignin content higher in comparison to wood produced during earlier stages of growth (38).

Formation of Heartwood. The changes associated with formation of heartwood may be macroscopic (color) and microscopic. Microscopically, formation of heartwood is observed to be closely associated with death (loss of protoplasm and nucleus) of the parenchyma cells (21, 33). This is preceded by a change of physiological activity in such cells (21, 86), or differentiation in morphology and volume of their nuclei (17, 33).

It has been suggested that death of parenchyma cells occurs as a result of the accumulation of toxic excretory products of metabolism (109). Such excretions are translocated through parenchyma cells toward the center of tree (pith), around which the cylinder of heartwood is formed and gradually expanded (93). Formation of heartwood is also associated with aspiration of bordered pits, formation of tyloses, and deposition of extractives. The amount of starch present in parenchyma cells declines in older sapwood and is completely metabolized (disappears) when sapwood is transformed to heartwood (41, 50, 68).

Deposition of extractives that impregnate the walls of wood tissues are responsible for changes of color. Change of color may be a conspicuous outward sign, but physiologically heartwood is not limited to species in which it can be macroscopically seen. In all species, parenchyma cells sooner or later die.[2] The formation of heartwood is a natural aging process (17).

Environmental factors may influence this process, however, It is suggested, for example, that heartwood is formed when the water (sap)–air relationship attains a certain "threshold" value in the tree (97). Exposure of living cells to aeration results in desiccation and death of cells, and to the accompanying changes mentioned above. It is hypothesized that aeration could result in part by broken live and dead branches, which should provide entrance of air to the center of the stem through branch piths–which are connected to the central pith of the stem. If this hypothesis is accepted, then such aeration should influence the width of sapwood. Decrease of sapwood as a result of artificial pruning is in support of this point of view (90).

It is interesting to note that experiments with pine (*Pinus radiata*) by use of radioactive carbon (C^{14}) have shown that heartwood extractives probably originate from materials already present in sapwood cells prior to conversion to heartwood, and that they are not translocated to the heartwood boundary from more actively metabolizing tissues (105). Different views are also held, however: it has been suggested that the phenolic materials are formed at the sapwood-heartwood boundary (68), or the precursors are formed in the cambium and are capable of diffusion to the heartwood boundary through the sapwood—by way of ray cells (17, 41). This is an added indication that the mechanism of transformation of sapwood to heartwood is not clearly understood.

The size (diameter) of heartwood increases with age but is also influenced by species and growth conditions. Usually, heartwood formation starts within 5–30 years of tree growth; however, some species (e.g., sugar maple) may form heartwood only after 100 years (41); delayed formation occurs also in alder (17). Slow-growing trees have more heartwood, and trees with a large crown have more sapwood. By forming heartwood, the tree regulates the size of sapwood, and keeps it to a "desired" optimum level (7). The periodicity of heartwood formation is also not known—namely, it is not known if one or more sapwood growth rings are transformed into heartwood each year (41, 45); there seems to be no regularity and, annually, the sapwood-heartwood boundary does not follow the boundary of a growth ring. At any rate, the width of sapwood is relatively constant within a species (under similar conditions of comparison). For this reason, such width is sometimes used as a macroscopic characteristic of diagnostic value.

Juvenile, Adult, and Overmature Wood. Juvenile wood or core wood is the wood comprising the growth rings that are near the pith. In accordance with the previous discussion, these rings are characterized by less pronounced latewood, shorter cells, larger microfibrillar angles, and lower crystallinity and cellulose content in comparison to adult wood that is produced later.[3] In trees, juvenile wood forms a central cylinder or core (hence *core wood*), the longitudinal axis of which theoretically coincides with the pith (Figure 6-4 A).

Practically synonymous to juvenile or core wood is the concept of *crown-formed wood* (20, 48). This is defined as wood that is now, or was previously, within the confines of the living crown at the time of its production (56).

Juvenile wood does not occur only in fast-growing trees. Formation of such wood is an age effect; therefore, juvenile wood exists irrespective of rate of growth (79). Of course, the volume of juvenile wood is greater when the growth rings near the pith are wider. As previously mentioned, the duration of the juvenile period varies, but juvenile wood is always present, occupying the first growth rings. The number of such rings cannot be precisely determined, not only because of differences

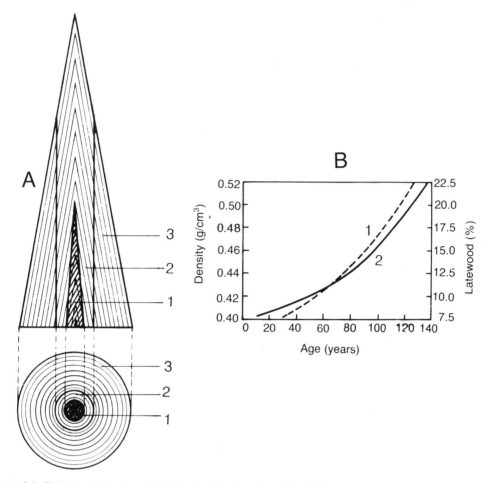

Figure 6-4. Variation of structure. (A) Vertical and horizontal variation (1, heartwood; 2, juvenile wood; 3, mature wood). (B) Variation of (1) density and (2) latewood with increasing age in spruce. (A, after Ref. 41; B, Ref. 67.)

among trees and species, but also because the patterns of variation of different criteria (latewood, cell length, etc.) do not coincide. A general characteristic of the juvenile period is that it is a period of rapid changes of wood structure. Juvenile wood has been primarily investigated in softwoods, especially in fast-growing American southern pines, where about 10 growth rings are often arbitrarily considered as juvenile wood (110). In general, about 10–20 growth rings are considered to comprise juvenile wood (47, 52, 59, 80, 89).

The juvenile period of wood development is followed by the formation of *adult* or *mature wood*, characterized by the attainment of a "typical" level of structural organization. Early stages of formation of adult wood may roughly coincide in many species with the beginning of transformation of the first growth rings into heartwood (24).

Overmature wood is produced by very old trees. As mentioned previously, such wood is atypical in structure in comparison to wood produced during the adult or mature period. It has very narrow rings with little latewood, lower cellulose and higher lignin content, and shorter cells with relatively thin walls.

Old age may also be associated with changes of inner heartwood. Such changes, gradually progressing from pith outward, may relate to chemical differentiation of extractives, formation of minute "compression failures" (25), or fungal attack. Thus, inner heartwood (near the pith) may, with increasing age, differ qualitatively from more recently formed heartwood.

Juvenile wood is the subject of extensive literature due to its great practical importance in wood utilization. The existing differences in anatomical structure have adverse effects on properties (59) (high longitudinal shrinkage, lower strength) (13), and therefore affect the quality of products. Strength is reduced by about 15–30% (80); greater longitudinal shrinkage causes distortions, checks and splits in lumber and other products, shorter fibers, and usually lower density (resulting from less latewood and lesser cell-wall thickness in comparison to mature wood) reduce not only the strength of wood but also certain mechanical properties as well as the yield of pulp. The importance of juvenile wood is greater in fast-growing trees, where the proportional volume of juvenile wood is high (110). Some figures are interesting: 15-year-old American southern pine (*Pinus taeda*) was estimated to have at least 40% juvenile wood in the harvested trunk; 40-year-old trunks may contain 25% juvenile wood; 50-year-old Douglas-fir of natural growth may be expected to have about 16% juvenile wood, but plantation trees of similar size and younger age may contain about 55% juvenile wood (89). The differences between juvenile wood and mature wood are less in hardwoods than softwoods, although there is also less research directed to hardwoods (13, 47) (Figure 6-5).

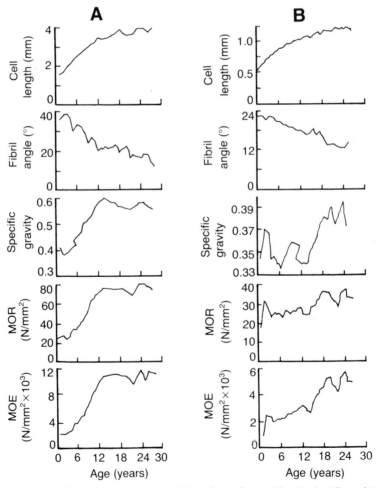

Figure 6-5. Change of anatomical and property characteristics of juvenile wood in (A) pine (*P. taeda*) and (B) cottonwood (*Populus* sp.). (After Ref. 13; reproduced by permission from the Society of Wood Science and Technology, USA.)

Horizontal Variation Within and Between Growth Rings. Within any one growth ring there are, of course, changes in cell-wall thickness, or in cell type and arrangement, associated with the formation of earlywood and latewood in trees growing in the temperate zones[4]. Earlywood and latewood differ also in cell size, ultrastructure, and chemical composition. In general, latewood cells—tracheids, fibers, vessel members—are longer than those of earlywood; the relationship is not linear from the beginning to the end of a growth ring, because short and long cells may be found in between (28, 92). Variation in length is accompanied by variation in the angle of microfibrils in the S_2 layer. In addition, latewood was found to have a higher cellulose and lower lignin content than earlywood in a number of softwoods and hardwoods investigated (19, 38, 106); however, the difference is due to the greater proportion of cellulose-rich secondary wall in latewood than in earlywood. In hemlock (*Tsuga heterophylla*), latewood was observed to have a higher degree of crystallinity than earlywood (58).

Variation *between rings* is mainly introduced by changes in ring width. As a rule, rings vary in width with increasing tree diameter (the normal tendency is for the rings to gradually narrow from pith to bark), but variations may also result from changes of the microenvironment in which each tree grows. Sudden release of a suppressed tree will be followed by a large increase in ring width.

The effects of changing ring width have been investigated in numerous studies in relationship to the proportion of latewood and cell length. The foregoing discussion should indicate, however, that such correlations must take into consideration the age of wood—whether it is juvenile, adult, or overmature. In many cases these factors are ignored, and therefore the results are conflicting and the validity of generalizations questionable.

With the assumption that consideration is limited to adult wood, the relationship between ring width and the proportion of latewood[5] is definite only in ring-porous hardwoods. In these woods, an increase of ring width—within limits—is associated with an increasing proportion of latewood (Figure 6-6). In diffuse-porous hardwoods, latewood is not distinct and, therefore, it is not feasible to identify a practical relationship. However, an increasing ring width may be associated with an increasing proportion of thick-walled latewood cells (fibers); wider rings were reported to contribute to a higher specific gravity in maple (*Acer saccharum*), and yellow poplar, for example (72). In softwoods, the relationship of ring width to the proportion of latewood varies among species, but generally the correlation is low (54, 97). However, there is apparently a moderate ring width, between 0.5 and 2.0 mm, where the proportion of latewood is more often higher than in very narrow or very wide rings (54).

With respect to the relationship of ring width to cell length, there is no general agreement; however, the balance of opinion appears to be in favor of a negative relationship—that is, within a tree, wider rings have shorter cells (27, 28, 92); extremely narrow rings may also have short tracheids (94).

Vertical Variation

The existence of a vertical variation of structural characteristics should be expected, since at different height levels wood is composed of growth rings of different structure. However, vertical variation is not simply a consequence of horizontal variation. Thus, the width of any one ring increases from the growing tip to a maximum near the lower part of the crown, then slowly decreases to the base. The proportion of latewood is higher at the base and decreases with height (55, 56)—but not always (Figure 6-7); cellulose content is changing in the same manner. Cell size also varies with height. This was first observed by Sanio (85) in Scots pine, and later by many workers in other species. Existing data indicate, in their majority, that, within a growth ring, length increases from the base upward for a certain distance; and, after reaching a maximum, it decreases progressively to the top of the tree, where length is less than at the base. Thus, in later-formed rings the longest cells occur higher up the stem than in earlier-formed rings (9, 27, 92). There is evidence of a positive relationship

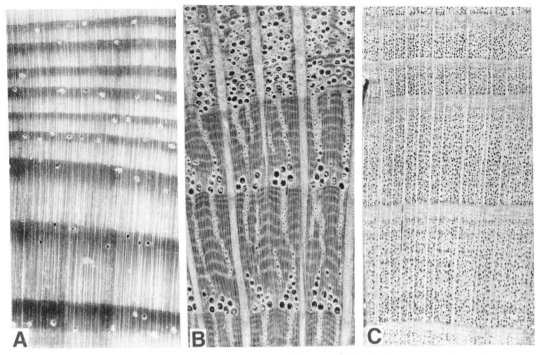

Figure 6-6. Effect of ring width on wood structure. (A) Black pine; (B) red oak; (C): beech (see text).

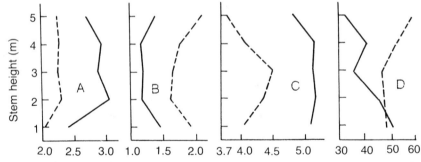

Figure 6-7. Vertical variation of structural characteristics in the stems of two young spruce trees (1, ─────── 30 years old; 2. ────── 16 years old). (A) Tracheid length (mm); (B) number of tracheids ($\times 10^3/mm^2$); (C) radial wall thickness (μm); (D) percentage of latewood (%). (Adapted from Ref. 108.)

between growth in height (activity of the apical meristem) and cell length (10, 28).

FACTORS AFFECTING VARIATION WITHIN A TREE

The factors involved in horizontal and vertical variation within a tree are *age*[6] of the cambium that produced wood, and *stimuli* to which the cambium is subjected during tree growth. Stimuli derive from external or *environmental influences* (e.g., mechanical stimuli that relate to distance from pith or from the base of the tree), and from internal *growth stresses* (43). These various factors interact, and their effects are often difficult to separate.

Environment, mainly in the form of wind action, induces mechanical stimuli, which influence structural characteristics of wood, such as cell-wall thickness (56). However, environmental factors affect the structure of the wood of a tree in a number of other ways. Environment includes a large diversity of factors, which act both below the ground (e.g., moisture, nutrients in the soil) and above the ground (e.g., light, temperature) (50, 81, 107). These factors create the microenvironment in which each individual tree grows. This microenvironment is not static throughout the life of a tree but subject to change, thus contributing to variation in wood structure within a growth ring and from ring to ring. It is difficult to assess individually all the factors of such microenvironments, and to relate quantitatively their influence on the structure of wood. Crown growth and development has been suggested as a unifying factor (55). Thus, the crown was observed to affect initiation and extent of heartwood; a large crown retards heartwood formation. Reduction of crown was found to reduce the amount of earlywood (14, 53, 64, 72). Active growth of the crown is associated with the formation of earlywood, while transition to latewood approximately coincides with the cessation of terminal elongation (55). Crown class (size) may influence tracheid length, diameter of earlywood tracheids, and height of rays (88).

The formation of latewood is, of course, a reflection of changing cell-wall thickness between the early and the latter part of a season of growth. What brings this change is not well understood, but certain environmental factors have been recognized to be influential. For example, the importance of soil moisture has been repeatedly observed; the width of earlywood depends on the moisture available at the beginning of the season, and the width of latewood on the amount of summer rainfall (40). Silvicultural practices affect moisture and other environmental factors (temperature, light), and thus influence the formation of earlywood and latewood. Fertilization was found to increase the amount of earlywood in longleaf pine (37, 72) and Norway spruce (74), but in slow-growing Scots pine an increase of latewood was observed (74). The type of transition from earlywood to latewood is generally characteristic of a species, but sometimes may change between rings as a result of annual environmental variations.

Such variations may also result in change of width, from ring to ring; this is accompanied by changes in proportion of latewood and cell length. The existing relationships (in general, and specifically for softwoods, ring-porous and diffuse-porous hardwoods) have already been explained.

The pattern of horizontal and vertical variation may not be the same in all radial planes of a tree. Such a condition arises because the effects of the environment may not be the same all around the periphery of a forming growth ring. Differences were found, for example, between the sunny and shaded side of trees; the sunny side having consistently shorter tracheids or fibers than the shaded one (60). Variation of wood structure in different radial planes (cardinal points) may be suspected especially in the presence of eccentric growth. Because of the possible existence of such a pattern of variability, sampling within a tree for studies of wood structure (or properties) is, as a rule, not limited to one radius.

Environmental factors (climate, fertilization, irrigation, tree spacing, etc.) are also thought to influence the development of growth stresses (1), and they may be related to changes of cell-wall thickness and microfibrillar angle (18). Development of stresses is also a species effect; for example, certain eucalypts are prone to a high degree of growth stresses[7].

VARIATION BETWEEN TREES

Variation of wood structure between trees of the same species exists because the microenvironment in which each tree grows is different. Differences occur in the same site, as well as between sites—in the same or different geographical localities, altitudes, and latitudes. Evaluation of structural variations between trees appears to be relatively difficult, however. The results are often conflicting (28); this may be due to different behavior between species, or to the large within-tree variation (Figure 6-7).

In a certain site, variation is very high—generally higher than between sites. The structure of wood of different trees varies in response to differences in microenvironments. Silvicultural measures (e.g., spacing of plantations, thinning, pruning, fertilization, etc.) will change this environment. As a result, adjacent trees may differ in pattern of ring width (from pith outward) and ring structure. Previous remarks regarding the relationship of ring width to the proportion of latewood (within a tree) should hold here, too. However, the relationship of ring width to cell length appears to be inverse in comparison to the relationship that exists within a tree; rapidly grown trees (e.g., dominant trees in a forest stand) have a greater average cell length than slowly grown trees (9, 27, 28).

Site is not correlated in a definite way with the proportion of latewood; investigations have shown that good sites and poor sites both decreased and increased latewood (37, 91, 99). Differences in the proportion of latewood (40, 54, 82) and cell length (9, 28, 82) have been observed between trees growing in different altitudes, and in different geographical localities within the range of a species. Increased latitude (associated to xeric conditions) was found to affect the structure of acacia wood; a positive relationship occurred between latitude and length of vessel members, proportion of fibers, and multiseriate rays, but the proportion of ray tissue and basic density were not correlated (104).

Aside from environmental effects, a significant factor for between-tree variation is that trees of the same species may differ in *genetic constitution*. It has been shown that certain characteristics of wood structure are inheritable (110), but separation of genetic and environmental effects is a difficult task.

QUALITY CONTROL

This chapter offers additional evidence that wood is a raw material of variable structure. This variability is not difficult to explain: wood is a product of biological processes, and growing trees are subject to internal and external influences during their life. Quality control, in the industrial sense, is making a product with standardized and desirable properties. Quality control may also be applied in the forest—the biological factory where wood is produced—by employing proper silvicultural and genetic practices. However, in contrast to the situation in a conventional factory, the expectations of control of tree growth—and especially of wood structure—cannot be too great. Trees are living organisms; they possess hereditary traits that are difficult to change, and there are practical limitations in efforts to regulate their microenvironment[8].

However, an understanding of the variability of wood and of the factors involved provides a desirable background for efforts to improve the quality of wood as a raw material in the forest—both through silviculture and genetics. Although the contribution of silviculture cannot be overemphasized, high hopes are placed in the improvement of trees and of their wood by genetics[9]. It is recognized that possibilities exist with respect to several characteristics, such as tracheid and fiber length, the proportion of latewood, specific gravity, cellulose and lignin content, quantity and quality of extractives, diameter of juvenile wood, figure, defects (e.g., knots, spiral grain, reaction wood), growth stresses, and others (49, 69, 110). The implementation of such goals is not easy, mainly because the processes of improvement require years to be completed[10], but this is undeniably a promising road to follow.

An understanding of the variability of wood is needed both on the part of foresters and technologists. With such knowledge, and the application of modern technology in wood-using industries (89), it becomes possible to practice rational wood utilization to the best advantage.

THE WOOD OF BRANCHES AND ROOTS

The structure of the wood of branches and roots differs in certain respects from that of the stem (tree trunk). However, because of limited or nonexisting importance of such wood from the wood utilization point of view, there is relatively little information on this subject. The major differences of branch and root wood from stem wood may be summarized as follows below.

The Wood of Branches

1. Branch wood is generally heavier than stem wood (32, 45).
2. Branches grow more slowly than stems, so their growth rings generally are more narrow than those of stems. Cross-sections of branches are usually eccentric; in softwoods, the rings of branches are wider on the lower side; whereas in hardwoods they are wider on the upper side. Many factors, such as distance from the stem, branch diameter, angle of branch to stem, and length of branch are related to eccentricity (71). As a rule, eccentricity is associated with the presence (in the wider rings) of reaction wood (i.e., compression wood in softwoods and tension wood in hardwoods) (42, 101) (see Chapter 7).
3. Branches retain the basic structural characteristics of ring-porous and diffuse-porous arrangement. However, vessels are smaller than in stems and more numerous per unit area of cross-section. Branch wood has more rays. In birch, branch wood was observed to be almost similar to juvenile wood in cellular proportions (5). Resin canals, if present in the stem, are also present in branches; however, in branches they are more numerous and smaller (32).
4. The cells of branch wood are shorter and narrower than those of stem wood. This applies both to tracheids of softwoods (9, 32, 97) and to fibers and vessel members of hardwoods (9, 63, 97). Length variation was investigated by Sanio (85) in Scots pine. He observed that the length of branch tracheids is related to the height at which branches arise from the main stem—and therefore, to the length of stem tracheids at that particular height. Branches, arising at heights where stem tracheids are relatively long, have longer tracheids; likewise, branches, arising at heights where stem tracheids are relatively short, have shorter tracheids. He also observed that the length of tracheids varies from the base to the tip of branches; it increases from the base outward and then falls. Similar observations were made in a Japanese pine: in 1-year-old branches, tracheid length increased from the base to a certain point, and then decreased gradually toward the top where it was generally at a minimum (94).

The Wood of Roots

In general, the wood of roots exhibits relatively large variability. An important factor is distance from the stem; with increasing distance there are greater differences in structure between root wood and stem wood. The structure of root wood depends also on whether roots are vertical or lateral, and on the soil conditions under which they grow.

1. Root wood is usually lighter in weight than stem wood (3), but may be heavier in softwoods (45).
2. In cross-section, roots are often deformed, especially those near the stem, and may attain an ellipsoid, I-beam, or T-beam shape (Figure 6-8). Such shapes are indicative of stresses—which derive from the weight of the tree, loads of snow, wind, and so on. In roots, pith is nonexistent or very small (70).

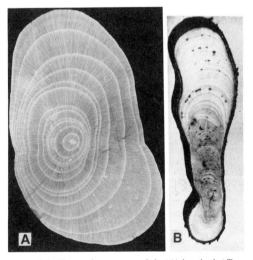

Figure 6.8. Eccentric root growth in (A) hemlock (*Tsuga canadensis*) and (B) birch. In B an extreme case of eccentricity and numerous pith flecks. (Photograph B from Ref. 5; reproduced by permission from the International Association of Wood Anatomists.)

Growth rings are relatively less pronounced or indistinct, and their width, while variable, tends to be narrower than in stem wood (45, 97). Wide rings may occur in roots growing in swampy areas (97). Roots are often eccentric. In several softwoods investigated, the greatest eccentricity was observed in lateral roots near the stem, and in vertical roots wedged between stones, while roots growing in sand were more or less round (8). Roots may possess false rings (70) and pith flecks (Figure 6-8). The bark-to-wood ratio may be higher in roots than in stems (31).

3. The ring-porous character is mostly not retained in root wood (Figure 6-9). The structure becomes diffuse-porous (83), or earlywood pores are not as distinctly different in size and arrangement as in stems (70). With increasing distance from the stem, latewood almost disappears; the same happens in softwoods. In diffuse-porous woods, the structure of root wood is looser compared to stem wood. Oak, black locust, and other hardwoods were reported to lack heartwood in roots (83, 97), but spruce and larch possess heartwood (16).

The number of rays in root wood may be smaller than in branch wood (32), but are about equal (32) or greater (61) in comparison to stem wood. The volume of rays was found to be greater in roots than in the stem and branches of hardwoods, although in the roots of softwoods ray volume was somewhat smaller (32). Rays may be different, e.g. aggregate in oak and birch (84).

Resin canals are less numerous in roots than in the stem, but their size is about the same (32). Stem and root canals are interconnected (73). In the primary xylem of roots of certain softwoods, primary resin canals may be present. The number and size of such canals vary: fir has one; spruce, larch and Douglas-fir, two; and pine, two to five (73, 97).

4. Cellular structure is very variable in roots. In softwood roots, the contrast in wall thickness between earlywood and latewood tracheids decreases with distance from the stem; gradually, latewood tracheids of roots have only smaller lumina. The radial walls of root tracheids in Scots pine were found to possess up to four rows of bordered pits. Tangential walls were also equipped with pits (97). Root wood of softwoods may differ from stem wood in additional characteristics, such as height and width of rays, presence of ray tracheids (they may be present in roots, while absent in stems), number of pits per cross-field, number of epithelial

Figure 6-9. Transverse section of root wood in red oak (*Quercus borealis*), showing pores (vessels) radiating from pith outward (5×). (Compare with stem wood in Figure II-4B, Appendix II). (Reproduced by permission from Pergamon Press.)

cells in resin canals, presence of axial parenchyma (31), and so on.

In ring-porous hardwoods, the ring-porous character is often altogether missing, as already mentioned, whereas in diffuse-porous hardwoods root wood appears looser (with more and larger vessels)—as in poplar—or may develop some pattern in vessel arrangement that does not exist in stem wood— as was reported in horse chestnut (70). In general, fibers are fewer in root wood, but axial parenchyma is abundant (35, 97). For example, root wood of black locust was found to contain 60% parenchyma and 27% vessels (97). Tyloses may be missing in root wood, as was observed in black locust, chestnut, ash, and other hardwoods (97). Oak and birch may have aggregate rays in root wood, while such rays do not exist in stem wood (46, 83). Other differences pertain to shape of lumina, perforation of vessel members, and size and structure of rays (homocellular, heterocellular) (70).

5. Roots may possess reaction wood but, in contrast to branches, this is a rare occurrence. With regard to compression wood, it has been observed that in some softwoods such wood is produced in the roots under normal (subterranean) conditions, but in other species only when roots become exposed (96, 103). Tension wood was found in the roots of poplar, birch, and other hardwoods. Fibers are gelatinous, or may be apparently normal but less lignified (44, 70).

6. It has been suggested that root woods may contain juvenile wood (70).

7. Finally, there are differences in cell size between root wood and stem wood. Tracheids, fibers, and vessel members have larger diameters in root wood (this statement does not apply to the earlywood vessels of ring-porous woods) (70). Tracheids and fibers in roots have been reported to be longer or shorter (27, 32, 62) in comparison to the stem, but the reports of shorter length are apparently not accurate (9). Size variation from pith outward has been investigated by Sanio in roots of Scots pine. He found that the length of tracheids first increases, then falls, and then rises again until a constant size is reached (85).

In general, variations in cellular structure of root wood may be affected by position (main, lateral roots) (62, 76), distance from the stem (101), and exposure to air and light (66). Exposed roots tend to produce wood that has structural characteristics resembling stem wood (8, 31). Inversely, stems that were experimentally grown in soil developed wood similar to root wood (61, 70).

Roots are primarily organs of storage and food translocation in the living tree, and these functions affect their structure. Thus, storage is served by abundant parenchyma, and food translocation by abundant pitting, more and larger vessels, and so on. Structural variations related to distance from the stem (therefore, distance from the synthesizing crown), and from pith outward, probably stem from differences in function, too. In birch, distance from pith was found to be an important source of variation; as distance increased, density and proportion of fibers increased, and the proportion of vessels and rays decreased (5).

It is obvious, that branch and root wood in particular possess considerable differences in structure from stem wood[11]. This would affect the properties of such wood, if it was to be utilized. Differences in structure also interfere with wood identification, with the result that identification of wood from samples deriving from branches, and particularly from roots, on the basis of keys for stem wood, is a difficult or impossible task. For this reason, special keys have been prepared for the identification of root wood (23).

REFERENCES

1. Anonymous. 1986. Growth stresses studied. HOUTIM/92. Pretoria, South Africa: National Timber Research Institute.
2. Baas, P., R. Schmid, and B. J. van Heuven. 1986. Wood anatomy of *Pinus longaeva* (bristlecone pine) and the sustained length-on-age increase of its tracheids. *IAWA Bull.* n.s. 7(3):221–228.
3. Bailey, I. W. 1920. The cambium and its derivative tissues. II. Size variation of cambial initials in gym-

nosperms and angiosperms. *Amer. J. Bot.* 7:355–367.
4. Bailey, I. W. 1923. The cambium and its derivative tissues. IV. The increase in girth of the cambium. *Amer. J. Bot.* 10:499–509.
5. Bhat, K. M. 1982. A note on cellular proportions and basic density of lateral roots in birch. *IAWA Bull.* n.s. 3(2):89–94.
6. Bhat, K. M., K. V. Bhat, and T. K. Dhamodaran. 1989. Fibre length variation in stem and branches of eleven tropical woods. *IAWA Bull.* n.s. 10(1):63–70.
7. Bamber, R. K. 1976. Heartwood: Its function and formation. *Wood Sci. Tech.* 10(1):1–8.
8. Bannan, M. W. 1941. Variability in wood structure in roots of native Ontario conifers. *Bull. Torrey Bot. Club.* 68:173–194.
9. Bannan, M. W. 1965. The length, tangential diameter and length/width ratio of conifer tracheids. *Can. J. Bot.* 43:967–984.
10. Bannan, M. W. 1966. Cell length and rate of anticlinal division in the cambium of the Sequoias. *Can. J. Bot.* 44:209–218.
11. Beckwith, J. B. 1965. Variation from pith to bark in some characteristics of the wood of red pine. Yale Forestry School (unpublished).
12. Bendtsen, B. A. 1978. Properties of wood from improved and intensely managed trees. *For. Prod. J.* 28(10):61–72.
13. Bendtsen, B. A., and J. Senft. 1986. Mechanical and anatomical properties in individual growth rings of plantation-grown eastern cottonwood and loblolly pine. *Wood and Fiber Sci.* 18(1):23–38.
14. Benić, R. 1956. Investigations on the proportion and some physical properties of the sapwood and heartwood of *Fraxinus angustifolia* Vahl. *Glasnik sumske pokuse* 12:13–104.
15. Bethel, J. S. 1964. The developing architecture of secondary xylem in conifers. *For. Sci.* 10:89–91.
16. Boettcher, P., and W. Liese. 1975. Zur Verkernung des Wurzelholzes von Fichte und Lärche. *Forstw. Cbl.* 94(4/5):152–160.
17. Bosshard, H. H. 1983. *Holzkunde*, Vol. 2. Basel: Birkhaeuser Verlag.
18. Boyd, J. D. 1980. Relationships between fibre morphology, growth strains and physical properties of wood. *Austral. For. Res.* 10:337–360.
19. Browning, B. L. 1963. The composition and chemical reactions of wood. In *The Chemistry of Wood*, ed. B. L. Browning, pp. 57–101. New York: Interscience.
20. Brunden, M. N. 1964. Specific gravity and fiber length in crown-formed and stem-formed wood. *For. Prod. J.* 14:13–17.
21. Chattaway, M. N. 1952. The sapwood-heartwood transition. *Austral. For.* 16:25–34.
22. Core, H. A., W. W. Moschler, and C. Chang. 1978. A note on prediction of wood properties in yellow-poplar. *Wood and Fiber* 9(4):258–261.
23. Cutler, D. F., P. J. Rudall, P. Gasson, and R. H. Gale. 1987. *Root Identification: Manual of Trees and Shrubs.* London: Chapman & Hall.
24. Dadswell, H. E. 1958. Wood structure variations occurring during tree growth and their influence on wood properties. *J. Inst. Wood Sci.* 1:2–23.
25. Dadswell, H. E., and W. E. Hillis. 1962. Wood. In *Wood Extractives*, ed. W. E. Hillis, pp. 3–55. New York: Academic Press.
26. Denne, M.P. 1988. Definition of latewood according to Mork (1928). *IAWA Bull.* n.s. 10(1):59–62.
27. Dinwoodie, J. M. 1961. Tracheid and fiber length in timber—A review of literature. *Forestry* 34:125–144.
28. Dinwoodie, J. M. 1963. Variation in Tracheid Length in *Picea sitchensis*. Carr. For. Prod. Res. Sp. Report No. 16, London.
29. Echols, R. M. 1955. Linear relation of fibrillar angle to tracheid length, and genetic control in slash pine. *Tropical Woods* 102:11–22.
30. Einspahr, D. E. 1986. Impact of biotechnology on wood properties. *Proc. 18th IUFRO World Congress, Div. 5*, pp. 347–358.
31. Esau, K. 1965. *Plant Anatomy*, 2nd ed. New York: John Wiley & Sons.
32. Fegel, A. C. 1941. Comparative Anatomy and Varying Physical Properties of Trunk, Branch and Root Wood in Certain Northeastern Trees. New York State College of Forestry. Tech. Bull. No. 55.
33. Frey-Wyssling, A., and H. H. Bosshard. 1959. Cytology of the ray cells in sapwood and heartwood. *Holzforschung* 13:129–136.
34. Fukazawa, K., J. Ohtani, and K. Takabe. 1986. The extent of juvenile wood in softwoods. Paper presented at the 18th IUFRO World For. Congr., Lubljana, Yugoslavia.
35. Gasson, P. 1987. Some implications of anatomical variations in the wood of penduculate oak (*Quercus robur* L.) including comparisons with common beech (*Fagus sylvatica* L.). *IAWA Bull.* n.s. 8(2):149–166.
36. Gerry, E. 1916. Fiber measurement studies—A comparison of tracheid dimensions in longleaf pine and Douglas-fir with data on the strength and length, mean diameter and thickness of wall of the tracheids. *Science* 40:360.
37. Goggans, J. F. 1961. The Interplay of Environment and Heredity as Factors Controlling Wood Properties in Conifers. North Carolina State College, School of Forestry, Tech. Report No. 11.
38. Hale, J. D., and L. P. Clermont. 1963. Influence of prosenchyma cell-wall morphology on basic physical and chemical characteristics of wood. *J. Polym. Sci.* 2:253–261.
39. Hejnowicz, Z., and A. Hejnowicz. 1959. Variations of length of vessel members and fibers in the trunk of *Robinia pseudoacacia*. Abstr. IX Intern. Bot. Congr., pp. 158–159, Montreal.
40. Hildebrandt, G. 1960. The effect of growth conditions on the structure and properties of wood. *Proc. Fifth World For. Congr.* 3:1348–1353.
41. Hillis, W. E. 1987. *Heartwood and Tree Exudates.* Berlin: Springer-Verlag.
42. Hoester, H. R. and W. Liese. 1966. Über das Vorkommen von Reaktionsgewebe in Wurzeln und Äste der Dikotyledonen. *Holzforschung* 20(3):80–103.

43. Jacobs, M. R. 1965. Stresses and Strains in Tree Trunks as They Grow in Length and Width. Austr. For. & Timber Bureau, Leaflet No. 96, Camberra.
44. Jagels, R. 1963. Gelatinous fibers in the roots of quaking aspen. *For. Sci.* 9:440–443.
45. Jane, F. W. 1970. *The Structure of Wood.* London: A. & C. Black.
46. Jeffrey, E. C. 1917. *The Anatomy of Woody Plants.* Chicago: The University of Chicago Press.
47. Kinninmoth, J. A. 1986. Wood from fast-grown short-rotation trees. *Proc. 18th IUFRO World Cong., Div. 5,* pp. 425–438. Ljubljana, Yugoslavia.
48. Knigge, W. 1960. The natural variability of wood as it affects selection of test material and structural applications of wood. *Proc. Fifth World For. Congr.* 3:1362–1367.
49. Koukos, P. K. 1987. Investigation of Wood Quality Characteristics of Six New Poplar Clones in Relation to Genotype, Age and Growth Rate. Diss., Aristot. Univ. Thessaloniki (Greek, English summary).
50. Kozlowski, T. T. 1971. *Growth and Development of Trees.* New York: Academic Press.
51. Ladell, J. L. 1963. Needle Density, Pith Size and Tracheid Length in Pine. Commonw. For. Inst., Paper No. 36, Oxford.
52. Ladrach, W. E. 1986. Control of wood properties in plantations. *Proc. 18th IUFRO World For. Congr., Div. 5,* pp. 369–380. Ljubljana, Yugoslavia.
53. Lappi-Seppälä, M. 1952. Über Verkernung und Stammform der Kiefer. *Communs. Inst. Forest. Fenniae* 40:1–27.
54. Larson, P. R. 1957. Effect of Environment on the Percentage of Summerwood and Specific Gravity of Slash Pine. Yale University, School of Forestry Bull. No. 63.
55. Larson, P. R. 1962. A biological approach to wood quality. *Tappi* 45:443–448.
56. Larson, P. R. 1962. Stem Form Development of Forest Trees. For. Sci. Monograph No. 5.
57. Larson, P. R. 1966. Changes in chemical composition of wood cell walls associated with age in *Pinus resinosa*. *For. Prod. J.* 16(4):37–45.
58. Lee, C. L. 1961. Crystallinity of wood cellulose fibers. *For. Prod. J.* 11:108–112.
59. Lewark, S. 1986. Anatomical and physical differences between juvenile and adult wood. *Proc. 18th IUFRO World Congr., Div. 5,* pp. 272–281. Lubljana, Yugoslavia.
60. Liese, W., and H. E. Dadswell. 1959. Über den Einfluss der Himmelsrichtung auf die Länge von Holzfasern und Tracheiden. *Holz Roh- Werkstoff* 17:421–427.
61. MacDonald, R. D. S. 1960. Comparative Studies on Stem and Rootwood with Special Reference to Some British Hardwoods. Sp. Subj. Report, Commonw. For. Inst., Oxford.
62. Manwiller, F. G. 1972. Tracheid dimensions in rootwood of Southern pine. *Wood Sci.* 52(2):122–124.
63. Manwiller, F. G. 1974. Fiber lengths in stems and branches of small hardwoods of Southern pine sites. *Wood Sci.* 7(2):130–134.
64. Marts, R. O. 1951. Influence of crown reduction on springwood and summerwood distribution in longleaf pine. *J. For.* 49:183–189.
65. Mork, E. 1928. Die Qualität des Fichtenholzes unter besonderer Rücksichtnahme auf Schleif- und Papierholz. *Der Papier Fabrikant* 26:741–747.
66. Morrison, T. M. 1953. Comparative histology of secondary xylem in buried and exposed roots of dicotyledonous trees. *Phytomorphology* 3:247–430.
67. Palovic, J. 1967. Die Abhängigkeit zwischen Rohdichte und Makrostruktur der Nadelhölzer und ihre Bedeutung in der Holztechnologie und Forstwirtschaft. *Holz Roh Werkstoff* 19:72–79.
68. Pandalai, R. C., G. M. Nair, and J. J. Shah. 1985. Ultrastructure of ray parenchyma cells in the wood of *Melia azedarach* L. (*Meliaceae*). *Wood Sci. Tech.* 19:201–209.
69. Panshin, A., and C. De Zeew. 1980. *Textbook of Wood Technology,* 4th ed. New York: McGraw-Hill.
70. Patel, R. N. 1965. A comparison of the anatomy of the secondary xylem in roots and stems. *Holzforschung* 19:72–79.
71. Patel, R. N. 1970. Anatomy of stem and root wood of *Pinus radiata* D. Don. *N. Z. J. For. Sci.* 1(1):37–49.
72. Paul, B. H. 1963. The Application of Silviculture in Controlling the Specific Gravity of Wood. USDA Tech. Bull. No. 1288.
73. Pauly, G. 1963. Les canaux sécréteurs du pin maritime. *Travaux Lab. Toulouse* 61(1):1–32.
74. Pechmann, H. 1958. Die Auswirkung der Wuchsgeschwindigkeit auf die Holzstruktur und die Holzeigenschaften einiger Baumarten. *Schweiz. Z. Forstw.* 109:615–647.
75. Preston, R. D. 1974. *The Physical Biology of Plant Cell Walls.* London: Chapman & Hall.
76. Poliquin, J. 1966. Changement Morphologiques et Physiologiques Reliés à l' Age dans le Bois de Racine de *Pinus silvestris* L. Thèse No. 3867 ETH, Zurich.
77. Poller, S. 1978. Studie über die chemische Zusammensetzung von Wurzel-, Stamm und Astholz zweier Kiefern (*Pinus silvestris*) unterschiedlichen Alters. *Holztechnologie* 19(1):22–25.
78. Ratanunga, M. S. 1964. A study of the fibre lengths of *Eucalyptus grandis* grown in Ceylon. *Ceylon For.* 6:101–112.
79. Rendle, B. J. 1960. Juvenile and adult wood. *J. Inst. Wood Sci.* 5:58–61.
80. Resch, H. 1986. Juvenile wood is topic of in-depth workshop. *For. Prod. J.* 36(1):4, 69.
81. Richardson, S. D. 1964. The external environment and tracheid size in conifers. In *The Formation of Wood in Forest Trees,* ed. M. H. Zimmermann, pp. 367–388. New York: Academic Press.
82. Rickson, F. R., and C. Heimsch. 1964. Length of tracheary elements of selected dicotyledons in relation to geographical distribution. *Amer. J. Bot.* 51:673.
83. Riedl, H. 1937. Bau und Leistungen des Wurzelholzes. *Jahrb. Wiss. Bot.* 85:1–75.

84. Rusch, J. 1973. Vergleichende anatomische Untersuchungen des Holzes von Wurzel und Stamm bei verschiedenen Laubbaumarten. Freiburg Univ. Forstwiss. Diss. (*Forstl. Umschau* 17/2:164, 1974).
85. Sanio, K. 1872. Über die Grösse der Holzzellen bei der gemeinen Kiefer (*Pinus silvestris*). *Jahrb. wiss. Bot.* 8:401–420.
86. Scaramuzzi, F. 1965. Istologia del legno di pino d'Aleppo e sue modificazioni durante la duramificazione. *Annali Acad. Ital. Sci. For.* 4:387–403.
87. Scaramuzzi, G. 1959. Variazioni dimensionali delle fibre nel fusto in *Populus X euramericana* cv. I-214. *Publ. Centr. Sper. Agr. For.* 2:87–118.
88. Schultze-Dewitz, G. 1961. Notes concerning the influence of the position of a tree on its anatomical structure. *Holzforschung* 15:89–91.
89. Senft, J. F. 1986. Practical significance of juvenile wood for the user. *Proc. 18th IUFRO World Congr. Div. 5*, pp. 261–269. Lubljana, Yugoslavia.
90. Smith, J. H. G., J. Walters, and R. W. Wellwood. 1966. Variation in sapwood thickness of Douglas-fir in relation to tree and section characteristics. *For. Sci.* 1:97–103.
91. Spurr, S. H., and W. Hsiung. 1954. Growth rate and specific gravity in conifers. *J. For.* 52:191–200.
92. Spurr, S. H., and M. J. Hyvärinen. 1954. Wood fiber length as related to position in tree and growth. *Bot. Review.* 20:561–575.
93. Stewart, C. M. 1966. Excretion and heartwood formation in living trees. *Science* 153(3740):1068–1074.
94. Sudo, S. Variation in tracheid length in Akamatsu (*Pinus densiflora*). [A series of papers published in *Mokuzai Gakkaishi* 1968 (14/1), 1969 (15/2, 15/6), 1970 (16/5), 1973 (19/2).]
95. Taylor, P. W. 1979. Property variation within stems of selected hardwoods grown in the mid-South. *Wood Sci.* 11:193–199.
96. Timell, T. E. 1980. Karl Gustav Sanio and the first scientific description of compression wood. *IAWA Bull.* n.s. 1(4):147–153.
97. Trendelenburg, R., and H. Mayer-Wegelin. 1956. *Das Holz als Rohstoff*. München: Hanser Verlag.
98. Tsoumis, G., and C. Passialis. 1976. Effect of growth rate and abnormal growth on wood substance and cell-wall density. *Wood Sci. Tech.* 10(4):33–38.
99. Tsoumis, G., and N. Panagiotidis. 1980. Effect of growth conditions on wood quality characteristics of black pine (*Pinus nigra* Arn.). *Wood Sci. Tech.* 14:301–310.
100. Upricard, J. M. 1971. Cellulose and lignin content in *Pinus radiata* D. Don. Within-tree variation in chemical composition, density and tracheid length. *Holzforschung* 25(4):97–105.
101. Vurdu, H., and D. W. Bendtsen. 1980. Proportions and types of cells in stems, branches and roots of European black alder (*Alnus glutinosa* L. Gaertn.). *Wood Sci.* 13(1):36–40.
102. Wardrop, A. B. 1951. Cell wall organization and the properties of the xylem. I. Cell wall organization and the variation of breaking load in tension of conifer stems. *Austral. J. Sci. Res. B* 4:391–414.
103. Westing, A. H. 1965. Formation and function of compression wood in gymnosperms. *Bot. Review* 31:381–480.
104. Wilkins, A. P., and S. Papassotiriou. 1989. Wood anatomical variation of *Acacia melanoxylon* in relation to latitude. *IAWA Bull.* n.s. 10(2):201–207.
105. Wilson, A. T. 1961. Carbon-14 from nuclear explosions as a short-term dating system: Use to determine the origin of heartwood. *Nature* 191:714.
106. Wilson, J. W., and R. W. Wellwood. 1965. Intra-increment chemical properties of certain western Canadian coniferous species. In *Cellular Ultrastructure of Woody Plants*, ed. W. A. Côté, pp. 551–559. Syracuse, New York: Syracuse University Press.
107. Zahner, R. 1963. Internal moisture stress and wood formation in conifers. *For. Prod. J.* 13:240–247.
108. Zenker, R. 1967. Die Rohdichte als Bindeglied zwischen Struktur und Eigenschaften des Holzes. *Holz Roh- Werkstoff* 25(1):37.
109. Zimmerman, M. H., and C. L. Brown. 1971. *Trees-Structure and Function*. Berlin: Springer-Verlag.
110. Zobel, B. 1981. Wood quality from fast-grown plantations. *Tappi* 64(1):71–74.

FOOTNOTES

[1] The length of the cambial initials does not vary in such species (see Chapter 5); therefore, an increase of cell length should not be expected. The observed slight change is attributed to an increasing degree of intrusive growth during the differentiation of cambial derivatives (39).

[2] The proportion of cells that remain alive in sapwood is about 10%, but may vary between 5 and 40% (41).

[3] Such variation results in changes of wood properties. For example, in softwoods and in the region of juvenile wood (from pith outward) it has been observed that density, transverse shrinkage, and strength increase, whereas axial shrinkage decreases (12). Figures 6-4B and 6-5 show the charge of density, bending strength (modulus of rupture), and elasticity (modulus of elasticity) from pith outward.

[4] In spite of differences in wall thickness between earlywood and latewood tracheids, the amount of cell-wall substance is reported to be about the same; in earlywood tracheids the cell-wall substance is spread over a larger diameter (15). The amount of cell-wall substance (density of cell walls) was also found not to differ substantially between growth rings of very large differences in width (98).

[5] The proportion of latewood is not always easy to measure; inaccuracies may arise because of difficulties to draw the line of demarcation between earlywood and latewood. For softwoods, a rule proposed by Mork is applicable to most species. According to this rule, latewood tracheids (as observed on microtome sections) are these in which the width of their common wall in the radial direction is equal or greater than the width of either cell cavity (26, 65).

[6] In a tree there are cambia of different ages (measured from their origination by apical meristems). For example, in a 100-year-old stem, the cambium at the ground level is 100 years old, but there is also a 1-year-old cambium at the

top. Cambia of all intermediate ages exist at various heights within a stem. Aside from other effects, cambial age was observed to relate to the size of cambial initials and their derivatives (4).

[7]Growth stresses may cause degradation of wood. In some species, they are so intense that trees, when felled, split apart with such force as to be dangerous to workers. Splitting may also occur during sawing of logs. In *Eucalyptus grandis*, growth stresses can render up to 20% of the sawn timber unusable for structural purposes because of the splitting and warping that they can cause (1).

[8]In this chapter, consideration of variability is limited to normal wood structure. An additional source of variability in wood is the occurrence of abnormalities and defects in growing trees. These are discussed in Chapter 7.

[9]In addition to conventional tree breeding, tissue culture is a promising tool that could increase the possibilities of improving wood quality (30).

[10]For this reason, evaluation at an early age (in seedlings) becomes very important. Outward criteria are desirable in practice, and research efforts have been directed in this respect. It was found, for example, that the length of internode (27, 28, 94), needle density, and pith size (51) can be useful indicators of cell length. It is also interesting to take note of the possibility to estimate the average tracheid length of a tree (mature wood) from a single ring near the pith (juvenile wood), and that comparisons between trees (or clones) may be based on 1-year-old branches (94).

[11]In addition to structure, differences exist in chemical composition (i.e., proportion of cellulose, hemicelluloses, lignin, and extractives between stem, branch, and root wood) (77).

7

Abnormalities in Wood

The preceding chapters have been exclusively devoted to the discussion of normal wood structure. Deviations from normal structure are not uncommon, since trees are living organisms and are subject to various influences throughout their life span. When wood is looked upon as a raw material, most abnormalities adversely affect its service value; these are commonly called *defects*. From the wood utilization point of view, defects are also certain normal characteristics of all trees—namely, knots (branches) and pith.

The line of demarcation between mere abnormalities and defects is not clear. Certain deviations from normal growth-ring structure (e.g., false rings, discontinuous rings, indented rings) are not known to exercise an adverse effect on the service value of wood and, therefore, cannot be classified as defects. Mild deviations cannot always be considered as defects either; for example, taper becomes a defect only when pronounced. Characterization of a structural feature as defect may also depend on the intended use of wood. Sometimes, defects for one purpose may be advantages for another. Thus, attractive figure may be produced from wood that is worthless for other technical uses, and knotty pine is in certain cases preferred over clear wood. The degree of adverse effect may also vary with intended use; for example, grain deviations are more objectionable if strength of a wooden structure is of primary importance.

This chapter will include (i) growth abnormalities caused by environmental influences on living trees, and (ii) natural growth characteristics (knots and pith).

GROWTH ABNORMALITIES

The abnormalities included here are (a) deviations from typical tree form, (b) spiral grain and other gain deviations, (c) abnormal arrangement of growth rings, (d) reaction wood (compression wood and tension wood), (e) disruption of inner wood tissues (compression failures, shakes, resin pockets), (f) abnormal color, and abnormalities (g) due to wounding, and (h) due to environmental pollution and atomic radiation.

Deviations from Typical Tree Form

Typically, a tree stands vertically, it is practically cylindrical in form (at least over parts of the stem), and has a circular cross-section. Abnormalities include deviations from the vertical position, from cylindrical form, and from circular cross-section.

Action of various environmental factors may cause the stems of trees to deviate from vertical. *Leaning, bending, crook, forking* and formation of *pistol-butted* stems are forms of such deviations (Figure 7-1). The factors involved act mechanically (wind, snow, soil movement), physiologically (light), or through destruction of the leader (frost, drought, people, animals, insects, fungi).

Deviations from cylindrical form is a natural tendency of trees; their stems tend to be conical

Figure 7-1. Pistol-butted trunk in fir (*Abies*).

due to the manner of tree growth, which takes place by superposition of conoid layers. Taper is a defect when the conical form is pronounced. The amount of taper changes with tree age (there is a natural tendency for stems to become more cylindrical with age), and it is also influenced by growth conditions and species. Taper is affected by the size of the crown—the greater the crown, the greater the taper. Thus, thinning promotes and pruning reduces taper, and open-grown trees are more tapered than trees grown in forest stands. Exposure to wind increases taper, and other factors such as species and site are also involved (54, 55).

A number of theories have been proposed to explain variations in taper. These include: the nutritional theory (stem form is a result of equilibrium between transpiration and assimilation; the water conduction theory (stem form results from attainment of equilibrium of water transportation between crown and root); the mechanistic theory (stem form is influenced by the weight of the tree, action of wind, and of snow and ice in the winter); and the hormonal theory (stem form is due to gradients of growth hormone translocation) (54).

Tree trunks may acquire a basal enlargement—called *butt-swell*. Butt-swell is more pronounced in trees growing in open forest stands. This is due to stresses developed at the base of the trunk as a result of wind action on a large crown (54), although moist sites (59) also increase butt-swell.[1]

Deviations from circular cross-section may be hereditary in nature, but they are usually caused by environmental influences, such as wind action. For example, the fluted stem form of hornbeam is considered to be hereditary (Figure 7-2). This is attributed to the presence of aggregate rays which do not grow as fast as the neighboring tissues (12); the result is a wavy pattern of growth rings and, therefore, a wavy circumference of the stem. Deviations due to wind action may be observed in regions where strong winds blow predominantly from the same direction. An affected tree develops an elongated cross-section—ellipsoid, oval, irregular, or I-beam-like—with the long diameter in the direction of the acting wind (67) (Figure 7-2). A noncircular circumference may also be caused by one-sided development of the crown.

All deviations from typical form, especially when pronounced, cause high manufacturing waste, and produce grain deviations that affect the properties (mainly strength and dimensional stability) of lumber and other wood products. In many cases, reaction wood (compression wood and tension wood) may also be present.

Spiral Grain and Other Grain Deviations

The bark of some trees is twisted in appearance, which indicates that the wood is likewise twisted. More often, however, this deviation may be hidden under normal bark. In this case, its presence is detected only after debarking. Drying and shrinkage of the exposed wood forms oblique checks which run in spirals over the length of the stem. This deviation is known as *spiral grain* (Figure 7-3).

Spiral grain is due to the spiral arrangement of wood cells in relation to stem axis. The severity of this deviation is highly variable. The magnitude of the angle between fibers and stem axis may vary from a few degrees to an exceptional 90°; it may vary from pith to bark and

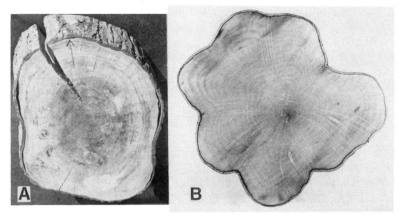

Figure 7-2. Deviations from circular cross-section. (A) I-beam-like form in a trunk of poplar (*Populus*) hybrid due to strong wind; (B) wavy circumference in hornbeam (*Carpinus*). (A, reproduced by permission from Pergamon Press.)

from bottom to the top of a tree; it may differ between stem and branches. The direction of spirality may also differ; it may be clockwise or counterclockwise, and may change from pith to bark. Initial direction of spirality and pattern of subsequent change may characterize a certain species.

Spiral grain is a very common defect, occurring in both softwoods and hardwoods. Trees with absolutely straight grain are rather scarce; this has led to views about spiral grain being "the normal pattern of growth" (35, 68). In fact, spiral grain has been looked upon as a physiological necessity related to transpiration (51).

However, the cause of spiral grain is not certain. The fact that under the same conditions some trees develop spiral grain and others do not, points to a genetic cause. It has been observed that spiral grain may be genetically transmitted through seeds (35, 69, 97). Several environmental influences have also been pro-

Figure 7-3. Spiral grain in poles; at right an extreme case in lodgepole pine, *Pinus contorta*.

posed as causal factors; they include twist due to wind action, unfavorable sites, crown development, and speculations about earth rotation and solar movement (69). An increase of spiral grain has been regarded as an active means by which trees try to increase their resistance to exterior influences, particularly to wind; spiral grain is thus designated as a "mechanomorphosis" (i.e., attainment of a form of growth which enables trees to better resist such an action) (91). The effect of site is not well understood. In a study of red alder, the potential of developing spiral grain was found to be less on a better site (where growth was faster) than on a poorer site (47). It is striking, however, that spiral grain presents an extreme variability; trees of the same species, growing in close proximity to each other, show great variation with respect to occurrence and degree of angle of deviation. The effect of crown development has been investigated in pine (*Pinus radiata*). It was observed that spiral grain may be influenced by tree spacing; closer spacing was found to produce beneficial effects, which were attributed to the faster upward movement of the crown. For the same reason, artificial pruning appeared to reduce spirality in the lower, branch-free portion of the stem (39).

With regard to the structure of wood possessing spiral grain, some observations were made in Scots pine (57). It was found that when the angle of deviation becomes large, the length of tracheids (and the height of rays) do not follow the normal pattern of increase with age; the walls of the first earlywood tracheids become thicker with increasing twist. Other structural characteristics were found not to differ significantly between normal and twisted wood.

Spiral grain is a serious defect. The strength of wood may become considerably less than normal, depending on the angle of deviation and the type of loading; most of all, the ability of wood to absorb shock is reduced (102) (see Chapter 11). Spiral grain affects adversely the dimensional stability, machining, drying and finishing characteristics of wood. Poles possessing spiral grain have a tendency to twist after installation; left-hand spiral poles deflect more than right-hand poles (11).

Occurrence of spirality, when not shown on the bark, may be detected by extracting increment cores; the angle of deviation is judged by splitting such cores in the tangential direction and observing the flow of a drop of ink. On debarked logs and poles, the angle may be measured on the basis of surface checks. Measurement of internal changes of angle or direction of spirality require splitting and detailed examination of disks (cross-sections). An "index" of spiral grain has been suggested for comparative measurements of such material. This index is an expression for the average amount of spiral grain at any level in a stem; it is a weighed average of a series of observations (10). On wood products (lumber, veneer, etc.), spiral grain may be detected by splitting; observing the alignment of structural elements (rays, vessels, resin canals, etc.) or checks on tangential surfaces; marking the surface with a scribe; or following the direction of a drop of ink; continuous, noncontact measurement in sawmills has also been suggested (63).

In some cases, the direction of fiber alignment alternates at intervals; for a number of years or cell layers (6) the fibers spiral in a given direction, then the direction is reversed, returning again to approximately the original slope. This condition, known as *interlocked grain* (Figure 7-4), may be considered a variation of spiral grain, and is a defect for similar reasons. However, it may be a valued decorative characteristic, imparting a pleasant ("rib-

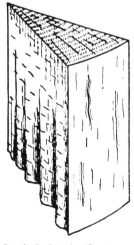

Figure 7-4. Interlocked grain (Courtesy of U.S. Forest Products Laboratory).

bon" or "striped") figure to radial surfaces; the attractive appearance of certain tropical woods (e.g., mahogany) is largely due to interlocked grain. It should be noted that interlocked grain was observed to relate to changes of anatomical structure. In an Asian species (*Anacardium officinale*), the zone where reversal of grain takes place was characterized by numerous small vessels occurring in groups and clusters, shorter and thin-walled fibers, abundant parenchyma and wider rays filled with extractives; the tendency to spirality was more pronounced in vessels (6).

Deviated grain may be produced from entirely straight-grained trees. This occurs when sawing or otherwise machining logs at an angle (rather than parallel) to growth rings when logs come from strongly tapered trees, and from trees with irregular circumference or eccentric growth. The defect is known as *diagonal grain*, and it affects adversely the properties of wood (particularly strength) in a manner analogous to spiral grain. Diagonal grain may be best observed and measured on radial surfaces.

Grain deviations in wood products are described by the cumulative term *cross-grain*; the term spiral grain is reserved for standing trees and logs. In general, slope of cross-grain is measured either by the ratio between a 1 cm (or 1 in.) deviation of the grain from the edge or axis of a piece and the distance within which this deviation occurs, or by measuring the angle in degrees (Figure 7-5).

Abnormal Arrangement of Growth Rings

Irrespective of whether a stem presents exterior deviations or looks perfectly normal, its cross-section may reveal various inner abnormalities. First to be noticed are abnormalities pertaining to the arrangement of growth rings. Deviations from the normal may include *eccentricity*, *false rings*, *discontinuous rings*, *indented rings*, and *double-* or *multi-pith* formation.

Eccentricity is the eccentric location of the pith. In extreme cases, the pith may be located near the edge of the cross-section (Figure 6-8). Usually, the presence of eccentricity may be detected from the outside because of its association with an oblong, oval, or otherwise noncircular circumference. It is possible, however, to find an eccentric arrangement of rings in an apparently circular stem.

Eccentricity may be caused by one-sided development of crown, which results in a better nutrition of one side of the stem. In many cases, however, it is a result of deviation of the stem from its vertical position. In addition to the eccentric arrangement of rings and the resulting unequal width, eccentricity is often associated with the production of reaction wood. Later, it will be explained that, as a rule, such wood is located in the side where the rings are wider.

False rings form in response to various environmental influences which cause intraseasonal disturbances of growth, with the result that more than one growth ring is laid down during a single growing season. Such rings may be formed due: to defoliation caused by late frosts, or insect or fungal attack during the growing season; to resumption of growth in late summer or early fall because of exceptionally favorable conditions; to a dry spring followed by summer rainfall; or to one or more periods of drought followed by heavy precipitation (84).

False rings may be detected in all species, but are more common in trees growing in dry regions. It has been observed in Arizona that pine trees (*Pinus radiata*), barely existing on

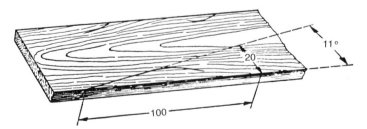

Figure 7-5. Measurement of spiral grain (20 cm in 1 m—i.e., 1:5 or 11°).

rocks, may form multiple false rings, which represent growth during several wet periods within a single year and cessation of growth during the intervening droughts (9). In the same region, when there is little winter snowfall, trees may begin to produce a layer of latewood before the end of the usually dry spring period. With the onset of summer rains, they gradually resume normal earlywood production, which is again followed by latewood. It is possible for this cycle to be repeated more than twice; therefore, more than two false rings may be produced.

False rings, sometimes several per year, have also been observed in second-growth bald-cypress. In this case, they are attributed to soil-moisture fluctuations (5). The climatic conditions of some regions (e.g., southern Greece, and Israel) contribute to the formation of false rings in local species (32).

False rings do not necessarily extend throughout the entire stem (84). When present, they may be recognized by careful inspection with a hand lens or microscope. Their outside boundary is characteristically hazy (diffuse), whereas in normal rings latewood terminates abruptly (Figure 7-6A and B). There exists, however, a wide variation of false rings. Some are very pronounced and are easily recognized, while others are difficult to differentiate, even with the microscope. It should be visualized that two or more false rings may lie within the boundary of one normal ring.

The structure of false rings presents cellular deviations. In softwoods, such deviations relate mainly to changing wall thickness, but in hardwoods they may appear in distribution, number, and diameter of vessels, and other cell elements (103).

Discontinuous rings are, by definition, rings that do not form a complete circle around the pith (Figure 7-6A and C). Aside from local cambium injury, such rings may be due to cambial inactivity resulting from lack of nutrition. The latter seems to be the case in strongly leaning trees, which tend to develop one-sided crowns, but the occurrence of discontinuous rings in vertical and apparently normal trees makes the explanation of this phenomenon difficult. Local cessation of ring formation may

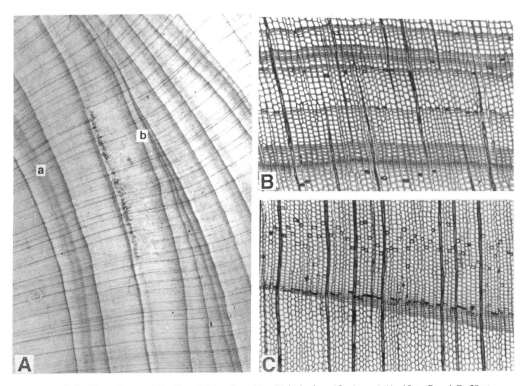

Figure 7-6. Discontinuous (Ab, C) and false rings (Aa, B) in juniper (*Juniperus*) (A, 10×; B and C, 50×).

last for 1, 2, or more years. The cambium seems to possess the ability to suspend its activity for many years and then to resume again. In a spruce it has been observed that no daughter cells had been produced for 10 years; in roots such interruptions of cambial activity extended in some parts up to 50 years (97). In leaning redwood trees, the cambium was locally dormant for most of the period of lean with occasional production of growth rings; in one case, 52 rings failed to be formed completely around the stem (31). There is no break of continuity of wood tissues during formation of discontinuous rings, since the cambium initials do not die—they merely remain inactive (73).

While discontinuous rings and false rings are structural abnormalities, they cannot be classified as defects because they have no known adverse effect on the service value of wood. Their practical importance is rather related to age and growth; they may cause foresters to underestimate or overestimate the growth potential of trees by falsely estimating their age.

Indented rings (Figure 7-7A) are occasionally found in some trees of softwood species, such as spruce, Douglas-fir, and pine. The cause of this phenomenon is not known. Such rings were reported to occur in trees growing in high elevations (116), but this is not a general condition; they may also be present in warm regions (e.g., *Pinus taeda*, an American southern pine). Anatomically, wood with indented rings differs from normal wood. In Sitka spruce, tracheids and rays differed from those of normal wood in both morphology and arrangement, and trabeculae (see Chapter 3) were commonly found to occur in tracheids (70). In Norway spruce, it has been observed that at the points of disturbed growth the number of rays showed an increase of about 40–50% and their volume was also 15–20% higher in comparison to rays of normal regions; the height of rays was somewhat diminished (116). In Europe, spruce wood containing indented rings is sought for making musical instruments, because of its reputedly desirable acoustical properties. There are no outward signs of its presence, however; therefore, its recognition in standing trees is not possible. The grain deviations associated with indented rings are not sufficient to affect the strength properties of wood (105).

Finally, *double-* or *multi-pith* formation is often observed in cross-sections of a stem at the base of a forking (Figure 7-7B); it may result from inclusion of a branch within the main stem, or common growth of two or more very closely spaced seedlings or sprouts.

Reaction Wood: Compression Wood and Tension Wood

The existence of reaction wood has been repeatedly mentioned in this text. Such wood may be found in both softwoods and hardwoods, and is known as *compression wood* and *tension wood*, respectively. As a rule, compression wood is formed on the lower (leeward) side of leaning softwood stems, whereas tension wood is found on the upper (windward) side of lean-

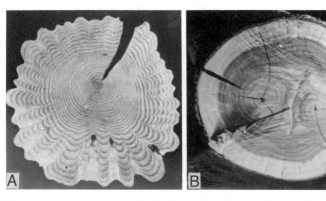

Figure 7-7. (A) Indented rings in spruce, and (B) double-pith in mulberry. (Photograph A from Ref. 28, reproduced by permission from VEB Fachbuchverlag.)

ing stems of hardwoods. The terms compression wood and tension wood imply, therefore, a direct relationship between these abnormal tissues and respective mechanical stresses developed on each side, as a result of leaning (such a relation cannot be always proved). Both compression wood and tension wood are cumulatively referred to as *reaction wood*, to denote that they are formed to counteract stimuli that have caused a stem to be displaced.

Compression wood and tension wood possesses certain structural similarities. As a rule, they are both associated with eccentric growth; they are contained in the side of the stem where the growth rings are wider; and they appear, on a cross-section, in the form of crescents. At the same time, however, they differ in many respects.

Compression wood is darker in tone than the surrounding normal wood. It has a reddish-brown color that makes it conspicuous, especially in species not possessing colored heartwood or too contrasting earlywood and latewood (Figure 7-8). There may be various degrees of intensity in which such abnormal wood is present, but when pronounced its growth rings seem to be composed entirely of latewood. In the case of mild occurrence, the recognition of compression wood becomes macroscopically difficult; observation of trans-

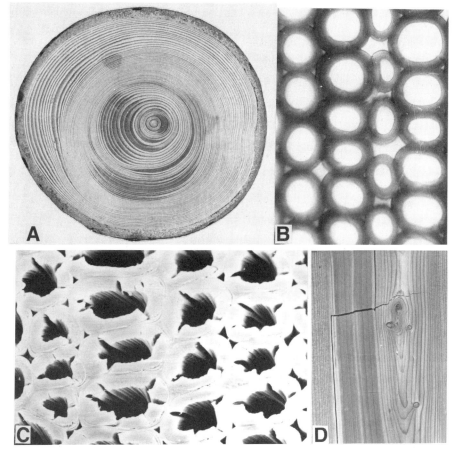

Figure 7-8. Compression wood. (A) Macroscopic appearance in fir (*Abies*) showing changing directions of formation during the lifetime of the particular tree (darker areas). (B) Compression wood in a light microscope with characteristic rounded tracheids and intercellular spaces (spruce, 350×). (C) A SEM photograph showing, in addition, spiral checks in cell walls (spruce, 580×); see also Figure 4-10A. (D) Splitting of spruce lumber due to differential (higher) longitudinal shrinkage of compression wood. (Photograph C by A. Hirzel, ETH Zurich, courtesy of L. J. Kucera; D courtesy of Building Research Establishment, U.K.)

versely sawed, thin sections, illuminated underneath, can reveal the presence of compression wood (101).

Microscopically, compression wood differs from normal in the following respects: On a cross-section, the shape of compression wood tracheids is circular; these cells round off at an early stage of cell development (at the beginning of formation of the primary wall) (15). Due to their circular shape, the tracheids are not tightly connected and they leave intercellular spaces among them (Figure 7-8B). The cell walls of compression wood are thick, even in earlywood when compression wood occupies that part of a growth ring. Wall checks are present on cross-sections. On longitudinal sections (radial or tangential), these checks appear to be arranged in spirals that usually form an angle of about 40–60° with the cell axis (Figures 4.10A and 7-8C). Checks occur even in mild compression wood, but they are absent from normal wood. Compression wood tracheids are significantly shorter than normal (10–40%)[2] (25), and their tips are abnormal (flattened, L-shaped) (114).

With regard to ultrastructure, compression wood tracheids lack the inner (S_3) layer of the secondary wall (Figure 7-8C). The outer (S_1) layer is usually thicker than normal. The microfibrils of the S_2 layer are arranged in flat spirals (angles are usually 40–60%, but smaller angles, 20–25° have also been measured); evidence of orientation is provided by inclined checks which are confined to the S_2 layer (99). These checks are not a result of drying; they exist in the living tree (14). They break up the S_2 in the form of thicker or thinner ribs that branch off toward the S_1 layer. Spiral thickenings, when present in normal wood tracheids, may be present in compression wood (yew), replaced by spiral checks (Douglas-fir, spruce, larch), or thickenings and checks may coexist; the microfibrils of thickenings gradually change direction in the course of transition from normal to compression wood (114).

Compression wood contains more lignin and less cellulose than normal wood. Cellulose content of normal wood, both in softwoods and hardwoods, is 40–45% (see Chapter 4). Lower values indicate the presence of compression wood (and inversely, higher values indicate the presence of tension wood). There are differences of opinion, however, regarding the manner of lignin distribution. According to some authors, lignin is placed in the walls of tracheids in a characteristic radial pattern, forming heavy bands next to the lumen and radially arranged plates (34); others report that the "extra" lignin is deposited between the S_1 and S_2 layers (19).

Compression wood differs also from normal wood with respect to physical and mechanical properties. It has a higher density (specific gravity) and longitudinal shrinkage, and erratic strength (105). The density of compression wood of several softwoods investigated at the U.S. Forest Products Laboratory was found to be up to about 40% higher than normal (75); such an increase may be explained on the basis of the thicker walls of its tracheids. The higher longitudinal shrinkage of compression wood is attributed to the larger angle of microfibrils in the microfibrils in the secondary wall; such shrinkage may be as high as 6–10% (20), whereas the longitudinal shrinkage of normal wood is only a few tenths of 1%. Shrinkage of compression wood in the radial and tangential directions is small, up to about half of normal (72).

With regard to strength, comparisons of normal and compression wood on unit weight basis indicate that the latter is weaker than normal, but differences exist according to type of loading. Compression wood has relatively low stiffness, bending strength, and toughness for its weight, and breaks with characteristic brash failures (105). In normal wood most strength properties increase rapidly when the moisture held in cell walls decreases. Strength also increases with increasing density (Chapter 11). These normal relationships do not apply to compression wood (76), the effects of which depend also on the degree of its intensity.

Therefore, compression wood is a serious defect. Its abnormal shrinkage may cause checking, warping, and other deformations. Failures of loaded wooden members may often be traced to the presence of compression wood. In addition, compression wood may present problems in chemical wood utilization; it produces less cellulose and makes pulp of inferior strength, especially when cooked with the sulfite process.

Tension wood is more difficult to identify macroscopically. Like compression wood, it usually appears in the form of crescents,[3] but may also be present in irregular patches; it is denser than the surrounding normal wood and shiny (silky); however, in contrast to compression wood, it is as a rule lighter in color.[4] These characteristics may be better observed in darker-colored (than in lighter-colored) woods. They are also more pronounced on the ends of freshly cut logs (or stumps), after they are exposed and somewhat dry (Figure 7-9A), but they are usually obscured by subsequent cross-cutting (67). In some cases of doubt, the application of suitable "cellulosic" stains—such as chloroiodide of zinc or phloroglucinol—may be helpful. When brushed on the surface, these stains produce a differential staining of tension and normal wood. Examination with the microscope is the most accurate method of identifying tension wood, however.

Microscopically, tension wood differs from normal wood mainly with respect to the structure of its fibers. In typically developed tension wood the walls of the fibers are abnormally thick, sometimes filling the entire lumen or reducing it to a mere slit. In unstained material,

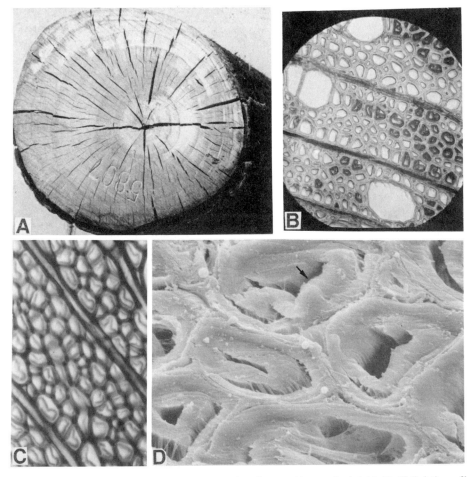

Figure 7-9. Tension wood. (A) Macroscopic appearance of tension wood in a poplar hybrid. (B, C) Gelatinous fibers—the gelatinous layer stained with (B) chloroiodide of zinc, and (C) safranin and fast green (poplar hybrid; B, 220× and C, 300×). (D) Tension wood in the electron microscope (tree-of-heaven, *Ailanthus*, 2000×). (Photograph A reproduced by permission from Pergamon Press; D courtesy of Y. G. Eom.)

an inner layer of these walls appears glassy or gelatinous; hence, such fibers are called *gelatinous fibers*. Application of chloroiodide of zinc, phloroglucinol, or safranin and fast green, on transverse microtome sections of tension wood produces a differential staining of the gelatinous layer (Figure 7-9B, C). The best contrast is achieved by chloroiodide of zinc, which stains this layer violet and the rest of the wall yellowish-brown. Staining with safranin and fast green has shown tension wood fibers to be limited in lignification, although a gelatinous layer may be missing (1, 21). Some species, such as ash and basswood, consistently show a reduction in degree of lignification, but no great anatomical modification (107). Under polarized light, tension wood fibers show characteristic spiral markings (slip planes), when viewed in longitudinal sections or in macerated material (16, 19). Tension wood appears to have a greater-than-normal proportion of fibers; vessels are smaller than normal and reduced in number, and rays appear somewhat increased in number (45, 98). With respect to size, tension wood fibers have been reported to be longer (71), shorter (25), or without consistent difference (79) from normal wood fibers; these discrepancies may be due to the source of material compared, or to differences among species.

The ultrastructure of tension wood fibers varies. The gelatinous layer may be present on top of the three layers (S_1, S_2, and S_3); it may replace the S_3 layer, or it may replace both the S_2 and S_3 layers (106). Microfibrils in the gelatinous layer tend to be oriented parallel to cell axis (77). Electron photomicrographs taken from transverse ultrathin sections of tension wood, after embedding in methacrylate, show that the gelatinous layer is porous; it exhibits a honeycombed texture in which cellulose lamellae appear to be firmly connected at irregularly spaced intervals. In the scanning electron microscope (SEM), the gelatinous layer (of dry wood) appears detached and cell cavities may be reduced to a slit (Figure 7-9D).

Tension wood contains more cellulose and less lignin than normal wood. The higher cellulose content is due to the presence of the gelatinous layer, which is highly cellulosic, or to limited lignification of tension wood fibers. Tension wood has a higher degree of crystallinity in comparison to normal wood, whereas compression wood was found to be less crystalline (108).

Depending on its intensity, tension wood has a higher density and longitudinal shrinkage than normal wood (75). The 5–10% (2–20%) higher density (73, 98) may be explained on the basis of the thicker walls of gelatinous fibers. The abnormal shrinkage, which may be up to 1.5% or higher (20, 98) (normal is a few tenths of 1%), is difficult to explain. On the basis of the parallel orientation of the microfibrils in the gelatinous layer, longitudinal shrinkage should be less than that of normal wood. It is believed that this abnormality is related to the chemistry of the cell wall rather than to its physical characteristics, although closing of the spiral checks previously mentioned has been proposed as a possible cause (16). On the basis of the honeycombed texture of the gelatinous layer, as revealed with the electron microscope, it has been suggested that high longitudinal shrinkage or swelling may result from shrinkage or swelling of isotropic interlamellar substances. Changes of the volume of these substances in a transverse direction would result in longitudinal contraction or expansion of tension wood fibers (80).

The strength of tension wood was found to be higher, comparable, or lower than normal—depending on the type of loading. In beech (*Fagus sylvatica*), specimens containing tension wood were found to be exceptionally weak in compression parallel to the grain and in toughness (17). Work pertaining to other species is not in agreement with these findings (73), and reports are often conflicting (115). As in the case of compression wood, normal relationships of strength to moisture content and density do not apply when tension wood is present (115). The proportion of gelatinous fibers is an important factor, but it is probable that other cellular changes associated with tension wood may be involved (45). Because of such erratic strength, tension wood in wooden structures should be viewed with concern analogous to that of compression wood, especially if strength of the structure is of primary importance (see also Chapter 11).

Other properties are influenced by the presence of tension wood. Nailing is difficult. In machining, saws become pinched and overheated; surfaces sawn longitudinally develop a "woolliness" or "fuzziness" (this may also be due to interlocked grain), so that proper finishing of wood products becomes impossible. Tension wood causes warping, corrugations, and checks in veneer. Matches bow and may break easily upon striking. In drying, the presence of tension wood may cause irreversible collapse (20). In chemical pulping, tension wood causes difficulties in cooking and beating. Pulping conditions for normal wood are too severe for tension wood. Pulp prepared from wood with a high proportion of tension wood has lower strength properties. However, more pulp is produced because of the higher cellulose content of tension wood. Tension wood is less objectionable in mechanical pulping because its lesser lignification makes grinding easier (21).[5]

In the extensive literature devoted to compression wood and tension wood, there is considerable discussion regarding the causes and function of these abnormal tissues in living trees (92, 108). It is generally accepted that displacement of stems from the typical, vertical position is the main cause, and that reaction wood is responsible for the tendency of displaced stems to recover. Increased displacement was observed to contribute to an increase of the amount of reaction wood both in softwoods and in hardwoods (76, 108)—but there is no general agreement on such a quantitative relation (2, 13, 92). Production of reaction wood ceases when a formerly displaced stem becomes again vertically oriented (108).

From the observed distribution of reaction wood in leaning stems—namely, that compression wood occurs on the lower side (which is under compression) and tension wood on the upper side (which is under tension)—it was originally proposed that these result from mechanical stresses. However, experimental work has proved that stresses of compression and tension do not always cause compression wood and tension wood, respectively. For example, by forming "loops" with young stems of softwoods and hardwoods, it was observed that compression wood and tension wood are not confined to parts subject to compression and tension. Compression wood was formed on the lower sides of such loops and tension wood on the upper sides (38, 107) (Figure 7-10). These and other related investigations demonstrate an effect of gravitational stimuli, and prove that the terms *compression wood* and *tension wood* are not literally correct, if they are taken to imply the cause of the respective tissues.

Formation of reaction wood has also been attributed to fast growth. Both the amount of such wood and the speed of movement of leaning trees toward their vertical position were reported to depend on vigor of growth (98, 108).

It was further observed that formation of reaction wood is associated with changes in the chemistry of the sap. By application of growth hormones, compression wood was produced artificially (89, 107, 108, 112).

The mechanism by which reaction wood assists displaced stems to recover is not well understood, however, and the different reaction

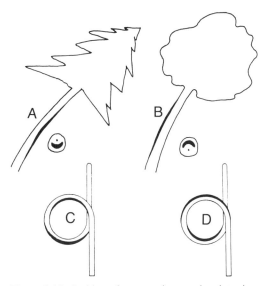

Figure 7-10. Position of compression wood and tension wood in trees displaced from the vertical (A, C: softwood; B, D: hardwood); the position of reaction wood is shown with bands of black. (C and D after Ref. 38; see text).

of softwoods and hardwoods to the same stimulus is still a largely puzzling biological phenomenon (107, 108).

From the forestry point of view, it should be noted that the formation of reaction wood may be controlled, within limits, in the forest through application of silvicultural measures that protect trees from becoming displaced from the vertical. Such measures include proper spacing of plantations, establishment of wind barriers, and timely thinnings of proper intensity in order to avoid abrupt exposure of tall and slender trees. Such trees may lean under their own weight, or due to the action of wind, snow, or sleet loads. Experiments have shown that neither strong nor prolonged displacement is required for a tree to form reaction wood; displacements as small as 2°, and for as short a period as 24 h or less (14, 48, 108), have produced these abnormal tissues. Reaction wood is not confined to learning trees, however. It may occur in trees subject to swaying due to wind—without being permanently displaced. On the other hand, leaning trees, especially when young and vigorous, may again become vertical; and leaning leaders of stems are gradually erected (98).[6] For all these reasons, reaction wood may be found in perfectly straight trees.[7] Utilization of such trees may create problems associated with compression wood and tension wood, although the presence of these tissues is not suspected.

Formation of reaction wood is largely unavoidable on steep slopes, where many trees attain a pistol-butted form. In pistol-butted hemlock, the interesting observation was made that compression wood occurred along the downhill radius at the 50 cm (1.5 ft) level, but this was reversed at the 2 m (6 ft) level. The stem appeared to have grown past the vertical, while tending to become erect (49). In this process, some leaning trees attain an inverted S-like form (12).

It is important to note that there is evidence for the inheritance of reaction wood, and that the evidence for the inheritance of stem form and straightness, which contribute to the incidence of such wood, is very strong (92).

Reaction wood occurs also in branches, where it is of practical importance when branches are looked upon as potential knots. Compression wood is formed on the lower side of branches of softwoods, and tension wood on the upper side of the branches of hardwoods. Reaction wood may be present in roots, but this has been little investigated (see also Chapter 6).

An extensive survey of reaction tissue (tension wood) in branches and roots of a large number of temperate and tropical hardwoods (including shrubs and herbs) (37) has shown the presence of such tissue in many but not all of the species investigated. On the basis of the material examined, about 50% of the species were found to possess tension wood in branches, but only 25% had it in root wood. Gelatinous fibers were observed both in wood and in bark (phloem). In wood, a positive relationship was found between the proportion of libriform fibers and the frequency of occurrence of tension wood. In a European hardwood (boxwood, *Buxus sempervirens*), tension wood fibers were observed to be rounded like compression wood tracheids of softwoods, whereas some softwoods possessed gelatinous fibers in wood (fir, spruce, Scots pine) or phloem (juniper) (37, 41). These observation show that the generally accepted concept, that compression wood is limited to softwoods and tension wood to hardwoods, is not without exceptions.

A note may be added here about the so-called *opposite wood* (i.e., wood formed in the opposite side of compression or tension wood). In softwoods (fir, larch, pine, hemlock), opposite wood differed from both normal and compression wood in a way that normal wood had intermediate characteristics; for example, the tracheids of opposite wood were longer (shorter in compression wood and intermediate in normal wood). They had square outlines in earlywood (angular in normal wood and round in compression wood), and a thick S_3 layer (thin in normal wood, absent in compression wood) (92). Opposite wood did not differ in chemical composition (cellulose, hemicelluloses, lignin) from normal wood, but had longer crystallites, a higher degree or crystallinity, and a better distribution of the orientation of microfibrils in the longitudinal (cell) direction in comparison to compression wood (shorter crystallites, lower degree of crystallinity, larger angle of microfibrils) (90). As a result, opposite wood is considered to have higher axial tensile strength and lower compressive strength, while the opposite occurs in compression wood—and normal wood is intermediate (90, 92). Like compression wood, opposite wood is variable depending on the extent of tree lean (87). There is little information about opposite wood in hardwoods; in beech, the fibers were shorter than the fibers of normal wood in contrast to the longer than normal fibers in tension wood (16, 73).

Disruption of the Continuity of Inner Wood Tissues

This category includes *compression failures*, *shakes*, and *resin* or *pitch pockets*.

Compression failures [also called *slip planes* or *compression creases* (110)] are ruptures of wood stressed in axial compression (compression parallel to the grain). Such ruptures may occur on the compression (inner) side of trees that are leaning due to wind storms or the accumulation of snow on their crown; they may extend to the center of the stem and beyond (66). The injured segment of the stems develops swellings or ridges (due to local faster growth), which may serve for recognition of this defect (Figure 7-11A). In the parts where swellings are formed, compression wood may be associated with compression failures (66). These failures are more common in young and slender trees exposed after intensive thinnings; such trees can be easily affected by wind or other loads.

Compression failures may also develop from careless felling or rough handling in logging operations, or in members of wooden structures as a result of application of heavy loads. On wooden surfaces, compression failures appear as localized transverse wrinkles, but often they cannot be easily distinguished with the naked eye. When suspected, but not clearly visible, it has been suggested that they can be made to stand out sharply by application of carbon tetrachloride to the surface of the wood (58). Under the microscope, compression failures

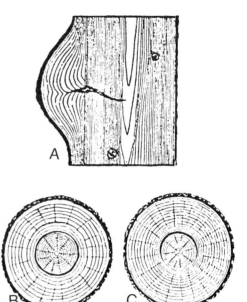

Figure 7-11. (A) Compression failure (with local swelling); (B) ring shake; and (C) cup shake. (After Ref. 50)

appear as zones or slip planes across the cell walls.

Shakes are also ruptures of the wood of living trees (Figures 7-11B and C, and 7-12); they may widen after the tree is felled due to shrinkage of the wood, but they originate when the tree is alive. Shakes are of two types—*ring* or *cup shakes* and *heart shakes* or *heart checks*. Ring or cup shakes are ruptures occurring parallel to growth rings, either on the boundary of adjacent rings or within a growth ring. Shakes extending all around a periphery are called *ring shakes*, whereas local separations are called *cup shakes*. The cause of shakes is uncertain. They are often found within wide rings or at the boundaries of a sudden change of ring width. This leads to the hypothesis that abrupt change of wood structure in trees subject to the action of bending or twisting stresses may be the reason for their formation (97). Internal growth stresses are also considered as possible causes of this defect (111). Furthermore, shakes have been related to compression wood (66), and to injury by woodpeckers (43) or other agents (86). Anatomically, shakes were observed (in eucalypts) to occur within bands containing abnormally large amounts of axial parenchyma

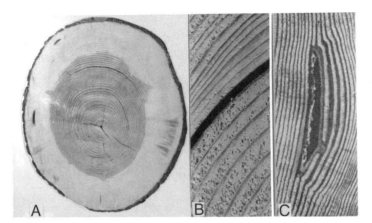

Figure 7-12. (A) Checks, shakes, and resin pockets in Balkan whitebark pine (*P. leucodermis*); (B) shake in fir (*Abies*); and (C) resin pocket in pine.

cells (normal or traumatic) and/or vessels (109). Observations with an electron microscope have shown that separation occurs usually in the middle lamella (less in cell walls), and microorganisms (bacteria, fungi) may be present in the shakes (62).

Heart shakes or *heart checks* are ruptures starting from the pith. They resemble checks, and they run radially from the pith outward mostly along rays. There may be one such check or crack on a diameter of the log, or many radiating from the pith in a star–or spider formation. This defect usually occurs in the lower part of the stem of older trees, and is attributed to growth stresses (40).

Resin pockets occur in softwoods possessing resin canals. They are oblong, lens-shaped ruptures of varying size, usually lying within the boundary of a growth ring and filled with resin (Figure 7-12C). Microscopic examination shows these cavities to be lined with parenchymatous, epithelial-like cells. The cause of resin pockets is uncertain; they are attributed to swaying or excessive bending of the trees concerned (30).

Abnormal Color

Deviations from normal color are not infrequent in the wood of living trees. They may be due to parasitic agents, or wounding, although for some the cause is uncertain. Parasitic agents (bacteria, fungi) are dealt with elsewhere (Chapter 15). Nonparasitic colorations may appear in the form of spots, or may occupy the central portion of the stem in the form of heartwood, or presented as peripheral bands inside heartwood (27).

Mineral streaks, floccosoids, and *included sapwood* appear in the form of spots. *Mineral streaks* are dark-colored streaks or patches observed in the wood of maples, sycamore, the hickories, and other hardwoods. Chemical analyses show an abnormal concentration of mineral matter (82)—hence, the name. The cause of mineral streaks is uncertain, but they have been attributed to bird (sapsucker) attack (61). The affected wood is harder than normal, is said to have a dulling effect on wood-working tools, resists impregnation with preservatives, and may cause checks in drying. However, the strength of wood is apparently not impaired (23).

Floccosoids are local discolorations in the form of whitish spots that frequently occur in the wood of western hemlock. They are not readily visible until the wood is dried or planed. Microscopic examination shows floccosoids to be associated with deposits included in tracheids; these deposits consist mainly of conidendrin—a normal extractive of hemlock wood (3).

Included sapwood appears as light-colored wood inside heartwood in the form of irregular streaks or in a peripheral ("moon-ring" or "target-ring") pattern (Figure 7-13). It is at-

Figure 7-13. Moon-ring (included sapwood) in oak. (After Ref. 27; reproduced by permission from Paul Parey.)

tributed to wounding (when streaks are formed) or to frost, which alter the behavior of the parenchyma cells of sapwood; they lose their normal physiological potential for transformation to heartwood. Included sapwood has been observed in association with resin pockets in larch, spruce, and eastern red cedar (*Juniperus virginiana*), and in peripheral form it was reported to occur in softwoods and hardwoods, such as western red cedar (*Thuja plicata*), Douglas-fir, European oaks (especially *Quercus petraea*), and eucalypts (27). Included sapwood is a defect due to color differentiation, and has a lower durability in comparison to heartwood (36).

Local change of color may also result from tapping of trees (pine, sugar maple), or from extraction of increment cores (especially in hardwoods) (8, 60). Both tapping and extraction of cores may provide entrance to parasitic agents, but such wounding may simply cause change of color near the wound, as a result of entrance of air and oxidation of organic compounds contained in wood.

Among colorations that occupy the central portion of stems is the *red-heartwood* (*Rotkern*) of European beech (*Fagus sylvatica*); this is a subject of considerable discussion in European literature. The macroscopic appearance of such wood may suggest fungal attack; it is often irregular in circumference and not uniform in color, and may be composed of overlapping zones which are darker in their boundaries (Figure 7-14). However, the inability to detect fungi, and the fact that it is present in almost all trees after a certain age, support the opinion that this is not a pathological form of heartwood. There is no clear explanation of this occurrence, although it has been attributed to injury (breakage of branches), and to moisture content of the wood falling below a certain level (117). The relation between dehydration and discoloration has been demonstrated by artificial induction of discoloration. Dehydration was found to be related to fast growth and simultaneous reduction of tree crown, and dis-

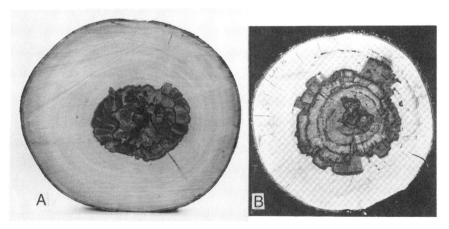

Figure 7-14. Discolorations. (A) Red-heartwood (Rotkern) in European beech; (B) sweetgum (see text). (Photograph A from Ref. 65; reproduced by permission from Springer Verlag; B from Ref. 22)

coloration was attributed to potential entry of oxygen into the dehydrated core through broken branches (96). At any rate, the physiological processes involved in red-heartwood formation are essentially the same as in formation of normal heartwood (i.e., death of parenchyma cells, formation of tyloses, and deposition of extractives) (24). Red-heartwood has a somewhat greater resistance to fungal attack when compared to the surrounding sapwood (117); it is, however, difficult to impregnate with preservatives, probably due to plugging of its vessels by tyloses.

Discolorations that in some cases are variegated, resembling the red-heartwood of beech, occur in sweetgum (*Liquidambar styraciflua*), and are attributed to branch stubs, and wounding by birds, insects, fire, and other agencies; it is strongly suggested that a considerable amount of these discolorations is a result of subsequent entrance of fungi (22). Heartwood similar in appearance to that of European beech (circular, star-shaped, or irregular) was also reported to occur in Japanese beech. This was related to damage of trees by insects, and subsequent introduction of air into the stem—causing oxidation of cell contents. Such heartwood was artificially produced by drilling holes to simulate insect attack (113). In addition, similar in appearance to red-heartwood of beech is the occurrence of *brown-heartwood* (*Braunkern*) in European ash (*Fraxinus excelsior*). Such wood varies with respect to time of appearance, size, and tone of color. Wet, especially stagnant sites, and bark lesions appear to contribute to its formation. However, the wood is sound with strength equal to that of unstained wood (7).

The heartwoods of beech and ash were found to have few extractives, the presence of which is limited to parenchyma cells (7, 24), where they are retained as a wall coating or drop-like inclusions. They do not impregnate the cell walls of all other wood tissues, as is the case with normally colored heartwood (7).

European beech and other hardwoods are also reported to form *frost-heartwood* (*Frostkern*) as a result of unusually low temperatures. In beech, frost-heartwood has a gray-reddish color, and appears suddenly in certain years of exceptionally cold winter weather affecting relatively young trees (8, 97). Such heartwood has few or no tyloses; it may be easily impregnated with preservatives, but possesses little durability when untreated.

Wood of abnormal color that is formed as a result of mechanical or physiological injury is sometimes called *protection wood* (42), in the sense that it is intended to provide protection against incoming invasion by destructive organisms (bacteria, fungi). Such protection (not always effective) is exercised by local concentration of extractives. These protective materials discolor wood in the form of heartwood, or they may be deposited in a zone surrounding a wound or an approaching attack—for example, in the boundary of a gradually expanding decay of inner heartwood.

It is important to appreciate that discoloration can exist without decay; but decay is always associated with discoloration.

Abnormalities Due to Wounding

In addition to being a factor in abnormal coloration of wood, wounding may cause cellular abnormalities. In general, trees react to wounding by forming a mass of traumatic cells, called *callus*. This is composed of thin-walled parenchyma cells, irregular in size and shape. In hardwoods, it may contain deformed fibers and vessel members, and gelatinous fibers. [The extent of such irregular tissues was observed, in walnut, to be greater when wounding occurs in the fall in comparison to spring (88).] Callus tissue originates from living parenchyma cells of the xylem and phloem, and from newly formed derivatives of the cambium (29). Gradually, depending on the extent of wounding, the injury may heal.

Wounding may cause *traumatic resin canals* to form in softwoods, and *traumatic gum canals* in hardwoods. Traumatic resin canals may form even in fir, redwood, hemlock, and other species where normal resin canals are absent. As a rule, traumatic resin canals are axial. On a transverse section, they appear to be arranged in more or less tangential rows (Figure 7-15A and B), and may be present in any part of the growth ring; whereas normal canals are rela-

ABNORMALITIES IN WOOD 101

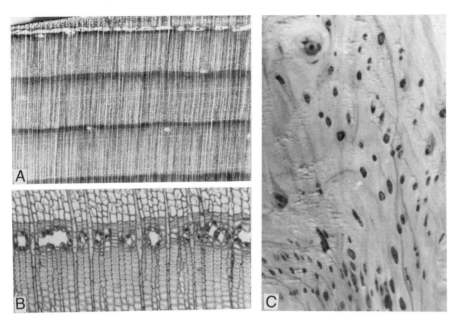

Figure 7-15. Traumatic resin canals: (A) in spruce (upper part, in a continuous row); (B) microscopic appearance in fir (40×); (C) portion of a lumber board from a fir tree attacked by mistletoe. (Photograph C reproduced by permission from Pergamon Press.)

tively scattered and are usually confined to the outer half of the growth ring. The epithelial cells of traumatic resin canals are thick-walled. As in the case of normal resin canals, traumatic ones are also *schizogenous* in origin (see Chapter 5). Traumatic gum canals form in cherry and sweetgum. These are either schizogenous (when they are formed by separation of cells, as in sweetgum) or *lysigenous* (when they develop from disintegration of cells, as in cherry); the latter process is also called *gummosis*. Schizogenous canals have epithelial cells, but lysigenous canals do not (29, 73). In eucalypts, gum canals are referred to as gum veins or *kino veins* (Figure 7-16A). Anatomically, they are associated with irregular parenchyma cells, and their cavities contain dark-colored inclusions (kino is a polyphenolic, medicinal, tanning and dyeing substance) (94).

Aside from the above effects, destruction and removal of parts of the bark, cambium, and wood results in exposure of the underlying tissues, which will shrink and check, and may be attacked by fungi and insects. In the process of healing, portions of the bark may be embedded in the wood; these are called *bark pockets* or *stone pockets*.

Both organic and inorganic agents may be involved in the wounding of trees. *Organic* agents that may cause wounding include people, animals, insects, and plants. People may cause extensive damages through careless logging, or by wounding trees in various other ways—including tapping of pines and sugar maples, and extraction of increment cores. Animals such as porcupine, bear, deer, squirrels, and rabbits may eat bark and wood. Woodpeckers may dig holes in living trees, although they seem to prefer dead trunks and utility poles. Wounding by insects is the cause of traumatic gum canals in cherry, for example, and of *pith-* or *parenchyma flecks* in this and other hardwood species (53) (Figure 7-16B). Attack by certain insects has also been associated with the formation of compression wood (92, 108). Mistletoe may damage trees to the extent that they become technically worthless. It is well known that mistletoes cause formation of tumors, in which the root system of this parasite is interwoven among the growth rings;

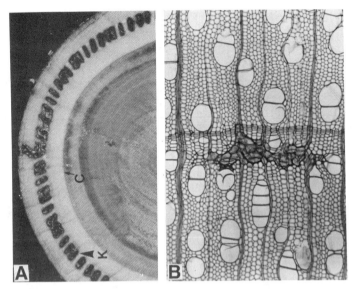

Figure 7-16. (A) Kino veins (K) in eucalypt (C, cambium); (B) pith fleck in birch (50×). (Photograph A from Ref. 94, reproduced by permission from the International Association of Wood Anatomists; B courtesy of H. Sachsse, reproduced by permission from Springer Verlag)

when the roots die, characteristic holes remain in the wood (Figure 7-15C).

Inorganic agents include sun, frost, lightning, hail, and fire. Sun may cause barkburns to relatively young trees having smooth and thin bark; such burns may be accompanied by local destruction of the cambium. Frost may cause frost-checks or frost-cracks in the lower part of stems parallel to their axis; these may be recognized macroscopically because the callus tissue is lining the boundaries of the wound in the form of lips (Figure 7-17). Frost checks are attributed to the contraction of the outer layers of sapwood due to cold, while the interior core remains unchanged (64); the initial superficial checks extend deeper by way of rays, which constitute planes of weakness in that respect. Lighting, hail (74), and forest fires may also cause wounds.

In addition to previously mentioned reactions of living trees to wounding, *barrier zones* are sometimes formed as a protective response to mechanical injury or infection. Such zones consist of bands of axial parenchyma, which

Figure 7-17. Frost cracks: (A) oak; (B) maple. (From Ref. 83; reproduced by permission from DRW-Verlag Weinbrenner.)

are considered specialized tissues and not irregular cells, as in callus tissue; they are formed by the surviving cambium as boundaries between dead, discolored, and clear, living sapwood. Barrier zones are often recognized as wood defects, because they are zones of structural weakness and may result in ring shakes; they are considered the most serious defect in the wood of some eucalypts (93, 95).[8]

Abnormalities Due to Environmental Pollution and Atomic Radiation

Environmental pollution (acid rain, etc.) is considered the cause of extensive damage of forests especially in Europe. The affected (dead or dying) trees show some differentiation in wood anatomy, which is mainly expressed in reduction of ring width. Other anatomical characteristics are not substantially altered. In softwoods (spruce, fir, pine): a change of the proportions of tracheids and ray tissue was not apparent; the number and dimensions of resin canals of spruce increased slightly; and sometimes there was an increase of the proportion of latewood (and increase of density)—associated with the reduction of ring width; cell dimensions, and cell-wall structure and percentage remained practically unchanged. Information regarding hardwoods is sparse: changes were reported in the proportion of vessels (increased) and fibers (decreased). The chemical composition of wood was also found unaltered—except in some inorganic elements (reduction of Ca and Mg, increase of metals). In general, the properties of wood (including permeability and gluability) were not affected from a practical point of view, beyond the effects of ring width (primarily to density), and there was no greater predisposition to fungal attack (4, 52).

High level ionizing radiation in the environment, even in amounts below that accepted as lethal to plants, was found to significantly affect the structure of the wood of pines (*Pinus rigida* and *P. echinata*). Tracheids were reduced in length, lumen diameter, and wall thickness; and cellular disorganization of tissues and suppression of cell multiplication was observed. The effects do not seem to be permanent, however; normal cell formation resumes following discontinuation or reduction of radiation to sublethal doses (33).[9]

NATURAL GROWTH CHARACTERISTICS

As already mentioned, when wood is looked upon as raw material, certain natural growth characteristics of all trees—namely, knots (branches) and pith—become defects.

Knots

A knot is an inclusion of the basal part of a branch within the stem of a tree. Because branches are indispensable members of a living tree, knots are largely unavoidable. Knot-free wood is produced only in the lower portion of the stem, in which the branches fall off naturally or are artificially removed by pruning.

There are two kinds of knots, depending on whether a branch was dead (dry) or alive at the time of inclusion. Dead branches are enclosed in living stems in the same manner as foreign objects (wire, nails, stones, bullets, etc.) may be incorporated. The knots resulting from such branches are called *encased* or *loose* (Figure 7-18B). When wood dries in the form of lumber or veneer, such knots may fall off, leaving knot-holes. On the other hand, branches enclosed while living produce knots, which are intimately connected with the surrounding wood through continuous growth rings; such branches give rise to *intergrown* or *tight* knots (Figure 7-18A).

Knots may differ in shape and size. Shape depends primarily on the cross-sectional shape of the enclosed branch, although it is also influenced by the direction of sectioning in relation to branch axis. There are *round*, *oval*, and *spike* knots (Figure 7-18C); the latter are produced when branches are cut along their axis. When considered by size, knots are classified as *pin*, *small*, *medium*, and *large*. Finally, on the basis of the quality of their wood, knots are distinguished as *sound*, *unsound*, and *decayed*. Knot characteristics are important criteria in formulating lumber grading rules (Chapters 11 and 17).

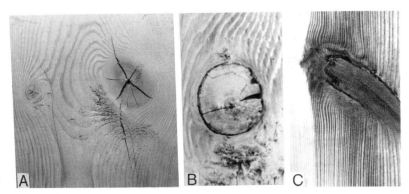

Figure 7-18. Knots: (A) intergrown (tight); (B) encased (loose); (C) spike (radial-longitudinal section). (Reproduced by permission from Pergamon Press.)

Knots affect adversely the appearance and properties of wood. Although knotty lumber may have some appeal, knots in general are aesthetically and technically undesirable.[10] Their adverse effect is due to the usually abnormal structure and higher density of their wood, and also to the association of knots with grain deviations and checking. The strength of wood may be considerably reduced by knots, depending on their kind, size, and location, and on the type of loading (Chapter 11). Knots also affect the machining, drying, and gluing properties of wood.

The degree of knotiness may be influenced by the conditions under which a tree grows; therefore, it may be controlled, up to a certain point, by silvicultural practices. Natural pruning may be induced by proper spacing—closer spacing prompts such pruning, particularly in nontolerant species in which lack of light causes the lower branches to dry and break off. When desirable results are not produced naturally, artificial pruning may be performed. It is necessary, however, to start as early as possible in order to reduce the diameter of the central knotty core.

Pith

The pith is an essential part of the living trees, but from a wood utilization point of view it is a defect. Its presence in wood products may reduce both strength and durability. This is primarily due to the different cellular (parenchymatous) structure of the pith in comparison to the surrounding wood. In addition, the wood near the pith is juvenile (see Chapter 6), often contains knots, and in older trees may develop radial checks (*heart shakes*).

REFERENCES

1. Barefoot, A. C. 1963. Abnormal wood in yellow poplar. *For. Prod. J.* 13:16–22.
2. Barger, R. L. and P. F. Folliott. 1976. Factors affecting occurrence of compression wood in individual ponderosa pine trees. *Wood Sci.* 8(3):201–208.
3. Barton, G. M. 1963. Conidendrin in floccosoids of western hemlock. *For. Prod. J.* 12:304.
4. Bauch, J. 1986. Characteristics and response of wood in declining trees of forests affected by pollution. *IAWA Bull.* n.s. 7(4):269–276 (and other papers in this bulletin).
5. Beaufait, R. W., and T. C. Nelson. 1957. Ring counts in second growth baldcypress. *J. For.* 55:588.
6. Bhat, K. V., and K. M. Bhat. 1983. Anatomical changes associated with interlocked grain in *Anacardium occidentale* L. *IAWA Bull.* n.s. 4(2/3):179–182.
7. Bosshard, H. H. 1983. *Holzkunde (1)*. Basel: Birkhaeuser Verlag.
8. Boyce, J. S. 1961. *Forest Pathology*, 3rd ed. New York: McGraw-Hill.
9. Brady, F. L. 1939. An effect of starvation of pine trees. *Tree-Ring Bull.* 5(3):20–23.
10. Brazier, J. D. 1965. An assessment of the incidence and significance of spiral grain in young conifer trees. *For. Prod. J.* 14:308–312.
11. Burger, H. 1953. Das "Arbeiten" imprägnierter Leitungsmasten. *Mitt. Schweiz. Anstalt f. forstl. Versuchswesen* 29(2):177–188.
12. Büsgen, M., and E. Münch. 1926. *Bau und Leben unserer Waldbäume*. Jena: Verlag Fischer.
13. Cano-Capri, J., and L. F. Burkhart. 1974. Distribution of gelatinous fibers as related to lean in south-

ern red oak (*Quercus falcata* Michx.). *Wood Sci.* 7(2):135–136.
14. Casperson, C. 1962. Über die Bildung der Zellwand beim Reaktionholz. Teil l: Zur Anatomie des Reaktionholzes. *Holztechnologie* 3:217–223.
15. Casperson, C. and A. Zinsser. 1965. Über die Bildung der Zellwand bei Reaktionholz-II. Zur Spaltenbildung im Druckholz von *Pinus silvestris* L. *Holz Roh-Werkstoff* 23:49–55.
16. Chow, K. Y. 1946. A comparative study of the structure and chemical composition of tension wood in beech (*Fagus sylvatica* L.). *Forestry* 20:52–77.
17. Clarke, S. H. 1937. The distribution, structure and properties of tension wood in beech (*Fagus sylvatica* L.). *Forestry* 11:85–91.
18. Cockrell, R. A. 1974. A comparison of latewood pits, fibril orientation, and shrinkage of normal and compression wood of giant sequoia. *Wood Sci. Tech.* 8(3):197–206.
19. Côté, W. A., and A. C. Day. 1965. Anatomy and ultrastructure of reaction wood. In *Cellular Ultrastructure of Woody Plants*, ed. W. A. Côté, pp. 391–418. Syracuse: Syracuse University Press.
20. Dadswell, H. E. 1958. Wood structure variations occurring during tree growth and their influence on properties. *J. Inst. Wood Sci.* 1:2–23.
21. Dadswell, H. E., and A. B. Wardrop. 1960. Recent progress in research on cell wall structure. *Proc. Fifth World For. Congr.* 2:1279–1288.
22. Davis, T. C., and H. O. Beals. 1977. Internal Discoloration of Sweetgum. Ag. Expt. Station, Circular 234. Auburn, Alabama.
23. Desch, H. E. 1981. *Timber, Its Structure, Properties and Utilization*, 6th ed. (revised by J. M. Dinwoodie). London: McMillan.
24. Dietrichs, H. H. 1964. *Chemisch-phyisiologische Untersuchungen über die Splint-Kern-Umwandlung der Rotbuche (Fagus sylvatica L.)—Ein Beitrag zur Frage der Holzverkernung*. Hamburg Kommissionsverlag.
25. Dinwoodie, J. M. 1961. Tracheid and fibre length in timber—A review of literature. *Forestry* 34:125–144.
26. Douglass, A. E. 1940. Examples of spiral compression wood. *Tree-Ring Bull.* 6(3):21–22.
27. Dujesiefken, D., and W. Liese. 1986. Vorkommen und Enstehung des Mondrings bei Eiche (*Quercus* spp.). *Forstw. Cbl.* 105:137–155.
28. Erteld, W., H.-J. Mette and W. Acterberg. 1963. *Holzfehler*. Leipzig: VEB Fachbuchverlag.
29. Esau, K. 1965. *Plant Anatomy*, 2nd ed. New York: J. Wiley & Sons.
30. Frey-Wyssling, A. 1938. Über die Entstehung von Harztaschen. *Holz Roh- Werkstoff*: 329–332.
31. Fritz, E. 1940. Problems in dating rings of California coast redwood. *Tree-Ring Bull.* 6(3):19–21.
32. Gindel, J. 1944. Aleppo pine as a medium for tree-ring analysis. *Tree-Ring Bull.* 10(1):6–8.
33. Hamilton, J. R. 1963. Characteristics of tracheids produced in a gamma and gamma-neutron environment. *For Prod. J.* 13:62–67.
34. Harlow, W. M. 1952. The chemistry of the cell walls of wood. In *Wood Chemistry*, Vol. 1, ed. L. E. Wise and E. C. Jahn, pp. 99–131. New York: Reinhold.
35. Harris, J. M. 1989. *Spiral Grain and Wave Phenomena in Wood Formation*. Berlin, New York: Springer Verlag.
36. Hillis, W. E. 1987. *Heartwood and Tree Exudates*. Berlin, New York: Springer Verlag.
37. Hoester, H. R. and W. Liese. 1966. Über das Vorkommen von Reaktionsgewebe in Wurzeln und Äste der Dikotyledonen. *Holzforschung* 20(3):80–103.
38. Jaccard, P. 1938. Exzentrisches Dickenwachstum und anatomisch-histologische Differenzierung des Holzes. *Ber. Schweiz. Bot. Ges.* 48:491–537.
39. Jacobs, M. R. 1935. The occurrence and importance of spiral grain in *Pinus radiata* in the federal capital territory. *Commonw. For. Bur. Australia Bull. No.* 50.
40. Jacobs, M. R. 1965. Stresses and Strains in Tree Trunks as They Grow in Length and Width. Australian Forest and Timber Bureau Leaflet No. 96, Canberra.
41. Jacqiot, C., and Y. Trenanrd. 1974. Note sur la présence de tracheides à parois gélatineuses dans des bois résineux. *Holzforschung* 28(2):73–76.
42. Jorgensen, E. 1962. Observations on the formation of protection wood. *For. Chron.* 38:292–294.
43. Jorgensen, R. N., and S. L. Lecznar. 1964. Anatomy of Hemlock Ring Shake Associated with Sapsucker Injury. U. S. For. Service, N. E. For. Expt. Station, Res. Paper NE-21.
44. Isebrands, J. G., and R. A. Parham. 1981. Reaction wood anatomy and its effect on Kraft paper from short rotation intensively cultured trees. *IAWA Bull.* n.s. 2(2/3):59.
45. Kaeser, M., and S. G. Boyce. 1965. The relationship of gelatinous fibers to wood structure in eastern cottonwood (*Populus deltoides*). *Am. J. Bot.* 2:711–715.
46. Kellog, R. M., and S. R. Warren. 1979. The occurrence of compression wood streaks in western hemlock. *For. Sci.* 25(1):129–131.
47. Kennedy, R. W., and G. K. Elliot. 1957. Spiral grain in red alder. *For. Chron.* 33:238–251.
48. Kennedy, R. W., and J. L. Farrar. 1965. Tracheid development in tilted seedlings. In *Cellular Ultrastructure of Woody Plants*, ed. W. A. Côté, pp. 419–453. Syracuse: Syracuse University Press.
49. Kienholz, R. 1930. The wood structure of "pistol-butted" mountain hemlock. *Am. J. Bot.* 17:739–764.
50. Koenig, E. 1958. *Fehler des Holzes*. Stuttgart: Holz-Zenralblatt Verlags.
51. Krempl, H. 1965. Drehwuchs in Stamm- und Astholz bei der Fichte. *Holzforsch. Holzverwert.* 17:21–26.
52. Kucera, L. J. and H. H. Bosshard. 1989. *Holzeigenschaften geschädigter Fichten*. Basel: Birkhaeuser.
53. Kulman, H. M. 1964. Defects in black cherry caused by barkbeetles and agromizid cambium miners. *For. Sci.* 10:259–266.

54. Larson, P. R. 1963. Stem Form and Development of Forest Trees. Forest Science Monograph No. 5.
55. Larson, P. R. 1965. Stem form of young *Larix* as influenced by wind and pruning. *For. Sci.* 11:412–424.
56. Lhoneux de, B., R. Antoine, and W. A. Côté. 1984. Ultrastructural implications of gamma-irradiation of wood. *Wood Sci. Tech.* 18:161–176.
57. Liese, W., and U. Ammer. 1962. Anatomische Untersuchungen an extrem drehwüchsigem Kiefernholz. *Holz Roh- Werkst.* 20:339–346.
58. Limbach, J. P. 1951. Compression Failures in Wood Detected by the Application of Carbon Tetrachloride to the Surface. U.S. For. Prod. Lab. Report No. 1591.
59. Locard, C. R., J. A. Putnam, and R. D. Carpenter. 1963. Grade Defects in Hardwood Timber and Logs. U.S.D.A. Agr. Handbook No. 244.
60. Lorenz, R. C. 1944. Discolorations and decay resulting from increment boring in hardwoods. *J. For.* 42:37–43.
61. Lund, A. E. 1966. Preservative penetration variations in hickory. *For. Prod. J.* 16(2):29–32.
62. McGinnes, E. A., J. E. Phelps, and J. W. Ward. 1974. Ultrastructure observations of tangential shake formation in hardwoods. *Wood Sci.* 6(3):206–211.
63. McLauchlan, T. A., and D. J. Kusec. 1978. Continuous non-contact slope-of-grain detection. Proc. 4th Nondestructive Testing of Wood Symposium, pp. 67–76.
64. Mayer-Wegelin, H., H. Kubler, and H. Traber. 1962. Über die Ursache der Frostrisse. *Forstw. Cbl.* 81:129–137.
65. Mehringer, H., Bauch, J. and A. Früwald. 1988. Holz-biologische Untersuchungen an Buchen aus Waldschadengebieten. *Holz Roh- Werkstoff* 46:447–455.
66. Mergen F. and H. I. Winer. 1952. Compression failures in the boles of living conifers. *J. For.* 50:677–679.
67. Moulopoulos, C., and G. Tsoumis. 1960. Growth abnormalities in Euramerican hybrid poplars cultivated in Macedonia-Greece. Annals, School Agr. For., pp. 193-237. Aristotelian University, Thessaloniki, Greece (Greek, English summary).
68. Northcott, P. L. 1957. Is spiral grain the normal growth pattern? *For. Chron.* 33:335–352.
69. Noskowiak, A. F. 1963. Spiral grain in trees—A review. *For. Prod. J.* 13:266–275.
70. Ohtani, J., K. Fukazawa, and T. Fukumorita. 1987. SEM observations on indented rings. *IAWA Bull.* n.s. 8(2):113–124.
71. Ollinmaa, P. J. 1956. On the anatomic structure and properties of tension wood in *Betula*. *Acta For. Fenn.* 64(3):1–263.
72. Onaka, F. 1949. Studies on Compression Wood and Tension Wood. Wood Res. Bull. No. 1, Kyoto, University, Japan.
73. Panshin, A., and C. De Zeew. 1980. *Textbook of Wood Technology*, 4th ed. New York: McGraw-Hill.
74. Pechmann v., H. 1949. Die Auswirkung eines Hagelschlages auf Zuwachsentwicklung und Holzwert. *Forstw. Cbl.* 68:445–456.
75. Pillow, M. Y. 1950. Presence of Tension Wood in Mahogany in Relation to Longitudinal Shrinkage. U.S. For. Prod. Lab. Report No. D 1763.
76. Pillow, M. Y., and R. F. Luxford. 1937. Structure, Occurrence and Properties of Compression Wood. U.S.D.A. Tech. Bull. No. 546.
77. Preston, R. D. 1974. *The Physical Biology of Plant Cell Walls*. London: Chapman and Hall.
78. Rendle, B. J. 1955. Tension wood, a natural defect of hardwoods. *Wood* 20:348–351.
79. Sachsse, H. 1961. Anteil und Verteilungsart von Richtgewebe im Holz der Rotbuche. *Holz Roh-Werkstoff* 19:253–259.
80. Sachsse, H. 1964. Der submikroskopische Bau der Faserzellwand beim Zugholz der Pappel. *Holz Roh-Werkstoff* 22:169–174.
81. Sachsse, H. 1965. Untersuchungen über den Einfluss der Astung auf die Farbkern-und Zugholzausbildung einiger Pappelsorten. *Holz Roh- Werkstoff.* 23: 425–434.
82. Scheffer, T. C. 1954. Mineral Stain in Hard Maple and Other Hardwoods. U.S. For. Lab. Report No. 1981.
83. Schirp, M., H. Kuebler and W. Liese. 1974. Untersuchungen an Baumscheiben über das Entstehen von Frostrissen. *Fortsw. Cbl.* 93(3):127–136.
84. Schulman, E. 1938. Classification of annual rings in Monterey pine. *Tree-Ring Bull.* 4(3):4–7.
85. Seth, M. K., and K. K. Jain. 1977. Relationship between percentage of compression wood and tracheid length in blue pine (*Pinus wallichiana* A. B. Jackson). *Holzforschung* 31(3):80–83.
86. Shigo, A. L. 1972. Ring and ray shakes associated with wounds in trees. *Holzforschung* 26(2):60–62.
87. Siripatadilok, S., and L. Leney. 1985. Compression wood in western hemlock, *Tsuga heterophylla* (Raf.) Sarg. *Wood Fiber Sci.* 17(2):254–265.
88. Smith, D. E. 1980. Abnormal wood formation following fall and spring injuries in black walnut. *Wood Sci.* 12(4)243–251.
89. Starbuck, C. J., and J. E. Phelps. 1986. Induction of compression wood in rooted cuttings of *Pseudotsuga menziesii* (Mirb.) Franco by indole-3-acetic acid. *IAWA Bull.* n.s. 7(1):13–16.
90. Tanaka, F., and T. Koshijima. 1981. Characterization of cellulose in compression wood and opposite wood of a *Pinus densiflora* tree grown under the influence of strong wind. *Wood Sci. Tech.* 15:265–273.
91. Thunell, B. 1951. Über die Drehwüchsigkeit. *Holz Roh- Werkstoff* 9:293–298.
92. Timell, T. E. 1986. *Compression Wood in Gymnosperms*. Berlin, New York: Springer Verlag.
93. Tippett, J. T. 1981. Barrier zones and the protection of cambia in perennial plants. *IAWA Bull.* n.s. 2(2/3):60.
94. Tippett, J. T. 1986. Formation and fate of kino veins in *Eucalyptus* L' Hérit. *IAWA Bull.* n.s. 7(2):137–143.
95. Tippett, J. T., and A. L. Shigo. 1981. Barrier zone

formation: A mechanism of tree defense against vascular pathogens. *IAWA Bull.* n.s. 2(4):163–168.
96. Torelli, N. 1984. The ecology of discoloured wood as illustrated by beech (*Fagus sylvatica* L.). *IAWA Bull.* n.s. 5(2):121–127.
97. Trendelenburg, R., and H. Mayer-Wegelin. 1955. *Das Holz als Rohstoff*, München: Hanser Verlag.
98. Tsoumis, G. 1952. Properties and effects of the abnormal wood produced by leaning hardwoods (tension wood). Report Yale For. School (unpublished).
99. Tsoumis, G. 1968. *Wood as Raw Material*. Oxford, New York: Pergamon Press.
100. Tsoumis, G. 1983. SEM observations on irradiated old wood. *IAWA Bull.* n.s. 4(1):41–45.
101. U.S. Forest Products Laboratory. 1962. A Simple Device for Detecting Compression Wood. Report No. 1390.
102. U.S. Forest Products Laboratory. 1987. Wood Handbook. U.S.D.A. Agr. Handbook No. 72 (revised).
103. Vasiljevic, S. 1955. Disposition, number and diameter of vessels of double (false) growth rings. *Coll. For. Bull.* 9:295–316. Univ. Belgrade.
104. Wagenfuehr, R., and C. Scheiber. 1974. *Holzatlas*. Leipzig. VEB Fachbuchverlag.
105. Wangaard, F. F. 1950. *The Mechanical Properties of Wood*. New York: J. Wiley & Sons.
106. Wardrop, A. B. 1964. Reaction anatomy of arborescent angiosperms. In *The Formation of Wood in Forest Trees*, ed. M. H. Zimmermann, pp. 405–456. New York, London: Academic Press.
107. Wardrop, A. B. 1965. The formation and function of reaction wood. In *Cellular Ultrastructure of Woody Plants*, ed. W. A. Côté, pp. 371–390. New York: Syracuse University Press.
108. Westing, A. H. 1965. Formation and function of reaction wood in gymnosperms. *Bot. Rev.* 31:381–480.
109. Wilkes, J. 1986. Anatomy of ring shake in *Eucalyptus maculata*. *IAWA Bull.* n.s. 7(1):3–11.
110. Wilkins, A. P. 1986. The nomenclature of cell wall deformations. *Wood Sci. Tech.* 20(2):97–109.
111. Wilson, B. F. 1962. A Survey of the Incidence of Ring Shake in Eastern Hemlock. Harvard Forest Paper No. 5.
112. Yamaguchi, K., T. Itoh, and K. Shimaji. 1980. Compression wood induced by 1-N-napthylphthalamic acid (NPA) and IAA transport inhibitor. *Wood Sci. Tech.* 14:181–185.
113. Yazawa, K. 1959. On the prevention of discoloration and decay and on the artificial formation of heartwood in beech. *Noyaju No Shinpo* 3(5). (English transl. No. 170/1962, Canada Dept. of Forestry).
114. Yoshizawa, N., Y. Okamoto, and T. Idei. 1986. Richting movement and xylem development in tilted young conifer trees. *Wood Fiber Sci.* 18(4):579–589.
115. Zenker, R., and W. R. Müller-Stoll. 1966. Einfluss der Zugholzfasern auf die Festigkeiteigenschaften des Pappelholzes in nassem und trockenem Zustand. *Holztechnologie* 7:17–25.
116. Ziegler, H., and W. Merz. 1961. Der "Hazelwuchs". *Holz Roh- Werkstoff* 19:1–8.
117. Zycha, H. 1953. Der rote Kern der Buche. *Holz Zentralbl.* 79:973–974.

FOOTNOTES

[1] A deviation similar in exterior appearance is due to internal decay. This is also attributed to stresses (40).

[2] Note that spiral checks were observed to form smaller angles (20 – 25°) in giant sequoia (*Sequoia gigantea*) (18), and cell length was negatively correlated to percentage of compression wood in blue pine (*Pinus wallichiana*) (85).

[3] Exceptional formation of alternating crescents, denoting a spiral arrangement of compression wood around the stem axis, has been observed in redwood, spruce, pine and fir (Figure 7-8A); in redwood, a case of eight full circuits has been reported (26, 31). Spiral tension wood was found in oak (*Quercus robur*) (78).

[4] Characteristically, in German tension wood is called *Weissholz* (white wood) and compression wood *Rotholz* (red wood). Tension wood is not always lighter in color, however. In certain tropical and Australian species, it is darker than normal (19).

[5] Pure reaction wood (compression wood, tension wood) is probably unacceptable in the pulping industry, but such a case (use of reaction wood alone) is practically nonexistent. Mixtures of normal and reaction wood which normally occur in trees will produce pulp with properties within acceptable ranges for many end products (44).

[6] Loss of leader (apical shoot) due to abiotic or biotic factors is also a cause for the formation of reaction wood, because the new leader deviates from the vertical position. Concerning biotic factors, it is interesting to mention that compression wood is believed to be caused in fir by saliva compounds of an insect (aphid) (92).

[7] Compression wood was observed in western hemlock to relate not to leaning but to presence of knots, and to extend toward the base of trees and a great distance from the knots, in the form of irregular streaks (46). Pruning was reported to increase the presence of tension wood in poplar (81).

[8] In a broader sense, barrier zones include callus tissue, gum (kino) veins, and traumatic resin canals (93).

[9] Wood exposed to gamma rays deteriorates, and becomes brittle depending on species and dosages (100). Lignin is the most resistant component.

[10] In some cases, knots have an aesthetic appeal (e.g., knotty pine), as do blue stain and "peckiness" (finger-sized pockets of decay) in baldcypress (*Taxodium distichum*) and incense cedar (*Libocedrus decurrens*); such "defective" wood is marketed at a high price and used decoratively (73).

II

PROPERTIES

Knowledge of the *properties* of wood, as of any other material (steel, cement, plastics), is a basic prerequisite for its rational utilization in making various products. In the chapters that follow, the following important properties are examined: *density (specific gravity)*, *hygroscopicity*, *shrinkage* and *swelling* (dimensional changes due to changing moisture content), and *mechanical*, *thermal*, *acoustical*, and *electrical* properties. A discussion of wood *deterioration* under the influence of fungi, insects, and other destructive factors is also included.

8

Density and Specific Gravity

Density is the mass contained in a unit volume of a material, and *specific gravity* is the ratio of the density of the material to the density of water. Specific gravity is also called *relative density*.

In the metric system, density is measured in grams per cubic centimeter (g/cm^3) or kilograms per cubic meter (kg/m^3). Density and specific gravity are numerically identical; for example, if the density of a material is 0.50 g/cm^3, its specific gravity is 0.50, because the density of water is 1 g/cm^3. (In both cases, the water is considered distilled and at 4°C or 39.2°F.)

In the English system, density is measured in slugs per cubic foot (10, 45) (1 $slug/ft^3$ = 0.5154 g/cm^3), but it is seldom used as such; instead, density is expressed as weight per unit volume [i.e., in pounds per cubic foot (lb/ft^3)], and this is sometimes called *weight density*. In units of this system, the density of water is 62.4 lb/ft^3 and, therefore, the specific gravity of a material may be found by dividing its weight density by 62.4; for example, 31.2 lb/ft^3 ÷ 62.4 lb/ft^3 = 0.50 (1 lb/ft^3 = 0.01602 g/cm^3, 1 g/cm^3 = 62.4 lb/ft^3).

Except for density in the English system, all other values—density (metric), specific gravity (metric, English)—are *numerically identical*, for the same piece of material and under the same conditions of measurement.

FACTORS AFFECTING THE DENSITY AND SPECIFIC GRAVITY OF WOOD

The density and specific gravity of wood are influenced by *moisture, structure, extractives,* and *chemical composition* (18, 26, 29, 34, 77).

Moisture

Wood is a hygroscopic material (i.e., it has the property to attract and retain moisture). Adsorption of moisture increases both its weight and volume, whereas loss of moisture results in their reduction (see Chapters 9 and 10). For this reason, and in order to obtain comparative values, the hygrometric condition of wood should be stated in each case. There are, however, certain conditions under which wood attains constant weight and volume, and these are oven-dry weight (i.e., weight at zero moisture content), oven-dry volume, and fully swollen volume.

Combinations of the above parameters (i.e., oven-dry weight[1]/oven-dry volume and oven-dry weight/fully swollen volume) define *oven-dry* density (specific gravity) and *basic* density, respectively. Oven-dry density is higher, because oven-dry volume is smaller than fully swollen volume. Both are comparative values, because their parameters are constant, as previously explained. A third concept is *air-dry* density, which also has a comparative value, if the air-dry condition has been attained under a controlled environment, leading to a constant moisture content, usually 12% and sometimes 15% (18). Air-dry density is calculated with oven-dry weight and air-dry volume (North America), or both weight and volume in air-dry condition (Europe). Other densities at x (i.e., not exactly known moisture contents) may be called *apparent*; they are not reproducible, and have only an indicative value in handling (shipping, etc.) wood, where weight is expressed in kg (lb) and volume in m^3 (ft^3). Figure 8-1 shows the relative magnitude of the above four densities in diagrammatic form.

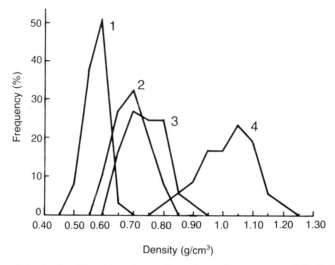

Figure 8-1. Variation of the density of beech wood according to base of measurement: 1, basic density (weight oven-dry/wet volume); 2, oven-dry density (wt od/volume od); 3, air-dry density (wt and volume air-dry); 4, apparent density (wt and volume wet). (After Ref. 17.)

Increasing moisture content increases the density of wood. As explained in the following chapters (Chapters 9 and 10), while the weight of wood increases with increasing moisture content (up to a maximum which can be quite high depending on oven-dry density, see Table 8-1, column 7), volume increases at first (up to maximum capacity of wood to swell, column 8), but remains constant thereafter irrespective of the moisture retained. As a result, apparent density becomes higher with increasing moisture content (Table 8-1, columns 2–6, Figure 8-2). In the range of 0 and 25% moisture content, density may be calculated from the relationship (33, 34)

$$r_u = r_o \frac{1 + u}{1 + 0.84\, r_o \cdot u} \quad (8\text{-}1)$$

where

r_u = density at u (or x) moisture content (g/cm^3)

r_o = oven-dry density (g/cm^3)

u = moisture content (%, decimal)

Table 8-1. Effect of Moisture on the Density of Wood[a]

		Moisture content (%)					Maximum mc %	Maximum swelling %
	r_o	10	20	30	50	100		
		Density (g/cm^3)						
Species	1	2	3	4	5	6	7	8
Balsa	0.13	0.139	0.148	0.156	0.180	0.240	732	8.1
Spruce	0.41	0.432	0.451	0.469	0.541	0.722	207	13.6
Pine, Scots	0.49	0.515	0.537	0.558	0.643	0.858	167	14.2
Larch	0.57	0.600	0.628	0.653	0.749	0.998	138	13.4
Beech	0.70	0.719	0.736	0.750	0.866	1.154	105	21.3
Oak, red	0.82	0.836	0.850	0.861	0.994	—	84	23.8

[a] The table reads as follows: column 1 is oven-dry density (wt. od/vol. od, r_o); columns 2–6 are densities at respective moisture contents, 10–100%. Except for column 1, which derives from experimental data, the remaining values are calculated. Maximum swelling was calculated from maximum shrinkage. The species are European—except balsa.

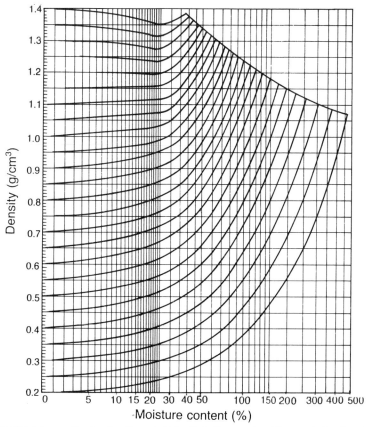

Figure 8-2. Effect of moisture on the density of wood. (After Ref. 33; reproduced by permission from Springer Verlag.)

Structure

Wood is mostly made of dead cells, which are composed of cell walls and cell cavities. The density of the material that constitutes the cell walls is practically constant, about 1.50 g/cm^3 on the basis of oven-dry weight (mass) and volume. The density of wood varies, depending on the amount of material (cell wall)[2] and voids (cell cavities) present in a certain volume. The variation is great especially in tropical woods, where oven-dry density ranges from about 0.1 g/cm^3 in balsa (*Ochroma lagopus*) to about 1.3 g/cm^3 in lignum vitae (*Guaiacum officinale*); temperate woods range from about 0.3 to 0.9 g/cm^3 (see Table 8-2).[3]

The density of wood is a measure of the quantity of cell-wall material contained in a certain volume, and is an index of void volume. The latter may be calculated from the relationship (34)

$$C = \left(1 - \frac{r_o}{r_w}\right) \times 100 \qquad (8\text{-}2)$$

where

C = proportion of void volume (% of total volume)
r_o = oven-dry density (g/cm^3)
r_w = density of cell-wall material (g/cm^3)

If $r_w = 1.50$, the relationship becomes: C (%) = 100 − 66.7 r_o. Void volume varies from about 95% in very light woods to about 10% in very heavy woods.

Differences in density and void volume derive from anatomical differences, such as differences in cell types (tracheids, vessel members, parenchyma cells) and their quantitative distribution, thickness of cell walls, and size of cell cavities.[4] Evaluation of the influence of such microscopic characteristics is difficult. For this reason, the relationship of specific gravity and structure is examined on the basis of factors which can be easily measured, such as width of growth rings and proportion of latewood.

Table 8-2. Density and Shrinkage[a]

Species	Density (g/cm³)	Shrinkage (%)			
		A	R	T	V
1. Baldcypress	0.46	—	3.8	6.2	10.5
Douglas-fir	0.48	—	5.0	7.8	11.8
Fir, white	0.39	—	3.2	7.1	9.8
Hemlock, western	0.42	—	4.3	7.9	11.9
Larch, western	0.52	—	4.2	8.1	13.2
Pine, loblolly	0.51	—	4.8	7.4	12.3
Pine, ponderosa	0.40	—	3.9	6.3	9.6
Pine, white eastern	0.35	—	2.3	6.0	8.2
Redcedar, western	0.33	—	2.4	5.0	6.8
Redwood, old growth	0.40	—	2.6	4.4	6.8
Spruce, Engelmann	0.35	—	3.4	6.6	10.4
Alder, red	0.41	—	4.4	7.3	12.6
Ash, white	0.60	—	4.8	7.8	13.4
Aspen, quaking	0.38	—	3.5	6.7	11.5
Basswood, American	0.37	—	6.6	9.3	15.8
Beech, American	0.64	—	5.1	11.0	16.3
Birch, yellow	0.62	—	7.2	9.2	16.7
Cherry	0.50	—	3.7	7.1	11.5
Cottonwood, eastern	0.40	—	3.9	9.2	14.1
Elm, American	0.50	—	4.2	9.5	14.6
Hickory, mockernut	0.72	—	7.8	11.0	17.9
Maple, sugar	0.63	—	4.9	9.5	14.9
Oak, northern red	0.63	—	4.0	8.2	13.5
Oak, white	0.68	—	5.3	9.6	18.9
Sweetgum	0.52	—	5.2	9.9	15.0
Sycamore, American	0.49	—	5.1	7.6	14.2
Walnut, black	0.55	—	5.5	7.8	12.8
Willow, black	0.39	—	2.6	8.1	14.4
Yellow poplar	0.42	—	4.0	7.1	12.3
2. Fir, white	0.44	0.1	3.8	7.6	11.7
Larch	0.60	0.3	3.3	7.8	11.8
Pine, black	0.55	0.3	4.1	7.7	12.5
Pine, Scots	0.53	0.4	4.0	7.7	12.4
Spruce	0.44	0.3	3.6	7.8	12.0
Alder	0.55	0.5	4.4	7.3	12.6
Ash	0.70	0.2	5.2	8.3	14.0
Aspen	0.46	—	3.5	8.5	12.8
Beech	0.74	0.3	5.8	11.8	17.6
Birch	0.73	0.6	5.3	7.8	14.2
Black locust	0.76	0.1	4.0	6.5	10.8
Chestnut	0.61	0.6	4.3	6.4	11.6
Elm, field	0.67	0.3	4.6	8.3	13.8
Hornbeam	0.82	0.5	6.8	11.5	19.7
Limetree (Basswood)	0.54	—	5.5	9.1	14.4
Oak, red	0.87	0.3	5.3	13.0	19.2
Oak, white (English)	0.69	0.4	4.0	7.8	12.2
Plane (Sycamore)	0.63	0.5	4.5	8.7	13.7
Poplar "I-214"	0.34	—	—	—	9.7
Walnut	0.69	0.5	5.4	7.5	13.9
Willow	0.52	0.5	3.9	6.8	11.5
3. Afara	0.60	0.2	4.7	5.5	10.4
Afzelia	0.81	—	2.6	4.3	—

DENSITY AND SPECIFIC GRAVITY 115

Table 8-2. (*Continued*)

Species	Density (g/cm³)	Shrinkage (%)			
		A	R	T	V
Aiélé	0.50	—	6.0	8.7	15.5
Amazakoué	0.78	0.1	4.2	8.0	13.3
Antiaris, Ako	0.55	—	3.5	5.8	—
Balsa	0.16	0.6	2.4	4.4	7.5
Beté	0.62	0.1	4.0	6.0	10.3
Bubinga	0.87	—	6.2	7.7	12.7
Dibetou	0.60	—	3.9	6.5	11.1
Difou	0.85	—	3.5	5.8	—
Idigbo	0.57	—	3.0	4.5	8.1
Iroko	0.62	0.1	3.8	5.5	9.4
Kosipo	0.70	—	4.3	6.0	11.0
Mahogany, African	0.51	0.2	3.2	5.7	9.1
Mahogany, American	0.60	0.3	3.2	5.1	8.6
Makoré	0.62	0.2	4.7	6.3	11.2
Meranti, red	0.67	—	4.5	8.2	—
Niangon	0.70	—	3.6	8.2	12.0
Obeche	0.38	0.2	3.3	5.6	9.1
Okoumé	0.43	0.2	3.8	5.7	9.7
Opepe	0.74	—	—	7.5	—
Padauk	0.70	—	3.0	4.7	8.5
Palissander	0.87	—	2.7	5.8	8.7
Ramin	0.60	—	4.0	9.4	14.3
Sapele	0.65	—	5.4	7.7	12.6
Seraya, white	0.53	—	—	—	—
Sipo	0.63	0.3	5.0	7.9	11.8
Teak	0.67	0.6	3.0	5.8	9.4
Tiama	0.59	—	5.1	7.6	12.4
Zebrano	0.78	—	5.1	8.0	13.5
4. Gum, blue southern	0.88	0.5	6.2	9.9	16.6
Jarrah	0.77	—	5.0	11.0	—
Gum, red river	0.91	—	5.0	8.0	—
Gum, manna	0.81	—	6.0	9.0	—

[a]Air-dry density (12–15% moisture content) and total shrinkage (from green to zero-moisture content). *A*, axial (longitudinal); *R*, radial; *T*, tangential; *V*, volumetric shrinkage. 1, North American; 2, European; 3, tropical; 4, Australian (eucalypts). Latin names of species are given in Appendix II and Index II. Data are based on Refs. 74 (North American), 17, 18, 34, 47, 48, 75 (European), 5, 16, 76 (tropical), and 4 and 76 (Australian).

Width of Growth Rings. The influence of ring width is different in softwoods, and in ring-porous and diffuse-porous hardwoods. In softwoods, the statistical correlation between ring width and density is low, but density tends to decrease with increasing ring width. In ring-porous hardwoods, density increases, up to a certain level, with increasing ring width (Figure 8-3), but in diffuse-porous hardwoods ring width is not a clear criterion of density (9, 12, 13, 44, 62).

It should be noted that rings of the same width do not always have the same influence on density, because their anatomical structure may differ depending on their position in the tree (juvenile, mature, overmature wood), or may vary due to growth conditions leading to formation of reaction wood (compression or tension wood).

Proportion of Latewood. Latewood is made of cells which have thicker walls and smaller cavities in comparison to earlywood. This results in a higher density of latewood as compared with earlywood [triple or more (13, 18, 62), see Table 8-3], and explains why the density of wood increases with increasing proportion of latewood. The relationship is clear

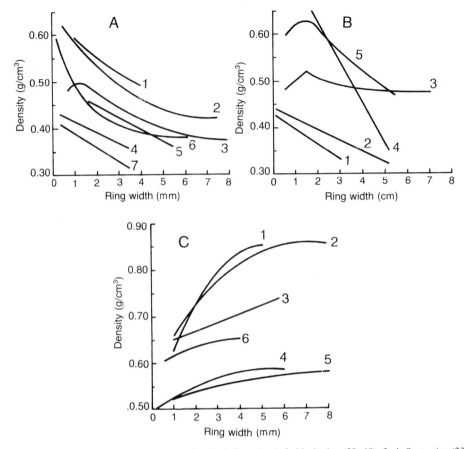

Figure 8-3. Relationships of density and ring width. (A) Softwoods: 1, 2, black pine (58, 68); 3, 4, Scots pine (23, 67); 5, Grecian fir (46); 6, hybrid fir (64); 7, European spruce (23). (B) Softwoods: 1, pacific silver fir (31); 2, western hemlock (38); 3, cedar of Lebanon; 4, Douglas-fir (22); 5, European larch (67). (C) Hardwoods (ring-porous): 1, white oak (*Qu. petraea*); 2, white oak (*Qu. farnetto*); 3, red oak (*Qu. borealis*) (21); 4, white oak (*Qu. robur*) (67); 5, European ash (33); 6, black locust (67). Note that numbers 4, 7(A), 1, 2(B), and 4, 5, and 6(C) indicate basic densities; the remaining are oven-dry densities.

in softwoods and ring-porous hardwoods (Figure 8-4). In diffuse-porous hardwoods, latewood is not clearly discernible; these woods usually have a narrow zone of cells with thicker walls at the end of growth rings, which is helpful to their delineation, but this zone is difficult to measure and relate to density.

In addition to proportion, the "quality" of latewood is important; features such as cell-wall thickness in relation to the size of cavities, and proportional representation of various cell types affect density.

In softwoods, differences in cell-wall thickness may be expressed with differences in color contrast between earlywood and latewood. In oak, ash, chestnut, and other ring-porous hardwoods, wide zones of latewood may have low density, because they contain relatively large vessels or a large proportion of parenchyma cells.

Proportion of latewood is the basic reason for the relationship between density and ring width, as previously mentioned, because latewood varies with ring width. The relationship is clear in ring-porous hardwoods, where an increase of width is associated with an increase of latewood proportion. In diffuse-porous hardwoods, there is no practical way to study the correlation. In softwoods, the proportion of latewood (and density) tends to decrease with increasing ring width, but the correlation is statistically low.

Table 8-3. Density of Earlywood and Latewood[a]

Species	Density r_o (g/cm^3)		Late / Early
	Latewood	Earlywood	
Spruce	0.55–0.87	0.31–0.36	1.53–2.49
Fir, white	0.62	0.28	2.21
Fir, Grecian	0.62	0.22	2.62
Pine Aleppo	0.67	0.35	1.91
Pine, hard	0.59	0.32	1.84
Pine, Scots	0.67–0.92	0.30–0.34	2.09–3.07
Pine, black	0.73	0.36	2.03
Pine, whitebark	0.53	0.35	1.51
Pine, umbrella	0.65	0.36	1.81
Douglas-fir	0.74–0.84	0.28–0.34	2.17–3.0
Larch	1.04	0.36	2.89
Oak, English	0.89–0.93	0.32–0.45	1.98–2.91
Oak, broadleaf	0.67	0.55	1.22
Ash	0.72–0.80	0.38–0.51	1.41–2.11
Black locust	0.68	0.49	1.39
Chestnut	0.56–0.65	0.42–0.49	1.33

[a]*Source*: Refs. 7, 17, 18, 34, 67, 69, and 79.

Extractives. Extractives are compounds of varying chemical composition (gums, fats, resins, sugars, oils, tannins, alkaloids, etc.) that are not part of the wood substance, but are deposited within cell walls and in cavities; their removal does not affect the cellular structure of wood. The proportion of extractives varies from less than 1% to 20% or more on the basis of oven-dry weight of wood. Higher amounts of extractives are a cause for the higher density of heartwood in comparison to sapwood; removal of extractives results in reduction of density.

Chemical Composition. The chemical components of cell walls differ in density. For example, in pine wood α-cellulose was found to have a density of 1.528 g/cm^3 and lignin 1.335 g/cm^3 (63). Differences of participation may, therefore, contribute to density differences. Cellulose variation is small, however

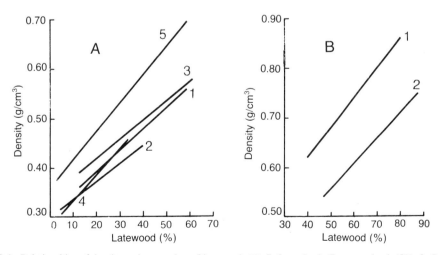

Figure 8-4. Relationships of density and proportion of latewood. (A) Softwoods: 1, European larch (51); 2, Scots pine (23); 3, hybrid fir (64); 4, European spruce (23); 5, Douglas-fir (22). (B) Hardwoods (ring-porous): 1, white oak (*Qu. farnetto*]); 2, red oak (*Qu. borealis*) (21). Note that numbers 1, 2, and 4(A) indicate basic densities; the remaining are oven-dry densities.

(40–45% in all woods), although the variation of lignin is greater (17–35%). Larger differences exist in compression and tension wood.

VARIATION OF DENSITY

The density of wood varies under the influence of the factors mentioned above. Variation exists within a tree, between trees of the same species, and between species (12, 45).

Variation Within a Tree

The density of wood varies in the trunk of a tree, and also between trunk, branch, and root wood.

Within a trunk—which is the main tree component with practical value—there is a *vertical* variation (from base to top of a tree) and a *horizontal* variation (from pith to bark).

In the *vertical* direction, there is a tendency for reduction of density with tree height, especially in softwoods (13, 19), but this does not occur in all cases (42) (Figures 8-5 and 8-6). The reduction is attributed to various factors—mechanical and biological. From a mechanical point of view, the trunk of a standing tree is considered a cantilever beam (supported at one end) (78). Under the influence of factors such as weight, wind, and snow acting on the crown, greater stresses develop at the base of the trunk, and result in local formation of wood of higher density (therefore, higher strength). It has been observed that supporting trees, so that they do not sway by the action of wind, leads to formation of lighter wood (27). In addition to mechanical factors, greater density at the base of a tree is contributed by the formation of heartwood. The proportion of heartwood is higher at the base, and its contribution to density differences is more pronounced when the color contrast between heartwood and sapwood is intense. The darker color of heartwood is caused by deposition of extractives; extractives are lighter than cell-wall material, but their presence contributes to a higher density, because they impregnate cell walls or are deposited in cell cavities.

Vertical variation is also influenced by the presence of juvenile wood around the pith. The proportion of such atypical wood is greater in the upper part of the trunk—in fact, at the top, all wood is atypical. Vertical variation is affected because the density of juvenile wood is different, higher or lower, in comparison to mature wood (see below).

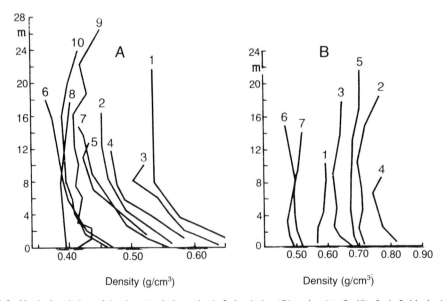

Figure 8-5. Vertical variation of density. (A) Softwoods: 1, 2, hard pine (*Pinus brutia*) (3, 48); 3, 4, 5, black pine (48, 58); 6, 7, Scots pine (23, 67); 8, 9, European spruce (23, 67); 10, hybrid fir (64). (B) Hardwoods: 1, chestnut; 2, 3, beech (67); 4, white oak (*Qu. petraea*); 5, European birch (23); 6, European alder (67). Note that numbers 6, 7(A), and 4(B) indicate basic densities; the remaining are oven-dry densities.

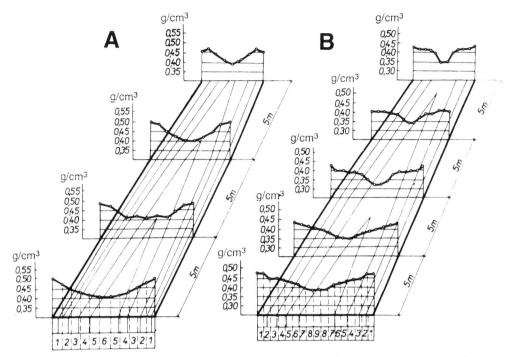

Figure 8-6. Vertical and horizontal variation of (air-dry) density. (A) European spruce; (B) European (silver) fir. Each subdivision at the base represents 20 years. (After Ref. 43.)

Taper contributes to vertical variation. In softwoods, where such observations have been made, the variation was higher (a greater reduction of density) in less tapering (more cylindrical) trunks (62). This effect is attributed to the presence of more juvenile wood in the upper part of the trunk, and to juvenile wood having a lower density in softwoods.

Horizontal variation is interrelated to vertical variation and, therefore, it is affected by the factors previously mentioned. In each horizontal level, the influence of age is more clear, however, considering that the wood produced in different stages of tree life (juvenile, mature, overmature) differs in density. In softwoods, density is low in youth (near the pith; Figure 8-6), ascends to a "typical" level, and finally is reduced at a great age, due to respective variation of cell-wall thickness and the proportion of latewood. Cellulose content varies in the same manner, but more lignin was found near the pith. In most softwoods, density is reduced after about 100 years of tree age (62). At a young age, the influence of distance from pith is greater than that of ring width, especially in fast-growing species (e.g., *Pinus radiata*), where an increase of density was observed even when ring width increased with age (15, 56, 72).

In hardwoods (oak, chestnut, beech), density was found to decrease with distance from pith, but available information is scarce with regard to these woods (2).

Interaction of ring width and proportion of latewood, and effects of age (juvenile wood, heartwood) in different parts within a trunk result in a complex distribution of density (17, 18, 64) (Figure 8-7); the complexity becomes higher when defects are present.

In *branches* and *roots*, density was found sometimes higher and other times lower both in softwoods (14, 52) and hardwoods (65, 80). Variability is affected by their distance from the stem and, especially in branches (66, 67), from the presence of compression or tension wood.

Variation Between Trees of the Same Species

Variation between trees of the same species is influenced by environmental conditions (soil, climate, tree spacing) and heredity. Environ-

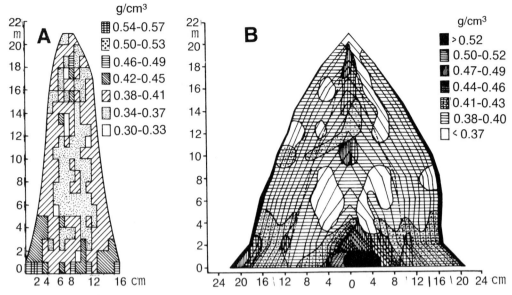

Figure 8-7. Diagrams of density distribution in tree trunks. (A) Scots pine, (B) fir (a natural hybrid between European silver fir and Grecian fir, *Abies alba* × *Abies cephalonica*). (1 m = 3.3 ft, 1 cm = 0.4 in.). (A after Ref. 18, reproduced by permission from Unione Tipografico-Editrice Torinese; B after Ref. 64.)

mental conditions change during the life of a tree and, therefore, they affect both between and within tree variation.

The effect of environment is basically expressed through changes of ring width and proportion of latewood. The evaluation of environmental factors (moisture, temperature, nutrients) is not easy. There is extensive literature on this subject (13, 19, 25, 37), which may be summarized as follows. The prevailing general view is that a combined influence of the environment, as expressed by site quality, does not exert a clear influence on density. High- and low-quality sites were not always related to production of high and low wood density, respectively; the opposite was also observed (71). Smaller differences were found between sites than between trees of the same site (20). The sociological position of a tree in a forest stand (crown class) was also observed not to exert a clear influence (13, 35, 36). Dominant coniferous trees tend to produce wood of lower density—but not always (73).

Density is affected by the space available for tree growth, and such space is determined by planting distances, intensity of thinning forest stands, and other silvicultural measures. A greater distance between trees favors the formation of wider growth rings and, therefore, exerts an indirect influence on density. Wider rings at a young age contribute more juvenile wood, which has a lower density in softwoods and a higher density in ring-porous hardwoods. Also, a greater distance between trees favors the development of a larger crown and more sapwood; the density of sapwood is lower in comparison to heartwood due to the presence of extractives and/or tyloses.

Density is affected by such factors as fertilization, thinning, pruning, and irrigation.[5] These factors induce faster growth with a tendency to lower density in softwoods (the reverse has also been observed) and higher density in hardwoods (2, 54, 61). In tropical species, it has been observed that drier regions produce wood of higher density (1).

Adverse growth conditions may influence density, when reaction wood is produced. Compression and tension wood have higher density in comparison to normal wood. The density of compression wood was found about 10–40% higher and that of tension wood 2–20% higher than normal wood (40, 44, 57, 62).

DENSITY AND SPECIFIC GRAVITY

Table 8-4. Variation of Density and Shrinkage[a]

Species	Density (g/cm³)	Shrinkage (%)			
		Axial	Radial	Tangential	Volume
Spruce	0.32–0.55[a]	0.1–0.5	1.4–6.1	4.7–12.1	6.8–16.5
Fir, white	0.35–0.52[a]	0.1–0.6	1.3–4.5	4.0–10.6	5.8–14.2
Fir, Grecian	0.36–0.49	—	—	—	9.5–14.5
Pine, hard	0.41–0.87	—	—	—	11.3–17.3
Pine, black	0.46–0.78	—	—	—	10.4–16.8
Oak, white (sessile)	0.53–0.89	0.1–0.4	3.0–7.7	5.9–13.7	8.9–19.6
Oak, white (English)	0.63–0.77[a]	0.1–0.4	3.9–6.5	7.8–13.8	11.8–19.0
Chestnut	0.51–0.65	—	—	—	9.4–14.2
Beech	0.66–0.78	0.1–1.1	3.0–9.8	8.6–18.2	12.3–21.8
Poplar (hybrids)	0.28–0.38	—	—	—	8.5–11.8

[a]Air-dry densities (12–15% mc); the remaining are dry densities. *Source*: Refs. 17, 18, 46, 48, and 69.

Heredity affects density in the sense that certain trees, under the same growth conditions, produce wood of high or low density, and that this property may be inherited (13, 19, 83). This is important for tree improvement efforts.

Density variation within and between trees results in considerable differences within the same wood species (17, 32) (see Table 8-4).

Variation Between Species

Density variation between species is basically due to differences in anatomical structure. Species differ with regard to cell types and their proportional participation. In addition, differences in extractives and chemical composition of cell walls may influence density. It should be noted, however, that woods of different species do not always differ in density (Table 8-2).

DETERMINATION OF DENSITY

Density is determined by measuring mass (weight) and volume, or by other methods. Dry (oven-dry) mass is measured by placing the chosen specimen of wood[6] in an oven at $103 \pm 2°C$ (214–221°F), until its weight becomes a constant; 12–48 h are needed for a specimen about 100 g in weight, depending on species and dimensions. The volume is determined by measuring the dimensions, if shape permits, or by water displacement (Archimedes' principle). In the case of dry volume, the specimen, immediately after its removal from the oven, is immersed first in molten paraffin for a few seconds to ensure a thin protective layer, and subsequently in water. Determination of wet volume does not need the above procedure; the wet specimen is directly immersed in water and, as in the previous case, its volume is measured by the weight of displaced water.[7] In both cases, the specimen is properly supported and carefully immersed to avoid touching the walls of the water receptacle (Figure 8-8).

With small specimens, mercury may be used instead of water to overcome the hygroscopicity of wood. The mass of displaced mercury (g) is converted to cubic centimeters (cm³) of volume by dividing the mass by the density of mercury (the latter is affected by temperature). Mercury is also used for measuring volume with Breuil's volume-meter (Figure 8-9). The specimen is immersed in mercury and the displaced volume of it is measured micrometrically (81).

Basic density (dry mass/fully swollen volume) may be determined by weighing the specimen of wood twice—in a dry condition (M_o) and after saturation with water (M_n). These two numbers are used for calculation of maximum moisture content (Y_{max}) and density from the following relationships (60):

$$Y_{max} = \frac{M_n - M_o}{M_o} \times 100 \quad (8\text{-}3)$$

Figure 8-8. Determination of the volume of a specimen by water displacement on an electric balance.

Figure 8-9. Breuil's mercury volume-meter.

or

$$Y_{max} = \frac{V_g - V_w}{M_o} \times 100$$

$$= \left(\frac{V_g}{M_o} - \frac{V_w}{M_o}\right) \times 100$$

$$= \left(\frac{1}{R} - \frac{1}{r_w}\right) \times 100 \quad (8\text{-}4)$$

If

$$r_w = 1.50 \text{ g/cm}^3,$$

$$Y_{max} = \left(\frac{1}{R} - 0.67\right) \times 100 \quad (8\text{-}5)$$

and

$$R = \frac{100}{Y_{max} + 67} \quad (8\text{-}6)$$

where

Y_{max} = maximum moisture content (%)
M_n = mass (weight) of water-saturated wood (g)
M_o = oven-dry mass (g)
V_g = fully swollen volume (cm^3)
V_w = volume of cell wall material (cm^3)
R = basic density (g/cm^3)
r_w = density of cell-wall material (1.50 g/cm^3)

The volume of wood may also be determined by measuring the volume of displaced water in a receptacle equipped with an attachment to read, from outside, the level of water before and after immersing the specimen (34); a common graduated glass tube may also be used for this purpose.

Density may be estimated by application of a simple method proposed by Paul (50): a prismatic specimen (e.g., 1×1×10 cm) is lengthwise immersed in water after its length is divided by pencil in 10 equal parts. The relationship of the length remaining immersed

DENSITY AND SPECIFIC GRAVITY

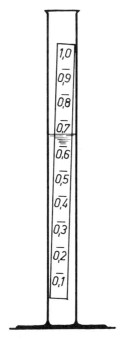

Figure 8-10. Estimation of density. (After Ref. 50.)

to the total length is an approximate measure of density (Figure 8-10). The method cannot be used with heavy tropical woods having density higher than 1 g/cm³, or specimens containing too much moisture, because they sink in the water.

The density of wood, and its variability within growth rings, may be determined directly by use of instruments (densitometers) scanning transverse surfaces or their photographs on film; they record the density by measuring the absorbed radiation (β- or x-rays) in different positions within a growth ring (Figures 8-11 and 8-12). Use of x-rays is faster and more accurate. Sound waves have also been tried (8, 11, 24, 49, 53, 54).

IMPORTANCE OF DENSITY

Density (the dry mass contained in a volume of wood) is directly related to other properties and, therefore, is important as an index of wood quality (12, 55, 59, 73, 77, 82). Density affects hygroscopicity, shrinkage and swelling, mechanical, thermal, acoustical, electrical, and other basic wood properties, as well as properties related to the industrial processing of wood (machining, drying, etc.). It should be noted, however, that the value of density as an index of quality refers to wood without defects, and that density is only an index. Differences in cellular composition or extractive content contribute in such a way that woods with the same density may exhibit differences in wood properties.

Density is also important as an index of quantitative production, because it is a measure of the mass contained in a certain wood volume. This aspect is of interest to industries

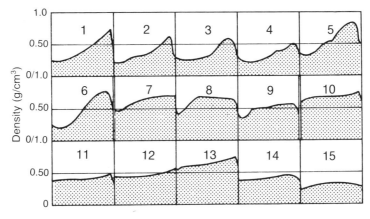

Figure 8-11. Variation of density within growth rings: 1, fir; 2, spruce; 3, ponderosa pine; 4, Scots pine; 5, larch; 6, Douglas-fir; 7, ash; 8, oak; 9, black locust; 10, maple; 11, alder; 12, cottonwood; 13, European beech; 14, aspen (trembling); 15, willow. (North American species, except 4 and 13; measurement of density with x-rays.) (Adapted from Ref. 11.)

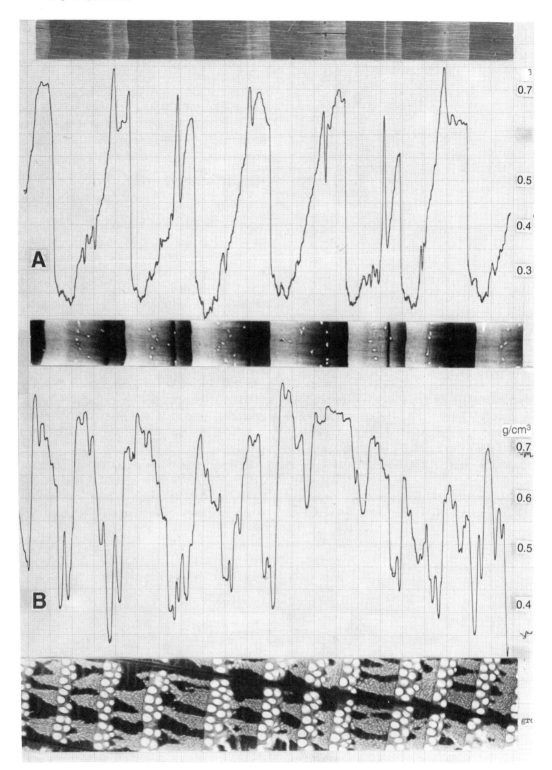

Figure 8-12. Variation of density between and within growth rings in (A) pine (maritime) and (B) chestnut. (X-ray densitometry: a, photograph; b, x-ray.) (Courtesy of H. Polge.)

making products such as pulp, paper and fiberboard, as well as to the production of wood in a forest.

REFERENCES

1. Batajas-Morales, J. 1987. Wood specific gravity in species from two tropical forests in Mexico. *IAWA Bull.* n.s. 8(2):143-148.
2. Bendtsen, B. A. 1978. Properties of wood from improved and intensively managed trees. *For. Prod. J.* 28(10):61-72.
3. Berkel, A. 1957. Untersuchungen über die Eigenschaften des brutischen Kiefernholzes—*Pinus brutia* Ten. *Istanbul Univ. Orman Fak. Derg.* 1:21-68.
4. Boas, I. H. 1947. *The Commercial Timbers of Australia.* C.S.I.R.O., Melbourne.
5. Bolza, E., and W. G. Keating. 1972. African Timbers—The Properties, Uses and Characteristics of 700 Species. Div. Build. Res., C.S.I.R.O., Melbourne.
6. Borgin, K., G. Tsoumis, and C. Passialis. 1979. Density and shrinkage of old wood. *Wood Sci. Tech.* 13:49-57.
7. Bosshard, H. H. 1964. Weisstannenholz. *Schweiz. Z. Forstwes.* 9/10:534-542.
8. Brazier, J. D. 1969. Some Considerations in Appraising Within-Ring Density. F.A.O., FO-FTB-69-4/2.
9. Brazier, J. D. 1977. The effect of forest practices on quality of the harvested crop. *Forestry* 50(1):49-65.
10. Brown, H. P., A. J. Panshin, and C. C. Forsaith. 1952. Textbook of Wood Technology, Vol. II. New York: McGraw-Hill.
11. Cown, D. J., and M. L. Parker. 1978. Comparison of annual ring density profiles in hardwoods and softwoods by X-ray densitometry. *Can. J. For. Res.* 8(4):442-449.
12. De Zeew, C. 1965. Variability in wood. In *Cellular Ultrastructure of Woody Plants*, ed. W. A. Côté, pp. 457-471, Syracuse, New York: Syracuse University Press.
13. Elliott, G. K. 1970. Wood Density in Conifers. Com. For. Bureau. Tech. Commun. No. 8, Oxford.
14. Enchev, E. 1972. Untersuchung einiger physikalischmechanischer Eigenschaften der Aeste von Nadelhölzern. *Holztechnologie* 13(4):232-236.
15. F.A.O./U.N. 1962. *Le Pin de Monterey (Pinus radiata* D. Don). Rome.
16. Farmer, R.H. 1972. *Handbook of Hardwoods.* London: HMSO.
17. Filipovici, J. 1965. *Studiul Lemnului.* Bucarest, Romania.
18. Giordano, G. 1971. *Tecnologia del Legno*, Vol. 1. Torino: Unione Tipografico-Editrice.
19. Goggans, J. F. 1961. The Interplay of Environment and Heredity as Factors Controlling Wood Properties in Conifers. N.C. State College, School of Forestry, Tech. Report No. 11.
20. Göhre, K. 1955. Einfluss von Wuchsgebiet, Standort, Rasse und Bewirtschaftung auf die Rohwichte des Holzes. *Arch. Forstwes.* 4(5/6):418-433.
21. Göhre, K. and E. Wagenknecht. 1955. Die Roteiche und Ihr Holz. Berlin: Deutscher Bauernverlag.
22. Göhre, K. 1958. *Die Douglasie und Ihr Holz.* Berlin: Akademie-Verlag.
23. Hakkila, P. 1966. Investigations on the Basic Density of Finnish Pine, Spruce and Birch Wood. Comm. Inst. For. Fenniae 61.5. Helsinki, Finland.
24. Heger, L., M. L. Parker, and R. W. Kennedy. 1974. X-Ray densitometry: A technique and an example of application. *Wood Sci.* 7(2):140-148.
25. Hildebrandt, G. 1960. The effect of growth conditions on the structure and properties of wood. *Proc. Fifth World For. Congr.*, 3:1348-1353.
26. Illston, J. M., J. M. Dinwoodie, and A. A. Smith. 1979. *Concrete, Timber and Metals.* New York: Van Nostrand Reinhold.
27. Jacobs, R. M. 1939. A Study of the Effect of Sway on Trees. Commonw. For. Bureau Bull. No. 26, Australia.
28. Keith, C. T. 1969. Resin content of red pine wood and its effect on specific gravity determinations. *For. Chron.* 45(5).
29. Kellogg, R. M. 1981. Physical properties of wood. In *Wood: Its Structure and Properties* (ed. F. F. Wangaard) pp. 195-223. Materials Education Council, The Pennsylvania State University, University Park, Pennsylvania.
30. Kellogg, R. M., and F. F. Wangaard. 1969. Variation in the cell-wall density of wood. *Wood Fiber* 1(3):180-204.
31. Kennedy, R. W., and G. W. Swann. 1968. Specific Gravity of Amabilis Fir (*Abies amabilis*) Determined from Increment Cores. W. For. Prod. Lab. Canada, Inform. Rep. VP-X-33.
32. Knigge, W., and H. Schulz. 1966. *Grundriss der Forstbenutzung.* Hamburg: P. Parey.
33. Kollmann, F. 1951. *Technologie des Holzes und der Holzwerkstoffe.* (1 Bd., 2. Auflage). Berlin: Springer Verlag.
34. Kollmann, F. F. P., and W. A. Côté. 1968. *Principles of Wood Science and Technology. I. Solid Wood.* Berlin: Springer-Verlag.
35. Koltzenburg, C. 1967. Der Einfluss von Lichtgenuss, soziologischer Stellung und des Standortes auf Holzeigenschaften der Rotbuche (*Fagus silvatica* L.). *XIV IUFRO Kongress*, 9:210-235.
36. Kommert, R. 1972. Untersuchung der Roh- und Raumdichte von Fichtenholz. *Holztechnologie* 13(4):229-231.
37. Kozlowski, T. T. 1971. *Growth and Development of Trees.* New York/London: Academic Press.
38. Krahmer, R. L. 1966. Variation of specific gravity in western hemlock tress. *Tappi* 49(5):227-229.
39. Lee, C. H., and H. E. Wahlgren. 1979. Specific gravity sampling with minimal tree damage. A study of red pine. *Wood Sci.* 11(4):241-245.

40. Moulopoulos, C., and G. Tsoumis. 1960. Growth abnormalities in euramerican hybrid poplars cultivated in Macedonia-Greece. *Annals, School Agr. For., Aristotelian University* 5:193-237 (Greek, English summary).
41. Murphey, W. K., W. J. Young, and B. E. Cutter. 1973. Effects of sewage effluent irrigation on various physical and anatomical characteristics of northern red oak. *Wood Sci.* 6(1):65-71.
42. Okkonen, E. A., H. E. Wahlgren, and R. R. Maeglin. 1972. Relationships of specific gravity to tree height in commercially important species. *For. Prod. J.* 22(7):37-42.
43. Palovic, J. 1966. Die Beziehung zwischen Rohdichte und Makrostruktur der Nadelhölzer und ihre Bedeutung in der Holztechnologie und Forstwirtschaft. Proc. Intern. Symposium, Eberswalde.
44. Panshin, A. J., C. De Zeew, and H. P. Brown. 1964. *Textbook of Wood Technology*, 2nd ed. New York: McGraw-Hill.
45. Panshin, A. J., and C. De Zeew. 1980. *Textbook of Wood Technology*, 4th ed. New York: McGraw-Hill.
46. Papamikail, P. 1962. Recherches sur les Propriétés Physiques et Méchaniques du Bois de Sapin de Chephalonie (*Abies cephalonica*) de la Forêt du Taygéte. Inst. Rech. Forest., Bull. No. 8, Athens (Greek, French summary).
47. Papamikail, P. 1966. La Qualité du Bois de Clones *Populus* x *euramericana* (Dode) Guinier cv? et *Populus* x *euramericana* (Dode) Guinier cv. "I-214." Inst. Rech. Forest., Bull. No. 13, Athens. (Greek, French summary).
48. Papamikail, P. 1970. Contribution a la Connaissance de la Qualité du Bois des Pins *Pinus brutia* Tenn. et *Pinus nigra var. pallasiana*. Inst. Rech. Forest., Bull. No. 40, Athens. (Greek, French summary).
49. Parker, M. L., and R. W. Kennedy. 1973. The status of radiation densitometry for measurement of wood specific gravity. Dept. Envir. Canad., For. Serv., W. For. Prod. Lab., Vancouver.
50. Paul, B. H. 1946. The Flotation Method of Determining the Specific Gravity of Wood. U.S. Forest Products Laboratory Report 1398.
51. Pearson, F. G. O., and H. A. Fielding. 1961. Some properties of individual growth rings in European larch and Japanese larch and their influence upon specific gravity. *Holzforschung* 15(3):82-89.
52. Phillips, D. R., A. Clark, and M. A. Taras. 1976. Wood and bark properties of southern pine branches. *Wood Sci.* 8(3):164-169.
53. Polge, H. 1966. Etablissement des Courbes de Variation de la Densité du Bois par Exploration Densitométrique de Radiographies d'Echantillions Prélevés a la Tarière sur des Arbres Vivants. Applications dans les Domaines Technologique et Physiologique. Thèse, Faculté des Sciences, University of Nancy.
54. Polge, H. 1978. Fifteen years of wood radiation densitometry. *Wood Sci. Tech.* 12(3):187-196.
55. Poller, S. 1967. Über die Bedeutung der Rohdichte in der Zellstoffindustrie. *Holz Roh- Werkst.* 25(1):39.
56. Rendle, B. J., and E. W. J. Phillips. 1957. *The Effect of the Rate of Growth (Ring Width) on the Density of Softwoods*. England: Princes Risborough.
57. Sachsse, H. 1961. Anteil und Verteilung von Richtgewebe im Holz der Rotbuche. *Holz Roh- Werkst.* 19(7):253-259.
58. Schalk, J. 1967. Über die Rohdichte und Festigkeit des Schwarzenkiefernholzes (*Pinus nigra* Arn.) und den Zusammenhang zwischen Rohdichte und Holzstruktur untersucht an belgischen Aufforstungsbestanden. *Forstw. Cbl.* Heft 24. Berlin: P. Parey.
59. Schultze-Dewitz, G. 1967. Internationales Symposium: Rohdichte von Holz und Holzwerkstoffen. *Holz Roh- Werkst.* 25(1):35-39.
60. Smith, D. M. 1955. A Comparison of Two Methods for Determining the Specific Gravity of Small Samples of Second Growth Douglas-fir. U.S. For. Prod. Lab. Report No. 2033.
61. Smith, D. M. 1968. Wood Quality of Loblolly Pine after Thinning. U.S. For. Prod. Lab. Res. Paper No. 89.
62. Spurr, S. H., and W.-Y. Hsiung. 1954. Growth rate and specific gravity in conifers. *J. For.* 52(3):191-200.
63. Stamm, A. J., and H. T. Sanders. 1966. Specific gravity of wood substance of loblolly pine as affected by chemical composition. *Tappi* 49(9):397-400.
64. Svarnas, D. 1964. Untersuchungen über die Holzeigenschaften der griechischen Tanne. Dissertation, Hann-Münden.
65. Taylor, F. W. 1977. A note on the relationship between branch- and stemwood properties of selected hardwoods growing in the mid-south. *Wood Fiber* 8(4):257-261.
66. Taylor, F. W. 1979. Variation of specific gravity and fiber length in loblolly pine branches. *J. Inst. Wood Sci.* 8(4):171-175.
67. Trendelenburg, R. H., and H. Mayer-Wegelin. 1955. *Das Holz als Rohstoff*. München: Hanser Verlag.
68. Tsoumis, G. 1958. Growth, Specific Gravity and Shrinkage of Wood of Black Pine, Beech, Oak and Chestnut. Thessaloniki (Greek, English summary).
69. Tsoumis, G. 1975. Density and Shrinkage of Greek Woods (unpublished).
70. Tsoumis, G., and C. Passialis. 1977. Effect of growth rate and abnormal growth on wood substance and cell wall density. *Wood Sci. Technol.* 11:33-38.
71. Tsoumis, G., and N. Panagiotidis. 1980. Effect of growth conditions on wood quality characteristics of black pine (*Pinus nigra* Arn.). *Wood Sci. Technol.* 14:301-310.
72. Turnbull, J. M. 1948. Factors affecting wood density in pines. *S. Afr. For. J.* 16:22-43.
73. U.S. Forest Products Laboratory. 1965. Proceedings of the Symposium on "Density a Key to Wood Quality."
74. U.S. Forest Products Laboratory. 1987. Wood Handbook. Ag. Handbook No. 72 (revised).
75. Vorreiter, L. 1958. *Holztechnologisches Handbuch*. Wien: G. Fromme Verlag.

76. Wagenführ, R., and C. Scheiber. 1974. *Holzatlas.* Leipzig: VEB Fachbuchverlag.
77. Wangaard, F. F. 1950. *The Mechanical Properties of Wood.* New York: J. Wiley & Sons.
78. Wangaard, F. F. 1974. A wood scientist's view of growth and wood quality. Proceedings "Third North American Forest Biology Workshop" (C. P. P. Reid and G. H. Fechner, eds.), pp. 227–241. Colorado State University.
79. Warren, W. G. 1979. The contribution of earlywood and latewood specific gravities to overall wood specific gravity. *Wood Fiber* 11(2):127–135.
80. Yang, K. C. 1980. Variation of physical and anatomical properties of the rootwood, stemwood and branchwood of trembling aspen. 1980 FPRS Annual Meeting.
81. Yao, J. 1968. Modified mercury immersion method in determining specific gravity of small, irregular specimens. *For. Prod. J.* 18(2):56–59.
82. Zenker, R. 1967. Die Rohdichte als Bindeglied zwischen Struktur und Eigenschaften des Holzes. *Holz Roh- Werkst.* 25(1):37.
83. Zobel, B. J. 1961. Inheritance of wood properties in conifers. *Silvae Genet.* 10:65–70.

FOOTNOTES

[1] Mass and weight are not the same. Mass is the quantity of substance contained in a body, whereas weight is a measure of the force by which a body is attracted to Earth. (Weight is expressed in Newtons, dynes, or pounds.) Mass is constant, but weight is a variable quantity, because the force of gravity varies at different latitudes and elevations—and in space. However, on Earth, and for practical purposes, weight is commonly used to signify the quantity of mass.

[2] The material that constitutes the cell walls is not compact. Research has shown that dry cell walls contain a small void volume, about 1–5%. Thus, theoretically, there is (i) *density of wood*, (ii) *density of cell walls*, and (iii) *density of cell-wall material* (30).

[3] These values are average, and often there is a wide range of variation; e.g., the density of balsa can vary from 0.04 to 0.32 g/cm^3 (26) (see also Table 8-4). Other heavy tropical woods are: *Quebracho colorado* 1.13 g/cm^3, *Brya ebenus (Inga vera)* 1.15 g/cm^3, *Mimuspa balata* 1.18 (0.87–1.24) g/cm^3, *Bulnesia sarmentii* 1.23 g/cm^3, and *Piratineria guyanensis* 1.38 g/cm^3, among others (74).

[4] Degradation of wood by decay fungi, chemicals or other agents attacking cell-wall material (see Chapter 15) may drastically reduce density. The density of oak wood from an old sunken ship was reduced to only 0.13 g/cm^3 (6), whereas oak wood has normally a density of about 0.65–0.90 g/cm^3 (see Table 8-1).

[5] It is interesting to note that sewage effluent has been tried as irrigation material. In red oak, this resulted in a significant increase of density, whereas in Douglas-fir density decreased at 50% effluent (in water) and increased at 100% effluent. However, such usage may have adverse effects on the soil, after long-term filtration through it (Ref. 41 and author's unpublished data).

[6] Trees are sampled by taking specimens from cross-cut disks or extracting increment borings. It was found that borings at breast height (1.30 m, 4 ft) give satisfactory results (39, 73).

[7] Accurate determination of mass and volume is difficult due to the presence of extractives. Volatile extractives are evaporated in the oven, and water-soluble extractives are dissolved (28).

9

Hygroscopicity

Hygroscopicity is the property of wood to attract moisture from the surrounding atmosphere and to hold it in the form of liquid water or water vapor.[1] This property originates from the chemical composition of wood: cellulose, hemicelluloses, pectins, lignin, and certain extractives are hygroscopic substances. As a result of hygroscopicity, wood always contains moisture. Hygroscopicity is an important property, because the moisture held in wood affects all other properties.

MOISTURE IN LIVING TREES

The moisture of wood in living trees varies from about 30 to 300% (11, 28). This variation is influenced by different factors, such as tree species, position of wood in the tree, and season of the year.

In softwoods, heartwood has a lower moisture content than sapwood; for example, Norway spruce heartwood was found to have 40–50% moisture and sapwood 160%, Scots pine 30–40% and 100–200%, black pine 35–40% and 100–150%, and Douglas-fir 35–65% and 85–130%, respectively (11, 39). In hardwoods, the differences are not so pronounced, and sometimes the situation may be opposite (Figure 9-1). A study of 27 American softwoods has shown heartwood to have an average moisture content of 55.4% and sapwood 148.9%; whereas in 34 hardwoods the respective values were 81.4 and 82.7% (30). Differences exist not only between species (28, 30) but also between trees of the same species.

Vertical variation of moisture is also more pronounced in softwoods, where an increase was observed from the base to the top of trees.

In hardwoods, the differences are comparatively small, and there is no definite tendency (increase or decrease) in the vertical direction (Figure 9-2).

The influence of season is not clear. Higher moisture contents were observed in the winter (3, 12, 16, 28) and lower in the fall (12, 28) than in summer (3); for example, in beech, moisture content averaged 92.2% in January and 66.6% in September (28), and in poplar 120–130% in the winter and 65–70% in the summer (3, 16). However, according to some observations, moisture was higher in the fall than in the spring, and others found that seasonal differences were not important (23, 36, 37). Tree age and site quality may also affect the variation of moisture (37).

Branchwood was found to have a high moisture content (Scots pine 120%, spruce 120%, fir 70%), decreasing from a lower to a higher position in the trunk, and from the base to the top of branches (28).

In abnormal situations, such as ''wetwood'' (2, 58) and ''black heartwood'' (8, 26), there is also an indicative abnormality in moisture content (see also Chapter 15). In poplar black heartwood, moisture contents up to 254% have been measured (33).

MOISTURE IN WOOD AS A MATERIAL

Irrespective of the moisture that wood may contain in a living tree, exposure to the atmosphere results in loss of moisture. The quantity that is finally held depends on the existing atmospheric conditions. For example, under the climatic conditions of Europe, the moisture

HYGROSCOPICITY 129

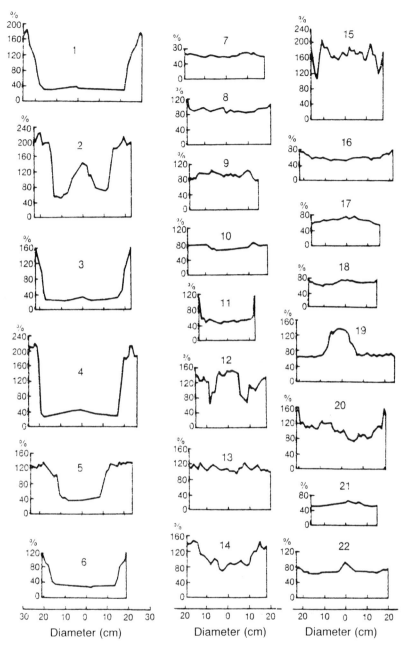

Figure 9-1. Horizontal moisture content variation of some European species at a height of 55 cm (about 2 ft) from the ground: 1, Norway spruce; 2, white (silver) fir; 3, Scots pine; 4, Balkan pine; 5, whitebark pine; 6, Douglas-fir; 7, white oak; 8, chestnut; 9, elm (field); 10, hackberry; 11, black locust; 12, willow; 13, basswood; 14, poplar (trembling); 15, poplar (black); 16, beech; 17, hornbeam; 18, maple; 19, plane (*Platanus*); 20, alder; 21, walnut; 22, birch. (1 cm = 0.4 in., 1 in. = 2.5 cm.) (Adapted from Ref. 28.)

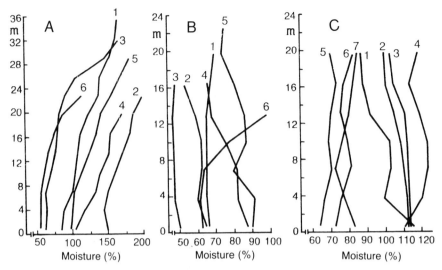

Figure 9-2. Vertical variation of moisture (1 m = 3.3 ft). (A) *Softwoods*: 1, Norway spruce; 2, white fir; 3, Scots pine; 4, black pine; 5, Balkan pine; 6, Douglas-fir. (B) *Hardwoods* (ring-porous): 1, white oak; 2, red oak; 3, ash; 4, chestnut; 5, elm; 6, black locust. (C) *Hardwoods* (diffuse-porous): 1, willow (white); 2, poplar (white); 3, poplar (trembling); 4, alder; 5, beech; 6, birch; 7, plane (*Platanus*). (Based on data from Ref. 28.)

content of wood kept under shelter (to avoid direct effects of rain, sun, or snow) is estimated to range between 6 and 26% (51). In the United States, the range is 6–24% (54), and in Greece 8–23% (48). Moisture content varies from place to place and with time (see Table 9-1).

Inversely, dry wood exposed to high atmospheric relative humidity, or in contact with liquid water, absorbs moisture; hence, its moisture content is raised. Saturation may raise it to very high values, depending on wood density. For example, the maximum moisture content of the very light tropical wood *balsa* may reach 1000%.

HOW MOISTURE IS HELD IN WOOD

Moisture is found in wood in two forms: as liquid water in cell walls, and as liquid and/or vapor in cell cavities. In saturated condition, there is only liquid water.

The basic reason for moisture entering into the mass of wood is the attraction of water molecules by the hydroxyls of its chemical constituents, mainly cellulose. As a result, a monomolecular layer of water is formed and held by these hydroxyls with strong hydrogen bonds. Formation of this layer results in pushing apart chains of cellulose molecules in the amorphous regions and between the crystallites of the microfibrils, so that wood starts to swell. Under the effect of secondary attractive forces, more water molecules enter and form a polymolecular layer (Figure 9-3). An additional part may enter by capillary condensation in cell-wall voids[2] and pit features (pit membrane openings, small pit mouths). After saturation of the walls, liquid water may also enter cell cavities (43, 56).

The above phases (monomolecular, polymolecular, capillary condensation[3]) are not clearly separated, but a distinction is made between water held in cell walls and in cell cavities: water held in cell walls is called "bound water" and that which is held in cavities "free water." The theoretical condition at which the walls are saturated but the cavities are empty is called the *fiber saturation point* (46). This is a useful term in wood utilization, as is explained later. The moisture held in wood at the fiber saturation point varies between species, but on the average it is about 30%.

DESORPTION–ADSORPTION

The quantity of moisture contained, at any one time, in wood exposed to the atmosphere is not

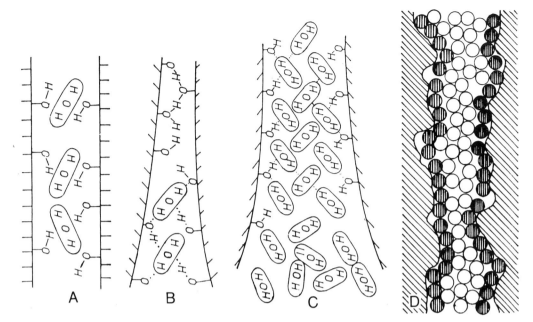

Figure 9-3. Water in wood. (A) Monomolecular layer of water held by free hydroxyls of cellulose chain molecules. (B) Gradual formation of the monomolecular layer by breaking the hydrogen bonds between adjacent cellulose molecules. (C) Polymolecular layer. (D) Schematic representation of monomolecular (dark circles) and polymolecular layer (white circles) of water. (After Refs. 5 and 21; reproduced by permission from VEB-Fachbuchverlag and Springer-Verlag.)

constant but subject to continuous change. Loss of moisture is called *desorption*, and gain is called *adsorption*.

Desorption begins by evaporation from an exposed surface of wood. If the walls are saturated, evaporation of water from the cavities needs little more energy than that required to evaporate water from a free surface (43). Reduced relative vapor pressure in the atmosphere (lower than unity) represents an attractive force holding water in the walls and, depending on its magnitude, causing lesser or greater desorption. To replace desorbed water, other water moves from the interior toward the surface of the wood. This movement is directed by expansion of air bubbles[4] contained in the cavities (Figure 9-4), and by diffusion and capillary rise. Diffusion takes place in the walls from po-

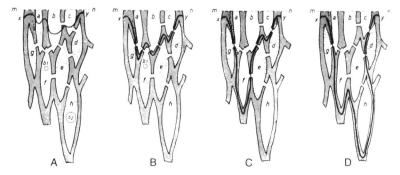

Figure 9-4. Schematic representation of the movement of water during desorption. (A) Cell cavities are full, but in f and h there are air bubbles (b_1 and b_2). Surface evaporation has reduced the level of water to the line xy, below the surface mn. (B, C, D) Gradual drop of the water level (mn) by emptying of cell cavities. Menisci formed in pits help the exit of water, which is also pushed out by air bubbles. If no bubbles exist (as in a, e, g), collapse may take place. (After Ref. 13.)

sitions of higher to positions of lower moisture content, but moisture also moves through cell cavities. If there is a difference in vapor pressure, water evaporates in the next cavity; it is held (and changed to liquid) in the next wall, and so on. Under zero relative vapor pressure, practically all moisture is lost. However, desorption is not complete, because certain hydrogen bonds are difficult to break even under vacuum and high temperature. With the standard method used to determine moisture content (placing wood in an oven at $103 \pm 2°C$), it is estimated that a small quantity of moisture, less than 0.5%, remains bonded (43).

The process of adsorption is reversed desorption. Dry wood, exposed to an atmosphere containing water vapor, adsorbs through its surface. At the beginning, a monomolecular layer is formed, then a polymolecular; and, if the relative pressure in the atmosphere is high (close to unity), capillary condensation may take place. Movement of moisture from the exterior to the interior of the wood is attained by diffusion, as previously described. Under relative vapor pressure lower than unity, no liquid water enters the cell cavities, because their diameter is large and does not allow reduction of pressure to levels needed for capillary condensation (53). However, if the relative pressure is close to unity, condensation may take place even with a small variation (fall) of temperature. In this way, liquid water enters the cavities, gradually accumulates, and moves inward through walls and adjoining cavities. When dry wood is immersed in water, there is active movement of water through cell cavities. Long immersion results in near complete saturation of the cavities.

MAXIMUM MOISTURE CONTENT

The maximum moisture that wood may contain, when both cell walls and cell cavities are saturated, depends on the space available in its mass. This space, in the form of empty cell cavities, is practically expressed by dry density; however, the space taken by water in the cell walls should be added. Maximum moisture content presents a large variation and, on the basis of density, it may be calculated, with approximation, by use of the following relationships (41):

$$Y_{max} = 100 \left(\frac{r_w - R}{r_w R} \right)$$

$$= 100 \left(\frac{1.50 - R}{1.50 R} \right)$$

$$= 100 \left(\frac{1}{R} - 0.67 \right) \quad (9\text{-}1)$$

and

$$Y_{max} = 100 \left(\frac{r_w - r_o}{r_w r_o} \right) + f$$

$$= 100 \left(\frac{1.50 - r_o}{1.50 r_o} \right)$$

$$= 100 \left(\frac{1}{r_o} - 0.67 \right) + 30 \quad (9\text{-}2)$$

where

Y_{max} = maximum moisture content (%)
R = basic density (g/cm^3)
r_o = dry density (g/cm^3)
r_w = density of wood substance (1.50 g/cm^3)
f = fiber saturation point (30%)

The above (second) equation gives maximum moisture content values of 76%, 183%, and 933% for woods having respective basic densities of 0.70 g/cm^3, 0.40 g/cm^3, and 0.10 g/cm^3. The relationship of maximum moisture content and density is shown in Figure 9.5. Maximum moisture content differs not only between species, but also in the same species, especially between heartwood and sapwood.

MOISTURE CONTENT UNDER CONSTANT ATMOSPHERIC CONDITIONS. EQUILIBRIUM MOISTURE CONTENT

Wood exposed to constant conditions of temperature and relative humidity,[5] for a sufficient time, desorbs or adsorbs moisture depending on its original hygrometric condition, and finally retains a certain quantity of moisture; this is

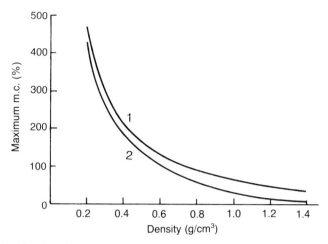
Figure 9-5. Relationship of maximum moisture content and wood density: 1, oven-dry density; 2, basic density.

called *equilibrium moisture content*. Equilibrium moisture content does not exist in an uncontrolled atmosphere, because both temperature and relative humidity are subject to continuous change. To achieve an equilibrium condition, wood (or any other hygroscopic material—e.g., cotton), must be placed in a closed container, where temperature and relative humidity can be kept constant. In practice, this is approached during kiln-drying of wood. Equilibrium moisture content is a measure of hygroscopicity, and is expressed as a percentage of the dry weight of wood.

The relationship between equilibrium moisture content and relative humidity, at a certain constant temperature, is given by a characteristic sigmoid curve. When relative humidity increases, equilibrium moisture content also increases; however, when temperature increases, equilibrium moisture content is reduced (20, 43). Differences exist between different species, but they are exhibited mainly at high relative humidities. A study of 136 different tropical and European species at a constant temperature of 25°C (77°F) and relative humidities of 30%, 60%, and 90%, has shown respective equilibrium differences of 3.3%, 5.4%, and 11.5% (9). Low values of equilibrium at high relative humidities may be attributed, as a rule, to a high content of extractives. It has been observed that removal of extractives results in an increase of equilibrium moisture content (27, 57).

Equilibrium moisture content values for various combinations of temperature and relative humidity may be estimated from Figure 9-6. The curves are based on a study of Sitka spruce (*Picea sitchensis*), but they are considered to be valid for most wood species. Important differences are found only at very high relative humidities.

Equilibrium moisture content also shows differences when wood loses moisture for the first time, or adsorbs moisture after drying, or desorbs moisture which has been previously adsorbed. Equilibrium is greater in desorption than in adsorption (Figure 9-7). The phenomenon is called *hysteresis*, and is a characteristic property of all cellulosic materials. It should be noted that, after the initial desorption of green wood, its hygroscopicity is permanently reduced at high relative humidities. In Figure 9-7, curve *a* is not repeated, in contrast to curves *b* and *c*.

In relatively large wood specimens, where achievement of equilibrium is practically difficult, and when control of conditions is not accurate, desorption and adsorption take place about simultaneously. The equilibrium of wood that has been rewetted follows an intermediate curve, between desorption and adsorption, called an "oscillating sorption" curve.

The magnitude of hysteresis may be expressed by the ratio of adsorption to desorption equilibrium moisture contents at the same relative humidity. The ratio is practically con-

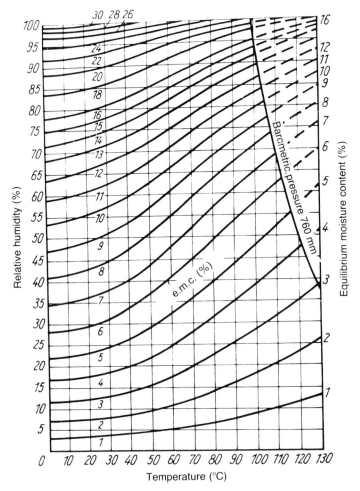

Figure 9-6. Isohygric curves: dependence of equilibrium moisture content on temperature and relative humidity. (After Refs. 19, 20 and 45; reproduced by permission from Springer Verlag.)

stant, on the average about 0.80 (0.74–0.88) (43, 53).

Various theories have been proposed to explain moisture hysteresis. A simple explanation is that the phenomenon is due to linkage of free hydroxyls of the wood constituents when there is no moisture or very little moisture in wood. Thus, during the adsorption that follows, the number of available hydroxyls is smaller (39, 43).

It may be concluded that equilibrium moisture content is affected by wood species, especially at high relative humidities, and by hysteresis; however, these effects are small and practically unimportant. Also, equilibrium is independent of the condition of the surface of wood (rough, smooth, painted, etc.), air velocity, and growth-ring orientation (transverse, radial, tangential). These factors do not affect the final equilibrium, although they do affect the rate of moisture exchange between wood and atmosphere.

FIBER SATURATION POINT

The term fiber saturation point has been previously defined; it represents the maximum moisture that wood may adsorb from atmospheric vapors assuming that no condensation occurs. The value of fiber saturation point may be found by extrapolation of adsorption curves determined at a constant temperature, and relative humidities reaching close to 100% (98% and up).[6]

Equilibrium moisture-content differences between species, at high relative humidities, are

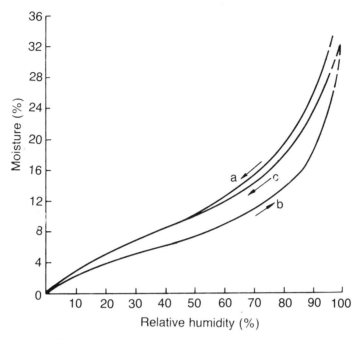

Figure 9-7. Moisture sorption isotherms for basswood (32°C, 90°F): *a*, initial desorption; *b*, adsorption; *c*, desorption following drying and saturation. (After Ref. 42)

expressed by differences in fiber saturation points. The moisture that different species retain at the fiber saturation point varies approximately between 20 and 40% (38, 40, 44, 47, 57). (There are higher and lower values—see Figure 9-8—but for practical purposes, the moisture content at that point is taken to be 30%.) Differences may be observed in the same species, depending on method of determination and other factors, such as extractive content (heartwood, sapwood), earlywood and latewood, compression and tension wood (1), density and temperature. The presence of extractives reduces the fiber saturation point (the opposite has also been observed) (46, 57); rise of temperature also has a reducing effect, and woods of higher density show a lower fiber saturation point. The latter is, at least partially, caused by the bulking effects of a high extractive content in most such woods (40).

The concept of fiber saturation point is useful from a practical point of view, because most properties change when the moisture content of wood is below this point. This relationship allows experimental determination of the fiber saturation point by measurement of a certain property (shrinkage, mechanical property, electric resistance, etc.) (22, 43, 55) at successively higher or lower moisture contents, until this property stops or starts to change (see Figures 10-2 and 11-5). However, the fiber saturation point cannot always be measured with accuracy, and for this reason the term "region of fiber saturation" is also used.

MOISTURE UNDER VARIABLE ATMOSPHERIC CONDITIONS

It is apparent from the foregoing discussion, that the moisture content of wood does not remain constant during storage, transportation, or use; it is subject to change, because the atmospheric conditions of temperature and relative humidity are continuously changing from hour to hour, day to day, month to month, and so on. Aside from temperature and relative humidity, a third meteorological factor, air velocity, may exercise a considerable influence, because it affects the rate of moisture evaporation.

The moisture that wood holds at a certain moment depends on species, source (heartwood, sapwood), thickness, surface condition, and direction of moisture movement. These factors affect the rate of moisture movement in the mass of wood. Higher density, greater

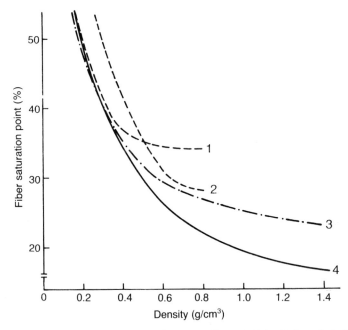

Figure 9-8. Dependence of fiber saturation point on density. (According to 1. Feist-Tarkow, 2. Kellogg-Wangaard, 3. Cudinov, 4. Vorreiter; after Refs. 7 and 40.)

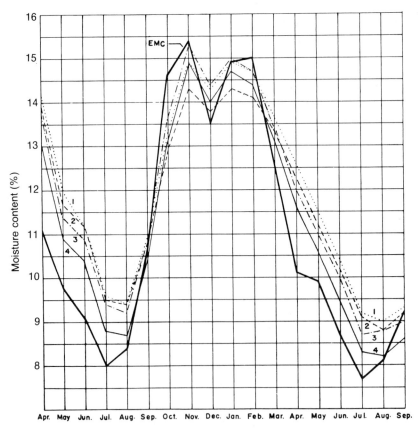

Figure 9-9. Relationship of actual mean monthly values of wood moisture to theoretical emc values for a period of 18 months: 1, beech; 2, white oak; 3, chestnut; 4, black pine (specimen thickness 2.5 cm or 1 in.).

thickness, and a nonhygroscopic cover (e.g., paint) have a retarding influence; whereas movement of moisture through transverse surfaces (parallel to grain) is 10–15 times faster in comparison to movement from radial or tangential surfaces (43).[7]

In spite of the above, the values of moisture content of wood exposed to variable atmospheric conditions, under shelter, are within certain limits, which may be estimated. These values cannot be greater than the maximum or lower than the minimum values of equilibrium moisture content determined by the existing temperature and relative humidity. Experiments with beech, oak, chestnut, and pine specimens (1 cm, 2.5 cm, and 5 cm thick) have shown that their actual mean monthly values did not differ significantly from respective theoretical values of equilibrium moisture content (Figure 9-9); the differences increased with thickness, but all differences, within a yearly period of observation, ranged from zero to 2.8% (49, 50).

These findings permit the conclusion that the probable values of moisture content of air-dry wood, exposed to variable atmospheric conditions under shelter, may be calculated as equilibrium moisture content values. Such values, based on mean monthly meteorological data of temperature and relative humidity, and the isohygric curves of Figure 9-6, are useful in establishing the limits of yearly (or other) variations of moisture content at a certain time in different locations. For example, a study based on data from 238 meteorological stations in Europe (51) has shown that the moisture content of wood ranges in July from 6.0 to 24.9% and in January from 11.0 to 26.8%. Isohygric curves for January are shown in Figure 9-10 (see also Table 9-1).

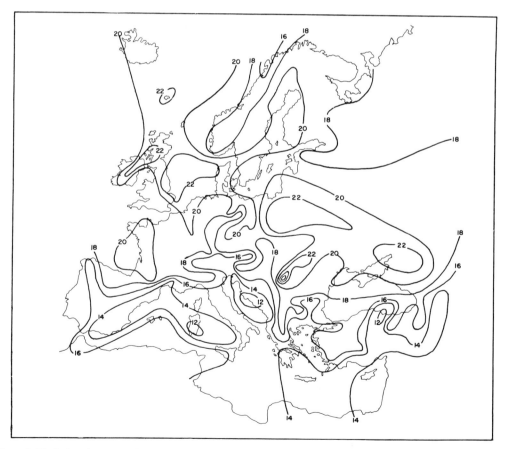

Figure 9-10. Isohygric curves of moisture content variation of air-dry wood under shelter in Europe for the month of January. They are based on equilibrium moisture content values determined from mean monthly values of temperature and relative humidity data taken from 238 meteorological stations for a 6-year period.

138 II / PROPERTIES

Table 9-1. Moisture Content Variation of Air-Dry Wood (Under Shelter, in Various Countries)[a]

Country	No.[b]	mc (%)	Country	No.[b]	mc (%)
Austria	7	12.6–22.5	Malta	1	12.2–15.2
Belgium	1	16.8–20.0	Netherlands	1	15.9–17.9
Bulgaria	3	11.9–17.5	Norway	6	12.8–19.8
Czechoslovakia	7	13.3–21.8	Poland	17	13.8–22.7
Cyprus	1	8.5–15.9	Portugal	1	11.8–17.5
Denmark	3	13.1–21.9	Romania	7	11.6–26.0
Finland	6	12.7–21.2	Spain	10	7.5–19.3
France	14	10.5–20.3	Sweden	9	12.5–20.5
Germany (W/E)	24	12.2–21.8	Switzerland	4	12.2–19.0
Greece	8	8.3–17.8	Turkey	23	6.0–17.7
Hungary	5	11.5–19.4	Yugoslavia	5	9.0–24.4
Iceland	2	16.8–20.3	U.K.	4	13.0–24.4
Ireland	3	16.5–21.4	U.S.S.R.	23	10.0–24.9
Italy	10	10.7–20.0			
Luxemburg	1	14.8–20.7	Africa, N.	9	6.1–18.0

[a] Equilibrium moisture content values based on mean monthly air temperature and relative humidity recorded in meteorological stations for the months of July (low values) and January (high values) (52). Yearly ranges according to other sources are as follows: Germany 13–22%, Greece 8–23%, U.S. 6–24%, Yugoslavia 8.7–26.3% (Refs. 20, 48, 49, 51, 54).
[b] Number of meteorological stations.

The moisture content of wood in heated rooms may be estimated in the same manner. For example, under the assumption that temperature is kept at 22°C (71.6°F) and relative humidity at 40%, the minimum moisture of the wood of flooring, furniture, or other interior woodwork will be about 7.5%. For other combinations of temperature and relative humidity, moisture content values may be determined from Figure 9-6. Figure 9-11 shows average moisture contents for interior use of wood in various areas of the United States.

Figure 9-11. Recommended average moisture content for interior use of wood in various areas of the United States. (Courtesy of U.S. Forest Products Laboratory.)

DETERMINATION OF MOISTURE CONTENT

The moisture content of wood may be found by different methods, which may be distinguished into direct and indirect methods. Direct methods involve separation of moisture from the wood, whereas in indirect determination a property related to moisture is measured. Main direct methods are (i) drying and weighing and (ii) distillation. Indirectly, the moisture is mainly measured by (iii) electric moisture meters (10, 14, 34, 40).

Drying and Weighing

According to this method, moisture content is determined from the relationship:

$$Y = \frac{M_x - M_o}{M_o} \times 100 \qquad (9\text{-}3)$$

or

$$Y = \left(\frac{M_x}{M_o} - 1\right) \times 100 \qquad (9\text{-}4)$$

where

Y = moisture content (%)
M_x = initial mass (weight) of wood (g)
M_o = oven-dry mass (g)

The instruments needed are an accurate balance, and an aerated oven to keep a constant temperature. The size and shape of wood specimens may vary. For example, in sampling lumber, the specimen has a thickness of 1–1.5 cm (about 0.5 in.) and is taken at least 30 cm (1 ft) from one end; if desired to find how moisture is distributed within the mass of wood, the specimen is prepared accordingly (see Chapter 18). In other products (poles, ties, furniture, etc.), specimens are taken with a borer. Irrespective of the sampling method, it is necessary to use sharp tools; otherwise, moisture will evaporate during cutting. Also, knots should be avoided, and wood splinters caused by sawing should be removed with sandpaper.

The specimen is weighed as soon as it is prepared, and then placed in the oven (at 103 ± 2°C) until its weight becomes constant. The required time varies depending on specimen size, wood species, number of specimens in the oven, and aeration of the oven. It is usually sufficient for the specimen to remain overnight, but experience will soon solve this problem. After removal from the oven, the specimen is not weighed immediately but is placed in a desiccator, above $CaCl_2$ or P_2O_5, until its temperature is reduced (Figure 9-12). If the specimen is weighed hot, the air above it is heated, becomes lighter, and the weight that is indicated on the balance is false—smaller than actual.[8]

Distillation

This method is applied with woods containing extractives (resins, oils, etc.) or treated with preservatives (18, 31). These substances evaporate in the oven; therefore, use of the previous method will not give accurate results.

The specimen is taken as in the previous method and reduced to small pieces (smaller than a match stick); sawdust may also be used. A representative specimen, 20–50 g in weight, is immediately weighed on an accurate balance (M_x) and then placed in a glass container with a liquid immiscible to water, such as toluol or xylol, in a saturated aqueous solution. Distillation is carried out in a special laboratory apparatus and lasts about 1.5 h. The moisture (water) of the specimen is collected in a graduated tube (Figure 9-13) and, as it is known, its volume in cm^3 measures its weight in grams (g). The oven-dry weight of the wood (M_o) is found by subtracting the weight of the water collected in the graduated tube from the initial weight (M_x).

Electric Moisture Meters

Measurement of moisture with electric moisture meters is based on the change of the electric properties of wood with change of its moisture (see Chapter 14). There are two types of instruments based on (a) electric resistance to passage of direct current and (b) dielectric properties of wood in a high-frequency electrical field. The second type includes two meters: (i) electric capacitance (based on the relation-

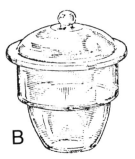

Figure 9-12. (A) Oven and (B) desiccator.

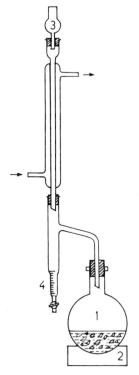

Figure 9-13. Determination of moisture content by distillation: 1, container with wood particles in toluol; 2, heating; 3, drying agent; 4, graduated tube to collect moisture (water). Arrows show entrance and exit of cooling water.

ship between moisture content and dielectric constant), and (ii) power loss of high (radio) frequency current (based on the relationship between moisture content and loss factor). Resistance and power loss are the two most widely used types of electric moisture meters. Resistance meters (Figure 9-14) are preferred; these

Figure 9-14. Electric moisture meter (resistance type). (Courtesy of Lignomat.)

instruments are more accurate due to the sensitivity of electrical resistance to moisture, and they are less affected by the density of the wood (25, 43). Measurements of moisture with electric meters presents advantages and disadvantages in comparison to the previous methods (10):

Advantages
1. Faster measurement of moisture.
2. Easy measurement of many specimens (or positions).
3. Nondestructive measurement.
4. Light and portable instruments[9] (except capacitance meters).
5. Cost of measurement small.

Disadvantages
1. Relatively smaller accuracy.
2. Measurements above the fiber saturation point are not accurate.
3. Resistance meters are not accurate for moistures lower than 6%.[10]
4. Measurements are affected by various factors, such as wood species, density, temperature of wood, and atmospheric conditions.

The disadvantages of electric moisture meters may be controlled by proper use, according to the instructions provided by the manufacturer; knowledge of the factors that affect the accuracy of readings is also necessary. Proper use may give an accuracy of ± 1–2% in comparison to weighing and drying; otherwise, the differences may be quite large (10, 14, 17, 32).

Factors affecting the electric properties of wood are the basic factors affecting the accuracy of electric moisture meters. In addition to the above, other factors that may affect the readings of a meter are grain direction, moisture distribution in the tested wood, included chemicals (preservatives, etc.), thickness of wood, type and contact of electrodes, and care of the operator (14, 15, 32).

The effect of species is due to variation of anatomical structure and density, and requires a respective correction of the instrument. Density is mainly effective with high-frequency meters. Temperature is important due to its relation to electric resistance (rising temperature causes a reduction of resistance; see Chapter 14). Grain direction is also important for resistance meters, and mainly for moisture contents higher than 20%; the electric current should flow parallel to the grain (14).

High surface moisture (due to rain, dew, etc.) gives high readings. Unequal distribution of moisture cannot be measured, and the readings are different from an average value. This is the case especially with resistance meters, because they measure the maximum moisture content at the contact points of the electrodes—and the region of maximum moisture is also the path of minimum electrical resistance (see Chapter 14). Unequal distribution of moisture creates problems with power-loss meters, too (see below). Wood treated with preservative salts or fire retardants has lower resistance and higher dielectric constant and loss factor than untreated wood of the same moisture content and, as a result, meter readings are too high (14).

The thickness of tested wood is not important if the distribution on moisture is uniform; however, there is no uniformity during drying of green or wet wood. In order to have readings that are close to an average moisture content, the electrodes of resistance meters should enter to a depth of $\frac{1}{4}$–$\frac{1}{5}$ of wood thickness (14, 25). The electrodes of such meters have the form of needles that may vary in length allowing varying depths of penetration. In high-frequency-type meters, the electrodes make contact only with the surface of the wood; the depth of penetration depends on the design of the electrode; however, there are problems of penetration of the electric field—and in the (usual) case of nonuniform distribution (gradient) of moisture, readings are affected by the condition of an outer shell (25). In general, the electrodes should not be placed close to the edges of tested wood, but at a distance at least equal to its thickness.

The importance of proper handling is obvious. Electric moisture meters have important advantages, but these are realized only by proper use. In addition to previous remarks, attention should be given to the following items: replacement of batteries in due time, clean electrodes, checking the instrument prior to use, proper correction of readings (according to manufacturer's instructions or comparative measurements by other methods), correct reading, and careful handling to avoid damaging the mechanism of the instrument (14, 15, 34).[11]

In general, and irrespective of the method of moisture content determination, it is necessary to make proper selection of specimens (or measurement positions) and sufficient number of measurements. In addition, the specimens should be representative; for example, measurement of specimens, which are more exposed to heat, moisture, or aeration, leads to conclusions that are not accurate and should not be generalized.

IMPORTANCE OF HYGROSCOPICITY

Hygroscopicity is an important property of wood. It affects all other basic wood properties (density, shrinkage, mechanical properties, etc.), as explained in other chapters. In addition, it is important in industrial processing of wood, such as machining, drying, preservative treatment, gluing, finishing, chemical processing, and others.

Hygroscopicity is a disadvantage of wood as a material from a dimensional stability viewpoint. Sound knowledge of the relationships of moisture to properties and processing is necessary for a rational wood utilization, because such knowledge allows application of proper measures that help to avoid adverse consequences in practice.

REFERENCES

1. Ahlgren, P. A., J. R. Wood, and D. A. I. Goring. 1972. The fiber saturation point of various morphological subdivisions of Douglas-fir and aspen. *Wood Sci. Technol.* 6(3):81–84.
2. Bauch, J. 1986. Characteristics and response of wood in declining trees from forests affected by pollution. *IAWA Bull.* n.s. 7(4):269–276.
3. Bendtsen, B. A., and L. W. Rees. 1962. Water-content variation in the standing aspen tree. *For. Prod. J.* 12(9):426–428.
4. Blankenstein, C. 1962. *Holztechnologisches Taschenbuch.* München: Hanser Verlag.
5. Blazej, A., L. Suty, and P. Krkoska. 1979. *Chemie des Holzes.* Leipzig: VEB Fachbuchverlag.
6. Bramhall, G., J. F. G. Mackay, and M. Salamon. 1980. Kiln Drying of Lumber: Fifteen Questions and Answers. West. For. Prod. Lab., Sp. Publ. No. SP-3R, VOT 1X2, Canada.
7. Cudinov, B. S. 1981. Zum Einfluss der Holzdichte auf der Fasersättigungspunkt. *Holztechnologie* 22(2):104–108.
8. F. A. O./U. N. 1956. Les Peupliers dans la Production du Bois et l' Utilisation des Terres. Rome.
9. Forest Products Research Laboratory, England. 1965. The Movement of Timbers. Leaflet No. 47. Princes Risborough, U.K.
10. Forest Products Research Laboratory, England. 1972. Report of a Seminar on Moisture Content Determination of Wood. Timberlab Papers No. 24–1970, Princes Risborough, U.K.
11. Giordano, G. 1971. *Tecnologia del Legno*, Vol. 1. Torino: Unione Tipografico-Editrice.
12. Gläser, H. 1955. *Die Ernte des Holzes.* Wirtschafts- u. Forstverlag Euting KG, Newwied.
13. Hawley, L. F. 1931. Wood-Liquid Relations, U.S.D.A. Bull. No. 248.
14. James, W. L. 1963. Electric moisture Meters for Wood. U.S. For. Prod. Lab. Res. Note FP1-08.
15. James, W. L. 1966. Effects of Wood Preservatives on Electric Moisture Meter Readings. U.S. For. Prod. Lab. Res. Note FPL-0106.
16. Jensen, R. A., and J. R. Davis. 1953. Seasonal Moisture Variations in Aspen. Minn. For. Notes No. 19.
17. Johnston, D. D., and R. H. Wynands. 1958. Determination of moisture in timber—A comparison of electrical resistance and oven–drying methods. *Wood* 23:458–462.
18. Keith, C. T. 1969. Resin content of red pine wood and its effect on specific gravity determinations. *For. Chron.* 45(5):1–6.
19. Keylwerth, R. 1949. *Holz-Zentr.* 75, No. 175 (23.9.49).
20. Kollmann, F. 1951. *Technologie des Holzes und der Holzwerkstoffe.* Berlin: Springer Verlag.
21. Kröll, K. 1951. Die Bewegung der Feuchtigkeit in Nadelholz während der Trocknung bei Temperaturen um 100°C. *Holz Roh- Werkst.* 9:176–181.
22. Krpan, J. 1954. Untersuchungen über den Fasersättigungspunkt des Buchen-, Eichen-, Tannen- und Fichtenholzes. *Holz Roh- Werkst.* 12(3):84–91.
23. Linzon, S. N. 1969. Seasonal water content and distribution in eastern white pine. *For. Chron.* 45(1):1–6.
24. Loos, W. E. 1965. Determining moisture content and density of wood by nuclear radiation techniques. *For. prod. J.* 15(3):102–106.
25. Mackay, J. F. G. 1976. Effect of moisture gradients on the accuracy of power-loss moisture meters. *For. Prod. J.* 26(3):49–52.
26. Moulopoulos, C., and G. Tsoumis. 1960. Growth abnormalities in Euramerican hybrid poplars cultivated in Macedonia-Greece. Annals School Agr. For. 5:193–

237, Aristotelian University (Greek, English summary).
27. Nearn, W. T. 1955. Effect of Water-Soluble Extractives on the Volumetric Shrinkage and Equilibrium Moisture Content of Eleven Tropical and Domestic Woods. Penn. State Ag. Expt. Station Bull. 598.
28. Nikolov, C., and E. Enchev. 1967. *Moisture Content of Green Wood.* Sofia, Bulgaria.
29. Noack, D., and W. Kleuters. 1960. Über die Bestimmung des Holzfeuchtikeitgehaltes mit Hilfe radioaktiver Isotope (b-Strahlen). *Holz Roh- Werkst.* 18(8):304-306.
30. Peck, E. C. 1953. The Sap or Moisture in Wood. U.S. For. Prod. Lab. Report No. 768.
31. Resch, H., and B. A. Ecklund. 1963. Moisture content determination for wood with highly volatile constituents. *For. Prod. J.* 13(2):481-482.
32. Rijdijk, J. F. 1969. Die Genauigkeit von Holzfeuchtigkeitmessungen mit elektrischen Feuchtigkeitgeräten. *Holz Roh- Werkst.* 27(1):17-23.
33. Roosen, P. 1955. La Teneur en Eau des Tiges de Peupliers Euramericains en Belgique. Lab. For. de l' Etat, Gembloux.
34. Salamon, M. 1971. Portable Electric Moisture Meters for Quality Control. For. Prod. Lab. Canada, Information Report VP-X-80.
35. Samborski, M. R., G. Tsoumis, and F. F. Wangaard. 1953. Moisture Absorption in Certain Tropical-American Woods. Yale Forestry School, Tech. Report No. 8.
36. Sauter, J. J. 1966. Über die jahresperiodischen Wassergehaltsänderungen und Wasserverschiebungen im Kern- und Splintholz von *Populus. Holzforschung* 20(5):137-142.
37. Schröder, J. G., and D. R. Philips. 1972. Seasonal moisture content of loblolly and slash pine. *For. Prod. J.* 22(4):54-56.
38. Siau, J. F. 1984. *Transport Processes in Wood.* Berlin, New York: Springer Verlag.
39. Skaar, C. 1979. Moisture sorption hysteresis in wood. *Symposium on Wood Moisture Content-Temperature and Humidity Relationships*, pp. 23-35. U.S. For. Prod. Laboratory.
40. Skaar, C. 1988. *Wood-Water Relationships.* Berlin, New York: Springer Verlag.
41. Smith, D. M. 1954. Maximum Moisture Content Method for Determining Specific Gravity of Small Wood Samples. U.S. For. Prod. Lab. Report No. 2014.
42. Spalt, H. A. 1958. The fundamentals of water sorption by wood. *For. Prod. J.* 8:288-295.
43. Stamm, A. J. 1964. *Wood and Cellulose Science.* New York: The Ronald Press.
44. Stamm, A. J. 1971. Review of nine methods for determining the fiber saturation points of wood and wood products. *Wood Sci.* 4(2):114-128.
45. Stamm, A. J. and W. K. Loughborough. 1935. Thermodynamics of the Swelling of Wood. U.S. Forest Products Laboratory Report (Published in the *Journal of Physical Chemistry*, Vol. 39, No. 1, January 1935.)
46. Tiemann, D. H. 1947. *Wood Technology.* New York: Pitmann.
47. Trendelenburg, R., and H. Mayer-Wegelin. 1955. *Das Holz als Rohstoff.* München: Hanser Verlag.
48. Tsoumis, G. 1955. The moisture of wood under the climatic conditions of Greece. *Technica Chronica* 371/372:158-163 (Greek, English summary)
49. Tsoumis, G. 1957. Moisture content variation of seasoned wood and its relationship to equilibrium moisture content. Dissertation, Yale Forestry School.
50. Tsoumis, G. 1960. Untersuchungen über die Schwankungen des Feuchtigkeitsgehaltes von Lufttrockenem Holz. *Holz Roh- Werkst.* 18:415-422 (English translation, No. 6791/1964, For. Prod. Lab., Australia).
51. Tsoumis, G. 1964. Estimated moisture content of airdry wood exposed to the atmosphere under shelter, especially in Europe. *Holzforschung* 18:76-81.
52. U.S. Department of Commerce, National Climatic Data Center. Monthly Climatic Data for The World (1955-1960). Ashville, North Carolina.
53. U.S. Forest Products Laboratory. 1946. Kiln Certification. A.N.C. Bull. 21.
54. U.S. Forest Products Laboratory. 1987. Wood Handbook. U.S.D.A. Ag. Handbook No. 72. (Revised)
55. Wangaard, F. F. 1957. A new approach to the determination of fiber saturation point from mechanical tests. *For. Prod. J.* 7(11):410-416.
56. Wangaard, F. F. 1979. The hygroscopic nature of wood. In *Symposium on Wood Moisture Content—Temperature and Humidity Relationship.* U.S. For. Prod. Lab.
57. Wangaard, F. F. and L. A. Granados. 1967. The effect of extractives on water-vapor sorption by wood. *Wood Sci. Technol.* 1:253-277.
58. Ward, J. C. 1980. Bacteriological, Chemical and Physical Properties of Wetwood in Living Trees. U.S. For. Prod. Lab.
59. Wengert, E. M., and P. H. Mitchell. 1979. Phsychrometric relationships and equilibrium moisture content of wood at temperatures below 212°F (100°C). In *Symposium on Wood Moisture Content—Temperature and Humidity Relationships*, pp. 4-11. U.S. For. Prod. Lab.

FOOTNOTES

[1] Wood may attract and hold other liquids and gasses, but water has a greater practical importance. The terms "moisture" and "water" are used here without distinction.

[2] Void spaces, invisible even with an electron microscope, exist in cell walls, but their proportion is small, about 1-5% (see previous chapter).

[3] Condensation in capillaries occurs at a lower relative vapor pressure (relative humidity) in comparison to condensation of large volumes of water vapor. In addition, there is a rise of water in capillaries due to the phenomena of cohesion (between water molecules) and adhesion (between water molecules and wood substance), and to surface tension exercised by menisci; the smaller the diameter

of capillaries, the higher the rise of water and the greater the curvature (and tension) of the menisci.

⁴Bubbles expand under the influence of high atmospheric temperature, or tension exercised by menisci. In the wood of living trees, cell cavities contain air and only in certain cases (e.g., in water-conducting sapwood, and in abnormal "wetwood" and "black heartwood"—see Chapter 15) they may be filled with water. In saturated wood, air bubbles may form when the rate of desorption is slow; the fast exit of water, such as in kiln-drying, may cause "collapse" by action of the tension of menisci (43) (see Chapter 10).

⁵Relative humidity is the ratio of the mass of water vapor contained in a certain volume of space to the maximum possible at the existing temperature ($\times$ 100). For example, relative humidity 25% means that the air contains $\frac{1}{4}$ of the maximum possible vapor, and 100% that the air is saturated. Absolute humidity is a measure of the mass of vapor in a unit of volume. Related to relative humidity is the term relative vapor pressure. This is the ratio of the pressure of the vapor contained in a certain volume of space at a certain temperature, to saturated vapor pressure at the same temperature. Relative vapor pressure multiplied by 100 is equal to relative humidity.

⁶The control of very high relative humidities is difficult due to surface condensation of vapors, with the result that liquid water may gradually enter cell cavities.

⁷The same factors affect the movement of moisture when wood is immersed in water. Different species differ in the rate of adsorption, and this may have practical importance in their utilization (35).

⁸If wood is kept too long in an oven, its weight decreases due to oxidation and loss of "water of constitution" (i.e., water that participates in the composition of cell-wall material); this results in a gradual chemical decomposition.

⁹In addition to portable (hand-held) meters, electric meters are sometimes installed "on-line" to monitor the changing moisture of lumber being dried in a kiln (see Chapter 18).

¹⁰Power loss meters are more accurate at low moisture contents (down to 1%) (6). The upper limit of accurate measurement with electric moisture meters is 25%.

¹¹In addition to electric moisture meters, moisture may be measured indirectly by other methods, such as radiation, but these are of no practical importance (10, 24, 29, 40).

10

Shrinkage and Swelling

Shrinkage is reduction, and *swelling* is an increase of the dimensions of wood due to changes of its moisture content. Such dimensional changes occur when the moisture of wood fluctuates below the fiber saturation point; changes of moisture above this point, irrespective of their magnitude, have no effect on dimensions.

Wood is anisotropic with regard to shrinkage and swelling (i.e., the reduction or increase of its dimensions, for the same change of moisture content, is different in different directions of tree growth). Specifically, the change of dimensions is least in the longitudinal direction (along the tree trunk), much greater in the radial direction (from pith to bark), and still greater in a direction tangential to growth rings (Figure 10-1 and Table 10-1).

FACTORS AFFECTING SHRINKAGE AND SWELLING

Shrinkage and swelling of wood are affected by many factors, such as moisture, density, anatomical structure, extractives, chemical composition, and mechanical stress. Density, structure, extractives, and (to a very limited extent) chemical composition affect differences between wood species.

Moisture

The magnitude of shrinkage and swelling is affected by the amount of moisture, which is lost or gained by wood when its moisture fluctuates between zero and the fiber saturation point, and vice versa (Figures 10-2 and 10-3). The relationship is practically linear[1] and applied to all growth directions (longitudinal, radial, tangential) and, therefore, to volumetric changes.

It has been observed that the relationship between shrinkage or swelling and moisture change is affected by the size of the specimen used for their measurement (1, 5, 44). Large specimens do not give consistent results due to lack of uniformity of moisture content distribution. Thus, internal stresses develop which affect the change of dimensions, and this may be the reason why, in certain cases, shrinkage and swelling is observed beyond the fiber saturation point (Figure 10-2). For this reason, measurements of shrinkage and swelling should be made with small, standard size specimens (Table 10-2) and procedures for the results to be comparable.

Density

The magnitude of shrinkage and swelling is higher with higher density (i.e., woods of high density shrink and swell more) (5, 40, 42) (Figures 10-2 and 10-4A). This is due to the larger amount of wood substance (greater cell-wall thickness) in woods of higher density, and to the exterior change of cell dimensions. It has been observed that, when moisture is lost or gained, the size of the cell cavity remains practically unchanged (35, 37, 40, 44).

The relationship of density and shrinkage or swelling is directly connected to the relationship between moisture and shrinkage or swelling, because woods of higher density contain more moisture in their cell walls. For example, at the same moisture, say 15%, 1 m^3 (35 ft^3)

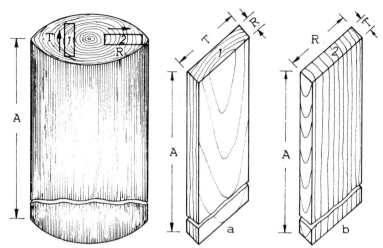

Figure 10-1. Axial (*A*), tangential (*T*), and radial (*R*) directions in round wood and lumber. See the different growth-ring arrangement and figure in *a* tangential (flat-sawn) and *b* radial (quarter-sawn) lumber.

Table 10-1. Shrinkage (%)[a]

Direction	A	B	C
Axial	0.4 (0.1–0.6)	—	—
Radial	4.3 (2.3–6.8)	3.7 (2.1–5.1)	4.9 (3.0–7.9)
Tangential	8.2 (6.0–11.8)	6.8 (4.4–9.1)	8.9 (6.2–12.7)
Volumetric	12.9 (8.5–18.8)	10.6 (6.8–14.0)	14.2 (10.2–19.2)

[a]Mean values and variation (in parentheses). A, European species (softwoods and hardwoods); B and C, North American (USA) species; B, softwoods; C, hardwoods.
Source: Refs. 16 and 45.

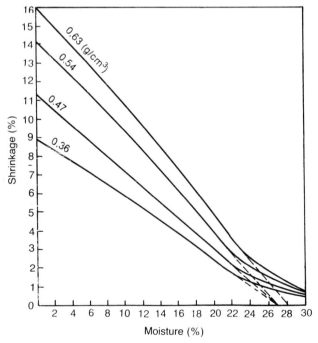

Figure 10-2. Relationships of volumetric shrinkage and moisture content of pine wood of varying density. (After Ref. 40).

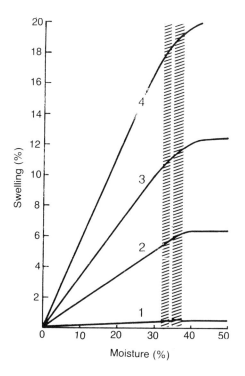

Figure 10-3. Relationships of swelling and moisture content in beech wood: 1, axial; 2, radial; 3, tangential; 4, volumetric swelling. The striated area is the "region" of fiber saturation point. (After Ref. 19; reproduced by permission from DRW-Verlag Weinbrenner.)

of a 0.80 g/cm^3 density wood contains 120 kg (260 lb) of water; whereas a 0.40 g/cm^3 density wood contains only 60 kg (130 lb) water.

Density affects the coefficient of anisotropy of shrinkage or swelling. The coefficient (ratio of tangential to radial shrinkage or swelling) becomes smaller with increasing density (5, 17) (Figure 10-4B). This means that in woods of higher density, the difference between tangential and radial shrinkage or swelling is smaller (i.e., such woods have a lower degree of anisotropy). The coefficient of shrinkage anisotropy increases with increasing moisture content of wood (18).

The relationship between shrinkage and density was observed to be absent in very old woods conserved in soil, sea, or glaciers for thousands of years. For example, in oak conserved for about 1000 years in the sea (sunken ship), density was reduced to 0.13 g/cm^3 and tangential shrinkage increased to 62.2%. As is suggested by marked reduction of density, such woods show deterioration of structure and changes in chemical composition (4).

Structure

The anatomical structure of wood is the basic reason for anisotropic shrinkage and swelling.

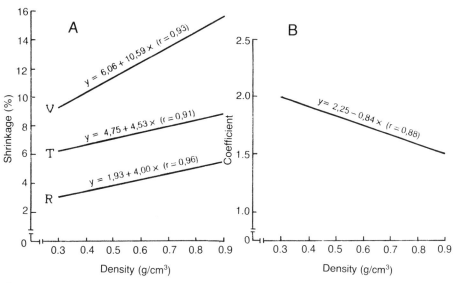

Figure 10-4. Relationships of (A) shrinkage (y) and density (x), and (B) coefficient of shrinkage anisotropy (tangential/radial shrinkage, y) and density (x). Based on 88 European, North American, and tropical species.

Table 10-2. Dimensions of Standard Specimens for Determination of Physical and Mechanical Properties of Wood in Various Countries (and proposed ISO standards)[a]

Test	USA[b] A	USA[b] B	Germany	France	U.K.	ISO
Prismatic specimens						
Density	5 × 5 × 15	5 × 5 × 15	—	2 × 2 × 2	—	2 × 2 × 2-3
Shrinkage, volume	5 × 5 × 15	5 × 5 × 15	—	2 × 2 × 2	—	—
Shrinkage, radial–tangential	2.5 × 2.5 × 10	2.5 × 2.5 × 10	3 × 3 × 1.5	—	—	—
Shrinkage, axial	—	—	3 × 3 × 10	—	—	—
Compression, axial	5 × 5 × 20	2.5 × 2.5 × 10	2 × 2 × 3[c]	2 × 2 × 6	2 × 2 × 6	2 × 2 × 3-6
Compression, transverse	5 × 5 × 15	5 × 5 × 15	5 × 5 × 15	—	—	—
Bending, static	5 × 5 × 76	2.5 × 2.5 × 41	2 × 2 × 36[d]	2 × 2 × 34	2 × 2 × 30	2 × 2 × 30-38
Bending, dynamic	5 × 5 × 76	5 × 5 × 76	—	—	2 × 2 × 30	2 × 2 × 30
Toughness	2 × 2 × 28	2 × 2 × 28	—	2 × 2 × 30	—	—
Hardness	5 × 5 × 15	5 × 5 × 15	—	minimum height 3 cm	—	5 × 5 × 15
Special Shape						
Tension, axial	2.5 × 2.5 × 45	—	5 × 1.5 × 45	—	—	(see ISO 3347)
Tension, transverse	5 × 5 × 6.3	—	—	2 × 2 × 7	—	(see ISO 3346)
Shear	5 × 5 × 6.3	—	3 × 6 × 8	—	2 × 2 × 2	—
Cleavage	5 × 5 × 9.4	—	—	2 × 2 × 4.5	2 × 2 × 4.5	—

[a] All dimensions in cm and clear wood specimens (without defects).
[b] Equivalents in inches: 5 × 5 × 15 cm = 2 × 2 × 6 in., 2.5 × 2.5 × 10 cm = 1 × 1 × 4 in., etc.
[c] Usual dimension; other specimens are 2.5 cm in cross-section and up to 12 cm high, or orthogonal and not square in cross-section.
[d] Also 3 × 3 × 54 cm and 4 × 4 × 72 cm.

Source: American (ASTM), German (DIN), French (AFNOR), British (B.S.), and ISO (international) standards.

The role of structure is discussed further in this chapter under "Reasons for shrinkage and swelling." The effect of density also relates to structural differences (cell types, cell-wall thickness, etc.).

Extractives

A large extractive content contributes to reduction of shrinkage and swelling.[2] The reduction is proportional to the space occupied by the extractives in the cell walls. Removal of extractives increases shrinkage and swelling (8, 25) (Figure 10-5).

Chemical Composition

The influence of the chemical composition of cell walls is small, because there are no great differences between different wood species, especially with regard to cellulose content. Lignin exercises a restraining effect on shrinkage and swelling. Also, lignin may contribute to the greater shrinkage of high-density woods, because it has been observed that lignin content is reduced with increasing density (5). Hardwoods shrink more than softwoods, at similar wood densities, and this is attributed to lower lignin content of hardwoods (34) (on the average, about 20% less; see also Chapter 4).

The relatively low shrinkage of certain woods, such as teak (see Table 8-1), is due to extractives, but has also been attributed to low content of hydrolyzed hemicelluloses (10).[3]

Mechanical Stresses

Mechanical stresses originating from externally applied loads, or developing when wood loses or adsorbs moisture,[4] if sufficiently large (above the proportional limit; see Chapter 11), may cause a permanent deformation of wood cells.[5] Such deformation results in secondary changes (reduction or increase) of shrinkage and swelling. Large compression results in a shrinkage greater than normal, when cross-sectional cell dimensions are permanently reduced. Inversely, under the influence of large tension stresses, shrinkage becomes smaller than normal.

REASONS FOR ANISOTROPY OF SHRINKAGE AND SWELLING

The differential shrinkage and swelling in different growth directions is attributed mainly to cell-wall structure. It should be remembered that there is a primary and a secondary cell wall. The primary wall is very thin, and the secondary wall is composed of three layers (S_1, S_2, S_3) with different orientation of microfibrils. In the outer (S_1) and the inner (S_3) layers, the microfibrils are arranged almost transversely to cell length, whereas in the middle

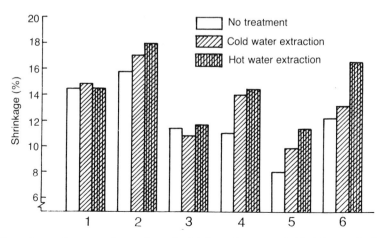

Figure 10-5. Effect of extractive removal on volumetric shrinkage: 1, spruce; 2, black locust; 3, pine; 4, yew; 5, mahogany; 6, a tropical wood (*Carapa guyanensis*). (After Ref. 25.)

(and thicker) layer (S_2) they are almost parallel to cell length (Chapter 4).

When moisture is adsorbed, the middle layer (S_2) tends to swell in proportion to the number of microfibrils (i.e., in proportion to its thickness), but the other two layers (S_1 and S_3) exercise a restraining effect due to the differing orientation of their microfibrils. The small axial (longitudinal) shrinkage is due to the orientation of microfibrils in the S_2 layer. If these microfibrils were precisely parallel, longitudinal shrinkage would be nill. Its small magnitude (see Table 8-1) is due to small deviations from parallelism. Larger deviations (e.g., in compression or juvenile wood) contribute to greater shrinkage and swelling (23, 24) (Figure 10-6). In compression wood, longitudinal shrinkage up to 10% has been measured (see Chapter 7).

The reasons for the difference between radial and tangential shrinkage and swelling are not well known. In part, this is attributed to the presence of rays, which, due to their radial orientation, exercise a restraining influence to the radial shrinkage and swelling (22). It has been observed that in certain species (oak, beech) removal of rays increases radial shrinkage, and ray tissue shrinks very little along its length (ray length). However, in other species, rays and wood (basswood), or wood with and without rays (pine), showed similar shrinkage. The restraining effect of the rays is attributed to the direction of microfibrils in the walls of their parenchyma cells; such microfibrils are, in their greater part, parallel to cell length, as it happens in the other cell types (12, 48). A decisive influence of rays is doubted, however, because, if such an influence did exist, woods with a higher ray content should exhibit a greater difference between radial and tangential shrinkage—which is not the case. However, a relationship between shrinkage and the number of rays has been observed, especially in hardwoods (7).

Another factor that is considered to produce a lower radial shrinkage, at least in softwoods, is the deviations of microfibrils caused by the presence of a greater number of pits in the radial walls of axial tracheids. However, the magnitude of such deviations does not explain the existing differences between radial and tangential shrinkage (6, 7, 37).

The difference in density between early- and latewood is also considered a cause of anisotropy (28, 37). Latewood shrinks and swells more due to its higher density—up to 3.5 times (6) (see Table 10-3); therefore, it has a predominant effect causing the attached earlywood to follow the same trend. Because growth rings have a tangential arrangement, tangential shrinkage becomes greater. This explanation is applicable especially to woods with pronounced density differences between earlywood and latewood; however, many species (diffuse-porous) exhibit a higher tangential than radial shrinkage, although they have no such differences.

According to another point of view, shrinkage and swelling are influenced by the middle lamella, which is made of lignin and pectic substances. It has been observed that partial removal of lignin causes an increase of shrinkage (5, 47). This is easy to explain, because the spaces occupied by lignin are emptied and its place is taken by water. The effect is similar to

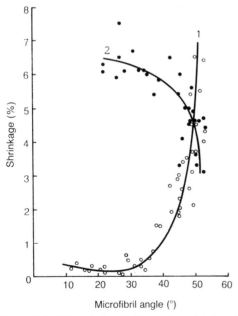

Figure 10-6. Effect of microfibrillar angle to shrinkage in (jeffrey) pine wood: 1, axial; 2, tangential shrinkage. (After Ref. 23; reproduced by permission from the Forest Products Research Society, USA.)

Table 10-3. Shrinkage of Earlywood and Latewood[a]

	Shrinkage (%)			
	Earlywood		Latewood	
Species	T	R	T	R
Fir, white	5.8	2.4	8.8	6.3
Pine, Scots	8.1	2.9	11.3	8.2
Douglas-fir	5.7	2.9	10.9	9.9
Larch	7.1	3.2	12.3	10.2

[a] T = tangential; R = radial shrinkage.
Source: Ref. 7.

that of extractives, but it should be noted that the remaining pectic substances may swell easier. The increase of shrinkage caused by the removal of lignin was found to be greater in the radial direction, and this is explained to mean that the middle lamella contains more lignin in the radial walls—which was experimentally verified (5). The effect of lignin is greater in woods of low density; whereas in woods of higher density, shrinkage is also influenced by cell-wall thickness. Wall thickness has been observed to be greater tangentially; the presence of rays also contributes more cell wall in that direction. In any case, the above factors cannot explain but a small part of the existing anisotropy (5, 40).

In conclusion, it may be stated that the difference between radial and tangential shrinkage and swelling is due to a combination of factors and that—in different species, specimens, or conditions—the influence of one or more of these factors is expressed in a different degree.

DETERMINATION OF SHRINKAGE AND SWELLING

Shrinkage is calculated on the basis of dimensions in the green condition (above fiber-saturation point); inversely, swelling is calculated on the basis of dimensions in the dry condition. The values are expressed per unit of the initial dimension (green or dry) or in percent of the original dimension. Calculations are made by use of the following relationships:

$$b = \frac{l_1 - l_2}{l_1} \quad (10\text{-}1)$$

or

$$b\% = \frac{l_1 - l_2}{l_1} \times 100 \quad (10\text{-}2)$$

and

$$a = \frac{l_1 - l_2}{l_2} \quad (10\text{-}3)$$

or

$$a\% = \frac{l_1 - l_2}{l_2} \times 100 \quad (10\text{-}4)$$

where

b = shrinkage (cm/cm or %)
a = swelling (cm/cm or %)
l_1 = green (initial) dimension (cm)
l_2 = dry (final) dimension (cm)

These relationships may be written:

$$b = 1 - \frac{l_2}{l_1} \quad (10\text{-}5)$$

and

$$a = \frac{l_1}{l_2} - 1 \quad (10\text{-}6)$$

and, finally,

$$b = \frac{a}{1 + a} \quad (10\text{-}7)$$

and

$$a = \frac{b}{1-b} \quad (10\text{-}8)$$

It should be noted that shrinkage and swelling, calculated on the basis of the same dimensions, are not equal,[6] and that one may be calculated from the other (see Table 10-4).

Also,

$$l_1 - l_2 = bl_1 = al_2 \quad (10\text{-}9)$$

That is, the change of dimensions ($l_1 - l_2$) may be calculated from dimensions in the green condition and shrinkage, or from dimensions in the dry condition and swelling.

The linear dimensions (axial, radial, tangential) are measured with a micrometer when accuracy is desired. Volume may be calculated from linear dimensions, when the sample has a regular geometric shape, or by immersion in water (Chapter 8). For practical purposes, volumetric shrinkage (and swelling) may be calculated by addition of the partial shrinkages or from the tangential and radial values only, i.e.,

$$b_v = b_t + b_r + b_l \simeq b_t + b_r$$

and

$$a_v = a_t + a_r + a_l \simeq a_t + a_r \quad (10\text{-}10)$$

If the assumption is made (approximately, it is true) that the relationship between shrinkage or swelling and moisture is linear, it is possible to estimate the following dimensional changes—which are important from a practical point of view.

a. Shrinkage corresponding to a certain change of moisture below the fiber saturation point. This is calculated from the relationship:

$$b_d = \frac{bY_d}{f} \quad (10\text{-}11)$$

where

b_d = shrinkage corresponding to moisture content change Y_d
b = total shrinkage
f = fiber saturation point
(b_d, b, and Y_d in percent or decimals).

Example: For b = 8% (0.08), Y_d = 14% − 8% = 6% (0.06), f = 30% (0.30), and b_d = 1.6% (0.016 cm/cm of green dimensions).

b. Change of dimensions corresponding to a certain change of moisture. This is calculated from the relationship:

$$l_d = \frac{bY_d l_1}{f} \quad (10\text{-}12)$$

where

l_d = change of dimension
b = total shrinkage
Y_d = change of moisture content
f = fiber saturation point
l_1 = initial (green) dimension
(b, Y_d, f in percent or decimals, and l_1 in decimal meters or centimeters).

Example: If b = 8% (0.08), Y_d = 6% (0.06), f = 30% (0.30), l_1 = 0.08 m (8 cm), and l_d = 0.128 cm.

IMPORTANCE OF SHRINKAGE AND SWELLING

Dimensional changes associated with shrinkage and swelling may cause degradation of products. Degradation may result from a simple reduction or increase of dimensions, anisotropy, or differential shrinkage or swelling in the mass of wood under the influence of differences in the distribution of moisture or density. As a result, various defects may develop, such

Table 10-4. Correspondence of Shrinkage and Swelling Values (%)

Swelling	0.1	0.2	0.4	0.6	0.8	1.0	2.0	5.0	10.0
Shrinkage	0.09	0.19	0.39	0.59	0.79	0.99	1.96	4.76	9.09

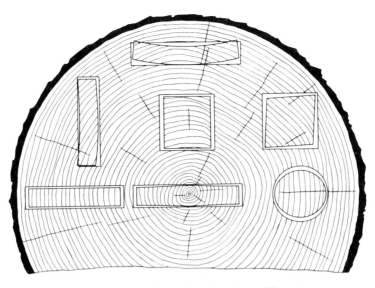

Figure 10-7. Change of shape of various cross-sections and warping due to differential radial and tangential shrinkage. (Courtesy of U.S. Forest Products Laboratory.)

as opening or tightening of joints, change of cross-sectional shape, warping, casehardening, honeycombing, collapse, and loosened or raised grain. All these may appear in wood of normal structure, but their magnitude may be seriously influenced by the presence of abnormalities, such as reaction wood (tension and compression wood), knots, etc.

Opening or *tightening* of joints may be caused by respective reduction or increase of dimensions. Increase of dimensions may result in difficulties of opening or closing doors, windows or drawers, whereas a decrease will cause openings in floors, loosening of furniture joints, etc.

Change of cross-sectional shape of wooden objects may result from differential radial and tangential shrinkage and swelling. Figure 10-7 shows the initial and final (after drying) shape of different cross-sections, depending on the position of the wood in the tree trunk.

Warping may also result from differences in radial and tangential shrinkage and swelling. For example, lumber boards, longer in the tangential direction (Figure 10-7, top), exhibit a natural tendency to warp. Warping may also be caused by differences in moisture-content distribution. Various types of warping are shown in Figure 10-8.

Checking is a result of unequal change of dimensions due to moisture gradient between the surface and the interior of a piece of wood. Checks appear after drying (air- or kiln-drying) at the ends due to faster loss of moisture, or in

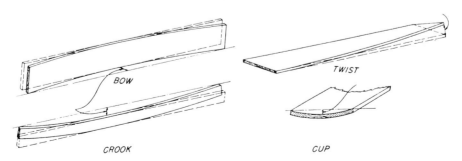

Figure 10-8. Types of warping. (Courtesy of U.S. Forest Products Laboratory.)

other positions of lumber and other wooden objects; they are superficial or deep, and usually follow the direction of rays, which constitute planes of weakness.

Casehardening, *honeycombing*, and *collapse* occur mainly in kiln-drying of lumber. *Casehardening* is a term applied to the presence of stresses in lumber, although no abnormality is seen. The defect is detected during machining, especially resawing; the sawn boards have a tendency to cup toward each other. A related defect, called *reverse casehardening*, causes resawn boards to cup toward the outside. *Honeycombing* is internal checking, which may develop in woods with wide rays, such as oak and other species. *Collapse* is the severe distortion of cells, causing a corrugated appearance of the surface of lumber (Figure 10-9). All these defects result largely from improper drying procedures, and specifically from drying wet and thick lumber at a high temperature and low relative humidity.

Briefly, the development of checking, casehardening, honeycombing, and collapse may be explained as follows.

Consider a lumber board, for example, with an overall moisture content above the fiber saturation point. Rapid evaporation dries a layer adjacent to the surface and this outer shell tends to shrink when its moisture content falls below the fiber saturation point. Its shrinkage is obstructed by the still wet core, and as a result tension stresses develop in the shell. When such stresses exceed the strength of wood, the shell ruptures; this is surface *checking*.

If the stresses developed are not high enough to cause rupture, but have nevertheless exceeded the so-called proportional limit (see Chapter 11), the result is a permanent deformation of the shell, or *tension set*. This means permanent stretching of the diameter of the cells; as a result, the shell will shrink less than normal. As drying proceeds, the moisture content of the core gradually falls below the fiber saturation point. The core also starts to shrink, but cannot shrink normally because of the restricting effect of the shell which, being itself under tension, exercises compression to the core. Excessive compression may cause *collapse* in some cases, but collapse may also result from a quick exit of moisture, which "deflates" cells in a manner similar to the collapse of a rubber tube or balloon when air or water is quickly expelled.

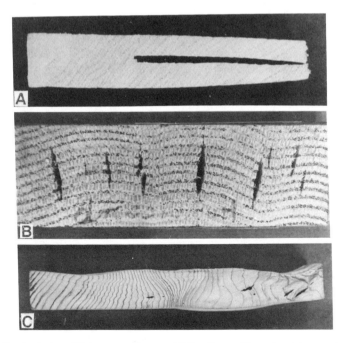

Figure 10-9. (A) Casehardening, (B) honeycombing, and (C) collapse. (Reproduced by permission from Pergamon Press.)

As the core is restricted by the shell and so unable to shrink normally, it falls under tension—exercising compression to the shell. Tension in the core may cause *honeycombing*.

If no collapse or honeycombing takes place, the condition of the shell being under compression and the core under tension is *casehardening*.

Changes in moisture content and differential shrinkage and swelling—in combination with improper machining—are the cause of two more defects, namely, *raised grain* and *loosened grain*. Raised grain refers to rough surfaces of lumber, caused by latewood being raised above the earlywood. Loosened grain involves a partial separation of growth rings; the loosened portions protrude from the surface of the wood. The main reasons for these defects are dull planer knives and greater shrinkage and swelling of latewood.

CONTROL OF SHRINKAGE AND SWELLING

The adverse effects of shrinkage and swelling have initiated efforts to control this property with the intention of securing dimensional stability to wood and its products. The difficulties encountered stem from the fact that it is practically impossible to control the hygroscopicity of wood. These efforts have led to certain methods which result in a considerable reduction of the *rate* or *magnitude* of shrinkage and swelling, but their practical application is often hindered by reasons related to effectiveness, cost, method of application, and effects on wood properties.

The methods include: *mechanical modification* of wood, *water-repellent coating*, *bulking treatment*, and *reduction of hygroscopicity* (30, 40, 43).

Mechanical Modification

Redistribution of the mass of wood, after mechanical processing, reduces its anisotropy which is the main reason for defects caused by shrinkage or swelling. In plywood, there is a drastic decrease of shrinkage and swelling in the direction of its length and width—to magnitudes approaching longitudinal values of solid wood; anisotropy is practically erased by the placement of adjacent layers with the grain of wood at right angles. In particleboard and fiberboard, there is a similar effect with regard to shrinkage and swelling (see Chapters 24 and 25).

Water-Repellent Coating

Coating the exterior surface of wood with paint, varnish, or similar substances does not affect the quantity of moisture that wood may retain when exposed to the atmosphere. These substances do not affect the equilibrium moisture content and, therefore, the final magnitude of shrinkage and swelling. None is entirely impermeable to water, but some may give considerable protection to atmospheric variation (aluminum paint 90–95%, various varnishes 50–85%) (27) through retardation of the rate of moisture exchange between wood and atmosphere; shrinkage and swelling under changing atmospheric conditions are, therefore, also retarded.

Aside from the exterior, coating may be applied to the interior surface of wood (inside cell cavities and cell walls) with water-repellent substances; a typical solution is made of alkydic resin (10–15%) and paraffin wax (0.5–1%) in an organic solvent (commercial "white spirit," shellsol) (46). Other chemicals that protect wood against insects and fungi, as well as dyes, may be added to the solution. The wood, in air-dry condition and in the form of a final product (door, window, etc.), is usually immersed for a few (2–3) minutes, and the protective solution enters to a limited depth; a deeper penetration is achieved by application of "double vacuum" (vacuum—atmospheric pressure or little higher—a second higher vacuum; see Chapter 19). After treatment, the organic solvent evaporates, and the solid components cover the exterior and part of the interior wood surface.[7]

In general, the effectiveness of protection by coating is reduced with passage of time, more so in structures exposed to the weather (exterior doors, windows, etc.), due to breakage of the continuity of the coating and erosion of the wood surface (46).

Bulking Treatment

Shrinkage may be partially or totally eliminated by keeping the wood in partially or totally swollen condition; this may be accomplished by various chemicals (salts, sugars, polyethylene glycol, synthetic resins, etc.), which are deposited in the cell walls.

Use of water-soluble *salts* (salts of sodium, barium, magnesium, etc.) (3) or *sugars* (sucrose) may reduce shrinkage up to 70% or more. Wood impregnated with a saturated solution of magnesium chloride starts shrinking only when the relative humidity of the surrounding atmosphere becomes lower than 32% (27, 40). However, these substances have the disadvantage that wood is always wet at high relative humidities, the salts or sugars appear on the surface, and the strength of such wood is reduced. Also, salts cause oxidation of tools and metallic connectors (nails, etc.). In order to control the higher hygroscopicity imparted by salts and sugars, two coats of varnish are usually used to seal these chemicals in the wood.

Polyethylene glycol is used in varying molecular weights (usually PEG 1000), and is introduced in the wood by immersing it in a 30–50% aqueous solution for a length of time depending on its thickness (27). Application of vacuum and pressure facilitates its entrance, especially in woods that are difficult to impregnate. The wood is usually treated in the green condition (31). After treatment and drying, the wood may be glued, dyed, or finished, and has a greater resistance to fungi; however, bending strength, abrasion resistance (40), and transverse compression (33) are slightly reduced (green strength is practically preserved) (30). Also, reduction in flammability, electrical resistance, and luster have been observed, but differences exist among species (30). Polyethylene glycol is used with carvings, gun stocks, tree disks (transverse sections), archaeological findings (old sunken ships), etc. (26), and practically eliminates shrinkage; in beech parquet flooring, shrinkage was reduced up to 93% (14). However, polyethylene glycol, like salts and sugars, may be depleted when treated wood comes in contact with water.

Nonleachable treatments are obtained by *synthetic resins*, various *monomers*, and *waxes*.

Water-soluble *synthetic resins* (thermosetting; see Chapter 21) are introduced in the wood and polymerized in situ. Phenol-formaldehyde is preferred; it may swell wood about 25% more than water and, therefore, the impregnation of cell walls is easier. The method has been used for many years in making two products of "improved wood," known as "Compreg" and "Impreg" (27, 30, 40). Both are made by impregnation of thin veneers. In "Compreg," after impregnation, the veneers are glued under high pressure and hot pressing (resulting to a $\frac{1}{3}$ to $\frac{1}{2}$ decrease of thickness and an increase of density to about 1.0–1.4 g/cm^3)—whereas in "Impreg," the resin is set in each veneer, and afterward the veneers are glued by use of adhesive and normal pressure. These products have a very good dimensional stability, but their cost is high and they have limited uses, such as rotors for helicopters and propellers for small airplanes, handles, wood patterns for automobiles, and textile shuttles.

The effects of monomers, such as *methacrylates*, *vinyl compounds*, *styrol*, etc. are similar (when used with a swelling agent). Their polymerization is attained by heat or atomic radiation. The product (Polymerholz, Lignomer, Wood-Polymer-Composite or WPC) shows very reduced shrinkage. In the case of styrol (styrene), a 60% shrinkage reduction has been reported (Figure 10-10). With acrylics, the reduction is 10–90% (in different species) (32). In addition to shrinkage, other properties are improved, such as mechanical (except toughness), resistance to climatic, chemical, and biotic factors (insects, fungi), as well as aesthetic appearance in certain species (20, 21, 32, 36). The method is promising due to the above improvements; however, application may be hindered by economic considerations, especially with regard to radiation polymerization. For this reason, its practical application is limited to items such as athletic equipment, flooring (parquet) (13), shoe forms, tool handles, and textile shuttles.

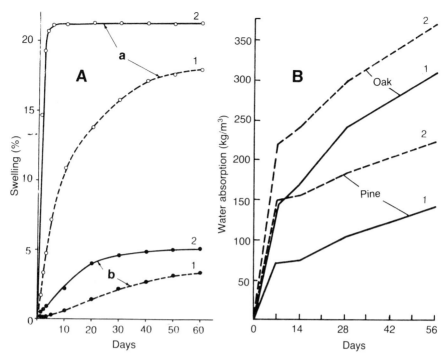

Figure 10-10. Effects of treatment (bulking) of wood. (A) Reduction of swelling of beech wood after impregnation with styrol: *a*, nontreated; *b*, treated wood; 1, exposed to air saturated with water vapors; 2, placed in water. (B) Comparative water absorption of oak and Scots pine specimens (3 × 3 × 50 cm, 1.2 × 1.2 × 20 in.) after treatment with urea-formaldehyde. Untreated oak absorbed 86% and oak 111% more water per unit volume in comparison to treated specimens. (Pine also showed a considerable increase of compression and bending strength but oak was practically unchanged): 1, treated; 2, untreated. (100 kg/m^3 = 6.25 lb/ft^3). (A, after Ref. 2, and B after Ref. 11; reproduced by permission from VEB-Fachbuchverlag and Springer Verlag.)

Wax is used to replace cell-wall water. The treatment is applied in two stages—first by substitution of moisture with the chemical "cellosolve" (ethylglyco-monoethyl ether) and, subsequently, by its substitution with wax; the substitution of cellosolve is achieved by immersion in molten wax and natural resin (cellosolve is distilled). In this manner, shrinkage may be reduced up to 80%, but the method has no practical application, because it cannot be used with larger dimensions of wood, and also because wax impedes subsequent gluing and finishing (31).

Reduction of Hygroscopicity

Hygroscopicity of wood, and therefore its shrinkage and swelling, may be reduced by *thermal* treatment, *replacement of hydroxyls* with less hygroscopic (hydrophobic) substances, and *cross-linkage of hydroxyls*.

The effect of high temperatures is mainly attributed to breakdown of hemicelluloses. The heating of oak wood at a temperature of 160°C (320°F) and a pressure of 6 atmospheres (the combination causes flow of lignin) reduced shrinkage by about 70% (9). High temperature (about 170°C, 340°F) and pressure have been used for the production of a product of high density (1.30–1.35 g/cm^3) called "Staypack." Another product, "Staybwood," is made by heating without presence of air. Shrinkage is reduced to about 50%, but the strength of wood is also reduced (40).

Replacement of hydroxyls is mainly done with acetyl groups. Acetylation may reduce shrinkage up to 75%, but the reduction is rather attributed to a combination of hydroxyl replacement and to keeping wood in a condition of partial swelling, because acetyl groups possess a considerably greater volume than the hydrophilic hydroxyls (40). The cost of treatment is high. Reduction of hygroscopicity is also affected by epoxides, isocyanates, etc., which react with wood hydroxyls. Epoxides may reduce shrinkage by 50–70%, but the treatment is expensive, and overtreatment results in loss of dimensional stability due to cell-wall rupture; isocyanates give similar results and have similar limitations (31).

Cross-linking of hydroxyls is mainly accomplished with formaldehyde vapors. Shrinkage is considerably reduced, up to about 90%, but the strength of wood is also reduced, due to hydrolysis of cellulose and hemicelluloses from the action of the acidic catalyst used (31, 40).[8]

The *conclusion* may be drawn that the methods described for control of shrinkage and swelling are either experimental or costly methods; therefore, they have no wide practical application, except in reconstructed products (plywood, particleboard, etc.). Thus, the simplest method of handling wood is its careful drying. Most dimensional problems with wood are caused by a high initial moisture content, with the result that shrinkage is high when wood is placed in the environment of use. This creates the need for careful drying, which aims at a moisture content nearing the expected average under the climatic conditions of use (see also Chapter 9). In this manner, the variation of moisture content and dimensions is the lowest possible.

REFERENCES

1. Bariska, M. 1966. Über den Einfluss der Teerölimprägnierung auf das Schwindverhalten von Buchenholz. *Holz Roh- Werkst.* 24(1):18–24.
2. Boiciuc, M., and C. Petrician. 1970. Dimensionsstabilisierung von Rotbuchenholz durch Anlagerung von Styrol. *Holztechnologie* 11(2):94–96.
3. Borgin, K. 1978. Progress report on evaluating the Thessaloniki process. *Int. J. Nautical Arch. Underwater Expl.* 7(4):314–317.
4. Borgin, K., G. Tsoumis, and C. Passialis. 1979. Density and shrinkage of old wood. *Wood Sci. Technol.* 13:49–57.
5. Bosshard, H. H. 1956. Über die Anisotropie der Holzschwindung. *Holz Roh- Werkst.* 14(8):286–295.
6. Bosshard, H. H. 1964. Weisstannenholz. *Schweiz. Z. Forstwes.* 9/10:534–542.
7. Bosshard, H. H. 1974. *Holzkunde*, Vol. 2. Basel: Birkhaeuser Verlag.
8. Brown, H. P., A. J. Panshin, and C. C. Forsaith. 1952. *Textbook of Wood Technology*, Vol. II. New York: McGraw-Hill.
9. Burmester, A. 1972. Quellung und Quellungsanisotropie von Holz in verschiedenen Feuchtigkeitsbereichen. *Holz Roh- Werkst.* 30(10):380–381.
10. Burmester, A. 1975. Zur Dimensionsstabilisierung von Holz. *Holz Roh- Werkst.* 33(9):333–335.
11. Erler, K., and D. Knospe. 1988. Vergütung von Kiefern- und Eichenholz durch Tränkung mit Harnstoffformaldehydharz. *Holz Roh- Werkst.* 46(8):327–329.
12. Harada, H. 1965. Ultrastructure of angiosperm vessels and ray parenchyma. In *Cellular Ultrastructure of Woody Plants*, ed. W. A. Côté, pp. 235–249. Syracuse, New York: Syracuse University Press.
13. Hills, P. R. 1969. The Use of Wood-Plastic Composites by the Flooring Industry. U.K. Atomic Energy Authority, H.M.S.O., London.
14. Kail, A. 1976. Untersuchungen zur Dimensionsstabilisierung von Fussbodenhölzern. *Holzforsch. Holzverwert.* 28(2):32–36.
15. Kellogg, R. M., and F. F. Wangaard. 1969. Variation in the cell-wall density of wood. *Wood Fiber* 1(3):180–204.
16. Kollmann, F. 1951. *Technologie des Holzes und der Holzwerkstoffe*. Berlin: Springer Verlag.
17. Kollmann, F. F. P., and W. A. Côté. 1968. Principles of Wood Science and Technology. I. Solid Wood. Berlin: Springer Verglag.
18. Kommert, R. 1980. Bemerkungen zur Quellungsanisotropie von Eichenholz. *Holztechnologie* 21(1):28–30.
19. Koenig, E. 1972. *Holz-Lexikon*, Vols. I and II. Stuttgart: DRW-Verlags.
20. Lala, M. K., H. K. Bali, and R. C. Cupta. 1980. Studies on wood-plastic composites. Part 3 (irradiated). *Holzforsch. Holzverwert.* 32(5):125–126.
21. Lutomski, K., and M. Lawniczak. 1977. Polymerholz und seine Widerstandsfähigkeit gegen biotische Einflusse. *Holz Roh- Werkst.* 35(2):63–65.
22. McIntosh, D. C. 1957. Transverse shrinkage of red oak and beech. *For. Prod. J.* 7(3):114–120.
23. Meylan, B. A. 1968. Cause of high longitudinal shrinkage in wood. *For. Prod. J.* 18(4):75–78.
24. Meylan, B. A. 1972. The influence of microfibril angle on the longitudinal shrinkage–moisture content relationship. *Wood Sci. Technol.* 6(4):293–301.
25. Nearn, W. T. 1955. Effect of Water Soluble Extractives on the Volumetric Shrinkage and Equilibrium Moisture Content of Eleven Tropical and Domestic Woods. Penn. State Ag. Expt. Sta., Bull. 598.
26. Netherlands National Commission for UNESCO. 1979. Conservation of Waterlogged Wood. Proc. Intern. Symposium, The Hague.
27. Palka, L. C. 1970. Current Trends in Dimensional Stabilization of Wood. W. For. Prod. Lab. Canada, Inform. Report VP-X-63.
28. Pentoney, R. E. 1953. Mechanisms affecting tangential and radial shrinkage. *For. Prod. J.* 3(2):27–32.

29. Raczkowski, J., and A. Krause. 1979. Swelling pressure of pine compression wood. *Drevasky Vyskum* 24(2):1-11.
30. Rowell, R. M., and R. L. Youngs. 1981. Dimensional Stabilization of Wood in Use. U.S. For. Prod. Lab. Res. Note FPL-0243.
31. Rowell, R. M., and W. B. Banks. 1985. Water Repellency and Dimensional Stability of Wood. U.S. For. Prod. Lab. Report FPL-50.
32. Schaudy, R., and E. Proksch. 1980. Untersuchung verschiedener Hölzer auf ihre Eignung zur Herstellung von dimensionsstabilen Holz-Kunstoff-Kombinationen. *Holzforsch. Holzverwert.* 32(2):25-35.
33. Schneider, A. 1970. Untersuchungen über Eigenschaftsänderung des Holzes durch Tränkung mit Polyäthylenglykol und über die Wirksamkeit verschiedener Tränkverfahren. *Holz Roh- Werkst.* 28(1):20-34.
34. Schroeder, H. A. 1972. Shrinking and swelling differences between hardwoods and softwoods. *Wood Fiber* 4(1):20-25.
35. Siau, J. F. 1984. *Transport Processes in Wood.* Berlin, New York: Springer Verlag.
36. Singh, Y. 1979. Properties of catalytically polymerized wood-plastic composites. *J. Inst. Wood Sci.* 8(3):127-128.
37. Skaar, C. 1988. *Wood-Water Relationships.* Berlin, New York: Springer Verlag.
38. Spalt, H. A. 1979. Water-vapor sorption by woods of high extractive content. In *Symposium on Wood Moisture Content—Temperature and Humidity Relationships*, pp. 55-61. U.S.Forest Products Laboratory.
39. Stamm, A. J. 1952. Surface properties of cellulosic materials. In *Wood Chemistry*, Vol. 2, ed. L. E. Wise and C. Jahn, pp. 691-814. New York: Reinhold.
40. Stamm, A. J. 1964. *Wood and Cellulose Science.* New York: The Ronald Press.
41. Tarkow, H. 1981. Wood and moisture. In *Wood: Its Structure and Properties*, ed. F. F. Wangaard, pp. 147-186. Materials Education Council, Penn. State University, University Park, Pennsylvania.
42. Trendelenburg, R. and H. Mayer-Wegelin. 1955. *Das Holz als Rohstoff.* München: Hanser Verlag.
43. U.S. Forest Products Laboratory. 1959. Dimensional Stabilization Seminar. Report No. 2145.
44. U.S. Forest Products Laboratory. 1961. Swelling of Wood. Report No. 1061.
45. U.S. Forest Products Laboratory. 1987. Wood Handbook. U.S.D.A., Ag. Handbook No. 72 (revised).
46. Voulgaridis, E. 1980. Physical Factors Affecting the Performance of Water Repellents Applied to Wood. Ph. D. Thesis, University of Wales, U.K.
47. Wardrop, A. B. 1957. The phase of lignification in the differentiation of wood fibers. *Tappi* 40:225-243.
48. Wardrop, A. B. 1964. The structure and formation of the cell wall in xylem. In *The Formation of Wood in Forest Trees*, ed. M. H. Zimmermann, pp. 87-134. New York/London: Academic Press.
49. Weatherwax, R. C., and H. Tarkow. 1968. Density of wood substance: Importance of penetration and adsorption compression of the displacement fluid. *For. Prod. J.* 18(7):44-46.

FOOTNOTES

[1]Linearity may be affected by densification of the water held in cell walls, but the effect is slight; the densification of such water was calculated to be 1.67% or 1.014 g/cm^3 (15, 49). It is noteworthy that during adsorption of vapors from the atmosphere, about half of the total swelling takes place in the relative humidity range of 0-85%, and the rest between 86 and 100% (10).

[2]This influence of extractives is exhibited at room temperature. At high temperatures, as applied in kiln-drying, woods with a high extractive content show high shrinkage and are subject to collapse (38).

[3]Such a relationship has also been ascribed to black locust, on the basis of measured low shrinkage (radial 2.8%, tangential 4.7%) (10), but there is no general agreement on the latter (see Table 8-1).

[4]Swelling creates strong pressure, which increases with increasing density (40, 41). Rocks may be split by placing wood in narrow cracks and wetting it. The pressure of pine compression wood was found 460% greater in the longitudinal direction and 40% greater in transverse direction in comparison to pine wood of normal structure (29).

[5]If wood adsorbs moisture while under compression, external swelling is constrained and swelling occurs instead internally, reducing the volume of the cell cavities.

[6]For example, for $l_1 = 16$ cm and $l_2 = 15.3$ cm, shrinkage $b = 0.04375$ cm/cm (4.375%) and swelling $a = 0.04475$ cm/cm (4.475%).

[7]The difficulty of controlling shrinkage and swelling is due to the very large interior surface of wood (area of contact with swelling agents). The interior surface of swollen spruce wood has been calculated to approximate 250 m^2 (2500 ft^2) per gram; generally, the values for wood are about 1000 times the microscopically visible surface (39).

[8]The treatments applied to wood in order to reduce shrinkage anisotropy modify other properties as well; hence, the product is sometimes called "modified" wood.

11

Mechanical Properties

The mechanical properties of wood are measures of its resistance to exterior forces[1] which tend to deform its mass. The resistance of wood to such forces depends on their magnitude and the manner of loading (tension, compression, shear, bending, etc.). In contrast to metals and other materials of homogeneous structure, wood exhibits different mechanical properties in different growth directions (axial, radial, tangential)—and, therefore, it is mechanically anisotropic.

BASIC CONCEPTS

Before discussing the mechanical properties of wood, it is useful to explain certain basic concepts regarding the mechanics of materials in general (65).

Force is any action that tends to move a body at rest, or change its shape or size, or, if the body is moving, to change the speed or direction of its movement. Under the action of exterior forces which tend to change its form, a body at rest exercises resistance. This resistance (i.e., the interior forces which develop inside its mass in reaction to the forces acting on the exterior) is called *interior stress* or simply *stress*. These stresses are equal and opposite to the exterior stresses.

A force (or load) is expressed in Newton (N) or pounds (lb). A force per unit area is called *stress*; it is expressed in N/mm^2, Pascal (Pa), or $lb/in.^2$ (psi), and is calculated from the following relationship[2]:

$$S = \frac{P}{A} \quad (11\text{-}1)$$

where

S = stress per unit area or unit stress (N/mm^2, Pa, $lb/in.^2$ or psi)
P = force or load (N, lb)
A = loaded area (mm^2, $in.^2$)

There are three basic stresses: *tensile*, *compressive*, and *shearing*. Tensile and compressive stresses are also called "normal stresses." A body is under *tensile* stress when the forces acting tend to increase its length. If the forces are acting in the opposite direction, the body is under *compressive stress* and tends to become shorter. *Shearing stresses* develop when the forces tend to cause a part of the stressed body to slide onto the adjacent part of the same body (Figure 11-1).

Under the influence of exterior forces which generate the above stresses, the stressed body tends to change shape and size. This change is called *deformation*. Under tension and compression, deformation is measured in mm (or in.), and is distinguished into *total deformation* and *strain*, which is *deformation per unit length* (i.e., the ratio of the total to the initial length of the body). That is,

$$D = L \times d \text{ or } d = \frac{D}{L} \quad (11\text{-}2)$$

where

D = total deformation (mm, in.)
L = initial length of the body (mm, in.)
d = strain (mm/mm, in./in.)

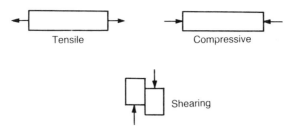

Figure 11-1. The three basic stresses.

In many materials, including wood, the relationship between stress and strain is linear if the stress is not too great (i.e., up to a certain point—Hooke's law). This point is called the *proportional* (or *elastic*) *limit*. Above the proportional limit, an increase of stress causes a greater than proportional deformation until the stressed body fails (Figure 11-2).

Elasticity is the property of a body to return to its initial condition (initial shape and size) when the load causing the corresponding stress and deformation is removed.[3] This occurs below the proportional limit. If loading is continued above this limit, part of the deformation is permanent.

The relationship between stress and strain defines the *modulus of elasticity* (Young's modulus). This is calculated from the relationship

$$E = \frac{S}{d} \qquad (11\text{-}3)$$

where

E = modulus of elasticity (N/mm², Pa, psi)
S = stress per unit area or unit stress (N/mm², Pa, psi)
d = unit deformation (mm/mm, in./in.)

Modulus of elasticity is valid only up to the proportional limit. A high modulus of elasticity indicates a stiff (difficult to bend) body (i.e., a body that can stand a high stress without great

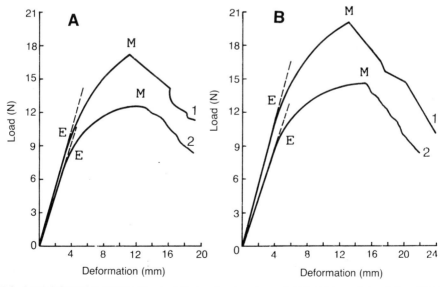

Figure 11-2. Load-deformation relationships in static bending (specimens 2 × 2 × 34 cm, 0.8 × 0.8 × 13.6 in.). (A) beech, (B) oak. 1, air-dry wood (12% mc); 2, green wood (beech 79%, oak 69% mc); E, proportional limit; M, maximum load.

deformation). In the case of wood, the modulus of elasticity is usually determined from loading in bending or axial loading.

Opposite to elasticity is the concept of plasticity. An ideally plastic body retains most of its deformation after removing the load. Such a body does not exist, and the same is true for ideally elastic materials.

Also noteworthy is the property of a body to present resistance to sudden or *impact loading* (i.e., *toughness*). The measure of this property is the energy which the loaded body can absorb. Energy is measured as *work*, which equals the product of force (load) and deformation. Work is expressed in Nm, Joules (J), or in. lb (in older literature as kpcm, kpm, or kpm/cm^2). The more work required to break a body means greater toughness. In most cases, bodies that generate little work are brittle. Brittle bodies cannot withstand large strains.

PROPERTIES

The mechanical properties of wood include its resistance to various types of loading. In the following, each property is considered separately and in relation to the influence of various factors.

Strength in Tension

The strength of wood in tension shows considerable differences if loading is axial (parallel to grain) or transverse. Strength in axial tension is much higher—up to 50 times and more (see Table 11-1). In the transverse direction, the influence of radial or tangential loading is not consistent (41).

The values of strength in axial tension of different temperate woods vary from about 50 to 160 N/mm^2 (50-160 MPa, 7250-23,200 psi), whereas in transverse tension the range is 1-7 N/mm^2 (1-7 MPa, 145-1015 psi). In certain tropical woods, axial tensile strength may reach 300 N/mm^2 (300 MPa, 43,500 psi).

Single cells (softwood axial tracheids, hardwood fibers) present a higher axial strength, 200-1300 N/mm^2 (200-1300 MPa, 29,000-188,500 psi) (41), and the strength of microfibrils is still higher. The strength of cellulose chains is theoretically estimated at 7500 N/mm^2 (7.5 GPa, 1,087,500 psi) (32). The successive reduction of strength from cellulose chains to wood is due to deviations of cellulose chains and microfibrils from parallelism to tree axis, and to low-strength substances between cellulose chains, microfibrils, and cells.[4] These substances are hemicelluloses, lignin, pectic substances, and extractives.

It has been observed that cell length is related to the axial tensile strength of wood—namely, that woods with longer cells (in general, softwoods in comparison to hardwoods) possess a higher strength. This may be attributed to the relationship between cell length and microfibrillar arrangement. It has been found, that, within a species, the angle between microfibrils and cell length is smaller (i.e., parallelism is greater) in longer cells and larger in shorter ones.

With the above values of axial tensile strength, especially the higher ones, this strength of wood compares favorably with metals and other materials. The comparison is more favorable for wood if strength is related to weight (density, specific gravity). Table 11-2 presents such a comparison (column 3), and shows that, weight for weight, wood is about equal to steel and superior to other construction materials. This characteristic of wood, a light but strong material, is also expressed by the so-called "breaking length," which is the length of a theoretical ribbon of material breaking under its own weight. Breaking length is measured in kilometers (km), and the range for different woods is 7-30 km (softwoods 11-30 km, hardwoods 7-30 km), construction steel 5.4 km, other steels 12-32 km, duraluminium 13.6 km, concrete 0.2 km, cast iron 1.8 km, PVC (plastic) 4 km, etc. (38, 39).

The high axial tensile strength of wood is seldom utilized. This is due to the development of shear stresses together with axial tensile stresses. The strength of wood in shear is comparatively very low, only 6-10% of axial tensile strength. Also, axial tensile strength is greatly reduced by the presence of knots, spiral grain, or other growth abnormalities. Axial tensile stresses develop in practice only in spe-

Table 11-1. Mechanical Properties (SI units)[a]

Species (North American)	Tension ∥	Tension ⊥	Compression ∥	Compression ⊥	Static bending MOR	Static bending MOE	Shear	Hardness (side)	Toughness R	Toughness T
			N/mm²					kN	Nm	
Baldcypress	—	1.9	44	6.2	73	9,940	6.9	2.3	—	—
Douglas-fir	130	2.3	51	6.0	83	13,660	8.0	3.2	22.6	40.8
Fir, white	—	1.8	37	4.1	64	9,520	6.4	2.1	14.7	22.6
Hemlock, western	—	2.1	43	4.7	69	10,280	8.1	2.4	15.8	23.8
Larch, western	134	3.0	56	6.8	96	13,520	9.7	3.7	23.8	38.5
Pine, loblolly	—	3.2	49	6.8	88	12,420	9.4	3.1	18.1	29.5
Pine, ponderosa	—	2.8	36	5.1	63	8,690	8.0	2.0	16.9	21.5
Pine, white, eastern	78	2.1	33	3.5	59	8,560	6.2	1.7	12.4	13.6
Red cedar, western	—	1.5	43	4.2	53	7,730	5.9	1.6	10.2	14.7
Redwood, old growth	—	1.7	42	5.9	69	9,250	1.5	2.1	10.2	15.9
Spruce, Engelmann	88	2.4	31	2.8	64	8,900	8.3	1.7	12.5	20.4
Alder, red	—	2.9	40	3.0	68	9,500	7.4	2.6	—	—
Ash, white	—	6.5	51	8.0	106	12,000	13.4	5.9	—	—
Aspen, quaking	—	1.8	29	2.6	58	8,100	5.9	1.6	—	—
Basswood, American	—	2.4	33	2.6	60	10,100	6.8	1.8	—	—
Beech, American	—	7.0	50	7.0	103	11,900	13.9	5.8	—	—
Birch, yellow	—	6.3	56	6.7	114	13,900	13.0	5.6	56.5	70.1
Cherry	—	3.9	49	5.9	85	10,280	11.7	4.2	—	—
Cottonwood, eastern	—	4.0	34	3.3	59	9,450	6.4	1.9	—	—
Elm, American	—	4.6	38	4.8	81	9,200	10.4	3.7	—	—
Hickory, mockernut	—	—	62	11.9	132	15,300	12.0	8.1[b]	70.3	74.8
Maple, sugar	—	—	54	12.5	109	12,630	16.1	6.4	41.8	40.7
Oak, northern red	—	5.5	47	8.6	99	12,560	12.3	5.7	53.1	49.1
Oak, white	—	5.5	51	9.1	105	12,280	13.8	6.0	38.4	35.0
Sweetgum	119	5.2	44	4.6	93	11,320	11.0	3.8	29.4	29.4
Sycamore, American	—	5.0	37	4.8	69	9,800	10.1	3.4	—	—
Walnut, black	—	4.8	52	7.0	101	11,600	9.4	4.5	—	—
Willow, black	109	—	28	3.0	54	7,000	8.6	—	23.8	26.1
Yellow poplar	155	3.7	38	3.9	70	10,900	8.2	2.4	24.9	23.8

[a] Data based on small, clear specimens in air-dry condition (Ref. 73). Original data in lb/in², except hardness (lb) and toughness (in. lb). Equivalent English unit data in Table 11-1A (Conversion equivalents are given at the end of this book).
[b] Pecan hickory.

Table 11-1. (Continued)[a]

Species (European)	Tension ‖	Tension ⊥	Compression ‖	Compression ⊥	Static bending MOR	Static bending MOE	Shear	Hardness (side)	Cleavage	Toughness
	N/mm²							kN	N/mm	J/cm²
Fir, white	78	1.4	33	4.7	67	9,600	5.5	1.7	3.1	5.9
Larch	105	2.2	54	7.3	97	13,530	8.8	3.4	3.3	5.9
Pine, black	102	2.0	39	—	103	11,760	9.8	2.6	5.6	3.9
Pine, Scots	102	2.9	54	7.5	98	11,760	9.8	2.4	3.6	6.9
Spruce	84	1.5	30	4.1	60	9,100	5.3	1.5	3.2	4.9
Alder	92	2.0	54	6.4	83	11,470	4.4	—	5.3	5.3
Ash	161	6.9	51	10.8	118	13,130	12.5	6.1	6.9	6.7
Aspen	108	2.7	47	2.6	77	10,700	7.7	2.6	7.2	3.9
Beech	130	3.5	46	7.9	104	13,130	12.3	4.8	9.1	7.8
Birch	134	6.9	50	10.8	144	16,170	11.8	—	5.2	8.3
Black locust	133	4.2	70	18.6	102	11,070	12.5	7.5	—	6.6
Chestnut	132	—	49	—	75	8,820	7.8	3.1	2.6	5.6
Elm, field	78	3.9	55	9.8	87	10,780	6.9	5.0	—	5.9
Hornbeam	153	3.8	54	16.7	140	14,700	16.9	7.6	10.2	8.3
Limetree (basswood)	83	4.9	51	1.8	104	7,250	4.4	—	2.6	4.9
Oak, red	108	3.3	42	11.5	116	11,560	12.7	4.6	9.1	—
Oak, white (English)	109	3.0	43	7.8	91	12,250	11.6	5.5	9.0	7.4
Plane (Sycamore)	—	5.2	45	6.4	97	10,290	9.8	4.8	—	6.9
Poplar, "robusta"	84	1.5	35	—	64	10,730	7.6	1.9	5.5	—
Walnut	98	3.5	71	11.8	144	12,250	6.9	5.3	4.6	9.3
Willow	83	2.4	26	3.4	53	9,800	6.7	1.6	6.4	6.9

[a] Data based on small, clear specimens in air-dry condition (Refs. 27, 39, and 74). Original data in kp/cm², except cleavage (kp/cm) and toughness (impact work, Bruchschlagarbeit, kpm/cm²). Note: Toughness values (J/cm²) could be converted to Nm (and vice versa) by multiplication (or division) by 4—considering a 2 × 2 cm cross-section of specimen. It should be noted, however, that differences in testing conditions (size of specimens, manner of loading) do not allow for accurate conversion.

Table 11-1. (Continued)[a]

Species (Tropical)	Tension		Compression		Static bending		Shear	Hardness (side)	Toughness
	∥	⊥	∥	⊥	MOR	MOE			
	N/mm²							kN	J/cm²
Afara	103	2.1	61	7.8	109	12,250	7.3	2.3	4.9
Afzelia	—	—	67	—	122	15,000	15.9	—	—
Aiélé	—	2.0	38	—	167	10,780	—	—	—
Amazakoué	130	3.9	79	—	132	16,760	12.4	3.8	11.0
Antiaris, Ako	—	—	41	—	69	9,310	10.3	—	—
Balsa	73	1.0	9	1.0	19	2,550	1.1	0.4	2.2
Beté	117	2.2	64	—	128	11,760	7.8	5.7	6.4
Bubinga	—	4.1	69	—	112	12,740	9.8	—	5.2
Dibetou	104	1.9	51	8.3	81	12,740	8.8	4.2	8.5
Difou	—	—	92	—	159	14,500	18.7	—	3.3
Idigbo	—	2.3	46	—	75	8,980	8.8	3.7	3.4
Iroko	78	2.5	68	—	111	11,270	10.8	3.2	2.5
Kosipo	—	1.9	49	—	88	7,940	5.9	—	6.1
Mahogany, African	60	2.0	45	6.9	85	9,800	7.8	3.7	3.9
Mahogany, American	—	2.4	49	8.3	83	8,330	9.7	3.1	5.2
Makoré	76	2.1	52	—	96	13,720	8.3	2.5	3.1
Meranti, red	—	—	52	—	90	11,660	—	—	—
Niangon	—	—	59	—	137	11,160	15.7	—	—
Obeche	48	1.3	39	—	72	6,660	5.4	2.0	2.9
Okoumé	57	1.8	38	5.3	71	9,100	5.9	1.7	2.4
Opepe	—	—	59	—	127	13,820	—	—	3.9
Padauk	—	2.2	73	—	135	12,450	7.7	4.3	6.4
Palissander	—	5.9	59	—	117	12,250	14.7	3.4	8.3
Ramin	—	—	70	—	127	15,910	10.6	—	—
Sapele	86	2.4	59	8.3	108	9,800	8.3	6.2	6.9
Seraya, white	—	—	49	—	90	11,560	—	3.6	—
Sipo	108	2.2	57	8.8	97	11,270	9.3	1.5	3.9
Teak	118	4.1	65	—	131	13,720	8.8	2.7	5.9
Tiama	—	2.1	47	—	76	9,900	9.6	4.2	4.3
Zebrano	—	3.4	49	—	98	11,270	7.7	2.6	5.8

[a] Data based on Refs. 10, 18, 25, and 75. The remarks of the previous footnote are also applicable here.

Table 11-1A. Mechanical Properties (English units)

Species (North American)	Tension ∥[a]	Tension ⊥	Compression ∥[a]	Compression ⊥	Static bending MOR[a]	Static bending MOE[b]	Shear	Hardness	Toughness R	Toughness T
			lb/in.² (psi)					lb	in.lb	
Baldcypress	—	270	6.36	730	10.6	1.44	1000	510	—	—
Douglas-fir	18.9	340	7.24	800	12.4	1.95	1130	710	200	360
Fir, white	—	300	5.81	530	9.8	1.49	1100	480	130	200
Hemlock, western	—	340	7.13	550	11.3	1.64	1250	540	140	210
Larch, western	19.4	430	7.64	930	13.1	1.87	1360	830	210	340
Pine, loblolly	—	470	7.13	790	12.8	1.79	1390	690	160	260
Pine, ponderosa	—	420	5.32	580	9.4	1.29	1130	460	150	190
Pine, white, eastern	11.3	310	4.80	440	8.6	1.24	900	380	110	120
Red cedar, western	—	220	4.56	460	7.5	1.11	990	350	90	130
Redwood, old growth	—	240	6.15	700	10.0	1.34	940	480	90	140
Spruce, Engelmann	13.0	350	4.48	410	9.3	1.30	1200	390	110	180
Alder, red	—	420	5.82	440	9.8	1.38	1080	590	—	—
Ash, white	—	940	7.41	1160	15.4	1.74	1950	1320	—	—
Aspen, quaking	—	260	4.25	370	8.4	1.18	850	350	—	—
Basswood, American	—	350	4.73	370	8.7	1.46	990	410	—	—
Beech, American	—	1010	7.90	1010	14.9	1.72	2010	1300	—	—
Birch, yellow	—	920	8.17	970	16.6	2.01	1880	1260	500	620
Cherry	—	560	7.11	690	12.3	1.49	1700	950	—	—
Cottonwood, eastern	—	580	4.91	380	8.5	1.37	930	430	—	—
Elm, American	—	660	5.52	690	11.8	1.34	1510	830	—	—
Hickory, mockernut	—	—	8.94	1730	19.2	2.22	1740	1820[c]	620	660
Maple, sugar	—	—	7.83	1470	15.8	1.83	2330	1450	370	360
Oak, red	—	800	6.76	1010	14.3	1.82	1780	1290	470	435
Oak, white	—	800	7.44	1070	15.2	1.78	2000	1360	340	310
Sweetgum	17.3	760	6.32	620	12.5	1.64	1600	850	260	260
Sycamore, American	—	720	5.38	700	10.0	1.42	1470	770	—	—
Walnut, black	—	690	7.58	1010	14.6	1.68	1370	1010	—	—
Willow, black	15.8	—	4.10	430	7.8	1.01	1250	—	210	230
Yellow poplar	22.4	540	5.54	500	10.1	1.58	1190	540	220	210

[a] 1000 psi.
[b] million psi.
[c] Pecan hickory. See footnote to equivalent Table 11-1.

Table 11-1A. (Continued)

Species (European)	Tension ‖[a]	Tension ⊥	Compression ‖[a]	Compression ⊥	Static bending MOR[a]	Static bending MOE[b]	Shear	Hardness	Toughness
			lb/in.² (psi)					lb	in.lb
Fir, white	11.3	200	4.78	680	9.7	1.40	800	370	210
Larch	15.2	320	7.83	1060	14.1	1.96	1280	750	210
Pine, black	14.8	290	5.65	—	14.9	1.70	1420	570	140
Pine, Scots	14.8	420	7.83	1420	14.2	1.70	1420	530	245
Spruce	12.2	220	4.35	590	8.7	1.32	770	330	170
Alder	13.3	290	7.83	930	12.0	1.66	640	—	190
Ash	23.3	1000	7.40	1570	17.1	1.90	1810	1340	240
Aspen	15.7	390	6.81	380	11.2	1.55	1120	570	140
Beech	18.9	510	6.67	1150	15.1	1.90	1780	1060	275
Birch	19.4	1000	7.25	1570	20.9	2.34	1710	—	295
Black locust[c]	19.3	610	10.15	1480	14.8	1.61	1810	1650	230
Chestnut	19.1	—	7.10	—	10.9	1.28	1130	680	200
Elm, field	11.3	565	8.00	1420	12.6	1.56	1000	1100	210
Hornbeam	22.2	550	7.83	2420	20.3	2.13	2450	1670	290
Lime tree (basswood)	12.0	710	7.39	260	15.1	1.05	640	—	170
Oak, red	15.7	480	6.10	1670	16.8	1.68	1840	1010	—
Oak, white (English)	15.8	430	6.23	1130	13.2	1.78	1680	1210	260
Plane (Sycamore)	—	750	6.52	930	14.1	1.49	1420	1060	240
Poplar, "robusta"	12.2	220	5.10	—	9.3	1.56	1100	420	—
Walnut	14.2	510	10.29	1710	20.9	1.78	1000	1170	330
Willow	12.0	350	3.77	490	7.7	1.42	970	350	245

[a]1000 psi.
[b]million psi.
[c]non-European (planted).
See footnote to equivalent Table 11-1.

Table 11-1A. (Continued)

Species (Tropical)	Tension ‖[a]	Tension ⊥	Compression ‖[a]	Compression ⊥	Static bending MOR[a]	Static bending MOE[b]	Shear	Hardness	Toughness
			lb/in.² (psi)					lb	in.lb
Afara	14.9	300	8.84	1130	15.8	1.78	1060	510	170
Afzelia	—	—	9.71	—	17.7	2.17	2300	—	—
Aiélé	—	290	5.51	—	24.2	1.56	—	—	—
Amazakoué	18.8	560	11.45	—	19.1	2.43	1800	840	390
Antiaris, Ako	—	—	5.94	—	10.0	1.35	1450	—	—
Balsa	10.6	145	1.30	145	2.8	0.37	160	90	75
Beté	17.0	320	9.30	—	18.6	1.70	1130	1250	225
Bubinga	—	590	10.00	—	16.2	1.85	1420	—	180
Dibetou	15.1	275	7.40	1200	11.7	1.85	1280	920	300
Difou	—	—	13.34	—	23.1	2.10	2710	—	120
Idigbo	—	330	6.67	—	10.9	1.30	1280	810	120
Iroko	11.3	360	9.90	—	16.1	1.63	1570	700	90
Kosipo	—	275	7.10	—	12.8	1.15	850	—	215
Mahogany, African	8.7	290	6.52	1000	12.3	1.42	1130	810	140
Mahogany, American	—	350	7.10	1200	12.0	1.21	1410	680	180
Makoré	11.0	300	7.54	—	13.9	1.99	1200	550	110
Meranti, red	—	—	7.54	—	13.0	1.69	—	—	—
Niangon	—	—	8.55	—	19.9	1.62	2280	—	—
Obeche	6.9	190	5.70	—	10.4	0.97	780	440	100
Okoumé	8.3	260	5.51	770	10.3	1.32	860	370	85
Opepe	—	—	8.60	—	18.4	2.00	—	—	140
Padauk	—	320	10.60	—	20.2	1.80	1120	950	225
Palissander	—	860	8.55	—	17.0	1.78	2130	750	290
Ramin	—	—	10.15	—	18.4	2.31	1540	—	—
Sapele	12.5	350	8.55	1200	15.7	1.42	1200	1360	245
Seraya, white	—	—	7.10	—	13.0	1.68	—	790	—
Sipo	15.7	320	8.55	1280	14.1	1.63	1350	330	140
Teak	17.1	590	9.42	—	19.0	1.99	1280	590	210
Tiama	—	300	6.81	—	11.0	1.43	1390	920	150
Zebrano	—	490	7.10	—	14.2	1.63	1120	570	200

[a]1000 psi; [b]million psi. See footnote to equivalent Table 11-1.

cific cases, such as rotors of helicopters and propellers of small airplanes.

The development of transverse tensile stresses is carefully avoided in wooden structures, because the strength of wood loaded in that manner is very small, and also because the formation of checks, due to shrinkage in drying, may practically reduce such strength to zero.

Strength in Compression

The strength of wood in compression is also different if loads are applied parallel or transverse to the grain. Axial compression strength is higher—up to about 15 times—and varies between 25 and 95 N/mm^2 (25–95 MPa, 3625–13,775 psi; Table 11-1), whereas transverse values vary between 1 and 20 N/mm^2 (1–20 MPa, 145–2900 psi). It has been observed, that in softwoods, tangential compression strength is higher than radial, whereas in hardwoods the situation is opposite (32).[5]

The strength of wood in axial compression is smaller in comparison to metals, but higher in comparison to most other construction materials, such as brick and stone. Also, wood differs from other materials (metals, minerals) because its strength in compression is about half that compared to its strength in tension. The difference is due to the structure of wood. The skeleton of wood is made of cellulose chain molecules which impart very high strength in axial tension. The other constituents (hemicelluloses and lignin) contribute to compression strength, but cellulose also supports compression loads.

The failure of wood due to axial compression may be traced to rupture of intercellular layers, cleavage or shearing, buckling or folding of cells, and rupture of cell walls (41, 44). Pits in the walls constitute positions of reduced strength. Inversely, stressing in transverse compression results in the change of the shape of cell cross-sections, and reduction of the size of cell cavities; with increasing loads, these changes advance gradually from the surface to the interior of the wood.

Stressing in transverse compression takes place, for example, in railroad ties, whereas axial compression occurs in columns. In the latter case, the ratio of length to least dimension in width of the wooden member is important. If this ratio is smaller than 11:1, the strength of the column depends entirely on the strength of wood in axial compression; if greater, the stiffness of the wood (modulus of elasticity) to resist buckling is also important (76).

Certain wood species are considered to possess a "warning" ability, expressed with a characteristic noise before they break when stressed in axial compression. Such woods are reported to be spruce, larch, pine, beech, birch, oak, and black locust (38, 70).

Strength in Shear

Shear may exist in longitudinal or transverse planes. Longitudinal shearing stresses are present when wooden members are stressed in bending. The strength of various woods in axial shear varies between 5 and 20 N/mm^2 (5–20 MPa, 725–2900 psi; see Table 11-1).

Strength in transverse shear acting on a cross-section is 3–4 times greater than in axial shear (41), but this is of no practical importance, since wood fails first in axial or rolling shear than in transverse shear.

Loads acting on a transverse plane may be a cause of rolling shear stresses; this action may (very seldomly) cause "rolling" of the fibers in an axial plane. The strength of wood in rolling shear is considerably lower in comparison to strength in longitudinal shear (62, 76).

Finally, oblique shear stresses may occur with axial tension or compression loads. Under the influence of such loads, oblique shear planes are formed in the cell walls and between cells. The highest shear stress occurs at an angle of about 45°, but the structure of wood as a whole (presence of various cell types, rays, earlywood and latewood, etc.) contributes so that the planes of rupture are formed at an angle of about 60–70° in relation to the axis of the stressed member (62, 76).

The strength of wood in axial shear has the greatest practical importance. Under the influence of shearing loads, wood usually fails in this manner.

Strength in Bending

Strength in static bending is an important mechanical property, because in most structures wood is subject to loads which cause it to bend. The typical case is of wood as a beam bent under external forces, which act transversely to its axis. Under their action three stresses develop—tension, compression, and shear (Figure 11-3). These stresses are axial. Tension stresses tend to lengthen the wood fibers, compression stresses tend to make them shorter, and shear stresses tend to make the upper part of the beam slide over its lower part.

In the usual case of a simple beam,[6] tension and compression stresses are, respectively, highest in the lower (convex) and the upper (concave) surfaces, gradually diminishing toward the center, and are zero in the neutral plane. Inversely, shear stresses are highest in the neutral plane and zero at the surfaces. The distribution of stresses along the length of the beam depends on the manner of loading (center, third-point, uniform).

The strength of wood in bending is usually expressed by the modulus of rupture, which shows the highest stresses in the outermost fibers of wood when the beam breaks under the influence of a load, which is applied gradually for a few minutes. Modulus of rupture varies between 55 and 160 N/mm^2 (55–160 MPa, 8000–23,200 psi; see Table 11-1), and this shows that bending strength is similar to the strength in axial tension. For this reason, modulus of rupture may be used as an index of strength in axial tension, when values of the latter property are not available.

The bending strength of wood is lower in comparison to metals, but higher than most nonmetallic materials. Wood often has the advantage of a lower modulus of elasticity and a more favorable relation of strength to weight.

Cleavage

The resistance of wood to cleavage refers to exterior forces acting in the form of a wedge. Due to its structure, wood has a low axial resistance to cleavage (i.e., it may be easily split). This is an advantage for certain uses (e.g., splitting fuelwood) and a disadvantage for others (e.g., wooden members splitting when nailed or screwed).

Different wood species possess a different resistance to cleavage. Softwoods and light hardwoods (fir, spruce, poplar) have low resistances in contrast to heavier hardwoods (maple, birch, sycamore, black locust, hornbeam). Also, the resistance is lower when the load is applied on a transverse surface and in a radial direction (due to the presence of rays).

The test of cleavage is similar to that of transverse tension (a similar specimen is used; see Figure 11-4), and cleavage values are expressed in N/mm (Newton per mm of width of the cleavaged surface).[7]

Toughness

Toughness (or energy in dynamic bending) refers to resistance against sudden loading in contrast to the previous cases, where the loads are static or slowly applied. This property is important for certain wood uses (e.g., tool handles, sport items, boxes and crates, etc.).

The energy absorbed by wood is higher with sudden rather than static loads. For example, a beam can support about double the load in the former case. Also, it has been observed that,

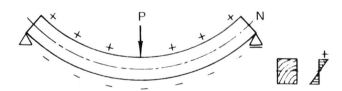

Figure 11-3. Stresses in a beam under the action of load P: + compressive and − tensile stresses; N, neutral plane. The distribution of compressive and tensile stresses is graphically shown (right) in relation to the cross-section of the beam.

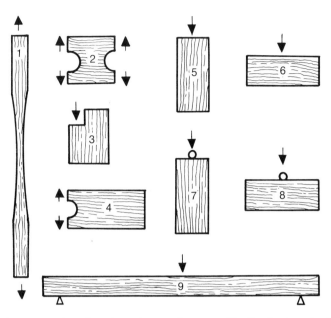

Figure 11-4. Schematic representation of specimen shapes and manner of loading for determination of the mechanical properties of wood. The shapes (especially 2, 3, and 4) are representative of American (ASTM) standards: 1, axial tension; 2, transverse tension; 3, shear; 4, cleavage; 5, axial compression; 6, transverse compression; 7, axial (end) hardness; 8, side hardness; 9, static bending. (The specimens are shown schematically, not in scale.)

with sudden loading, the deflection of a beam is about double in comparison to static loading (69, 76).

Elasticity

From an elasticity point of view, wood has an intermediate position in comparison to other materials. The values of the modulus of elasticity varies between 2500 and 17,000 N/mm^2 (2.5–17 GPa, 362,500–2,465,000 psi; see Table 11-1). Comparative data included in Table 11-2 (column 4) show that wood has a lower modulus of elasticity than other materials (it bends more under a certain load); however, if weight (density, specific gravity) is taken into consideration (column 5), wood is comparable to steel. Modulus of elasticity is different in the three growth directions (i.e., axial, radial, and tangential). The above values apply to the axial direction, whereas transverse values are only 300–600 N/mm^2 (300–600 MPa, 43,500–87,000 psi). There are no important differences between radial and tangential directions.

Modulus of elasticity is determined from static or dynamic bending tests—usually static. The values derived from dynamic tests are a little higher (on the average, about 10–15%) (8), and those from static bending (with a simple beam) are a little lower than those obtained from tensile tests, because part of bending derives from shear deformation (48). More accurate values may be determined from tests of axial tension, but such tests present practical difficulties. Modulus of elasticity values determined from axial compression are lower than in bending (35).

Aside from dynamic bending tests, modulus of elasticity may be determined dynamically by vibration of wood specimens with sound waves (41).

Hardness

Hardness is a measure of the resistance of wood to the entrance of foreign bodies in its mass. This resistance is higher—up to about double in the axial direction than sidewise, but the dif-

Table 11-2. Mechanical Properties of Wood and Other Materials in Relation to Specific Gravity[a]

Material	Specific gravity 1	Tensile strength 2	T/SG 3	MOE 4	MOE/SG 5
			N/mm²		
Wood					
Spruce	0.44	84	191	9,100	20,680
Oak, red	0.87	108	124	11,650	13,290
Concrete	2.5	4	1.6	13,800	5,520
Glass	2.5	50	20	72,400	28,960
Aluminum	2.8	250	89	69,000	24,640
Cast iron	7.0	140	20	82,800	11,830
Steel	7.9	450	57	207,000	26,200
Plastics					
PVC	1.3	60	46	5,800	4,460
Polysterene	1	70	70	3,450	3,450
			lb/in² (psi) × 10³		
Wood					
Spruce	0.44	12.2	27.7	1,320	3,000
Oak, red	0.87	15.7	14.2	1,680	1,930
Concrete	2.5	0.58	0.23	2,000	800
Glass	2.5	7.2	2.9	10,500	4,200
Aluminum	2.8	36.2	12.9	10,000	3,570
Cast iron	7.0	20.3	2.9	12,000	1,710
Steel	7.9	65.2	8.2	30,000	3,800
Plastics					
PVC	1.3	8.7	6.7	400	310
Polysterene	1	10	10	500	500

Sources: ACI Manual of Concrete Practice. 1985. Part 1 (Materials), and General Properties of Concrete, ACI Publication; Baumeister, T. (ed.). 1967. *Standard Handbook for Mechanical Engineers*. New York: McGraw-Hill; Ref. 9; Fellers, W. O. 1990. *Materials Science, Testing and Properties for Technicians*. Englewood Cliffs, New Jersey: Prentice Hall; Junival, R. C. 1967. *Stress-Strain-Strength*. New York: McGraw-Hill. U.S. Forest Products Laboratory 1987. *Wood Handbook*. Madison, Wisconsin (73).
[a]Indicative values; there is a wide variation, especially in metals (and concrete) according to composition and processing. For example, the range of tensile strength is reported for aluminum 70–550 N/mm² (10,000–80,000 psi), cast iron 120–400 N/mm² (18,000–60,000 psi), steel 350–2200 N/mm² (50,000–320,000 psi), and PVC 10–40 N/mm² (1500–9000 psi). Specific gravity and strength of woods: air-dry.

ference between radial and tangential surfaces is seldom important (39). Hardness is related to the strength of wood in abrasion and scratching with various objects, as well as to the difficulty or ease of working wood with tools and machines. It is an important property for various uses, such as floors, furniture, sport items, pencils, etc.

Some woods are relatively soft (poplar, willow, basswood, pine), others have a medium hardness (pine, fir, juniper, walnut), and some are hard (yew, oak, elm, black locust, ash, beech, sycamore, hornbeam, maple, birch, olive). Tropical woods include a range from very soft (balsa) to very hard species.

FACTORS AFFECTING MECHANICAL PROPERTIES

The mechanical properties of wood are affected by various factors—mainly, moisture, density, temperature, duration of loading, and defects.

Moisture

Moisture affects the mechanical properties when it changes below the fiber saturation point. When moisture is reduced, strength increases—and vice versa (Figure 11-5, Tables 11-3 and 11-3A). This increase is due to changes in the cell walls, which become more

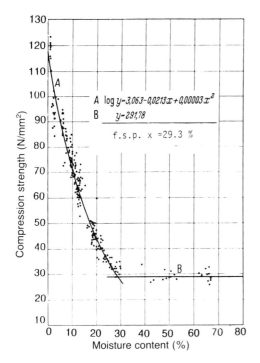

Figure 11-5. Effect of moisture on compression strength of Scots pine. Strength is reduced with increasing moisture content—up to the fiber saturation point. (After Ref. 42; reproduced by permission from Springer-Verlag.)

Table 11-3A. Effect of Moisture on Toughness

Species	% increase of green over air-dry values[a]	
	R	T
Douglas-fir	5.0	0
Hemlock, western	7.1	−19.0
Pine, loblolly	93.7	46.2
Redcedar, western	133.3	76.9
Hickory	12.9	9.1
Oak, white	114.7	119.3
Sweetgum	30.8	26.9
Yellow poplar	45.4	42.8

[a]Green toughness appears higher in all woods except Douglas-fir (no change) and western hemlock—in tangential loading. This is in contrast to the other mechanical properties, as shown in Table 11-3. *Source:* Ref. 73. North American species.

compact. Their structural units (i.e., the microfibrils) come closer together and the attractive forces between cellulose chain molecules become stronger (64). Also, increase is affected by shrinkage. With loss of moisture from cell walls, the mass of wood substance contained in a certain volume increases.[8]

The magnitude of moisture influence is different in different properties. According to studies of this relationship, a 1% change of moisture changes the strength in axial compression by 6%, bending strength (modulus of rupture) 5%, hardness 2.5–4% (more in the axial direction), modulus of elasticity (in static bending) 2%, etc. (76).[9] An exception is toughness, which is not increased with decreasing moisture but sometimes actually decreases as the wood dries (Table 11-3A). This is attributed to the greater deformation of green wood. Toughness is affected not only by the magnitude of load but also by the stiffness of wood. Dry wood may carry a greater load but bends less before it breaks (73, 76).

Due to the effect of moisture, and in order to have comparable results, mechanical properties are determined at a constant moisture content—that is, in green condition (above the fiber sat-

Table 11-3. Effect of Moisture on Strength Properties

Species	Tension ∥	Compression ∥	Static bending		Shear	Cleavage	Hardness	
			MOR	MOE			∥	⊥
	% increase of air-dry over green values							
Fir, white	35.3	64.2	45.7	16.0	33.3	10.7	73.0	32.6
Spruce	41.2	68.1	38.6	58.5	45.9	28.0	66.0	40.5
Pine, Scots	—	120.0	117.4	13.2	66.7	—	—	—
Birch	—	121.7	145.0	27.0	71.4	—	95.0	—
Chestnut	57.0	78.6	—	—	33.3	—	—	—
Beech	34.0	63.2	43.1	22.8	38.5	15.0	45.4	41.0
Oak, white (sessile)	43.1	56.0	28.1	28.9	25.0	35.3	28.0	19.8

Source: Ref. 27. European species.

uration point) or in air-dry conditions (usually 12% and sometimes 15%) (32, 41).

Strength, or other mechanical property values based on different moisture contents, may be corrected for the purpose of comparison (28, 36, 73). The following relationship is applied (28):

$$P = P_{12} \left(\frac{P_{12}}{P_g}\right)^{-\left(\frac{M-12}{M_p-12}\right)} \quad (11\text{-}4)$$

where

P = strength at M moisture content
M = moisture content (%)
M_p = fiber saturation point[10]
P_{12} = strength at 12% mc
P_g = strength of green wood (mc > M_p)

Also, correction may be based on the linearized form of the above relationship (7):

$$\log A_3 = \log A_1 + \left(\frac{M_1 - M_3}{M_1 - M_2}\right) \cdot \log \frac{A_2}{A_1} \quad (11\text{-}5)$$

where A_1, A_2 and M_1, M_2 are known values of strength and moisture, and A_3 is the strength value corresponding to moisture M_3.[11] This correction is based on observations according to which, within certain limits, the relationship of the logarithms of the values of a property and of the corresponding moisture content is linear (48).

In correcting strength values, it is assumed that moisture content is uniformly distributed in the mass of wood. Also, it is obvious that corrections refer to moisture contents below the fiber saturation point.

Density

Density is the best and simplest index of the strength of wood without defects. With increasing density, strength also increases. This is because density is a measure of the wood substance contained in a given volume. Greater density derives from a greater proportion of cells with thick walls and small cavities—and this results in greater strength of denser defect-free wood.

The relationship of density and strength varies with different properties and species (79), but in most cases it is linear (48, 54) (Figure 11-6). Differences between species are due to different cellular composition and different extractive content (strength is reduced by removal of extractives; Figure 11-7), with the result that woods of the same density may have different strength and vice versa. For the same reason, density is not an accurate measure but, as previously mentioned, an index of strength.

Density is related to strength within a species, especially when heartwood and sapwood are separately compared, by the following relationship (76):

$$\frac{S}{S'} = \left(\frac{g}{g'}\right)^n \quad (11\text{-}6)$$

where S and S' are values of strength corresponding to densities g and g'. The value of n varies between 1.25 and 2.50, depending on the property. For example, it is 1.25 for axial compression and modulus of elasticity, 1.50 for strength in static bending (modulus of rupture), and 2.50 for transverse compression and hardness.

Structure

The effects of density on mechanical properties derive from structural differences that produce density variations. Characteristics that lead to higher or lower density (ring width, proportion of latewood, etc.; see Chapter 3) affect, respectively, the strength of wood. Thus, in softwoods, where latewood tends to decrease with increasing ring width, fast-growing trees produce wood of lower strength; in ring-porous hardwoods, wider rings are related to higher proportion of latewood and higher strength, whereas in diffuse-porous hardwoods, there is no clear relationship due to a lack of distinct latewood. These are generalized relationships.

Cellular characteristics are fundamental features that affect mechanical properties, but such correlations are rare in literature. A study of teak (*Tectona grandis*) revealed the following

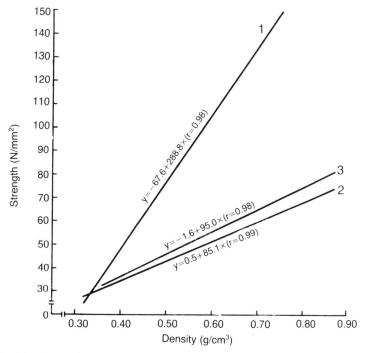

Figure 11-6. Relationships of strength (y) and density (x): 1, static bending (MOR); 2, axial compression; 3, hardness. Based on 41 (hardness 31) European and North American species.

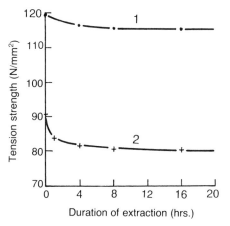

Figure 11-7. Effect of removal of extractives on axial tension strength: 1, birch (extraction with ethanol-benzol); 2, spruce (extraction with hot water). (After Ref. 55; reproduced by permission from VEB Fachbuchverlag.)

(19): An increased number of pits per fiber had a definite weakening effect on tensile strength. Fiber-wall volume and mean thickness of individual fiber walls had a rough relation to that property. Compressive strength was usually determined by the abundance of parenchyma cells, and length and aggregation of vessel members. Wood specimens with shorter vessel members were more resistant to compression. Bending strength was related to the frequency of fibers, distribution of soft elements, and aggregation of vessels.

Ultrastructural characteristics are also very important. For example, the microfibrillar angle of the S_2 layer has a profound effect. Large angles that characterize juvenile wood result in lower strength (Figure 11-8).

Temperature

In general, the strength of wood is reduced with increasing temperature (15, 16, 59). Reduction is influenced by such factors as moisture content of wood, level of temperature and duration of heating, manner of loading, species of wood, and dimensions of members of wooden structures. Reduction may also derive from defects (e.g., checks, which may result from moisture content changes accompanying changes of temperature).[12]

Figure 11-9 shows the effect of temperature

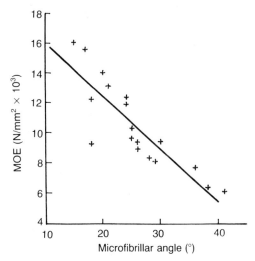

Figure 11-8. Relationship of microfibrillar angle and modulus of elasticity in spruce wood. (After Ref. 11.)

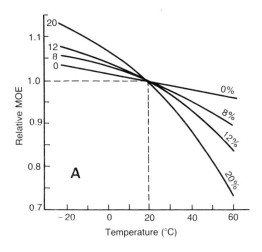

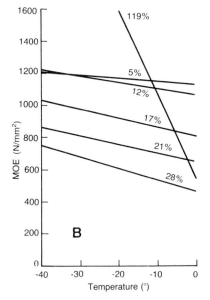

Figure 11-9. Effect of temperature and moisture on modulus of elasticity: (A) average of six species, (B) spruce. (After Refs. 8, 30, and 60.)

and moisture. Low temperatures are shown to increase the modulus of elasticity, while temperatures higher than room temperature (20°C, 68°F) decreased this property. The effect of moisture is respectively opposite; higher moisture contents have an increasing effect at low temperatures and a decreasing effect at higher temperatures. Above the fiber saturation point, low temperatures also have an increasing effect; a sharp linear increase of modulus of elasticity, between zero and −20°C (32 to −4°F) was observed in spruce wood with a 119% moisture content (30). Such an increase in frozen wood is attributed to additional stiffness provided by ice formed in the cell lumina. The effect of temperature and moisture is also depicted in Figure 11-10; increasing temperature reduced the modulus of elasticity, and the reduction was greater at higher moisture contents.[13]

The duration of heating is very important. Temperatures lower than 100°C (212°F) have no adverse effect when wood is exposed for a short period of time, but temperatures higher than 65°C (150°F) may have a permanent adverse effect with a long duration of heating. A temperature of 200°C (400°F) will reduce strength in a few minutes (28). Figure 11-11 shows a relationship between temperature, duration, and bending properties. The degrading effect of long duration is due to chemical decomposition of wood.

Various properties are affected by temperature in different ways (40, 59). Toughness is especially sensitive. At low moisture contents it is reduced, and at high moisture contents it increases with increasing temperature (34). Frozen wood, saturated with water, was found to have a higher strength in static bending (modulus of rupture), but toughness was considerably lower in comparison to air-dry wood at room temperature (40).

Heating in water results in smaller reduction of strength in comparison to steaming, but hot air has the smallest effect (64, 76). Wooden

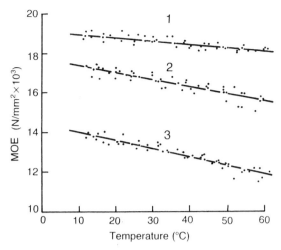

Figure 11-10. Effect of temperature and moisture on the modulus of elasticity of Scots pine wood (1, 0%; 2, 15%; 3, 30% moisture content). (After Ref. 29; reproduced by permission from VEB Fachbuchverlag.)

members of small dimensions are affected more than large members, and repeated applications of high temperatures are additive (34).

Duration of Load

The duration of load has an important influence on the strength of wood (i.e., on the magnitude of load that a wooden structure can support). For this reason, results of laboratory tests, based on loading of a few minutes or seconds, have only comparative value and are applicable in practice only after correction (79).

The magnitude of strength change in relation to time[14] is influenced, under similar other conditions, by the manner of loading (i.e., if the load is permanent or periodic). Under the action of a permanent load, wood (and other materials) exhibits the phenomenon of *creep* (i.e., the deformation increases with the passage of time). Periodic loads result in *fatigue*. In both cases, strength is reduced. Research has shown

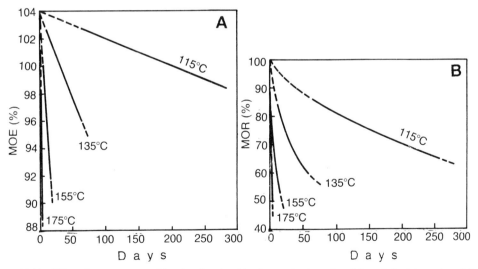

Figure 11-11. Effect of temperature and heating duration (days) on reduction of modulus of elasticity and modulus of rupture. (The relationships are based on four softwoods and two hardwoods, and small specimens exposed to dry heat.) Changes are shown in percent of the value at 23°C (73.5°F). (Courtesy of U.S. Forest Products Laboratory.)

that permanent loading reduces the strength to 50–75% of the values obtained from static tests of short duration (Figure 11-12). Periodic loads have a similar influence, but the residual strength (endurance limit) may be as low as 25% of the static values (32, 76).

In general, the behavior of wood to load duration is influenced by various factors which are related to wood (species, density, moisture) and to loading conditions (magnitude of load, duration, rhythm of change, etc.). The resistance to fatigue increases with density. Creep increases with moisture content and temperature (62).

Defects in Wood

Defects reduce the strength of wood. The degree of their influence depends on the kind, size, and position of the defect, and the manner of loading.[15] The most important defects that reduce strength are: knots, grain deviations, checks, compression and tension wood. Secondary attacks by fungi, insects, and other factors which cause deterioration of wood, also have an adverse effect.

Knots. The adverse effect of knots is mainly due to local grain deviations and checks caused by their presence. Checks are formed due to differential shrinkage and swelling of knots, because their density is higher, they usually contain compression or tension wood, and their fiber orientation is different in comparison to those of the adjacent wood. Intergrown (tight) knots cause greater deviations and more checking, whereas encased (loose) ones act, in addition, through absence of material or discon-

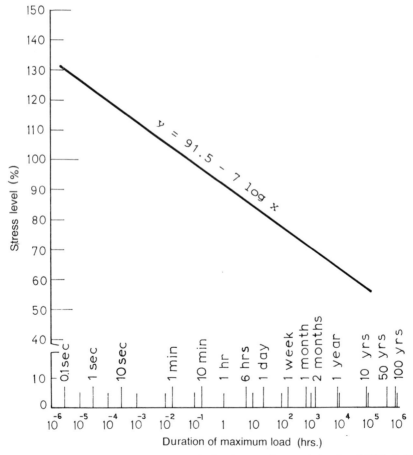

Figure 11-12. Effect of duration of loading on the strength of various woods on static bending (MOR). A 100% strength corresponds to laboratory testing conditions—duration of about 1 min. The woods are various softwoods and hardwoods in air-dry or green condition. (Adapted from Ref. 53.)

tinuity of tissues. Comparatively, the effect of intergrown and encased knots may be similar or different (38, 43, 75). Under the same conditions, this depends mainly on the manner of loading (e.g., the same effect of both knot types was observed in axial tension). In general, the diameter of knots has a greater effect than their number.[16] The effect of holes caused by falling out of encased knots is different from that of the same sized holes opened with tools (41).

The presence of knots greatly reduces the strength of wood in axial tension (20, 47, 73) (Figure 11-13), due to associated grain deviations. As previously mentioned, transverse ten-

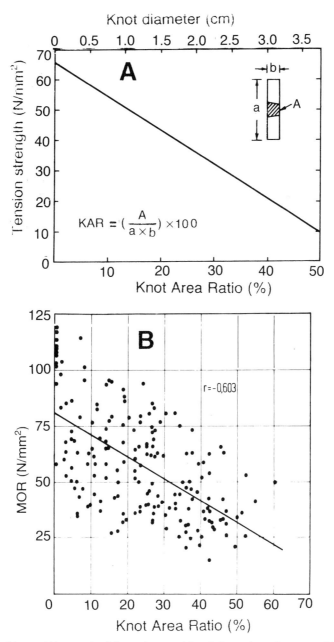

Figure 11-13. Effect of knots: (A) strength of Scots pine wood in relation to knot diameter and knot area ratio, (B) strength of Douglas-fir wood in static bending (modulus of rupture) in relation to the knot area ratio. (A, adapted from Ref. 17; B, after Refs. 20, and 31.)

sile strength is very low in comparison to axial strength. Strength in axial compression is reduced less, whereas strength in transverse compression may increase by the presence of intergrown knots (38, 70). (This is a practical advantage when wood is used for certain structures, such as railroad ties.) Bending strength (modulus of rupture) is considerably influenced by the position of knots. Knots found near the middle of the lower side of simple beams—supported at both ends—have the greatest adverse effect. The effect is smaller when knots are in the middle of the upper side, and very small when found between the two sides or near the ends of the beam. Strength in horizontal shear is affected little or not at all—in fact, it may increase, because knots interrupt the continuity of checks, which greatly reduce this strength (77). Also, hardness and the resistance of wood to cleavage increase. Modulus of elasticity is reduced (39, 41, 47). In general, knots (or other defects) in the edges of wooden members have a greater adverse effect in comparison to knots of the same size located in their interior (6, 47).

Grain Deviations. The influence of grain deviations is basically due to the difference between axial and transverse strength of wood. Because the former is higher (e.g., tension or compression strength), it is obvious that strength is reduced by loading at an angle (Figure 11-14); a greater angle results in a greater reduction. Tensile strength is affected more than compressive strength, while the reduction of bending strength is intermediate (5, 41, 62, 76) (Figure 11-15). Modulus of elasticity is also reduced by grain deviations, but the greatest effect is shown in toughness. It has been observed in beech that a 5° deviation reduces strength by 10%, and an angle of 10° up to

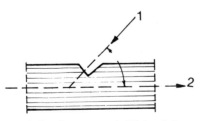

Figure 11-14. Loading at an angle (1) in relation to grain direction (2).

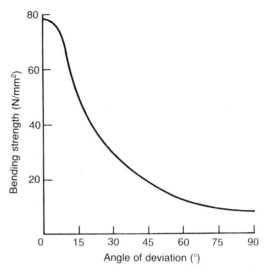

Figure 11-15. Effect of grain deviation on bending strength (MOR) of air-dry fir (*Abies*) wood. (After Ref. 5.)

50% (41). The effect of grain deviation to horizontal shear is minimal—in fact, such strength may increase (76).

The reduction of strength due to grain deviations may be calculated from the relationship (28, 41):

$$N = \frac{PQ}{P \sin^n \theta + Q \cos^n \theta} \quad (11\text{-}7)$$

where

N = strength at an angle θ from the fiber direction
P = strength parallel to grain ($\theta = 0°$)
Q = strength perpendicular to grain ($\theta = 90°$)

The following values are given to n: axial tension 1.5–2, axial compression 2–2.5, static bending 1.5–2, toughness 1.5–2, modulus of elasticity 2.

Checks. The influence of checks depends on their size, direction, and the manner of loading. Axial tensile strength is unaffected or very little affected when checks have the same direction with the forces exerting tension. Inversely, transverse tensile strength is greatly reduced. The effect of checks on compressive strength is small, comparatively smaller in transverse than axial compression, but the reduction of strength in horizontal shear is considerable. It is evident that in loading of beams, the influence is greater when checks are near the neutral plane, where horizontal shear stresses are greatest. The resistance to cleavage is greatly reduced when checks are present. Checks usually coexist with grain deviations and knots.

Compression Failures. Compression failures (see Chapter 7) reduce considerably the strength of wood in static bending and toughness, especially when found on the side stressed in tension.

Compression and Tension Wood. The presence of compression or tension wood has sometimes a positive and sometimes a negative influence on the strength of wood, whereas in certain cases no important differences are found (14, 43, 56, 76). The results are influenced by the manner of loading, species of wood, and extent of these abnormalities (43, 56). Compression wood has lower modulus of elasticity, and lower static bending strength and toughness (76), whereas tension wood was found (in beech) to have very low strength in axial compression but higher strength in axial tension and toughness (13). The relationships of strength to density and moisture, which apply to wood of normal structure, do not apply to compression or tension wood of the same species. In general, these abnormalities should be considered to have a negative effect on strength, because they may cause checks or warping with changing moisture content due to their higher axial shrinkage (see Chapter 10); therefore, they should be avoided in wooden structures bearing loads, especially when pronounced. Sudden breakages, which sometimes occur in wood (ladders, matches, etc.), are due to compression or tension wood—but also to grain deviations, or a combination of such defects. Compression wood breaks with characteristic brush failures.

Other abnormalities, such as juvenile wood near the pith, as well as pith itself, reduce the strength of wood (43).

Decay and Other Biological Deterioration. Attack of wood by fungi, insects, and marine borers may minimize the strength of wood, but incipient stages of attack also have adverse effects. Toughness is especially reduced by incipient decay. Plant parasites, such as mistletoe, also reduce strength. In any case, the use of attacked wood should be avoided where strength is of primary concern.

DETERMINATION OF MECHANICAL PROPERTIES

The mechanical properties of wood are determined from small specimens (without defects), or structural members (beams, lumber boards, etc.), or wooden structures (e.g., trusses). Small specimens present the possibility of wider sampling (more specimens) and the systematic study of the effects of various factors (moisture content, density, growth-ring structure, physical or chemical treatment, etc.) on mechanical properties. However, such effects are difficult to transfer to full-size members due to variation of wood structure and presence of defects. Engineers prefer testing full-size members.

The following discussion refers mainly to small specimens, which may be studied to determine the strength of wood of standing trees, logs, or lumber (1, 21).

Sampling

There is no internationally accepted method of sampling. The American standard procedure (ASTM D143) (1) provides for a cruciform pattern with matched (double) specimens to be tested in green and air-dry conditions (Figure 11-16); a modified pattern is proposed for small-diameter (< 30 cm, 12 in.) logs. A tree trunk is cross-cut to logs 4 ft (1.20 m) in length, and specimens are taken from the third and fourth log—from the tree stump. Cardinal points (north, south, east, west) are noted in order to study their effect. Specimens are 5 × 5 cm (2 × 2 in.) or 2.5 × 2.5 cm (1 × 1 in.) for smaller trees, length and shape differ according to test.[17] In Europe there is no suggestion for such a procedure, specimens are small

182 II / PROPERTIES

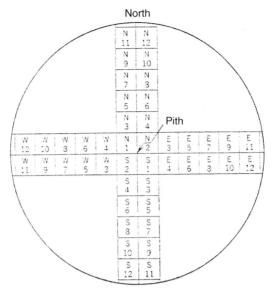

Figure 11-16. The cruciform pattern of sampling to study property variation within a tree trunk (bolt); matched (double) specimens are tested in green and air-dry condition. (After Refs. 1 and 76.)

(see Table 10-2), and shapes differ even among countries (2, 3, 4, 21, 48). A proposal for international standardization (ISO 3129) specifies a (single) cruciform pattern for small logs (18 cm, about 7 in. or less), and sampling from a central "plank" for larger logs. An (impractical) template has also been suggested, sampling in analogy to the volume of wood at different distances from the center (pith) of a log (48).

Systematic sampling, such as by application of the American standard, has been extensively used in the past to determine the properties of various species, but is now practically abandoned; current practice is to sample at random from logs or lumber, and according to the purpose of tests (evaluation of certain factors affecting strength).

Application of standards is desirable to permit comparison of results of research, considering that wood is a material of international commerce. For this reason, standardization efforts were initiated by the FAO (Food and Agricultural Organization of the United Nations) years ago (24); although certain proposals have been made, these as well as recent proposals of the International Organization of Standardization (ISO) (36) have not been generally accepted.[18]

Irrespective of the method of sampling, sample size (number of specimens investigated) is an important criterion in assessing wood properties. Sample size depends on property variation and is determined by statistical considerations.

Testing—Results

Tests for the determination of the mechanical properties of wood (or other materials) are carried out with special testing machines which allow loading of a specimen with a measured load applied gradually or suddenly. Loading is performed with vertical movement of a steel plate, hydraulically or mechanically (with screws), or with a falling weight. The specimens are held with special accessories or supports, and deformation is measured with special instruments (Figure 11-17).

The results are related to loads, deformations, and size of the specimens, and are calculated as follows (1, 12, 71, 76).

Strength in *tension* (*axial* or *transverse*)[19] is determined from the relationship

$$S = \frac{P}{A} \qquad (11\text{-}8)$$

where

MECHANICAL PROPERTIES   183

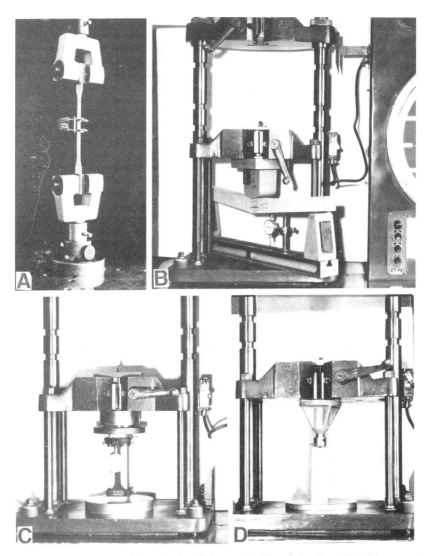

Figure 11-17. Mechanical tests: (A) axial tension (parallel to grain), (B) static bending, (C) axial compression, (D) end hardness. (Photograph A courtesy of Istituto Legno Firenze.)

S = ultimate stress in axial (or transverse) tension (N/mm², Pa, psi)
P = ultimate load (N, lb)
A = minimum cross-section of the specimen (mm², in.²)

Tests of *axial compression* give the following properties:

a. *Proportional fiber stress* in axial compression from the relationship

$$S' = \frac{P'}{A} \qquad (11\text{-}9)$$

where

S' = stress at proportional limit (N/mm², Pa, psi)
P' = load at proportional limit (N, lb)
A = cross-section of the specimen (mm², in.²)

b. *Modulus of elasticity* in axial compression from the relationship

$$E = \frac{P' L}{A D} \qquad (11\text{-}10)$$

where

E = modulus of elasticity (N/mm², Pa, psi)
P' = load at proportional limit (N, lb)
L = distance of specimen supports (mm, in.)
A = cross-section of the specimen (mm², in.²)
D = deformation at proportional limit (mm, in.)

c. *Strength in axial compression* from the relationship [same as Eq. (11-8)]

$$C = \frac{P}{A} \quad (11\text{-}11)$$

where

C = ultimate stress in axial compression (N/mm², Pa, psi)
P = ultimate load (N, lb)
A = cross-section of the specimen (mm², in.²)

This equation is usually applied to express the strength of wood in axial compression.

Strength in transverse *compression* is determined as stress at proportional limit from Eq. (11-8), with A = loaded surface, and strength in axial *shear* from Eq. (11-9), with A = sheared surface.[20]

Tests of *static bending* give the following properties:

a. *Stress at proportional limit* in static bending from the relationship

$$S' = \frac{1.5 \, P' \, l}{bd^2} \quad (11\text{-}12)$$

where

S' = stress at proportional limit (N/mm², Pa, psi)
P' = load at proportional limit (N, lb)
l = length between beam supports (mm, in.)
b = width of beam (mm, in.)
d = length of beam (mm, in.)

b. *Modulus of rupture* from the same relationship, with P = ultimate load instead of P'. Note should be made that this is an approximate stress due to the assumption that linear behavior is sustained to level of failure.

c. *Modulus of elasticity* in static bending from the relationship

$$\text{MOE} = \frac{P' \, l^3}{4 D b d^3} \quad (11\text{-}13)$$

where MOE = modulus of elasticity (N/mm², Pa, psi) and D = deflection of the neutral plane at the proportional limit (mm, in.) measured at half span. The other parameters are as previously defined.

d. *Work*[21] *to proportional limit* from the relationship

$$W' = \frac{P'D}{2 \, lbd} \quad (11\text{-}14)$$

where W' = work to proportional limit (J/mm³, in.·lb/in.³), and the other parameters are as previously defined.

e. *Work to maximum load* from the relationship

$$W = \frac{A}{lbd} \times c \quad (11\text{-}15)$$

where

W = work to maximum load (J/mm³, in.·lb/in.³)
A = area of the load-deflection curve included between the point of zero deflection and the deflection corresponding to maximum load (see Figure 11-2); the area is measured with a planimeter (mm², in.²)
l, b, d = as previously defined
c = correction factor depending on the scale of the load-deflection diagram; it is the load-deflection equivalent (load × deflection) in a unit of scale (1 mm², 1 in.²)

In addition, the test of static bending may be utilized to determine the stress at the neutral plane in axial shear from the relationship

$$S = \frac{0.75P}{bd} \quad (11\text{-}16)$$

where S = maximum shear stress at the neutral plane (N/mm², Pa, psi), and P, b, and d are as previously defined.

Cleavage resistance is determined from the relationship

$$C = \frac{P}{W} \quad (11\text{-}17)$$

where

C = cleavage resistance (N/mm, lb/in.)
P = load causing cleavage of the specimen (N, lb)
W = width of specimen (mm, in.)

Toughness is determined by tests intended to break the specimen with one stroke of a hammer acting in the form of a pendulum. The test is carried out in different ways. According to the American method (1, 76), toughness is measured from the change of the initial angle under which the pendulum is released—the change being due to the presence of the breaking specimen. Initial angles are set at 30°, 45°, or 60°, and the position of the hammer, which is placed at the end of the pendulum, may change according to wood species. Indicative values of toughness, with this method, are calculated from the initial and final angles of the pendulum from tables or from the relationship

$$T = Wl\,(\cos A_2 - \cos A_1) \quad (11\text{-}18)$$

where

T = toughness, work per specimen[22] (Nm, in.·lb)
W = weight of the hammer (N, lb)
l = distance between the axis supporting the specimen and the center of the hammer (mm, in.)
A_1 = initial angle (°)
A_2 = final angle, after breakage of the specimen (°)

With another method (German standard DIN 52189, ISO 3348), the machine measures directly the work in toughness, and coefficient a is calculated from the relationship

$$a = \frac{W}{F} \quad (11\text{-}19)$$

where

a = toughness coefficient (J/mm², in·lb/in.²)[23]
W = work in toughness (J, in.·lb)
F = cross-section of the specimen (mm², in.²)

The dynamic behavior of wood may also be determined by an impact test, entailing the fall of a certain weight (50 lb = 22.3 kg) from an increasing height. Loading is continued until the specimen breaks, and the height of drop (mm, in.) causing complete failure is recorded. Use of this test has declined due to the long time period required and the arbitrary assumptions made in deriving relationships.

The determination of (static) *hardness* presents difficulties because wood is a heterogeneous material. The usual method is one proposed by Janka (see also ISO 3350), according to which hardness is measured by the load needed to insert a steel ball with a diameter of 1.128 cm (0.444 in.) into the transverse, radial, or tangential surface of the specimen and to a depth equal to half its diameter. In this position, the perimeter of the indentation made into wood defines an area of 1 cm². Hardness is measured in N/mm² (lb in North America).

Other methods of measuring hardness include the following (41).

Brinell Method. A steel ball with a 10 mm diameter is pressed into wood with a 5 N load (10 N for very hard and 1 N for very soft woods). The maximum load is placed in 15 s; it is kept constant for 30 s, and is reduced to zero in 15 s. Hardness is calculated from the relationship

$$H = \frac{2P}{\pi D(D - \sqrt{D^2 - d^2})} \quad (11\text{-}21)$$

where

H = hardness (N/mm^2)
P = maximum load (N)
D = ball diameter (mm)
d = diameter of indentation (mm)

Chalais-Meudon Method (French Standard) (3). A steel ball, 30 mm in diameter, is pressed with a maximum load of 20 N into the radial surface for 5 s. "Hardness index" is calculated from the relationship

$$N = \frac{1}{t} \quad (11\text{-}22)$$

where

$t = 15 - 0.5\sqrt{900 - b^2}$
(t = depth and b = width of indentation, mm)

Finally, *dynamic hardness* is determined by the fall of a steel ball (density 7.8 g/cm^3, diameter 25 ± 0.05 mm) from a height of 50 cm. Hardness (H) is calculated from the relationship

$$H = \frac{4mh}{d^2} \quad (11\text{-}23)$$

where

m = mass of the ball (kg)
h = height of fall (m)
d = average diameter of indentation (cm)

Allowable Stresses

The values of the various mechanical properties, determined from the laboratory tests previously described, cannot be used directly for engineering design of wooden structures, because they apply only to the specific test conditions (i.e., specimens without defects and short test duration). For practical design purposes, laboratory data are corrected and the corrected values are called *allowable, working*, or *design stresses* (17, 33, 34, 51, 65). Use of allowable stresses permits a better utilization, because it is possible to calculate the necessary dimensions of the members of a given structure without the waste of material and the sacrifice of function and safety. In this manner, the results of studies of the mechanical properties of wood find their practical applications.

Allowable stresses are calculated from laboratory data of the strength of small, clear specimens in the green condition by successive corrections that take into consideration the following factors: variability of the strength of wood, time (duration) of loading, moisture content, size of a structural element, grade (quality) of wood, and safety (accidental overloading) of a structure.

Allowable stresses are usually needed for certain properties—namely, bending, tension, compression, and shear. The specific properties calculated are modulus of rupture (strength in static bending), ultimate stress in axial tension and compression, transverse compression, and horizontal shear (65, 76). An estimation of the magnitude of bending deformation (deflection) may be made from tables of modulus of elasticity. Mean values of this modulus are not corrected for certain factors (e.g., duration of load and accidental overloading). Minimum values may be taken from the variation of this property (65).

The stresses derived from laboratory tests are usually expressed as mean values; however, mean values do not give a picture of strength variability of the specimens tested. Such variability, as expressed by frequency curves, approaches a normal statistical distribution (48, 65). Most values are found around the mean value, but certain specimens have much lower or higher strength (Figure 11-18).

The extremely low values are obviously very important for the safety of a wooden structure. This does not mean that the calculation of allowable stresses should be based on such low values, because the probability of their appearance is small, and it is possible to exclude members with very low strength on the basis of their weight, or macroscopic criteria of wood density (ring width, proportion of latewood). In any case, the need to correct (reduce) mean values is clear.

The amount of reduction of a mean value de-

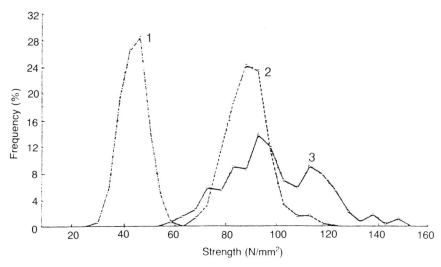

Figure 11-18. Frequency distribution of mechanical property values of fir (a natural hybrid between European silver fir and Grecian fir, *Abies alba* × *Abies cephalonica*): 1, axial compression; 2, static bending (MOR); 3, axial tension. (Adapted from Ref. 67.)

pends on its standard deviation and the desired level of statistical probability. A minimum value that excludes the lower 5% of the total laboratory values is calculated by subtracting 1.65 times the standard deviation (s); this assumes that the chance of getting a value lower than that is 5%. Assuming chances of 2.5% or 1%, respective factors are $1.96 \times s$ and $2.33 \times s$.

The effect of time is calculated on the basis of observations that the load-carrying capacity of wood is reduced with duration of loading. After a 10-year period, the strength of clear wood is reduced, on the average, to about 60% (Figure 11-12); laboratory values are reduced by a factor of 0.625.

A correction for moisture content is applied to correct from green to air-dry wood. Correction factors differ among properties and levels of moisture below 19%.

Size is a factor applied in beams, for example. A correction is applied to account for the reduction of strength of beams of greater depth in comparison to standard laboratory test specimens.

Considering that wood is practically never free from defects in structural applications; further reductions are affected by the grade of wood. Grading is usually visual, and grading rules are based on kind, size, number, and position of defects; growth-ring characteristics (width, number of rings per cm or in.) that affect density are also sometimes taken into consideration. Corrections based on grade constitute the so-called *strength ratio*, which is the ratio of the strength of wood with defects over that of clear wood.[24]

The final step is to apply a factor of safety, which varies depending on property, and is different in softwoods and hardwoods. Thus, laboratory values are reduced many-fold in calculating allowable stresses.[25] Detailed information regarding allowable stresses is contained in specialized publications (21, 33, 65, 73).

Other Methods for Determination of Mechanical Properties

In addition to previously mentioned testing procedures, the determination of the mechanical properties of wood may be done by other methods, such as nondestructive evaluation (NDE)—and this is an important advantage. Nondestructive evaluation is applied with special machines (Figure 11-19), sound waves, etc. (26, 31, 45, 49). Use of nondestructive testing is based on empirical relationships established between modulus of rupture and

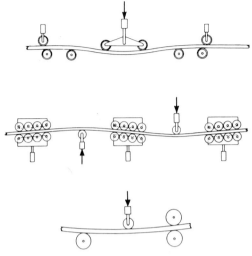

Figure 11-19. Schematic representation of machine stress grading of lumber. (After Ref. 46; reproduced by permission from Springer-Verlag.)

modulus of elasticity (Figure 11-20). This relationship allows for the determination of the strength of a wooden member (e.g., a lumber board) with a certain degree of accuracy by simply passing it through a machine which loads it in bending.[26] Smaller deflection predicts higher strength, and vice versa. A small computer calculates the allowable stress, which is automatically marked on the wood with different grade identification, such as color. In this manner, wood is "stress-graded" continuously and without destruction of material. Also, machine grading is advantageous because it may be applied irrespective of wood species—which contributes to better utilization. Machines are used in some countries (U.S., Australia, Britain, etc.) (61, 66) but not in large scale. Grading is generally done with visual criteria (defects, ring width, etc., see Table 11-14), although this is tiresome (because all the faces of wood should be checked) and usually conservative.

In laboratories, strength tests are also conducted with individual wood cells, mainly softwood tracheids (37, 57). It has been observed that such strength is related to the strength of pulp and paper (22, 37, 52, 77) (see Chapter 26).

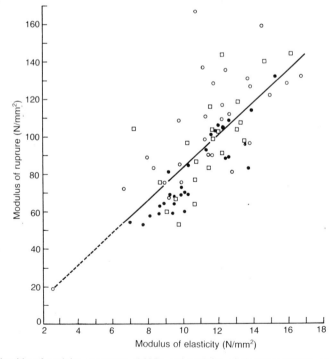

Figure 11-20. Relationship of modulus or rupture (MOR) and modulus of elasticity (MOE). Data from Table 11-1: European woods (□), North American woods (●), tropical woods (○). The straight line is based on all woods and expressed by the equation $y = -3.24 + 8.7x$ ($N = 80$, $r = 0.745$). The wood at extreme left (down) is balsa.

Table 11-4. EEC Recommended Grades for Structural Use
(Permissible Limits for the S10, S8, and S6 Visual Stress Grades)

Characteristic	Grade				
	S10	S8		S6	
		EITHER	OR	EITHER	OR
Knots:					
Margin KAR	Equal to or less than 1/5	Equal to or less than 1/2	More than 1/2	Equal to or less than 1/2	More than 1/2
Total KAR	Equal to or less than 1/5	Equal to or less than 1/3	Equal to or less than 1/5	Equal to or less than 1/2	Equal to or less than 1/3
Slope of grain	1 in 10			1 in 6	
Rate of growth: Average width of annual rings	Not more than 6 mm			Not more than 10 mm	
Size and number of fissures:	Surface fissures less than 300 mm in length may be ignored				
Not through the thickness	Not more than 1/4 of the length but 600 mm maximum			Not more than 1/4 of the length but 900 mm maximum	
	On any running metre, not more than one fissure of the maximum length				
Through the thickness	Only permitted at the ends with a length of not more than the width of the piece			Not more than 600 mm if at the ends with a length of not more than 1½ times the width of the piece	
Wane	1/4 of the thickness, 1/4 of the width, full length; 1/3 in any 300 mm length, provided that at ends it is confined to one arris only			1/3 of the thickness, 1/3 of the width for full length; 1/2 in any 300 mm length, provided that at ends it is confined to one arris only	
Distortion:	Any piece which is bowed, twisted or cupped to an excessive extent having regard to the end use shall be rejected				
Approximate limits of distortion*					
Bow (per 2 m length)	In a thickness of 38 mm should generally not exceed 30 mm In thickness of 75 mm and greater should generally not exceed 10 mm (Intermediate sizes may be determined by interpolation)				
Spring (per 2 m length)	In a width of 63 mm should not exceed 10 mm In widths of 250 mm and greater should not exceed 5 mm (Intermediate sizes may be determined by interpolation)				
Cup	Should generally not exceed 1 mm per 25 mm of width				
Twist (per 2 m length)	Should generally not exceed 1.5 mm per 25 mm of width				
Pitch pockets and inbark:					
Not through the thickness	Unlimited, if shorter than the width of the piece. Otherwise the same limits as for size of fissures				
Through the thickness	Unlimited, if shorter than half the width of the piece. Otherwise the same limits as for size of fissures				
Insect damage	Worm holes and pin holes are permitted to a slight extent in a few pieces No active infestation is permitted Wood wasp holes are not permitted				

Source: Timber Bulletin for Europe. 1982. 26 (Suppl. 16).

*Distortion will largely depend on the moisture content of the timber at the time it is measured. Where for a particular reason other limits than those indicated are required, these should be subject to contract between supplier and purchaser.

ADDENDUM

In the above discussion, the mechanical properties of wood have been examined in reference to sampling and laboratory testing of small clear specimens in order to calculate allowable stresses for engineering design of wooden structures; also, direct, nondestructive evaluation (testing) of structural members has been briefly mentioned.

Other, relatively new wood-engineering concepts include "in-grade testing," "proof-loading," and "reliability-based design" (7, 8).

In-grade testing is testing samples of a given species, size, and grade of full-size structural lumber (or other wood products). The samples are selected at random from the production line of a factory (sawmill, etc.), or factories of a region, and they are tested on site using portable equipment to determine strength and stiffness (modulus of rupture, modulus of elasticity). In-grade testing has at least two purposes: it establishes design stresses and quality control to see that a minimum property is met.

Proof-loading entails testing an initial random sample (in bending, tension, compression, torsion, etc., according to loading expected in service) in order to estimate the variability of the material. On the basis of the results (calculation of modulus of rupture, etc.), a load is determined to guarantee a minimum strength performance of the structure by exclusion of the lower 5% or 10% of the strength values. This load is then applied to all remaining pieces; those that break are removed. The main disadvantage of the method is the possibility that proof loading may damage a member during testing; however, no data are available to prove the validity of the damage. The advantage is that one can guarantee a minimum strength value for each piece of structural product.

Reliability-based design of wooden structures is based on the probability distribution of expected load effects and of the resistance of wood products (lumber, etc.). The area of overlap of the two frequency distribution curves (i.e., the area of high loads and low resistance) is proportional to the probability of failure. By controlling the distance between the means of resistance and load effect in the design process, the probability of failure can be controlled. The scope of this procedure is to reduce the variability of the material,[27] and to overcome the limitations of wood waste and required high level of safety, which may result from designing on the basis of allowable stresses. The subject of reliability-based design (probability of failure) is quite complex, and is still in the developmental stage.

REFERENCES

1. American Society for Testing and Materials. 1984. *Annual Book of ASTM Standards*, Vol. 04.09. Philadelphia.
2. Armstrong, F. H. 1955. The Strength Properties of Timber. For. Prod. Res. Bull. No. 34, HMSO, London.
3. Association Francaise de Normalisation (AFNOR). Normes-Bois. Paris.
4. Blankenstein, C. 1962. *Holztechnisches Taschenbuch*. München: Hanser Verlag.
5. Baumann, R. 1922. Die bisherigen Ergebnisse der Holzprüfungen in der Materialprüfung Anstalt an der Tech. Hochschule Stuttgart. Forsch. Gebiete Ingenieurw. H231, Berlin.
6. Boatright, S. W. J., and G. G. Garrett. 1979. The effect of knots on the fracture strength of wood. *Holzforschung* 33:68–77.
7. Bodig, J. 1984. Personal communication.
8. Bodig, J., and B. A. Jayne. 1982. *Mechanics of Wood and Wood Composites*. New York: Van Nostrand Reinhold.
9. Bolz, R. E., and G. L. Tune. 1970. *Handbook of Tables for Applied Engineering Science*. Cleveland, Ohio: Chemical Rubber Co.
10. Bolza, E., and W. G. Keating. 1972. African Timbers—The Properties, Uses and Characteristics of 700 Species. Div. Build. Res. Melbourne, Australia: CSIRO.
11. Brazier, J. 1986. Growth features and structural wood performance. *Proc. 18th IUFRO World For. Congress, Div. 5*, pp. 37–49.
12. Brown, H. P., A. J. Panshin, and C. C. Forsaith. 1952. *Textbook of Wood Technology*, Vol. II. New York: McGraw-Hill.
13. Clarke, S. H. 1937. The distribution, structure and properties of tension wood in beech (*Fagus sylvatica* L.). *Forestry* 11:85–91.
14. Cockrell, R. A., and R. M. Knudson. 1973. A comparison of static bending, compression and tension parallel to grain and toughness properties of compression wood and normal wood of a Giant Sequoia. *Wood Sci. Technol.* 7(4):241–250.
15. Comben, A. J. 1955. The effect of high temperature kiln-drying on the strength properties of timber. *Wood* 20:311–313.
16. Comben, A. J. 1964. The effect of low temperatures on the strength and elastic properties of timber. *J. Inst. Wood Sci.* 13:44–55.
17. Curry, W. T. 1965. Progress in stress-grading of boards. *Timber Trades J.* 254(11).
18. Dams, K. 1968. Afrikanische Exporthölzer. Stuttgart: DRW-Verlag.
19. Datta, P.C., and B. Basu. 1983. Strength elements of wood structure and their biochemical control. Abstract *IAWA Bull.* n.s. 4(1):5–6.
20. Dawe, P. S. 1964. The effect of knots on the tensile strength of European redwood. *Wood* 29(1):49–51.
21. Deutsches Institut f. Normung (DIN). 1979. Holznormen (Taschenbuch 31). Berlin: Beuth Verlag.
22. Dinwoodie, J. M. 1965. Tensile strength of individual compression wood fibers and its influence on properties of paper. *Nature* 205(4973):763–764.
23. Dinwoodie, J. M. 1975. Timber—A review of the structure-mechanical property relationship. *J. Microscopy* 104(1):3–32.

24. F. A. O./U. N. 1954. Report of the Third Conference on Wood Technology. Rome.
25. Farmer, R. H. 1972. *Handbook of Hardwoods*. London: HMSO.
26. Fewell, A. R. 1984. Timber stress grading machines. BRE Inform. Paper IP 17/84.
27. Filipovici, J. 1965. *Studiul Lemnului*. Bucarest, Romania.
28. Galligan, W. L. 1975. Mechanical properties of wood. In *Wood Structures*, pp. 32–54. New York: Am. Soc. Civil Eng.
29. Ganowicz, R., R. Guzenda, and R. Plenzer. 1980. Einfluss der Temperatur auf den Elastizitätsmodul parallel zur Faserrichtung des Holzes. *Holztechnologie* 21(1):5–8.
30. Geissen, A. 1976. The effect of temperature and moisture content on the elastic and strength properties of wood in the freezing range. Ph.D. Dissertation, University of Hamburg.
31. Glos, P., and H. Schulz. 1980. Stand und Aussichten der maschinellen Schnittholzsortierung. *Holz Roh- Werkst.* 38:409–417.
32. Giordano, G. 1971. *Tecnologia del Legno*, Vol. 1. Torino: Unione Tipografico-Editrice.
33. Hoyle, R. J. 1978. *Wood Technology in the Design of Structures*, 4th ed. Missoula, Montana: Mountain Press.
34. Illston, J. M., J. M. Dinwoodie, and A. A. Smith. 1979. *Concrete, Timber and Metals*. New York: Van Nostrand Reinhold.
35. Ingenieurschule f. Holztechnik, Dresden. 1965. Taschenbuch der Holztechnologie. Leipzig: VEB Fachbuchverlag.
36. International Organization for Standardization, Switzerland. Standard Nos. 3129–3133, 3345, 3346, 3348–3351.
37. Kersavage, P. C. 1973. Moisture content effect on tensile properties of individual Douglas-fir latewood tracheids. *Wood Fiber* 5(2):105–117.
38. Knigge, W., and H. Schulz. 1966. *Grundriss der Forstbenutzung*. Hamburg, Berlin: P. Parey.
39. Kollmann, F. 1951. *Technologie des Holzes und der Holzwerkstoffe*. Berlin: Springer Verlag.
40. Kollmann, F. 1960. Die Abhängigkeit der elastischen Eigenschaften von Holz von der Temperatur. *Holz Roh- Werkst.* 18(8):308–314.
41. Kollmann, F. F. P., and W. A. Côté. 1968. Principles of Wood Science and Technology, I. Solid Wood. Berlin: Springer Verlag.
42. Krpan, J. 1954. Untersuchungen über den Fasersättigungspunkt des Buchen-, Eichen-, Tannen- und Fichtenholzes. *Holz Roh- Werkst.* 12(3):84–91.
43. Kucera, L. 1973. Holzfehler und ihr Einfluss auf die mechanischen Eigenschaften der Fichte und Kiefer. *Holztechnologie* 14(1):8–17.
44. Kucera, L., and M. Bariska. 1982, 1985. On the fracture morphology of wood (I, II). *Wood Sci. Technol.* 16:241–259; 19:19–34.
45. Kufner, M. 1977. Maschinelle Schnittholzsortierung und ihr möglicher Einfluss auf die Holzverwertung. *Holz Roh- Werkst.* 35:173–178.
46. Kufner, M. 1978. Elastizitätsmodul und Zugfestigkeit von Holz verschiedener Rohdichte in Abhängigkeit von Feuchtigkeitsgehalt. *Holz Roh- Werkst.* 36(11):435–439.
47. Kunesh, R. H., and J. W. Johnson. 1972. Effect of single knots on tensile strength of 2-by 8 inch Douglas-fir dimension lumber. *For. Prod. J.* 22(1):32–36.
48. Lavers, G. M. 1969. The Strength Properties of Timbers. For. Prod. Res. Bull. No. 50, HMSO, London.
49. Lee, I. D. G. 1958. *Non-Destructive Testing of Wood by Vibration Methods*. London: Timber Development Association.
50. National Forest Products Association. 1980. *Design Values for Wood Construction*. Washington, D.C.
51. Noack, D. 1971. Tropische Hölzer in tragenden Bauwerken. Grundsätzliche Überlegung zur Feststellung der zulässigen Spannungen von Bauholz. *Holz-Zbl.* 55:795–798.
52. Page, D. H., F. El-Hosseiny, K. Winkler, and R. Bain. 1972. The mechanical properties of single wood pulp fibers, Part I: A new approach. *Pulp Pap. Mag. Can.* 73(8):72–77.
53. Pearson, R. G. 1972. The effect of duration of load on the bending strength of wood. *Holzforschung* 26(4):153–158.
54. Pearson, R. G., and R. C. Gilmore. 1980. Effect of fast growth rate on the mechanical properties of loblolly pine. *For. Prod. J.* 30(5):47–54.
55. Pecina, H. 1981. Eigenschaftsänderungen durch chemische Einflussnahme auf die Grundbestandteile des Holzes. *Holztechnologie* 22(1):45–51.
56. Perem, E. 1960. The Effect of Compression Wood on the Mechanical Properties of White Spruce and Red Pine. For. Prod. Lab., Canada, Tech. Note No. 13.
57. Popper, R. 1985. Computergestreurtes System zur Ermittlung der mechanischen Eigenscahften von Einzelfasern unter variablen Klimabedingungen. *Holz Roh- Werkst.* 43:193–197.
58. Rajput, S. S., N. K. Shukla, and R. R. Sharma. 1980. Mechanical tests for wood. Comparison of test results of large and small size specimens. *Holforsch. Holzverwert.* 32(5):117–120.
59. Rusche, H. 1973. Festigkeitseigenschaften von trockenem Holz nach thermischer Behandlung. *Holz Roh- Werkst.* 31(7):273–281.
60. Schulzberger, P. H. 1953. The Effect of Temperature on the Strength of Wood. Aeron. Res. Cons. Comm. Report ACA-46, Melbourne.
61. Serry, V. 1972. European lumber industry moves toward machine grading. *World Wood* 13(3):23.
62. Silvester, D. 1967. *Timber—Its Mechanical Properties and Factors Affecting Its Structural Use*. Oxford, New York: Pergamon Press.
63. Singer, F. L., and A. Pytel. 1980. *Strength of Materials*, 3rd ed. New York: Harper and Row.
64. Stamm, A. J. 1964. *Wood and Cellulose Science*. New York: The Ronald Press.
65. Sunley, J. G. 1968. Grade Stresses for Structural Timbers. For. Prod. Res., Bull. No. 47, HMSO, London.
66. Sunley, J. G., and W. M. Hudson. 1964. Machine grading of lumber in Britain. *For. Prod. J.* 14(4):155–158.
67. Svarnas, D. 1964. Untersuchungen über die Holz-

eigenschaften der griechischen Tanne. Dissertation, Hann. Münden.
68. Tebbe, J. 1987. Investigation of the national and international standard of machine stress grading. Proc. EEC Seminar on Wood Technology, pp. 95–97, Munich.
69. Tiemann, D. H. 1947. *Wood Technology.* New York: Pitmann.
70. Trendelenburg, R., and H. Mayer-Wegelin. 1955. *Das Holz als Rohstoff.* München: Hanser Verlag.
71. Tsoumis, G. 1951. A Study of Strength and Related Properties of *Fraxinus americana* L. through Standard Tests of Small Clear Specimens in the Green Condition. Thesis, University of Michigan.
72. U.S. Forest Products Laboratory. 1964. Proceedings of the Symposium on Nondestructive Testing of Wood. FPL-040, Madison, Wisconsin.
73. U.S. Forest Products Laboratory. 1987. *Wood Handbook.* U.S.D.A. Ag. Handbook No. 72 (revised).
74. Vorreiter, L. 1949. *Holztechnologisches Handbuch*, Band I. Wien: Verlag Fromme.
75. Wagenführ, R., and C. Scheiber. 1974. *Holzatlas.* Leipzig: VEB Fachbuchverlag.
76. Wangaard, F. F. 1950. *The Mechanical Properties of Wood.* New York: J. Wiley & Sons.
77. Wangaard, F. F., and G. E. Woodson. 1973. Fiber length–fiber strength interrelationship for slash pine and its effect on pulpsheet properties. *Wood Sci.* 5(3):235–240.
78. Wengert, E. M. 1979. Relationship between toughness of hardwoods and specific gravity. *Wood Sci.* 11(4):233.
79. Wood, L. W. 1960. Relation of Strength of Wood to Duration of Load. U.S. For. Prod. Lab. Report No. 1916.

FOOTNOTES

[1] In wood, interior forces resulting from shrinkage and swelling may also be active (see Chapter 10).

[2] Newton and Pascal are units of the International System of Units (SI). In older European literature, force (load) is expressed in kp (kilopond) and stress in kp/cm^2. Equivalents are given at the end of this book.

[3] The technical meaning of the term "elasticity" is different from the common concept of the word. In everyday usage, an elastic body exhibits a large deformation under the action of a certain load and practically returns to its original dimensions when this load is released. In mechanics, the criterion of elasticity is not the magnitude of deformation but the accomplishment of full recovery. From this point of view, steel and glass are elastic materials. In beams, the deformation is called *deflection* (63).

[4] The failure of wood in axial tension takes place in latewood tracheids axially between the S_1 and S_2 layers, and in earlywood tracheids by transverse breakage of the cell wall (23).

[5] Members with a diagonal growth-ring arrangement, as it appears on cross-sections, have lower strength in comparison to members with a radial or tangential arrangement (76).

[6] There are three common types of beams: simple (supported at the ends), cantilever (supported at one end), and continuous (supported at the ends and in–between).

[7] The cleavage test tends to be replaced by fracture toughness, which is a measure of resistance to propagation of cracks. Available data are very limited, however.

[8] Exit of moisture without adverse consequences (e.g., splitting) is prerequisite to the increase of strength. Otherwise, the strength is reduced.

[9] The strength of wood may be more than double when its moisture content is reduced from fiber saturation point to zero. Table 11-3 shows that the strength of pine and birch wood increased by about 120% in axial compression and 145% in static bending after air-drying (12–15% mc). Theoretically, the strength of oven-dry wood should again double.

[10] The value used is less than 30% (21–27%), varying for different species, and on average 25% (28).

[11] For example, suppose that the laboratory values A_1 and A_2 were determined in the green condition (M_1) and at 9.5% mc (M_2). What strength (A_3) corresponds to 12% mc (M_3)? The value M_1 is the mc at the fiber saturation point of the species under consideration.

[12] It is difficult to determine the influence of temperature (heat) on strength and other wood properties due to the simultaneous change of its moisture.

[13] This is of practical importance in kiln-drying. For example, the reduction of strength in transverse tension during the initial stages of drying, and when the moisture content of the wood lies above the fiber saturation point, may have a decisive influence on the development of checks. Strength in transverse tension may be reduced to $\frac{1}{3}$–$\frac{1}{2}$ in comparison to the values of this property at room temperature (76).

[14] The branch of mechanics dealing with the change of properties of materials in relation to time is called "rheology."

[15] The size of a wooden member is also a factor. For example, lumber of structural size with defects is affected less by high moisture than clear wood. It is also worth noting that the strength of a structural member is reduced when its size increases.

[16] The effect of knots on strength is estimated from the relationship of the diameter of a knot (or knots) and the width of the wooden member (knot ratio) (23, 34), or from the knot area ratio (20) as shown in Figure 11-13. Knots with a diameter up to half the width of a member may reduce its strength up to 50%, depending on their location and type of loading (23, 34).

[17] A comparative study found no important differences between specimens 5 × 5 cm (2 × 2 in.) and 2 × 2 cm (0.8 × 0.8 in.) in cross-section. Properties investigated included: static bending, axial and transverse compression, shear, and hardness (58).

[18] International standardization is not easy, mainly because

separate methods have been used for many years, and change of results from one method to another cannot be done with accuracy due to many factors involved, such as size and shape of specimens, and manner and speed of loading.

[19] Half the specimens tested in transverse tension have a radial and half a tangential surface of fracture.

[20] In transverse compression, the specimens are loaded on a radial surface; and in shear, half of them have a radial and half a tangential shearing surface. In static bending, the specimens are loaded on the middle of the tangential surface nearest to the center (pith) of the tree trunk.

[21] Work measures the energy that wood absorbs under static bending loads, and is an index of toughness—a property measured by dynamic loading.

[22] Toughness values are chiefly useful for comparative purposes; they are not readily adapted to expression as work per unit volume for other size specimens.

[23] In French standards (NF 351-009), the coefficient of toughness K is calculated from the relationship (39).

$$K = \frac{W}{F^{10/6}} \qquad (11\text{-}20)$$

[24] Structural grades differ in various countries. Three grades are recognized in most European countries (as proposed for the European Community; see Table 11.4), and four grades—75, 65, 50, 40—in Britain. In the United States, for a given range of lumber sizes, there may be up to five grades (select structural, No. 1, No. 2, No. 3, appearance); these apply to softwoods and some hardwoods (yellow-poplar, aspen) (33, 73). Details of specifications that define grades cannot be presented in this book, especially concerning grades in the United States, where complicated rules exist. The reader may seek such information in other, specialized publications (33, 50, 73)—for example, U.S. or German specifications (DIN 4074). Table 11.4 provides an example of how grades are formulated, and what is the importance of wood defects. Worth noting is the concept of strength grouping, which is included in the specifications (codes) of some countries; three or more strength classes are recognized, and each class defines minimum requirements for all relevant strength parameters. In this manner, stress grades from different wood species (and countries) are combined in one group (7).

[25] Allowable stresses for clear wood, called *basic* stresses (65), are sometimes calculated, but their value is theoretical. Instead of applying separate factors (duration of loading, size, etc.), basic stresses are calculated by dividing the statistical minimum value (mean $-1.65s$, etc., where s = standard deviation) by a factor of 2.25 for most properties, except strength in axial compression where the factor is 1.4. The result covers also the possibility of slight overloading. Calculation of basic stresses involves reduction of laboratory values to $\frac{1}{3}$–$\frac{1}{4}$, whereas allowable stresses are as low as $\frac{1}{10}$ or more depending on property, species, and grade (7).

Allowable stresses vary in different countries (and wood species); therefore, reference to numbers is not practical in this book. A comment may be made that, in spite of extensive tests in laboratories and existing data on individual species, some specifications (e.g., DIN 1052) make no distinction among softwoods, and only major hardwoods (oak, beech) are included—grouped together.

[26] "Bending-type" machines are the conventional type. Other machines are based on vibration, resonant pulse technique, and irradiation principles; these can be considered as equal or even superior to bending machines, but they have not been adequately developed as yet, and, in addition to suitability, economic considerations are also involved in selecting a machine (68).

[27] Reduction of variability is also sought by use of laminated wood (beams, etc.) produced by gluing lumber or veneer; the variability of such structures can be reduced due to possible reduction of growth-ring variability and distribution of defects (see Chapter 23).

12

Thermal Properties

Effects of heat on the properties of wood are discussed throughout this book. Heat affects hygroscopicity, shrinkage and swelling, mechanical, acoustical, electrical, and other properties. This chapter is devoted to thermal properties of wood—and namely: expansion and contraction (dimensional change), thermal conductivity, thermal capacity, specific heat, combustion, and heating value.

Expansion and Contraction

As with other materials, when wood is heated its dimensions increase; and inversely, when cooled, its dimensions decrease. This phenomenon is called thermal *expansion* and *contraction*, respectively. From a practical point of view, these changes are seldom important, because they are very small in comparison to shrinkage and swelling—which are simultaneously present, because changes of temperature often change the moisture content of wood.[1]

Expansion is measured by the *coefficient of thermal expansion*, which refers to oven-dry wood, and measures the elongation of unit length when its temperature increases by 1°C.[2] Due to wood structure, the coefficient of expansion is different in different directions of growth—much smaller in the axial (longitudinal) direction than transversely to growth rings, and smaller radially than tangentially. According to some measurements in European woods, the coefficients of expansion range as follows ($\times 10^{-6}$): axial 3.7 (fir) to 9.5 (ash), radial 6.3 (spruce) to 22.0 (beech), and tangential 29.0 to 72.7 (pine; Table 12-1). In comparison, the coefficient of glass is 9×10^{-6}, steel 11×10^{-6}, and aluminum 24×10^{-6} (8, 18, 19). The thermal expansion of wood has no practical importance, however, because, in common usage, temperature changes are within relatively narrow limits; and also because such changes result in changes of moisture and, therefore, in simultaneous dimensional changes due to shrinkage or swelling, which are much greater than thermal expansion or contraction.

The relationship between expansion and temperature is nearly linear in all three growth directions (Figure 12-1). Also, the relationship of expansion and density is linear, but the influence of density is very small in the axial direction (Figure 12-2). It has been observed that when green wood is heated in water, steam, or preservatives, it expands tangentially and contracts radially (20, 22).[3] This may result in the formation of checks (e.g., when logs are heated prior to veneer production). Temperatures much below zero, which cause wood to contract quickly, may also result in surface checking—and in frost cracks in living trees.[4]

Expansion of a sample of wood in any growth direction and for a temperature change from t_1 to t_2 (°C, °F) may be calculated from the relationship

$$d = \frac{l_2 - l_1}{l_1} \qquad (12\text{-}1)$$

where

d = expansion (cm/cm, in./in.)

Table 12-1. Thermal Expansion

Species	Coefficient of expansion × 10⁻⁶		
	A^a	R	T
Fir	3.7	15.8	58.4
Pine, black	4.0	18.4	72.7
Pine, Scots	4.2	15.0	29.0
Spruce	5.4	6.3	34.1
Ash	9.5	—	—
Beech	5.4	22.0	34.8
Chestnut	6.5	—	32.5
Oak	4.9	11.1	54.4

$^a A$ = axial; R = radial; T = tangential.
Source: Refs. 19 and 34.

l_1 = original length (cm, in.)
l_2 = final length (cm, in.)

This relationship may be used to calculate expansion in percent of the original length (by multiplication of the second part × 100), and thus becomes similar to Eq. (10-1) applied in the determination of shrinkage. Volumetric expansion is calculated in the same manner.

If the coefficient of expansion is known, expansion may be calculated from the relationship (18)

$$d = a l_1 d_t \qquad (12\text{-}2)$$

where

d = expansion (cm, in.)
a = coefficient of expansion (cm/cm/°C, in./in./°F)
l_1 = original length (cm, in.)
d_t = change of temperature (°C, °F)

The above relationship allows the following calculations:

$$a = \frac{d}{l_1 d_t} \qquad (12\text{-}3)$$

and because $d = l_2 - l_1$ and $d_t = t_2 - t_1$

$$l_2 = l_1 \left[1 - a(t_2 - t_1) \right]$$

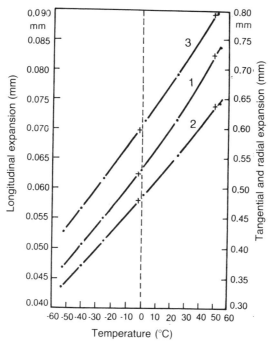

Figure 12-1. Relationship of expansion and temperature of birch wood (r_o = 0.593 g/cm³) in the range of −55°C to +55°C (−47°F to +131°F): 1, axial; 2, radial; 3, tangential expansion. (After Ref. 41.)

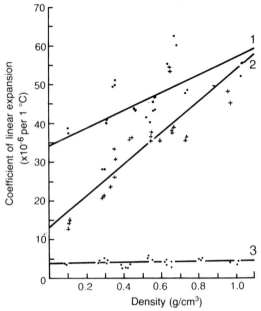

Figure 12-2. Relationship of thermal expansion and (oven-dry) density. (Based on eleven temperate and tropical woods): 1, tangential; 2, radial; 3, axial expansion. (After Ref. 5; reproduced by permission from Springer-Verlag.)

Thermal Conductivity

The meaning of thermal conductivity is the opposite of the thermal insulation capacity of wood and other materials. Wood is low in thermal conductivity because its structure is porous. Thermal conductivity is expressed by the *coefficient of thermal conductivity* (κ). This is a measure of the quantity of heat in calories, which will flow during a unit of time (s) through a body 1 cm thick with a surface area of 1 cm^2, when a difference of 1°C is maintained between the two surfaces; κ is measured in cal·cm/s·cm^2·°C, or kcal·m/h·m^2·°C.[5]

Thermal conductivity is influenced by various factors, such as wood structure, density, moisture, temperature, extractives, and defects (checks, knots, cross grain) (28, 33).

Axially, thermal conductivity is about twice higher in comparison to the transverse (radial, tangential) direction. Coefficients of thermal conductivity of different woods, at a temperature of 20°C (68°F), were calculated, on the average, as follows (16):

axial	0.191–0.284	kcal·m/h·°C
radial	0.104–0.151	"
tangential	0.090–0.140	"

The higher axial thermal conductivity is mainly due to the fibrous morphology and axial placement of most wood cells (microfibrillar deviations reduce such conductivity) (39, 40). Between radial and tangential direction, important differences do not exist. Radially, thermal conductivity is a little higher (5–10%) (2, 26, 38), and this is attributed to the wood rays. However, a higher tangential conductivity has also been observed in woods with pronounced density differences between earlywood and latewood. Finally, the ratio of axial to transverse thermal conductivity was found to relate to the ultrastructure of wood—namely, to the angle of microfibrils. A greater angle, as in compression wood, results in a smaller difference between axial and transverse thermal conductivity (39).

Thermal conductivity increases with wood density, moisture content, and temperature (18, 19, 31, 37) (Figure 12-3). The relationships are linear. The effect of density is expressed by the following relationship—valid for transverse conductivity, density 0.2–0.8 g/cm^3 and 25°C temperature (19):

$$\kappa = 0.177\, R_{12} + 0.0205 \qquad (12\text{-}4)$$

THERMAL PROPERTIES

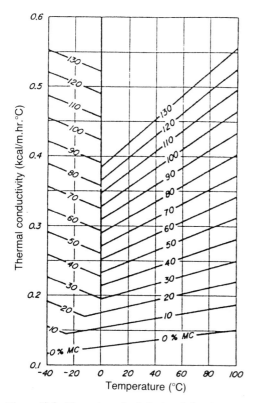

Figure 12-3. Thermal conductivity in relation to temperature and moisture content (birch wood, radial direction, basic density 0.515 g/cm³). (After Ref. 15.)

where

κ = coefficient of thermal conductivity (kcal·m/h·°C)
R_{12} = air-dry density at 12% moisture content

When moisture is increased or reduced below the fiber saturation point by 1%, thermal conductivity is increased or reduced from 0.7 to 1.18%, and on the average 1.25% (18); above the fiber saturation point, increase or reduction is somewhat higher, and in general wood with a moisture content higher than 40% has an approximately $\frac{1}{3}$ higher conductivity than dry wood. The combined effect of density and moisture is expressed by the relationship (6):

$$\kappa = 0.165 + r_o(1.39 + 0.038M) \quad (12\text{-}6)^6$$

where

r_o = oven-dry density (specific gravity)
M = moisture content (%),

and the effect of temperature by the relationship

$$\kappa = \kappa_o(1 + bt) \quad (12\text{-}7)$$

where

κ_o = thermal conductivity at 0°C
b = a proportionality coefficient, related to the effect of density and moisture content
t = temperature (°C)

$$\text{Coefficient } b = \frac{0.3}{1 - 270\frac{r_o}{r_w}} \quad (12\text{-}8)$$

where

$$r_w = 1.50 \text{ g/cm}^3$$

Thermal conductivity is also affected by extractives; woods with a high extractive content (dark-colored) have a higher conductivity (38). The same effect is contributed by oleoresin (e.g., in pine wood). In general, thermal conductivity is mainly influenced by changes of density and moisture (36).

On the basis of the above, and considering that the meaning of thermal conductivity is opposite to that of the thermal insulating capacity of wood, the following practical conclusions may be drawn:

- Wood possesses a greater insulating capacity in the transverse direction. This is an advantageous property, considering common usages of wood (e.g., in the construction of doors, partitions, wooden houses, etc.).
- Light woods are better thermal insulators.
- Dry wood is a better insulator.
- Wood is a better insulator at lower temperatures (the effect is small).

In general, wood is better in comparison to other materials from a thermal-insulation point of view. Its thermal conductivity is lower in comparison to metals, marble, glass, concrete, etc. (4) (see Table 12-2).

Wooden structures of special construction

Table 12-2. Thermal Properties of Wood and Other Materials[a,b]

Material	1	2	3	4	5
Aluminum	173 (1400)	0.22	2.7 (168)	0.2912 (0.1255)	570
Steel	38.3 (310)	0.12	7.8 (486)	0.0408 (0.0176)	80
Concrete	1.55 (12.6)	0.25	2.2 (137)	0.00228 (0.00121)	5.5
Brick	0.62 (5.0)	0.20	1.6 (100)	0.001949 (0.00084)	3.8
Wood					
Douglas-fir	0.10 (0.80)	0.39	0.49 (31)	0.000515 (0.00022)	1.0
White oak	0.15 (1.20)	0.39	0.72 (45)	0.000515 (0.00022)	1.0

Source: Courtesy of F. F. Wangaard (adapted to SI units).
[a] 1, Thermal conductivity: kcal·m/h·m²·°C (Btu·in./h·ft²·°F); 2, specific heat: kcal/kg°C; 3, density: g/cm³ (lb/ft³); 4, diffusivity: m²/h (in.²/s); 5, diffusivity ratio—these values mean that, under the same conditions, brick heats 3.8 times as fast as wood, concrete 5.5 times, steel 80 times, and aluminum 570 times as fast.
[b] Respective calculations of insulating efficiency (opposite to conductivity) show that the insulating efficiency of wood is about 6 times greater than brick, 16 times greater in comparison to concrete, 400 times than steel, and 1700 times than aluminum. In all cases, wood is considered air-dry and values refer to transverse direction (across the grain).

(e.g., with a middle layer of undulating paperboard or low-density fiberboard) reduce thermal conductivity—that is, increase thermal insulation due to the thermal-insulating capacity of air (21).

Specific Heat

Specific heat of a body is the quantity of heat needed to increase the temperature of its unit mass by 1°C. Specific heat is measured in cal/g °C or kcal/kg °C. Considering that the average specific heat of water is 1 (i.e., 1 cal is needed to increase the temperature of 1 g of water—from 15 to 16°C), specific heat of a material, including wood, may be defined as the ratio of the quantity of heat needed to increase its temperature by 1°C to the quantity needed to increase the temperature of an equal mass of water by 1°C (32).

The specific heat of wood is higher in comparison to metals and other common materials, which means that relatively greater quantities of heat are required to increase its temperature (see Table 12-2). This property, in conjunction with the low thermal conductivity of wood, is an advantage because it makes it suitable for handles, matches, and other products. Also, it is important in various industrial processes, such as drying, preservative treatment, and gluing.

Specific heat is not considerably affected by species and density of wood, but increases when its temperature and moisture content increase (30, 31) (Figure 12-4).

Thermal Diffusivity

Thermal diffusivity is a measure of the rate of change of the temperature of a material, when the temperature of its surroundings changes. Wood has a much lower diffusivity than steel (see Table 12-2). This difference is the reason why wood feels "warm" to touch in comparison to steel. Diffusivity is lower in wood of lower moisture content and higher density, but the effect of the latter is smaller (40).

Combustion

Wood may burn. This property makes it suitable for heating purposes, but is a disadvantage from a technical utilization point of view, because it limits its wider use, and results in its replacement by other materials in certain constructions and products.

Wood burns under the action of high temperatures, which result in its chemical decomposition and production of flammable gases. Gradually, the following changes take place (12):

a. Evaporation of moisture (up to 100°C).
b. Evaporation of volatile substances (95–150°C and higher).

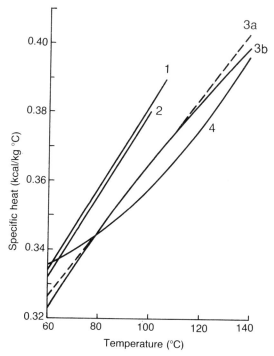

Figure 12-4. Influence of temperature on specific heat (oven-dry wood and bark): 1 and 2, various woods; 3 and 4, pine (*Pinus glabra*—3a, latewood; 3b, earlywood; 4, bark). (After Ref. 17.)

c. Superficial carbonization and slow exit of flammable gases (150–200°C).
d. Faster exit of flammable gases, followed by ignition and glow (200–370°C).
e. Fast ignition of flammable gases and formation of glowing charcoal (370–500°C).[7]

The flammability of wood (speed of ignition and burning) is affected by various factors, such as *species, moisture content, temperature, dimensions*, and *type of wooden structure*.

With regard to species, extractive content (mainly resin) is of major importance. Wood structure is also an important factor: species with long, open passages in their mass (e.g., hardwoods with open vessels, and without tyloses) are more flammable. Softwood tracheids have a negative effect, because of their short length and closed ends. In the axial direction, burning proceeds at about double speed in comparison to transverse direction (29). Moisture makes both ignition and progress of burning more difficult. With increasing temperature, wood becomes more flammable. At a temperature of about 250°C (500°F), a spark or flame is needed for its ignition, but at higher temperatures (500°C, or about 1000°F and higher), ignition is spontaneous (1, 13, 18).[8] Pieces of wood with relatively small dimensions ignite and burn easier, as well as constructions that favor the movement of air.

However, it should be noted that:

a. Wooden members of large dimensions (e.g., thick beams) burn with difficulty, and their strength is reduced gradually in comparison to metals, which fail (bend) under the influence of the high temperatures of a fire (up to 1000°C or 1800°F). Due to the low thermal conductivity and high specific heat of wood, only a thin, superficial layer is initially carbonized in heavy timbers; this layer has an insulating effect, and delays the progress of burning (13, 29).[9]
b. The flammability of wood may be considerably reduced by spraying or impregnation with suitable chemicals which delay its ignition and burning (see Chapter 19).

200 II / PROPERTIES

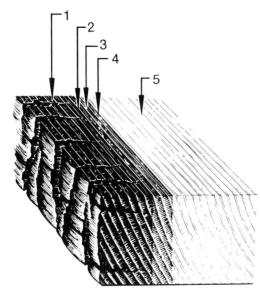

Figure 12-5. Progress of carbonization: 1, carbonized exterior layer; 2, its base; 3, pyrolysis zone; 4, its base; 5, unaffected wood. (Courtesy of U.S. Forest Products Laboratory.)

Table 12-3. Heating Values[a]

Species	Wood	Bark
Cypress, Mediterranean	5920	—
Douglas-fir	4580–5030	5610
Fir	4440–4650	5060
Larch	4050	—
Pine	4780–6790	5040–5980
Spruce	4500–4700	4900
Alder	4300–4440	4670
Ash	4250–5350	—
Beech	4500–4870	5340
Birch	4650–5190	5310–5480
Black locust	4500	—
Hornbeam	4060	—
Maple	4180–4670	4300
Oak	4390–5280	4140–4660
Poplar	4120–5350	4240–4670
Willow	4190–4260	—

[a]Maximum heating values (oven-dry wood) determined in a laboratory (kcal/kg); in practice, there are losses of combustible gases and heat consumed to evaporate moisture. See also Figures 12-6 and 12-7.
Source: Refs. 8, 14, and 18.

Heating Value

Burning wood produces heat. The quantity of heat produced from a mass of 1 g or 1 kg of wood, completely burnt, is called *heating value*. The maximum heating value of (oven-dry) wood is on the average about 4500 cal/g (4500 kcal/kg).[10] Differences are found between and within a species (Table 12-3, Figure 12-6). In general, hardwoods possess lower heating value in comparison to softwoods (respective average values are about 4350 kcal/kg and 4700 kcal/kg) (9, 18).

Heating value is influenced by various factors, such as *moisture content*, *extractives*, and *chemical composition* of wood.

Moisture reduces the heating value (24) (Figure 12-7). The heating value of air-dry wood is about 15% lower than that of oven-dry wood. The effect of moisture may be estimated from the relationship (18)

$$H = \frac{4500 - 600u}{1 + u} \quad (12\text{-}9)$$

where

H = heating value at u moisture content (kcal/kg)

u = moisture content (% od weight, decimal number)

or from the relationship (35)

$$H = H_a - (0.0114\, H_a \times u_1) \quad (12\text{-}10)$$

where

H = heating value at u_1 moisture content (kcal/kg)

H_a = heating value of od wood (kcal/kg)

u_1 = moisture content (% of green weight)

Extractives may have an important influence on heating values (11). For example, oleoresin has high heating value (8500 kcal/kg), and for this reason softwoods containing resin (e.g., pines) have a higher heating value. The influence of chemical composition derives from the fact that lignin has a higher heating value (about 6100 kcal/kg) in comparison to cellulose (4150–4350 kcal/kg) (18, 43).

Utilization of the heating value of wood is affected by the manner of its burning (27). In

THERMAL PROPERTIES   201

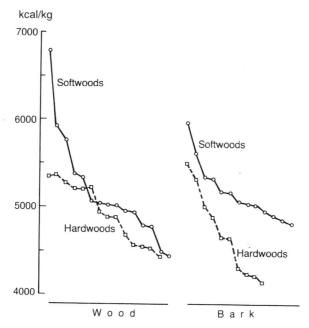

Figure 12-6. Graphical presentation of (maximum) heating values of softwoods, hardwoods, wood and bark. (Based on data from Ref. 14.)

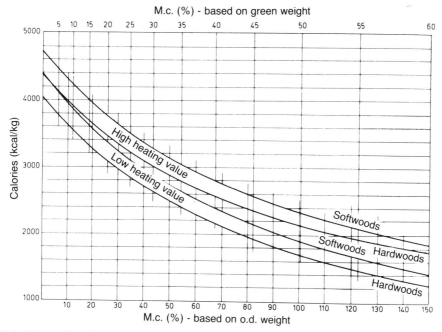

Figure 12-7. Relationship of heating value and moisture. (After Ref. 9; reproduced by permission from Unione Tipografico-Editrice Torinese.)

traditional fireplaces, only a small part of the heating value is utilized (5–20%), and in stoves 10–70%.

In addition to production of heat by direct burning, wood is utilized in the form of products, which are produced by pyrolysis (carbonization, destructive distillation, liquification), gasification, and hydrolysis (7, 23, 42) (see Appendix I).

REFERENCES

1. Beall, F. C., and H. W. Eickner. 1970. Thermal Degradation of Wood Components: A Review of Literature. U.S. For. Prod. Lab. Res. Paper FPL 130.
2. Blankenstein, C. 1962. *Holztechnologisches Taschenbuch.* München: Hanser Verlag.
3. Browne, F. L. 1963. Theories of the Combustion of Wood and its Control. U.S. For. Prod. Lab. Report No. 2136.
4. Bolz, R. E., and G. L. Tune. 1970. *Handbook of Tables for Applied Engineering Science.* Cleveland, Ohio: Chemical Rubber Co.
5. Christoph, N., and G. Brettel. 1977. Untersuchungen zur Wärmedehnung von Holz in Abhängigkeit von Rohdichte und Temperatur. *Holz Roh- Werkst.* 35(3):99–108.
6. Deliinski, N. 1977. Berechnung der instationären Temperatureverteilung im Holz bei der Erwärmung durch Wärmeleitung. *Holz Roh- Werkst.* 35:141–145.
7. Earl, D. E. 1974. *Charcoal.* Rome: FAO.
8. Filipovici, J. 1965. *Studiul Lemnului.* Bucarest, Romania.
9. Giordano, G. 1971. *Tecnologia del Legno*, Vol. 1. Torino: Unione Tipografico-Editrice.
10. Holliday, D., and R. Resnick, 1962/1966. *Physics*, Vols. 1 and 2. New York: J. Wiley & Sons.
11. Howard, E. T. 1973. Heat of combustion of various southern pine materials. *Wood Sci.* 5(3):194–197.
12. Hunt, G. M., and G. A. Garratt. 1967. *Wood Preservation*, 3rd ed. New York: McGraw-Hill.
13. Illston, J. M., J. M. Dinwoodie, and A. A. Smith. 1979. *Timber, Concrete and Metals.* New York: Van Nostrand Reinhold.
14. Ince, P. J. 1977. Estimating Effective Heating Value of Wood or Bark Fuels at Various Moisture Contents. U.S. For. Prod. Lab. FPL 13.
15. Kanter, K. R. 1957. The thermal properties of wood. *Derev. Prom.* 6(7):17–18 (Russian, cited by H. P. Steinhagen).
16. Knigge, W., and H. Schulz. 1966. *Grundriss der Forstbenutzung.* Hamburg: P. Parey.
17. Koch, P. 1969. Specific heat of ovendry spruce pine wood and bark. *Wood Sci.* 1(4):203–214.
18. Kollmann, F. 1951. *Technologie des Holzes und der Holzverkstoffe. I. Band.* Berlin: Springer Verlag.
19. Kollmann, F. F. P., and W. A. Côté. *Principles of Wood Science and Technology, I. Solid Wood.* Berlin: Springer Verlag.
20. Kubler, H. 1973. Role of moisture in hygrothermal recovery of wood. *Wood Sci.* 5(3):198–204.
21. Lewis, W. C. 1968. Thermal Insulation from Wood in Buildings: Effects of Moisture and its Control. U.S. For. Prod. Lab. Res. Paper FPL 86.
22. Maclean, J. D. 1952. Effect of temperature on the dimensions of green wood. *Proc. A.W.P.A.* 48:136–157.
23. Miller, R. H. P. 1951. Gasogens. U.S. For. Prod. Lab. Report No. 1463.
24. Murphey, W. K., and B. E. Cutter. 1974. Gross heat of combustion of five hardwood species at differing moisture contents. *For. Prod. J.* 24(2):44–45.
25. Oberli, H. 1966. Schwinden des lebenden Stammholzes bei Frost. *Schweiz. Z. Forstwes.* 117(7):503–511.
26. Ratcliffe, E. H. 1964. A review of thermal conductivity data. *Wood* 29(7):49–51; 29(8):46–49; 29(9):50–54.
27. Reineke, L. H. 1961. Wood Fuel Combustion Practice. U.S. For. Prod. Lab. Report No. 1666-18.
28. Rowley, F. B. 1933. The heat conductivity of wood at climatic temperature differences. *Heat Piping Air Cond.* 5:313.
29. Schaffer, E. L. 1975. Fire considerations. In *Wood Structures*, pp. 392–410. New York: Soc. Civil Engineers (ASCE).
30. Skaar, C. 1988. *Wood-Water Relationships.* Berlin, New York: Springer Verlag.
31. Sova, V., D. Brenndoerfer, and G. Zlate. 1970. Zur Bestimmung der thermischen Eigenschaften von Ahornholz (*Acer pseudoplatanus* L.) *Holz Roh- Werkst.* 28:117–119.
32. Stamm, A. J. 1964. *Wood and Cellulose Science.* New York: The Ronald Press.
33. Steinhagen, H. P. 1977. Thermal Conductivity Properties of Wood. A Literature Review. U.S. For. Prod. Lab., Gen. Tech. Report FPL-9.
34. Stevens, W. C. 1960. The thermal expansion of wood. *Wood* 25(8):328–329.
35. Tillman, D. A. 1978. *Wood as an Energy Resource.* New York: Academic Press.
36. U.S. Forest Products Laboratory. 1952. Computed Thermal Conductivity of Common Woods. Tech. Note 248.
37. U.S. Forest Products Laboratory. 1987. Wood Handbook. USDA, Ag. Handbook No. 72 (revised).
38. Vorreiter, L. 1949. *Holztechnologisches Handbuch, Band I.* Vienna: Verlag Fromme.
39. Wangaard, F. F. 1943. The effect of wood structure upon heat conductivity. *Trans. Am. Soc. Mech. Eng.* 65(2):127–135.
40. Wangaard, F. F. 1969. Heat transmissivity of southern pine wood, plywood, fiberboard, and particleboard. *Wood Sci.* 2(1):54–60.
41. Weatherwax, R. C., and A. J. Stamm. 1946. The Coefficients of Thermal Expansion of Wood and Wood Products. U.S. For. Prod. Lab. Report No. 1487.

42. Wendorff, G. B. 1980. Moteurs à gaz de bois. *Revue Bois Appl.* 12:25–28.
43. White, R. H. 1986. Effect of lignin content and extractives on the higher heating value of wood. *Wood Fiber Sci.* 19(4):446–452.

FOOTNOTES

[1] The effect of temperature is opposite to that of moisture: dimensions increase due to thermal expansion, and decrease due to shrinkage.

[2] Heat and temperature are different as follows: Heat is a form of kinetic energy which, when brought to or removed from a body, changes its temperature. Temperature is a comparative measure of the thermal condition of a body (i.e., an index that characterizes how much a body is warmer or colder in comparison to another). (Temperature does not measure the quantity of heat contained in a body, because the quantity of thermal energy needed to raise the temperature of two materials (e.g., steel and wood) to the same level, may be entirely different.) According to the kinetic theory of heat, an increase of the temperature of a body is accompanied by an increase of the mobility of its molecules, which are removed from one another, more so with increasing temperature. The result is expansion—and, inversely, contraction (10).

[3] Measurements have shown that tangential expansion ranged from 0.5% to more than 2%, and radial contraction was on the average 2%. These changes are permanent (attributed to release of growth stresses). Tangential expansion becomes higher with increasing temperature (20).

[4] According to another point of view, the formation of frost cracks is attributed to the exit of moisture from cell walls to cell cavities (i.e., to shrinkage of wood and not to its thermal contraction) (16, 25).

[5] In the English system, thermal conductivity (κ) is measured as the quantity of heat in British thermal units (Btu) which will flow in 1 h through a body 1 in. thick and 1 ft^2 when a difference of 1°F is maintained between the two surfaces.

$$1 \text{ kcal} \cdot \text{m/h} \cdot °C = 0.8062 \text{ Btu} \cdot \text{in.}/\text{h} \cdot \text{ft}^2 \cdot °F$$

and

$$1 \text{ Btu} \cdot \text{in.}/\text{h} \cdot \text{ft}^2 \, °F = 0.12404 \text{ kcal} \cdot \text{m/h} \cdot \text{m}^2 \cdot °C$$

[6] In the English system, the following formula is applied:

$$k = S(1.39 + 0.028M) + 0.165 \qquad (12\text{-}5)$$

where

k = thermal conductivity (Btu·in./h·ft^2·°F)
S = specific gravity based on oven-dry weight and volume at the current moisture content
M = moisture content (% of oven-dry weight)

For a moisture content of 40% or higher, $k = S(1.39 + 0.038M) + 0.165$ (37).

[7] According to other sources (1, 3), four stages are distinguished with regard to combustion products, as follows:

1. Below 200°C: inflammable gases, mainly water vapors and traces of CO_2, formic, and acetic acid.
2. Between 200 and 280°C: the same gases (few water vapors) and a little CO. Reactions are endothermic and almost all products are flammable.
3. Between 280 and 500°C: pyrolysis takes place with exothermic reaction. Most products are flammable (oxygen, methane, tar in the form of smoke, etc.).
4. Above 500°C: main residue is coal, which burns and is consumed.

Pyrolysis is macroscopically recognized by the color which becomes darker. Exit of gaseous products of pyrolysis has a cooling effect (reduces the temperature of the carbonized zone) (13, 29).

[8] Spontaneous ignition may take place without exterior application of heat—from the heat developed by slow oxidation, moisture adsorption, and fungal or other enzymatic action. Such conditions may exist in the interior of large piles of sawdust or wood particles (32, 38).

[9] There is considerable reduction of temperature toward the interior of burning wood. According to some observations in wood of Douglas-fir, the temperature at the base of the carbonized zone (see Figure 12-5) was 300°C, at a distance of 6 mm ($\frac{1}{4}$ in.) 180°C, and at a distance of 12 mm ($\frac{1}{2}$ in.) about 90°C (29). According to other research, the burning front advances at a rate of 0.64 mm/min in softwoods and medium density hardwoods, and 0.5 mm/min in heavy hardwoods (13).

[10] In the English system, the heating value (fuel value, calorific value, heat of combustion) is expressed in Btu/lb; 1 kcal/kg = 1.8 Btu/lb. Heating energy is also expressed in Joules (J): 1 cal = 4.184 J, 1 Btu = 1.055 × 10^3 J, 1 Btu = 252 cal, 1 cal = 0.00396 Btu.

13

Acoustical Properties

The acoustical properties of wood refer to (i) production of sound by direct striking, and (ii) behavior of wood to sounds produced by other sources, transmitted through the air, and affecting wood in the form of sound waves.

WOOD AS SOURCE OF SOUND

Wood is sometimes used as a source of sound. An example is the xylophone—a musical instrument made of wooden bars of varying size. Musical sounds are produced by striking the bars with suitable wooden or metallic mallets. In some old monasteries and village churches, wooden slabs are used, in place of bells, to produce sounds by rhythmical striking.

The pitch or tone of sound, whether low or high, depends on frequency of vibration. Frequency is affected by dimensions, density, and elasticity (modulus of elasticity) of a particular wooden member; smaller dimensions, lower moisture content, and higher modulus of elasticity produce sounds of higher pitch.

SOUND WAVES FROM OTHER SOURCES

When sound waves produced by another source reach wood, part of the acoustical energy is reflected and part enters its mass. Wood vibrates, and the original sound is intensified, or subjected to partial or total absorption.

Consonance—Resonance

Consonance or intensification of sound takes place when wood is used as a resonator. Its performance is affected by frequency of vibration, shape of the resonator, and condition of the surface of wood (e.g., a lacquered surface has a favorable effect). A resonator does not change the pitch of the original sound but can intensify it (make it louder), and increase its duration.

Wood is used as a resonator in stringed musical instruments; especially noteworthy is the case of violins. There is a preference for spruce wood (it has a high modulus of elasticity in relation to density)[1]—straight-grained, radially (quarter) sawn, homogeneous in structure, with narrow growth rings (up to 2 mm), low proportion of latewood (up to 25%), and from old trees (130–150 years in age, and a diameter greater than 40 cm or 16 in.) (5, 8). Spruce wood with indented growth rings (see Chapter 7) is considered very suitable for musical instruments (5). Stradivari, Amati, Guarneri, and other famous Italian violin makers used spruce wood, but their secret of making fine instruments is not clear; research has shown that, in addition to suitable wood, the quality of sound should be related to thickness and curvature of the resonator (body), and to treatment of the wood (chemical impregnation, drilling small holes, etc.) (4). In addition to spruce, resonators of musical instruments are made of fir and pine wood; hardwoods, such as maple and tropical woods, are also used.

Absorption of Sound

As mentioned earlier, a part of the acoustical energy reaching wood enters its mass. This energy may be absorbed (at least partially), due

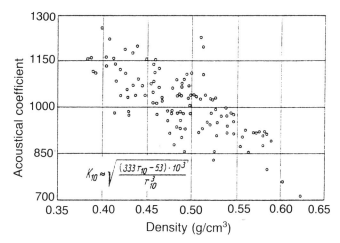

Figure 13-1. Relationship of acoustical coefficient and density in spruce wood (10% moisture content). (After Ref. 8; reproduced by permission from Springer-Verlag.)

to repeated refraction and reflection of sound waves. In this manner, friction of the molecules, which make up the mass of wood, results in change of acoustical to thermal energy.

The ability of wood to absorb sound is measured by the *coefficient of sound absorption*, which is an expression of the proportion (percentage) of the absorbed sound (1). Wood has an advantage in comparison to other materials, due to its porous structure, but has a relatively low coefficient of absorption, less than 10% (Table 13-1). The coefficient is affected by wood density and other factors, such as modulus of elasticity, moisture content, temperature, intensity and frequency of sound, and condition of the surface of wood: woods with lower density and modulus of elasticity, and at a higher moisture content and temperature, absorb more sound (Figure 13-1); absorption is greater with sounds of lower frequency, and is lower from lacquered wood.

The sound-insulating capacity of wood may be substantially improved (up to 90%) by providing void spaces inside walls of partitions. In wood products (e.g., fiberboard), low density and drilling holes increase the insulation.

SPEED OF SOUND

The speed of sound in the mass of wood varies, depending on direction (axial, transverse) and species of wood. In the axial direction, it is about 3500–5000 m (10,000–15,000 ft)/s. Transversely, the speed is lower, because modulus of elasticity, which affects the speed of sound, is lower in that direction. In other materials, speeds of sound are as follows: air 340 m (1100 ft)/s, cork 430–530 m (1400–1750 ft)/s, water 1440 m (4750 ft)/s, iron 5000 m (15,000 ft)/s, glass 5000–6000 m (15,000–20,000 ft)/s.

Speed is theoretically calculated from the relationship (3, 7)

$$V = \frac{E}{r_o} \quad (13\text{-}1)^2$$

where

V = speed of sound (m or ft/s)

Table 13-1. Sound Absorption (coefficient of absorption, %)

Material	Frequency (cps)		
	125	500	2000
Walls			
Wood	8	6	6
Brick	2	3	5
Floors			
Wood	5	3	3
Concrete	1	2	2
Glass	3	3	2
Boards, excelsior	13	36	70
Boards, insulation	39	52	59

Source: Ref. 1.

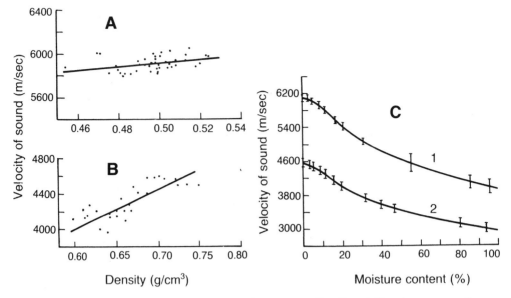

Figure 13-2. Relationship of sound velocity to (A, B) density (A, spruce; B, oak), and (C) moisture content (1, spruce; 2, oak). (After Ref. 6; reproduced by permission from Springer-Verlag.)

E = modulus of elasticity (N/mm^2, psi)
r_o = oven-dry density (specific gravity)

Therefore, the speed of sound depends on the elasticity of wood (or other solid material) and the quantity of vibrating substance (i.e., its density) (Figure 13-2A and B).

Axial and transverse speed, in a specimen of wood, differ because their respective moduli of elasticity are different; transverse modulus is considerably lower and, therefore, transverse speed is less. Moisture reduces the speed of sound (Figure 13-2C) because, with increasing moisture, modulus of elasticity decreases and density increases. Speed is also reduced with increasing temperature, because higher temperatures produce lower density due to thermal expansion of wood (2).

Generally, a smaller difference between axial and transverse speed is considered an indicator of suitability for musical instruments (14) although differences do exist. For example, for violin sounding boards (front piece), a light, straight-grained spruce with a high ratio of axial-to-transverse speed is preferred; for backs, a curly grained maple has been used (low ratio of axial-to-transverse speed).

DAMPING OF SOUND

Sound produced by wood is gradually subject to damping, due to decay of its vibrations. The acoustical energy is dissipated partly by radiation in the atmosphere and partly due to internal friction (resulting in the development of heat).

The damping capacity of wood varies with species, moisture content (reduced with increasing moisture), direction of vibration (longitudinal, transverse, torsional), and mode of vibration. Damping due to sound radiation depends mainly on the ratio of speed of sound to density of wood (12).

In musical instruments (e.g., sounding boards of pianos or violins), it is desirable to have low damping due to internal friction and high damping due to radiation of sound (7, 12).

EFFECT OF WOOD DEFECTS

The acoustical properties of wood are influenced by defects. This relationship is utilized by the application of acoustical methods to the detection of defects (e.g., decay, shakes, etc.) (9, 10). Other technical applications of wood acoustics (in architectural structures, etc.) have been considered inadequately investigated (13). An empirical application to detection of inner decay is made in standing trees prior to felling by striking their trunk with an axe and judging the tone of sound.

REFERENCES

1. Hoyle, R. J. 1975. Physical character of wood. In *Wood Structures*, pp. 1–31. New York: Am. Soc. Civil Engineers (ASCE).
2. James, W. L. 1961. Effect of temperature and moisture content on internal friction and speed of sound in Douglas-fir. *For. Prod. J.* 11(9):383–390.
3. Holliday, D., and R. Resnick. 1962/1966. *Physics*, Vols. 1 and 2. New York: J. Wiley & Sons.
4. Ille, R. 1976. Eigenschaften und Verarbeitung von Fichtenresonanzholz für Meistergeigen (II). *Holztechnologie* 17(1):32–35.
5. Knigge, W., and H. Schulz. 1966. *Grundriss der Forstbenutzung*. Hamburg: P. Parey.
6. Kollmann, F., and H. Krech. 1960. Dynamische Messung des elastischen Holzeigenschaften und der Dämpfung. *Holz Roh- Werkst.* 18:41–51.
7. Kollmann, F. F. P., and W. A. Côté. 1968. *Principles of Wood Science and Technology, I. Solid Wood.* Berlin: Springer Verlag.
8. Krzysik, F. 1967. Untersuchungen über den Einfluss der Rohdichte auf die Verwendungsmöglichkeit von Fichtenklangholz. *Holz Roh- Werkst.* 25(1):37.
9. Lee, I. D. G. 1958. *Non-Destructive Testing of Wood by Vibrational Methods.* London: Timber Development Association.
10. McDonald, K. A. 1978. Lumber Defect Detection by Ultrasonics. U.S. For. Prod. Lab. FPL 311.
11. Ono, T. 1983. On dynamic mechanical properties in the trunks of woods for musical instruments. *Holzforschung* 37(5):245–250.
12. Schniewind, A. P. 1981. Mechanical behavior and properties of wood. In *Wood: Its Structure and Properties*, ed. F. F. Wangaard, pp. 225–270. Materials Education Council, Pennsylvania State University, University Park, Pennsylvania.
13. Schultz, T. J. 1969. Acoustical properties of wood: A critique of the literature and a survey of practical applications. *For. Prod. J.* 19(2):21–29.
14. Vorreiter, L. 1949. *Holztechnologisches Handbuch, Band I.* Wien: Verlag Fromme.

FOOTNOTES

[1] Dynamic modulus of elasticity, internal friction, and density are considered important properties of wood for musical instruments (11).

[2] The following similar relationship is reportedly used in the musical instrument industry (Poland, U.S.S.R.) (8) as criterion for selection of suitable woods:

$$K = \sqrt{\frac{E}{R_{10}^3}}$$

where

K = acoustical coefficient
E = modulus of elasticity (N/mm^2, psi)
R_{10} = density (specific gravity)
 at 10% moisture content

Modulus of elasticity is determined from static bending tests, and its effect is about three times greater than that of density. The relationship between K and R is negative (Figure 13-1). For suitable spruce wood (Klangholz), K = 1000–2000 (ring width 1.5–2 mm, latewood 25%, density 0.45 g/cm^3) (8).

14

Electrical Properties

The most important electrical properties of wood are: *resistance* to passage of direct electric current, and *dielectric properties* under the action of alternating high-frequency electric current.

ELECTRICAL RESISTANCE

Electrical *resistance* is the property of a material to oppose the passage of electric current; it is the opposite of electrical *conductivity*. Electrical resistance is measured in ohm (Ω) units, and conductivity is the reciprocal ($1/\Omega$). The electrical resistance of a conductor of unit cross-sectional area and unit length is called *specific electrical resistance* or *resistivity*, and is expressed in ohm-meter or ohm-centimeter units. This property is characteristic of each material, and useful for comparison of various materials on the basis of their ability to conduct electric currents. High specific electrical resistance (resistivity) denotes a poor conductor. The reciprocal of specific electrical resistance is called *specific conductivity* ($1/\Omega$ cm), and it also characterizes materials on the basis of how well electric current flows through them. Good electrical conductors have high specific conductivity and low specific resistance (6).

The electrical resistance of wood is affected by various factors, such as *species, structure, density, temperature,* and *moisture*. The effect of moisture is greater than that of the other factors.

Oven-dry wood is an insulating material; it practically does not allow the passage of electric current through its mass. However, with increasing moisture content, electrical resistance is greatly reduced, and saturated wood behaves almost like water in this respect.

The changes of electrical resistance (and conductivity) with changing moisture content is great in the region between zero and the fiber saturation point; above that point, the change is relatively very small (Figures 14-1 and 14-2).[1] The specific electrical resistance of oven-dry wood varies between 3×10^{17} and 3×10^{18} Ω cm; it is reduced to 10^8 in air-dry wood, and to about 10^6–10^5 Ω cm at the fiber saturation point. Thus, in the region between zero and the fiber saturation point the resistance is reduced more than a billion times; whereas between the fiber saturation point and the maximum moisture content the reduction is only about 50 times.

Oven-dry wood is comparable to the best insulating materials. Its specific electrical resistance is similar to that of porcelain (3×10^{14}) and paraffin (1×10^{16}). Inversely, the resistance of good conductors is low: copper 1.72×10^{-6}, iron 10×10^{-6}, water (distilled) 5×10^{-5}.

The rest of the previously mentioned factors affect electric resistance as discussed below.

Wood species is of little importance. Existing small differences are due to chemical substances (mainly metallic ions) (8, 16) contained in the cell walls (and remaining in the ash) and to extractives.

Structure is effective in the sense that electrical resistance varies axially and transversely (Figure 14-2). Transverse resistance was found to be 2–8 times greater (2.3–4.5 in softwoods,

ELECTRICAL PROPERTIES 209

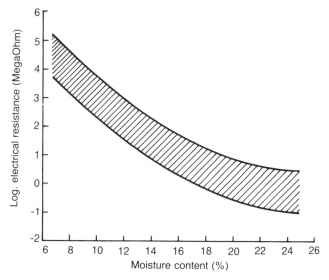

Figure 14-1. Relationship of the logarithm of electrical resistance of wood and moisture. The shaded band includes 90% of the values obtained from various species. (Courtesy of U.S. Forest Products Laboratory.)

and 2.5–8 times in hardwoods) (15). Between radial and tangential directions, differences do not exist or are small (radial resistance may be a little smaller) (14, 23).

The effect of density is unclear. It has been reported that, at the same moisture content, there is an analogy—that is, when density is doubled, resistance is halved (conductivity is doubled) (18). However, other research has shown that conductivity is not related to density, but rather to the chemical composition of wood (20)—namely, to lignin content (conductivity increases with increasing lignin content[2]).

The electrical resistance of oven-dry wood increases with decreasing temperature, and is about doubled for every 12.5°C drop of temperature (19). This behavior of wood is opposite to that of metals.

Practical application of the great effect of moisture on the electrical resistance of wood is found in electrical moisture meters (see Chapter 9). Also, due to its relatively low conductivity at low moisture contents, wood is used for utility poles, handles of electrical instruments, etc. Impregnation of wood with preservatives, especially water-soluble salts, for protection against decay or fire, considerably reduces its electrical resistance (2, 23) (Figure 14-3)—that is, increases its conductivity[3]. Inversely, impregnation with water-repellent substances of low electrical conductivity (e.g., phenolic resins) reduces its conductivity, especially at high relative humidities, because resin reduces the hygroscopicity of wood (14, 18).

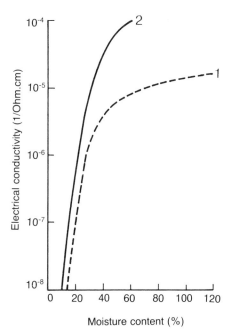

Figure 14-2. Relationship of electrical conductivity and moisture content in alder veneer: 1, transverse to grain; 2, parallel to grain. (After Ref. 5.)

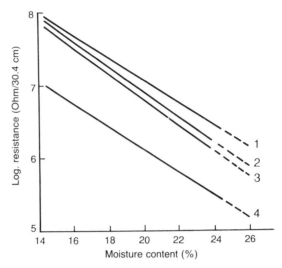

Figure 14-3. Relationship of electrical resistance and moisture in pine sapwood specimens 1. untreated, and treated with 2. creosote, 3. pentachlophenol, and 4. water-borne (CCA) preservatives. (After Ref. 11.)

DIELECTRIC PROPERTIES

A *dielectric* is a poor conductor of electric current, or an electrically insulating material, especially when considered from the view of its behavior in an electric field. Wood is dielectric when oven-dry or containing little moisture.

Dielectric properties of a material are the *dielectric constant* and the *power factor*. These properties present a practical importance in gluing wood with electrical methods and in making electrical moisture meters. Drying wood is also theoretically possible but practically inapplicable. Gluing with synthetic resins (and drying) may be performed because, when the direction of alternating current changes very rapidly (high-frequency current), dielectric materials are heated due to friction of the electrons of their molecules. This heat derives from the creation of an electric field and not from the electrical resistance of the dielectric. Moisture content may be measured because it affects the dielectric properties of wood.

Dielectric constant, also called electric permeability, is a measure of the insulating value of a material with regard to the passage of a high-frequency current. The dielectric constant of a vacuum is 1—other liquid or solid substances have a greater value—and the dielectric constant of water is 81. The dielectric constant of dry wood varies between about 2 and 3, increases with density, moisture content and temperature, and declines with increasing frequency of alternating current; it is about 1.5 times higher in the axial than in the transverse direction. At a density of 0.4–0.6 g/cm^3, the dielectric constant of oven-dry wood is 1.8–2.2. At a moisture content of 10, it varies from 2.7 to 3.5, at 20% from 4.0 to 5.4, and in green wood it is 81, (i.e., is equal to that of water (1, 7, 12, 17, 22). Some electric moisture meters are based on the relationship between dielectric constant and moisture. (The relationship is curved up to the fiber saturation point, and is linear above it.) The main disadvantage of such meters (also called capacitance hygrometers) is that their readings are considerably influenced by wood density.

The *power factor* is a measure of the rate at which electric energy is absorbed by a dielectric; the energy is presented in the form of heat. The factor has a fractional value from zero to 1; it increases with moisture content, wood density, and frequency of the current, and is higher parallel to the grain (7, 17). This property is important in gluing and drying of wood with high-frequency current (5). One type of electric moisture meter is based on the relationship between power factor and moisture (see Chapter 9).

OTHER PROPERTIES

Wood exhibits a *piezoelectric* effect. As with other materials, especially quartz crystals, electric polarization may take place in wood (equal and opposite electric loads develop on opposite sides of a piece) under the action of mechanical stresses (compression, tension). Also, deformation (elongation, reduction of size) may take place under the action of an electric field (4). The piezoelectric effect is influenced by grain distortions (such as found around knots), density, and degree of crystallinity. Research has indicated that this effect could be a strength predictor and/or weakest-point indicator—both of which are important for stress grading of structural timber (13).

Special note should be made regarding magnetic properties. *Magnetic susceptibility* is a property that measures the state of magnetization in relation to the intensity of a magnetic field; its value is positive or negative. Wood has a very low magnetic susceptibility—ranging from -0.2 to $+0.5 \times 10^6$ in various species (depending on density); whereas the respective values for the following materials are: paraffin -0.58×10^6, rubber $+1.1 \times 10^6$, and iron $+720 \times 10^6$ (18). Moisture has a positive effect, because the magnetic susceptibility of water is about double that of wood. Susceptibility is also higher parallel to the grain, and higher in cellulose than in lignin (24).

REFERENCES

1. Aleon, D. 1975. Propriétés diélectriques du bois et chauffage haute frequence. *Revue Bois Applic.* 1/2:29-37.
2. Chanmamedov, K. M. 1973. Zum Problem der elektrischen Leitfähigkeit von Holz. *Holztechnologie* 14(1):47-50.
3. Darvenitsa, M. 1980. *Electrical Properties of Wood and Line Design.* Australia: University of Queensland Press.
4. Fukada, E. 1968. Piezoelectricity as a fundamental property of wood. *Wood Sci. Technol.* 2:299-307.
5. Gefahrt, J. 1967. Die Verwendung von Hochfrequenzenergie in der Holzindustrie. *Holz Roh- Werkst.* 35(1):125-129.
6. Holliday, D., and R. Resnick. 1962/1966. *Physics,* Vols. 1 and 2. New York: J. Wiley & Sons.
7. Hoyle, R. J. 1975. Physical character of wood. In *Wood Structures,* pp. 1-31. New York: Am. Soc. Civil Engineers (ASCE).
8. Ingenieurschule f. Holztechnik, Dresden. 1965. *Taschenbuch der Holztechnologie.* Leipzig: VEB Fachbuchverlag.
9. James, W. L. 1975. Dielectric Properties of Wood and Hardboard: Variation with Temperature, Frequency and Grain Orientation. U.S. For. Prod. Lab. Res. Paper FPL 245.
10. Katz, A. R., and D. G. Miller. 1963. Effect of water storage on electrical resistance of wood. *For. Prod. J.* 13(7):255-259.
11. Katz, A. R., and D. G. Miller. 1963. Effects of some preservatives on the electrical resistance of red pine. *Am. Wood Pres. Assoc.* 59:204-217.
12. Kellogg, R. M. 1981. Physical properties of wood. In *Wood: Its Structure and Properties,* ed. F. F. Wangaard, pp. 187-244. Materials Education Council, Pennsylvania State University, University Park, Pennsylvania.
13. Knuffel, W., and A. Pizzi. 1986. The piezoelectric effect in structural timber. *Holzforschung* 40(3):157-162; 1988. II. The influence of natural defects. *Holzforschung* 42(4):247-252.
14. Kollmann, F. F. P., and W. A. Côté. 1968. *Principles of Wood Science and Technology. I. Solid Wood.* Berlin: Springer Verlag.
15. Lint, R. T. 1967. Review of the electrical properties of wood and cellulose. *For. Prod. J.* 17(7):54-61.
16. Panshin, A. J., and C. De Zeew. 1980. *Textbook of Wood Technology,* 4th ed. New York: McGraw-Hill.
17. Skaar, C. 1948. The Dielectric Properties of Wood at Several Radio Frequencies. N.Y. State College of Forestry, Tech. Bull. No. 69, Syracuse, New York.
18. Stamm, A. J. 1964. *Wood and Cellulose Science.* New York: The Ronald Press.
19. U.S. Department of Agriculture, Forest Products Laboratory. 1987. *Wood Handbook.* Agr. Handbook No. 72 (revised).
20. Venkateswaran, A. 1972. A note on densities and conductivities of wood. *Wood Sci.* 5(1):60-62.
21. Venkateswaran, A. 1974. A note on the relationship between electrical properties and thermal conductivity of wood. *Wood Sci. Technol.* 1(8):50-55.
22. Vermaas, H. F. 1973. Regression equations for determining the dielectric properties of wood. *Holzforschung* 27(4):132-136.
23. Vorreiter, L. 1949. *Holztechnologisches Handbuch,* Band I. Wien: Verlag Fromme.
24. Zürcher, E. 1988. Diagnosemethoden des Gesundheits- und Vitalitäts-zustandes der Bäume. *Vierteljahrsschr. Naturforsch. Ges. Zürich* 133/1:24-42.

FOOTNOTES

[1]From zero to about 7% moisture content, there is a linear relationship between logarithm of specific conductivity and moisture content; from 7% to the fiber saturation point, the logarithm of conductivity is linearly related to the loga-

rithm of moisture content; above the fiber saturation point, the relationship is curvilinear and the change small (24).

[2]Electrical conductivity is related to thermal conductivity, which is also influenced by the chemical composition of wood (21).

[3]Use of wood containing salts should be avoided where electrical resistance (or conductivity) is critical. For example, an increase of electrical conductivity in poles may lead to hazards for power lines (i.e., wood charring and pole-top fires) (3). Dielectric heating, such as in gluing laminated timbers, should also be avoided. Electric moisture meters may give erroneous readings for such wood. Storage of wood in seawater may increase its electrical conductivity due to the salt contained in such water (10).

15

Degradation of Wood

Various external factors—plant (bacteria, fungi), animal (insects, marine organisms), climatic, mechanical, chemical, thermal—may cause degradation of appearance, structure, or chemical composition of wood. Such degradation varies from simple discoloration to rendering wood completely useless, and may refer to living trees, timber, or products.

The resistance of wood to factors that may cause degradation is called *durability*, and may be expressed as the time during which wood preserves its usefulness without special protection, such as preservative treatment. If wood is not exposed to the above-mentioned factors, its durability is practically unlimited. A good example is given by furniture and other wooden items preserved in excellent condition for thousands of years in the pyramids of Egypt (tombs of the Pharaohs). Also, old wood, thousands and even millions of years old, has been found in excavations, on sea bottoms (ancient sunken ships), in rivers (bridge supports), lakes, and swamps (supports of prehistoric dwellings, small objects or remnants),[1] in caves and ancient mines, glaciers, and elsewhere (9, 79). Wood is not degraded or destroyed with the passage of time, but always under the action of exterior factors. It should be remembered that in living trees, the greatest part of wood is dead, but remains sound for hundreds and sometimes thousands of years (25).

PLANT FACTORS OF DEGRADATION

Bacteria

The effects of bacteria are relatively small in comparison to other factors that degrade wood, such as decay fungi, insects, or marine organisms. Bacteria appear in wood stored for an extended period of time in water (even seawater) (61) or in soil, because they have the ability to live in environments with little or no oxygen, since most are anaerobic. Bacterial activity is expressed by perforation or destruction of pit membranes (Figure 15-1) in sapwood (47), erosion of cell walls, and consumption of parenchyma cell contents (1, 13, 18, 46, 62). In this manner, the permeability of wood increases (7–10 times) (1), and strength (toughness, compression, bending) is reduced (1, 34). Also, discolorations, softening of surface layers and excessive shrinkage have been attributed to bacteria (16).[2]

It has been observed that the chemical constituents of wood are attacked more readily when separated from one another. Attack of wood is more difficult; this is probably due to the presence of extractives and interconstituent bonding, or to the acidity (pH) of wood. Wood flour is attacked more easily than solid wood. Lignin acts protectively for the carbohydrates against bacterial decomposition.

Bacteria are considered to be the cause of *wetwood*, which is present in living trees of various species, mainly fir and poplar (4, 13, 67, 77, 78). Wetwood is considered an irregular form of heartwood (biologically, it may not coincide with normal heartwood) (4).[3] In wetwood, parenchyma cells are dead, and storage substances (starch, etc.) do not exist. Wetwood has more extractives and a darker color. Initially, wetwood appears in roots or the lower part of the trunk and extends upward (according to some observations, it also starts from the base of cut or broken branches and extends in-

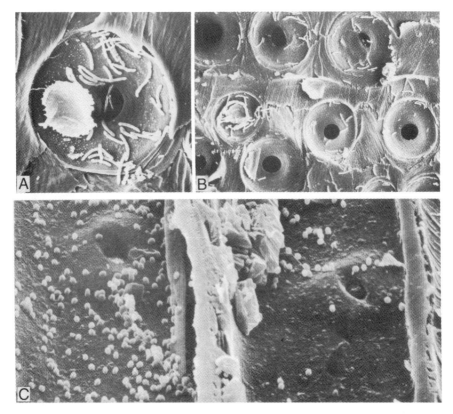

Figure 15-1. Bacteria (A, B) in pine pits after long storage in water (membranes are destroyed); (C) in "wet heartwood" of fir. (A, 2500×; B, 800×; C, 1500×.) (Photographs A and B after Ref. 56, reproduced by permission from DRW-Verlag Weinbrenner, Stuttgart; Photograph C after Ref. 13.)

ward) (4, 54). In transverse sections, it resembles colored heartwood (Figure 15-2), or appears in a zone between sapwood and heartwood, or in patches. Although it appears wet, its moisture content may be lower or about equal to that of sapwood, or there may be an irregular dry zone between sapwood and wetwood. The higher moisture content is attributed to bacteria (78). In poplar, wetwood is called *brown* or *black heartwood* and is common in hybrid poplars (50) (Figure 15-3). The properties of wetwood (density, mechanical properties, acidity, permeability) do not always show differences (35, 54, 77, 78), but its presence, in certain species, has been related to checks (and frost-checks in living trees), detachment of growth rings (shakes), and drying problems [collapse, honeycombing (77), wavy surfaces in poplar veneers (50)]. The darker color and the unpleasant odor that sometimes

Figure 15-2. "Wetwood" in European white (silver) fir *(Abies alba)*. (Courtesy of N. Torelli.)

Figure 15-3. "Black heartwood" in hybrid poplars. (Reproduced by permission from Pergamon Press.)

is present in wetwood practically disappear, or their intensity diminishes with drying lumber or other products, although color may constitute an aesthetic disadvantage (in veneer, matches, etc.). In round timber, the color of transverse sections becomes darker with exposure and passage of time (54). Bacteria may contribute to decomposition of wood after an initial fungal attack or may favor such an attack (33, 61).

Fungi

There are two kinds of fungi: stain and decay fungi.

Stain Fungi This category includes fungi that cause discoloration of wood. They usually attack softwoods (pine, spruce, etc.) and seldom hardwoods (poplar, beech, oak, ash, tropical species). Subject to attack are fallen trees, logs, products, and sometimes living trees; the attack is usually limited to sapwood, and seldom extends to heartwood (10).

Blue stain (or *sap stain*) of pines is the most common and serious consequence of attack by stain fungi. The wood becomes ash-bluish to blackish in its entire sapwood, or in irregular and usually wedge-shaped patches (Figure 15-4); the latter is due to the arrangement of rays, and occurs mainly in softwoods, because in most hardwoods, in addition to rays, there is abundant axial parenchyma. Color is imparted by hyphae or caused by colored substances produced by them (45). Stain fungi belong to *Ceratocystis (Ceratostomella)*, *Graphium*, and other genera. These fungi live mainly on storage substances found in parenchyma cells (mainly ray parenchyma), but sometimes their hyphae are also observed in tracheids. From cell to cell, they pass mainly through pits and sometimes by boring cell walls (Figure 15-4B and C).

Blue stain may appear very quickly in warm weather, sometimes a few hours or days after felling the trees or processing (sawing, etc.) green wood. In addition to temperature and moisture, the attack is favored by nutrients (starch, etc.) found in sapwood, and by other factors, such as absence of intense light and suitable pH (61). Dry wood may be attacked if rewetted. Blue stain may be prevented by moving the harvested wood outside the forest without delay, especially during warm weather, and processing (sawing, etc.) and drying the products. Prevention is also possible by spraying or by immersion in fungicides (10, 68, 70).

Consequences of attack. Stain fungi may cause considerable reduction of the market value of wood, mainly due to discoloration, which has an undesirable effect on the appearance of nonpainted products, such as flooring and certain types of furniture. The discoloration is not superficial and, therefore, cannot be removed by planing or sanding. Other consequences of attack are:

- Degradation of wood properties. The effects are not clear (there are disagreements in re-

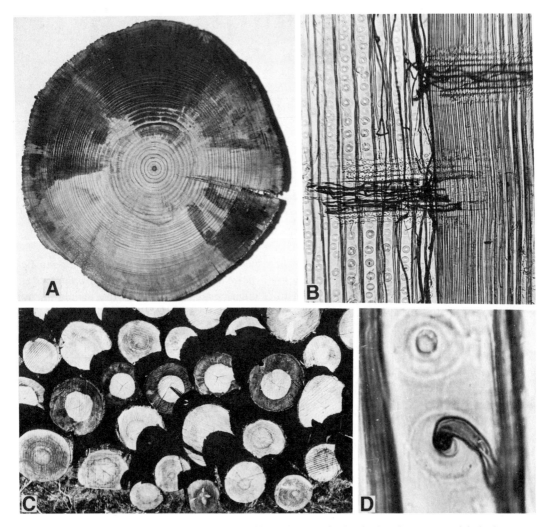

Figure 15-4. Blue stain, (A) macroscopic appearance (black pine); (B) hyphae in the microscope, mainly in the rays (black pine, 180×); (C) all around attack (peripheral portion outside of wetwood in fir); (D) a hypha is directed toward a pit mouth (black pine, 1080×).

search findings), although it is generally believed that there is no strength reduction of practical importance—except in toughness, which is reduced by 15–30% or more (61, 68). This means that wood with pronounced blue stain is unsuitable for products subjected to stresses, such as tool handles, athletic items, ladders, scaffolding, and machine parts.
- The permeability of wood by preservatives or water changes, but the effects are again not clear; retention and penetration were found sometimes higher and sometimes lower (10).
- Pulp production is not affected in quantity, but the pulp is darker in color and, therefore, bleaching becomes more expensive.
- Painting is not affected if the wood is dry; however, if not properly dried, or the hyphae are not dead, rewetting with a water-based paint may cause a change of color.
- Blue stain is not decay. Pronounced blue stain may be accompanied by incipient decay, but wood attacked by stain fungi does show a greater susceptibility to decay fungi (61). Blue stain may predispose wood to decay (16).

Aside from blue stain, discoloration may be caused by molding fungi growing on moist surfaces. Such fungi belong to *Penicilium, Trichoderma, Gliocladium*, and other genera (61). The effects vary according to specific fungus and the conditions of attack. The discoloration is usually superficial (can be removed) but sometimes may be deep, or there may be an erosion of cell walls. In addition to discoloration, molding renders wood unsuitable for certain uses, such as packaging. Also, gluing difficulties have been reported, as well as increased permeability (12, 52, 83).

Decay Fungi. Decay fungi constitute the most important factor that affects the durability of wood. In many cases, resistance to decay is considered synonymous to the durability of wood.

Factors of attack. Decay is not an innate property of wood; it appears only if the conditions are favorable for fungal development. The four main factors that are needed are: *food, moisture, air,* and *heat*. Suitable pH is also a factor. Fungi may grow without light, but some require light for undisturbed reproduction (61).

Food is the wood itself, which may be attacked due to its organic composition. If other conditions are favorable, fungal spores arriving on its surface germinate and produce fine (microscopic) tube-like filaments called *hyphae*. Hyphae grow by elongation of their tips and take their nourishment from cell walls or cell contents. Decay fungi may consume cell walls by secretion of enzymes, which possess the ability to dissolve many organic substances after their change to forms that may be assimilated.

The resistance of various woods to fungal attack varies, but no wood is immune (see Table 15-1). Differences are due mainly to variations in content of toxic extractives. A higher extractive content, as in heartwood, imparts a higher resistance. Sapwood is less resistant because of much lesser extractive content and the presence of nutritive substances, such as starch, in living (parenchyma) cells; after felling a tree, starch is gradually decomposed (see also Chapter 5).

Wood density is of secondary importance. Woods higher in density are not necessarily more resistant, because resistance is mainly related to chemical composition. Thus, the relatively lighter wood of conifers (e.g., pines) is more resistant in comparison to certain heavy hardwoods, such as beech, birch, and maple. However, under the same extractive content and the same conditions of exposure, heavier woods exhibit a higher resistance (i.e., they last longer). Aside from more wood substance, the consumption of which needs more time, the longer durability of very heavy woods is attributed to lack of oxygen; it is hypothesized that carbon dioxide concentrates around the hyphae. In general, the effect of density on durability, both between as well as within species, if existing, is small and in no case analogous to its effect on other wood properties, such as strength.[4]

The combined action of density and heartwood color may be expressed as follows: light woods with light color have a low natural durability in contrast to heavy and dark woods.

Two main categories of decay (rot) are recognized—*brown* and *white*. The terminology refers to the appearance of decayed wood. In an advanced stage, wood attacked by brown rot fungi is changed to a brown, checked mass that may be easily broken or reduced to dust by finger pressure. The checks are both parallel and perpendicular to the grain, similar to charcoal. In contrast, wood attacked by white rot fungi has a spongy or fibrous appearance, with white pockets or streaks separated by areas where wood appears firm and strong (Figure 15-5). Brown and white rots differ from a chemical point of view. Fungi that cause brown rot consume mainly carbohydrates (i.e., cellulose and hemicelluloses). Lignin is not consumed to a considerable extent, but its chemical properties (e.g., solubility) change. White rot fungi may decompose both carbohydrates and lignin. Sometimes, the same wood may be attacked by both types of fungi. Brown rots are more often present in softwoods and white rots in hardwoods (19, 43).

A third category of decay, called *soft rot*, is of lesser importance. Fungi that cause such decay (ascomycetes and deuteromycetes or *Fungi Imperfecti*) are similar to brown rot fungi with regard to chemical action. Wood becomes superficially soft, darker with progressing attack, and when dried its surface layers become checked (as in brown rot) and brittle. The un-

Table 15-1. Resistance of Woods to Fungi, Insects, and Marine Borers—and Permeability (Resistance to Preservative Treatment)[a-c]

Species (European, N. American)	F h	L s/h	A	MB	P s/h	Species (Tropical)	F h	L s/h	A	MB	P s/h
Cypress, Mediterranean	2	—	—	4	—/—	Acajou	3	5/1	—	—	2/4
Fir, white	5	—	5	5	—/—	Afara	1	5/1	—	3	—/4
Larch	2	—	5	4	2/3	Ako	—	5/5	—	—	1/1
Pine, black	4	—	5	5	1/3	Azobe	2	—/—	—	4	2/4
Pine, maritime	4	—	—	5	1/3	Bilinga	2	—/—	—	2	1/2
Pine, Scots	4	—	5	5	1/3	Dibetou	3	5/1	—	4	2/4
Spruce	5	—	5	5	—/3	Framire	2	5/1	—	4	2/4
						Iroko	—	5/1	—	4	—/4
Douglas-fir[d]	3	—	—	5	2/3	Kosipo	3	5/1	—	—	—/—
Pine, pitch[d]	3	—	—	—	1/3	Limba	4	5/5	—	—	—/2
Pine, Monterey[d]	4	—	5	5	1/3	Makoré	2	5/1	—	2	2/4
						Meranti, dark	3	5/—	—	—	2/4
Ash	5	5/—	5	—	2/2	Meranti, light	3	5/—	—	—	2/4
Beech	5	1/1	5	5	1/1	Niangon	3	5/1	—	4	3/4
Birch, white	—	1/1	—	1	—/—	Obeche	4	5/5	—	—	1/3
Black locust	2	5/1	—	—	—/4	Okoumé	4	5/5	—	—	—/—
Chestnut	3	5/1	—	4	—/4	Padauk	1	5/1	—	1	—/—
Elm	4	5/—	5	—	1/3	Palissander	3	3/3	—	—	—/—
Maple, field	5	5/—	5	—	1/1	Ramin	5	—/—	—	—	—/—
Oak, red	4	5/5	5	4	1/4	Sapele	3	5/1	—	4	2/3
Oak, white	2	5/1	5	4	—/—	Sipo	—	5/1	—	—	—/4
Plane, oriental	5	—/—	—	—	—/—	Teak	1	—/—	—	3	—/4
Poplar, hybrid	5	1/1	5	5	1/2	Tiama	4	5/1	—	4	3/4
Walnut	3	5/1	5	—	1/3						
Willow	—	1/1	5	—	1/3						

[a] F = Fungi, L/A = insects (L = Lyctidae, A = Anobiidae), MB = marine borers, P = permeability, s = sapwood, h = heartwood.
[b] Resistance to fungi, insects, and marine borers: 1, very durable (resistant); 2, durable; 3, moderate durable; 4, little durable; 5, non-durable.
[c] Permeability to preservatives: 1, permeable; 2, moderately resistant; 3, resistant; 4, very resistant.
[d] North American species.
Source: Ref. 29.

derlying wood is firm and sound. Soft rot appears in wood exposed to very high moisture, in water (including seawater) (47), or in contact with moist soil.

Microscopically, wood attacked by decay fungi shows holes in the cell walls and erosion in places that are not in direct contact with hyphae. Apparently, this takes place by translocation of enzymes. The morphology of degradation varies depending on the kind of fungus. In brown rots no thinning of cell walls is observed up to the last stage of attack. In contrast, in white rots there is a progressive thinning toward the middle lamella even at the incipient stage. Finally, soft rots are characterized by polygonal or cylindrical cavities with pointed ends in the cell walls. In most cases of brown rots, the diameter of the holes opened in cell walls is greater than the diameter of the hyphae.

Interesting observations have also been made with the electron microscope. White rot fungi remove cell-wall layers progressively; whereas in brown rots, different layers are attacked at the same time (61). Erosion of cell walls away from hyphae is characteristic, and differences in the morphology of holes and the manner of passage of hyphae through walls and pits may be observed. The warty layer appears more resistant to brown rot fungi, whereas white and soft rot fungi may decompose it (36). The selectivity of different fungi (brown, white, and soft rot) from a chemical point of view has been utilized in studies of distribution of chemical constituents in cell walls.

Moisture and *air*, in a favorable combina-

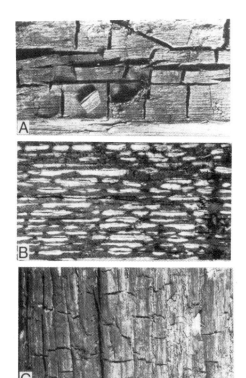

Figure 15-5. Decay: (A) brown, (B) white, (C) soft. (Photographs B and C courtesy of U.S. Forest Products Laboratory.)

tion, are prerequisite for growth and activity of fungi. These factors are in close relationship, because air occupies in the mass of wood spaces that are not occupied by water. Different species of fungi differ in moisture requirements; however, in general, wood must contain at least 20% moisture (based on oven-dry weight) for an attack to start. Conditions become favorable for continuous fungal activity when moisture increases toward the fiber saturation point, and become optimal a little beyond that point (moisture content about 35–50%). At that hygrometric condition, the cell walls are saturated and a thin layer of free water exists on their inner surface. While the rest of the cell cavity is occupied by air, this thin water layer facilitates the diffusion of enzymes from hyphae to walls (43).

At lower moisture levels, wood is not attacked but some species of fungi are not killed; they possess the ability to remain in hibernation a long time, reviving when conditions become again favorable. Some species form thick-walled spores that are resistant to desiccation; whereas others, such as *Serpula (Merulius) lacrymans*, conduct water to decaying wood through rhizomorphs, or release water during the decomposition of wood, and thus increase its moisture content. The drying capacity (relative humidity) of the air becomes an important factor.

Very high moisture content is also a restraining factor for most fungi, because it deprives them of needed air (oxygen). Wood may be protected by storage in water.

The requirements of decay fungi with respect to *heat* vary depending on species (73). Optimal temperatures range from about 20 to 25°C (70–80°F); soft rot fungi require higher temperatures in comparison to brown and white rot fungi. Fungal activity is retarded at temperatures lower than 10°C (50°F) and higher than 30°C (85°F), and ceases outside the range of 0°C (32°F) and 40°C (104°F) (61). Fungi are definitely killed only at very low and very high temperatures. High temperatures in combination with high relative humidities, as applied in kiln-drying of wood, are more effective than dry heat. Fungi may be killed, but the wood is not protected from a new attack.

Material Attacked. Fungi cause decay in living trees, timber, and wood products. From an economic viewpoint, the consequences of attack are more important in products, due to the cost of their replacement or repair.

In living trees, fungi usually attack heartwood. Sometimes, the attack extends to sapwood, but only if it is advanced in heartwood. This differential resistance may be explained on the basis of chemical composition and moisture content of wood. The resistance of older heartwood is reduced with age because the toxicity of extractives is progressively reduced. The outer (younger) sapwood in living trees presents a higher resistance because it contains more moisture. As previously mentioned, fungi cannot attack wood with very high moisture content, especially wood saturated with water, because they need air for growth and activity.

The heartwood of living trees may be attacked by many species of fungi. Some attack only softwoods, some only hardwoods, others both. Some fungi are limited to one or a few species of wood. Fungi enter as spores from wounds of the trunk, branches, or roots—and, when conditions are favorable, they germinate, grow, and gradually decompose the wood both structurally and chemically. Attacked trees may appear healthy, but careful observation will reveal exterior signs of attack, such as wounds, a swollen trunk base, resin secretion, or sporophores of fungi. However, such criteria appear only in advanced stages. At the incipient stage, the attack may not be apparent even after felling a tree. Macroscopically, the wood appears healthy, and a discoloration, in patches or in the form of heartwood, may be the only indication of attack. At this stage, a microscopic examination is needed to verify the existence of hyphae (Figure 15-6).

After a fungus is established, the attack extends—much faster axially than transversely, due to wood structure and direction of sap movement. Gradually, the wood changes exterior appearance, structure, and properties. At the beginning it becomes softer, and with progressing attack parts of it disappear. In an advanced stage, the trunk becomes hollow and the wood may be easily broken by hand. The probability of living trees being attacked is higher in coppice forests (sprouts), and increases with increasing age of the stump.

In a forest, the conditions of attack are more favorable in summer, both with regard to wood moisture and environmental temperature. The wood of trees killed by insects, harvested during this period, or previously fallen due to wind or snow, is subject to attack, especially if left in contact with the ground. For this reason, the logs should be properly piled to facilitate surface drying, or moved promptly to sawmills or other processing factories (45, 70). In contrast to living trees, fallen trees or logs—when left in contact with the ground—are preferably attacked in their sapwood. This may be explained on the basis of higher (favorable) moisture content and the presence of nutrients (starch, etc.) in sapwood, whereas heartwood is additionally protected by the presence of extractives.

Favorable conditions may also occur in wood products, such as stored green lumber or structures exposed to conditions suitable for growth of fungi (exterior doors, windows, etc.). In

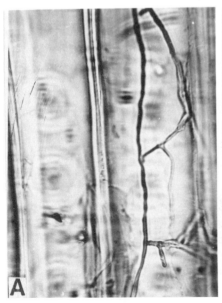
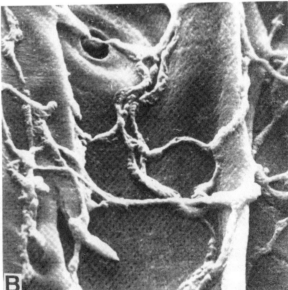

Figure 15-6. Hyphae of decay fungi. A. *Fomes pinicola* in fir, incipient stage of attack (light microscope, 1000×), and B. *Coniophora cerebella* in Scots pine (SEM, 1700×). (A. reproduced by permission from Pergamon Press; B. after Ref. 11, reproduced by permission from the Institute of Wood Science, U.K.).

such cases, the resistance of wood may be considerably increased by preservative treatment (Chapter 19).

Consequences of Attack. Fungal attack results in changes of color, structure, chemical composition, and properties of wood. More important are changes occurring at the incipient stage, because in advanced stages the wood becomes useless.

- The color almost always changes and becomes lighter with white rot and darker with brown rot. Sometimes, for example, in oak attacked by *Stereum guasapatum*, the color becomes darker at the incipient stage of white rot.
- Changes of structure and chemical composition were previously discussed. In brown rot, even when the cellulosic skeleton has been removed, the remaining lignin retains the initial form of cell walls.
- The density of wood decreases, according to the stage of attack; for this reason, the reduction of weight (in percent of the original and under the same moisture content) is used as a quantitative criterion to determine the degree of attack in research studies. In addition to same moisture content, a prerequisite to comparison is wood of the same species and of similar structure (i.e., about equal percentage of earlywood and latewood).
- The hygroscopicity of wood changes. Wood adsorbs water faster and, therefore, logs would sink faster during water transportation or storage. Also, wood adsorbs more moisture (water vapors) from the atmosphere at high relative humidities and less at low relative humidities—and, when periodically moistened, it retains the moisture longer. Preservatives enter more easily into the mass of wood, even at incipient stages of attack.
- Shrinkage remains about normal or becomes a little higher in white rot, but is considerably higher in brown rot, especially in the axial direction. Wood, at an incipient stage of attack, may collapse during kiln-drying.
- The mechanical properties of wood change even at the incipient stage, when hardness does not seem affected. The degree of degradation depends on wood species, species of fungus, duration and conditions of attack, and manner of loading (55, 80, 82). At the incipient stage, no important differences are found between fungi or woods (softwoods, hardwoods) in most properties, although toughness is considerably affected (3, 55, 61, 81) (Figure 15-7). Reduction of that property (up to $\frac{1}{3}-\frac{1}{2}$) takes place before the weight of wood is reduced (3) (the reduction is attributed to broken links of lignin and carbohydrates) (61).[5] Toughness may be almost nullified by a weight reduction of about 10% (81). Bending properties (modulus of rupture and elasticity) are less affected, and even smaller is the reduction of axial compression, transverse compression, and axial tension. Shear and hardness are little affected (81) (under the above-mentioned weight reduction). In general, mechanical properties are most sensitive to brown rot, and should not be considered unaffected even when the wood is hard and firm.[6]
- The thermal, acoustical, and electrical properties of wood are also affected due to their relationship to density (which is reduced).
- Wood with pronounced attack by certain fungi, especially *Armillaria mellea*, may be luminescent; this is attributed to hyphae.
- In brown-rotted wood, the yield of pulp is low and its quality poor; whereas in white-rotted wood both yield and quality are good—nearly equal to that of sound wood (19, 43).

Main Fungi. Fungi attack the wood of living trees, timber, or products. For the purpose of this book, the most important fungi are those that attack wood after harvesting.[7]

Fungi attacking wood exposed out-of-doors (10, 16, 31).
Softwoods
Lentinus lepideus Fr., *Lenzites sepiaria* (Wulf.) Fr., *Peniophora gigantea* (Fr.) Massee, *Polystictus abietinus* (Dickes) Fr.
Hardwoods
Daedalia quercina (Lihn.) Fr., *Polystictus versicolor* (Lihn.) Fr., *Stereum purpureum* Pers.

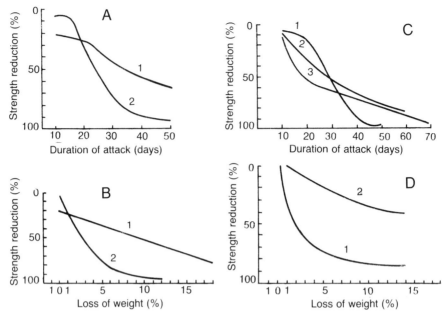

Figure 15-7. Effects of fungal attack on mechanical properties of wood. (A) Reduction of toughness in relation to duration of attack of beech wood by: 1, *Polystictus versicolor* (white rot); 2, *Polyporus vaporarius* (white rot). (B) Reduction of toughness in relation to duration of attack of different woods by *Merulius domesticus* (brown rot): 1, beech; 2, Scots pine; 3, spruce. (C) Reduction of toughness in relation to loss of weight of beech wood attacked by: 1, *Polystictus versicolor* (white rot); 2, *Daedalia quercina* (brown rot). (D) Reduction of: 1, toughness and 2, static bending in relation to loss of weight of beech wood attacked by *Chaetomum globosum* (soft rot). (A–C, after Ref. 55; D, after Ref. 3.)

Fungi attacking wood products (buildings, etc.)

Serpula ((Merulius) lacrymans (Wulf.) Fr., *Coniophora cerebella* (Pers.) Dubr., *Phellinus megaloporus* (Pers.) Heim.

Details about the characteristics of the above fungi, regarding the appearance of wood in incipient and advanced stages, species attacked, sporophores, etc., are available in specialized books of forest and wood pathology (10, 16), and they will not be discussed here. Generally, however, most of these fungi cause brown rots (except *Polystictus* and *Stereum*, which are white rots), and are of world-wide occurrence. Wood exposed out-of-doors refers to poles, posts, mine timbers, railroad ties, and generally wood in contact with the ground. In buildings, the attack is connected to similar conditions of use or leakage of water. The particular case of *Serpula (Merulius) lacrymans* was previously mentioned. This fungus is able to transport moisture to long distances through its hyphae (micellium), which, in an advanced stage, form a thick, cotton-like layer on the surface of the wood; they can even pass through walls, transporting moisture high above the ground.

ANIMAL FACTORS OF DEGRADATION

Insects

Like fungi, insects attack the wood of living trees, timber, and products. In living trees, insects bore through the bark or enter through debarked areas. The attack may be confined to the region between bark and wood, due to the abundance of nutritive substances in that region (23), or it may extend throughout the sapwood and heartwood.

The damage is greater in trees of low vigor, logs, and wood products. In general, insects attack wood for shelter, food, and oviposition. They cannot digest wood substance directly, however, and for this reason xylophagous (wood-destroying) insects cooperate with microorganisms (protozoa, bacteria, fungi) to change undigested cellulose and lignin to more available subproducts. Insects open bore holes and galleries (tunnels) of varying size, up to 2.5 cm (1 in.) in diameter, and some species change the interior of wood to dust, leaving a thin exterior layer. Some insects bear decay fungi into wood.

The biological cycle of insects includes four stages—*egg, larva, pupa* (nymph), and *adult*. Eggs are deposited on surfaces, in fissures of wood or bark, or inside cell cavities (pores). After their incubation (hatching), larvae start boring, and sometimes they fill the galleries with their excretions, which may include wood dust. Larvae develop into pupae, which remain in hibernation—the insect does not eat and, therefore, it is not destructive. Gradually, pupae are transformed to adults with final shaping of their morphological and anatomical characteristics (Figure 15-8). Adults are short-lived, certain species do not destroy wood and others fly actively. The larval stage is longer. In temperate climates, the biological cycle may be completed in 1–10 years, depending on wood species, insect species, and environmental conditions (8) (Figure 15-9).

The environmental conditions that affect the development and activity of insects are the same as in fungi (i.e., *temperature, moisture,* and *air*). Some insects are adaptable to a wide variation of temperature and moisture, others cannot tolerate a very high temperature or moisture. In general, however, the insects are favored by higher temperatures and (in contrast to fungi) lower moisture contents of wood.

Insects usually develop under the bark of dead trees or logs, and enter wood afterward (therefore, prompt debarking is a protective measure), but wood in service is also attacked. Infected wood may be sterilized at high temperatures, 50–60°C (120–140°F) and higher, as applied in kiln-drying (see Chapter 18). Toxic liquids (insecticides) and poisonous gases are also effective, and recommended for prod-

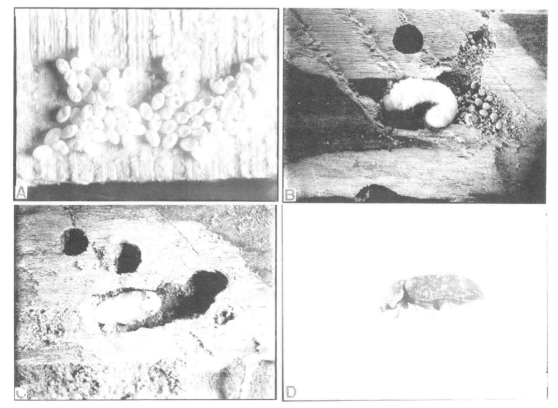

Figure 15-8. The four stages of the biological cycle of *Xestobium rufovillosum* (A, eggs; B, larva; C, pupa; D, adult). (A, 10×; B and C, 3×; D, 4×). (Courtesy of Building Research Establishment, U.K.; reproduced by permission.)

Figure 15-9. Schematic representation of the biological cycle of an insect attacking wood. (A) An egg drops in a fissure on the surface of the wood. (B) After hatching, the larva starts boring and opens galleries, from 1–10 years. (C) The larva is transformed to pupa near the wood surface. (D) After a few weeks, the adult insect exits flying, it pairs, lays eggs on another surface—and the cycle is repeated. (Courtesy of Timber Research and Development Association, TRADA, U.K.; reproduced by permission.)

ucts (furniture, etc.) that cannot be exposed to high temperatures. Preservative treatment, where applicable, affords protection, and painting or coating wood surfaces with varnish reduces egg-laying sites. Biological control has also been tried (74).

Main Insects

The insects attacking timber and products[8] belong to three classes: *Coleoptera, Hymenoptera* and *Isoptera* (8, 31, 41).

Coleoptera. These insects (Figures 15-8 and 15-10) are characterized by two hard outer wings (elytra), which cover and protect the real (functional) wings. The main Coleoptera are: *Anobium punctatum* De G. (26), *Xestobium rufovillosum* De G. (59), *Ptillinus pectinicornis* L., *Lyctus linearis* Goetze, *Hylotrupes bajulus* L., *Hesperophanes cinereus* Vill., *Bostrychus capucinus* L., *Xyloterus lineatus* Oliv., *Platypus cylindricus* Fab.

Some of these insects are well known. *Anobium punctatum* is the "common furniture beetle." The adult is dark-brown to blackish, about 2.5–5 mm (0.1–0.2 in.) long. Exit holes are circular, about 1.5 mm in diameter, and inside infested wood there is a network of tunnels short in length and packed with excrements ("bore-dust," "frass"). In heavy attack, all the interior is reduced to dust. Continuing attack is revealed by small piles of dust outside exit holes and new holes in the summer. Under magnification, the particles of bore-dust appear ellipsoid or lemon-shaped. The biological cycle is usually 2 years, but may extend to 10 years or more. Both hardwoods and softwoods are attacked.

Xestobium rufovillosum is the "death watch beetle." The adult is brown to chocolate brown, variegated, and about 8 mm long. Exit holes are circular, about 3 mm in diameter, and the tunnels are filled with coarse, bun-shaped particles ("pellets"), visible to the naked eye. The attack is favored by damp conditions, and severe damage may be hidden under a healthy-looking outer appearance. The woods attacked are preferably hardwoods; the biological cycle is 1–4 years (Figure 15-8).

Lyctus beetles are known as "powder post beetles." The adult is brown to black, about 5 mm long. Exit holes are circular with a 1.5 mm diameter, and the tunnels packed with very fine bore dust. Damaged woods are preferably larger-pored hardwoods. The biological cycle

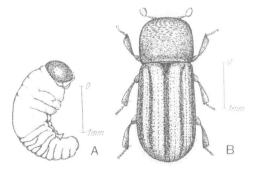

Figure 15-10. An insect of the class Coleoptera, *Xyloterus lineatus:* (A) larva; (B) adult. (After Ref. 2; reproduced by permission from Springer-Verlag.)

is 1–2 years, but in warmer environments it may be as low as 2–3 months.

Hylotrupes bajulus is the "house longhorn beetle." Adults are black, sometimes brownish, about 10–20 mm (up to $\frac{3}{4}$ in.) long; exit holes are 6–10 mm, tunnels are large, and pellets cylindrical. The biological cycle is 2–3, and sometimes 10–12 years. Softwoods and sometimes hardwoods are subject to attack.

More specific information about the above and other *Coleoptera* (beetles) is found in specialized publications (8, 41, 65). A general remark is that in almost all cases the damage is caused by larvae (sometimes also by adults), and mainly in sapwood.

Hymenoptera. These insects (Figure 15-11) have four (two pairs) of membranous wings, of which the back ones are shorter. The ovipositors of xylophagous species are usually saw-like or pointed. Main Hymenoptera are: *Urocerus (Sirex) gigas* L., *Camponotus herculeanus* L., and *C. ligniperda* Latr.

Urocerus (Sirex) gigas belongs to the "wood wasps." The adult is large, 18–35 mm (up to 1.5 in.) long (females are larger); exit holes are 5–10 mm and circular, tunnels are long and curved and packed with coarse dust. The biological cycle is 1–3 years. Woods attacked are mainly softwoods—sickly trees or freshly felled logs with bark, but sometimes the attack continues into timbers in buildings. Damage is caused by larvae, and the insect cooperates with a fungus in the process of attack.

Camponotus spp. belong to "carpenter ants." There is a similarity and sometimes confusion with termites ("white ants"). They are present in colonies that include winged males and females, queen, and wingless workers. All forms are black, and resemble ants. They attack mainly softwoods, trees and logs affected by decay fungi. The tunnels are longitudinal and peripheral—around growth rings, usually empty with smooth walls.

Isoptera. Isoptera (Figure 15-12) are the termites, which cause severe damage to timber and wood products (8, 41). Termites live in complicated colonies and not as solitary individuals. A colony is made of a "king,"

Figure 15-11. An adult (female) *Urocerus (Sirex) gigas* (Hymenoptera). (After Ref. 72; reproduced by permission from VEB-Fachbuchverlag.)

"queen" (laying many thousands of eggs every day and for many years), numerous "workers" (responsible for damage to wood), "soldiers" (engaged in defense of the colony), and a number of immature individuals in various stages of development. Only the king and queen are reproductive, but supplementary reproductive individuals (nymphs) also exist, which may replace the king or queen if killed or a new colony is started. Adult reproductives (alates) have four transparent wings of equal size. Workers and soldiers are sterile. In some families, true workers are absent, replaced by "pseudergates" (i.e., false workers—older nymphs).[9]

Termites are represented by many families, genera, and species that belong to two main groups: dry-wood termites (*Kalotermes, Cryptotermes, Neotermes*, etc.), and moist-wood or subterranean termites (*Reticulitermes, Coptotermes, Heterotermes*, etc.). In southern Europe, important species are *Kalotermes flavicollis* F. and *Reticulitermes lucifugus* Rossi; in North America an important species is *Reticulitermes flavipes* Kollar. Many other destructive genera and species exist in other parts of the world, especially in the tropics. Termites seldom attack living trees (e.g., *Eucalyptus* plantations and forest nurseries).

Subterranean termites establish colonies usu-

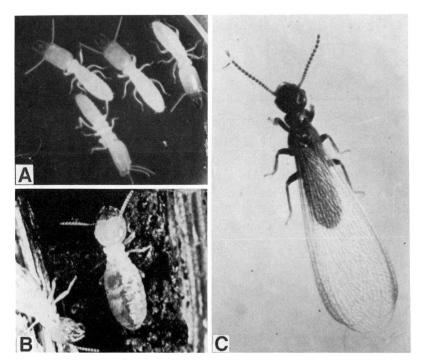

Figure 15-12. Termites *(Reticulitermes flavipes)*. The damage is caused by workers (A), and the colony is guarded by soldiers (B). Male and female reproductives (C) are the only forms with wings; they become king and queen in a new colony (A, B, and C 10×). (Courtesy of T. Amburgey.)

ally in damp soil, and they attack wood in contact with the ground; however, wood above ground may also be attacked, in which case they construct connecting earth tunnels. Dry-wood termites nest in the wood on which they feed, and do not invade a structure from the soil.

Termites do not open distinct galleries in wood. Presence of attack may be detected from powdered excretions ejected from surface holes opened by workers; eventually, the entire interior of wood may be destroyed and only a thin outer layer remains. The extent of attack depends on termite species, manner of wood use, and environmental conditions (temperature, humidity). Some wood species are resistant to termites, such as West Indian mahogany (*Swietenia mahogany*), Iroko (*Chlorophora excelsa*), Teak (*Tectona grandis*), Azobe (*Lophira procera*), Courbaril (*Hymenaea courbaril*), bald cypress (*Taxodium distichum*), and others.

Termites utilize cellulose as food, but not lignin—which is excreted. In some species, the consumption of cellulose is helped by symbiotic protozoa or bacteria living in their digestive tracts.

Subterranean termite attack may be prevented by such measures as proper design of wooden structures (avoidance of excessive moisture, etc.), use of mechanical barriers (metal caps, concrete slabs or barriers), and use of toxic chemicals (to produce a lethal or repellent layer of soil). Dry-wood termites may be avoided by use of woods with natural resistance, and care in introducing infected items (e.g., furniture) in a house. An existing termite attack may be controlled by application of chemical solutions (e.g., introduced into surface holes with a syringe), or fumigation (e.g., use of methyl bromide, which is also effective with other insects).

Termites are also called "white ants" (workers and nymphs are whitish), but they are not ants, because the latter belong to *Hymenoptera*.

Marine Borers

Certain marine organisms that belong to the phylum *Mollusca* and to the class *Crustacea* of the phylum *Arthropoda* attack wooden structures in seawater, such as wharf piling, boats (usually not in motion), and stored round timber, causing severe damage (7, 8, 15, 22). These organisms attack wood for the same reasons that insects attack it (i.e., for shelter, oviposition, and food), although some obtain their food from the sea (plankton). The extent of attack depends on various factors, such as kind of organism, temperature and salinity of the water, fungal symbiosis,[10] and wood species. All species of temperate woods are sooner or later destroyed, sometimes in a few months, if placed without proper preservative treatment (see Chapter 19) in regions where such organisms are abundant. Certain tropical woods present an increased natural resistance, and this is attributed to their relatively high silica (SiO_2) content (63, 66) or to toxic extractives (extractives are considered to be of greater importance) (5).

Molluscan borers belong to genera *Teredo*, *Bankia*, *Martesia*, and others, while borers of the phyllum *Arthropoda* include *Limnoria*, *Chelura*, *Sphaeroma*, and others (Figure 15-13). The most important borers that attack wood belong to *Teredo* (*T. navalis* L., *T. pedicellatus* Quatr., *T. utriculus* Gmel., etc.) and *Limnoria* (*L. tripunctata* Menzies, *L. carinata* Menzies, *L. lignorum* Rathke, etc.) (30).

***Teredo* spp.** These organisms are worm-like (hence, called "shipworms"). The front part of their body is equipped with a pair of shells with teeth-like projections that can bore into wood.[11] They may reach a length of 1 m (3 ft.) or more, but usually this is not greater than a few centimeters with a few millimeters width. The dimensions are influenced by such factors as extent of attack, natural resistance of wood, and species of organism. The biological cycle (egg–adult) is completed in about 1 year. The food is wood and plankton. Attacked wood has few and small holes 1–2 mm in diameter on the surface, but inside there are many galleries 2–5 cm (1–2 in.) in diameter (Figures 15-14

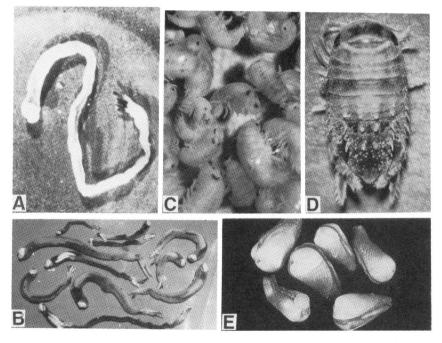

Figure 15-13. Marine borers: A, *Teredo*; B, *Bankia*; C, *Limnoria*; D, *Sphaeroma*; E, *Martesia*. (Borers are not in scale.) (Photograph A, courtesy of T. Amburgey; Photographs B–E, courtesy of CSIRO, Australia.)

228 II / PROPERTIES

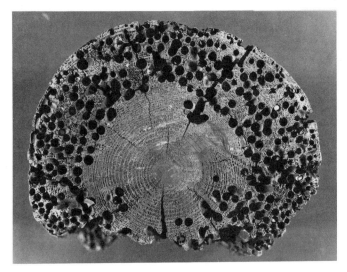

Figure 15-14. Piling attacked by *Teredo* (the outer layer appears almost intact). (Reproduced by permission from Pergamon Press.)

and 15-15). These galleries are empty (seldom containing remnants of organisms), and their walls have a lime-like (calcium) coating (71). Use of woods that have a natural resistance (e.g., certain tropical woods) and proper preservative treatment reduce the danger of attack, but protection of the wood with cement, metals, or plastics is also effective. Transfer of wooden structures (e.g., boats) from sea to nonsalty water will kill these organisms in 2–3 weeks.

Limnoria spp. The body of these organisms is composed of segments (like shrimp, but in shape they resemble lice); they are 3–5 mm long, about 1 mm wide, and are able to destroy wood with their fine teeth, which act like a "file." The biological cycle lasts 2–3 weeks. The galleries are shorter and smaller compared to *Teredo*, about 1 mm in diameter; their depth of entrance into wood is usually no more than 1–1.5 cm (0.5 in.), and they are interconnected so that the wood becomes spongy. The danger

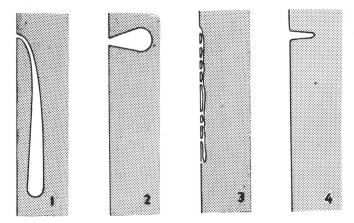

Figure 15-15. General morphology of marine borer galleries: 1, *Teredo*; 2, *Martesia*; 3, *Limnoria*; 4, *Sphaeroma*. (After Ref. 7; reproduced by permission from Springer-Verlag.)

of attack may be reduced by preservative treatment. Transfer of wooden structures from sea to nonsalty water or on the ground will stop the attack faster than in the case of *Teredo*.

CLIMATIC FACTORS

Climatic factors (temperature, relative humidity, rain, snow, air, sun, light) may cause degradation after long exposure of wood out-of-doors. Under such conditions, the wood is subject to repeated dimensional changes, which may result in detachment of growth rings and checking, in combination with color changes and differential surface erosion (earlywood–latewood, sapwood–heartwood). At an advanced stage, degradation may resemble fungal attack, which may also be present. The effect of climatic factors is greater on tangential than radial surfaces. Other degradation, due to shrinkage or swelling, includes change of dimensions, change of shape, warping, and checking (see Chapter 10).

In addition to color changes, light may cause reduction of strength, checking, surface erosion (the surface of wood becomes more porous and fibrous) (27), reduction of the degree of cellulose polymerization, and other chemical changes (breaking of chemical linkages, loss of chemical substances, degradation of the lignin of middle lamella) (20, 21, 49). These are physicochemical effects caused by ultraviolet and common solar radiation, and are influenced by other factors, such as moisture, temperature, and oxygen of the atmosphere (20, 27, 49, 58). Degradation of wood may also be caused by atmospheric pollution (sulfur dioxide, etc.) (61).

MECHANICAL FACTORS

Degradation may result from repeated mechanical stresses, when wood is used for products, such as railroad ties, flooring, and ladders. The resistance of wood varies, depending on species, moisture content, manner of loading, exposure conditions, and grain of the loaded surface. Products, such as railroad ties, are stressed in a different way (rolling loads and outdoor conditions that favor fungal attack) in comparison to floors and other products used under interior conditions and subject to wear by friction. Different species present a different resistance, depending on density, hardness, and other mechanical properties, such as transverse compression and horizontal shear. As with climatic factors, radial surfaces show a higher resistance in comparison to tangential surfaces, but the resistance of transverse surfaces is much higher. For the latter reason, floors subjected to heavy use (factories, etc.) are sometimes made with transversely exposed surfaces.

CHEMICAL FACTORS

Wood presents considerable resistance to many chemical substances, when the conditions of exposure are not severe (mild concentration) and the temperature is low. Due to such chemical inertness, wood is used for containers of chemicals and foodstuffs. The resistance of wood to mild acids is far superior to common steel, but its resistance to alkaline solutions is not so remarkable. This is due to the inertness of cellulose and lignin to acids, while hemicelluloses and lignin are easily attacked by strong alkalis (75).

The results of the action of chemicals are shown in loss of strength, and depend on wood species, kind and concentration of chemical, duration of exposure, and temperature. Low concentrations of chemicals—for example, 2% solutions of hydrochloric acid (HCl), sodium hydroxide (NaOH), and other acids and alkalis, acting under room temperature conditions—were shown to exercise little or no degrading effect on most of the species studied. Naturally, degradation increases with higher concentrations, and increasing duration and temperature of exposure. Studies with 10% concentrations, and temperatures of about 50°C (120°F), have shown that most species lost more than half (and some more than $\frac{3}{4}$) of the original strength. Alkalis are more destructive; they reduce modulus of elasticity, strength in bending, and transverse compression. In general, softwoods are more resistant than hardwoods, probably because they contain less hemicelluloses. It has been observed that differences among species are not related to their original chemical composition but rather to the retention of hemicelluloses (75).

The effects of chemical solutions also depend on the solvent. Nonswelling solvents have no important influence, but others (e.g., water) contribute to the reduction of strength due to the resulting swelling (28).

Preservative treatment has no practical adverse effects on the strength of wood under normal conditions of application (see Chapter 19). Fire retardants (in combination with kiln-drying) may reduce bending strength (modulus of rupture) up to 10%, but modulus of elasticity was found not substantially affected (28).

Aside from loss of strength, the color of wood may change for chemical reasons. After air- or kiln-drying, the color of various woods becomes darker due to oxidation of cell contents (44). Discoloration, in patches, may also result from the chemical reaction of wood tannins with nails or other metallic connectors or objects in the presence of moisture.

THERMAL FACTORS

High temperatures cause chemical decomposition of wood. The successive changes resulting from the action of increasing temperatures were discussed in Chapter 12. Long duration heating of wood, even at the temperature of a laboratory oven ($103 \pm 2°C$), results in chemical decomposition; such an effect may also be caused by short duration heating at high temperatures (64). Decomposition products are: carbon monoxide, formic and acetic acid, methane, tar, and others (6). Depending on temperature, duration of exposure, manner of heating, moisture content of wood, and other factors, the results vary; they include: loss of weight, microscopic differentiation of wood structure (24, 42), softening, reduction of hygroscopicity, increase of shrinkage, and reduction of strength (40, 42, 64) (Figures 11-9, 11-10, and 11-11). Lignin is the most resistant chemical constituent of wood. Hemicelluloses decompose at 200–260°C (400–500°F) and lignin at 280–500°C (540–930°F).[12] Softening of lignin is favored by moisture and pressure (32, 48).

Wood may be consumed or degraded by fire. It should be noted, however, that wood obtained from trees killed by a forest fire (as well as by insects, fungi, etc.) may not be different in quality from that of normally harvested healthy trees—if there is no subsequent attack. The trees are killed, because their foliage is burnt or their cambium is killed by the action of high temperatures, but it should be remembered that even the wood of living trees is mostly dead (see Chapter 3); also, the action of high temperatures is usually short (due to fast movement of the front of a fire), and bark provides protection especially in thick-barked species (69). It should also be noted that the resistance of wood to fire may be increased by fire-retardant treatment, and wood products of large dimensions (e.g., glued-laminated beams) may offer considerable protection—in some cases, better than steel. Wooden load-carrying elements fail only when their cross-sectional dimension is reduced to the extent that the load cannot be supported. Increase of cross-section to dimensions larger than those prescribed by engineering calculations may guarantee a better resistance, for a certain time, because the fire may affect only a superficial layer, a few millimeters thick (see Figure 12-5). This situation is more favorable than that of steel supports that fail totally at a critical temperature (38). Wood that is not carbonized in a fire practically preserves its original strength and may be reused (60).

REFERENCES

1. Adolf, A., E. Gerstetter, and W. Liese. 1972. Untersuchungen über einige Eigenschaften von Fichtenholz nach dreijähriger Wasserlagerung. *Holzforschung* 26(1):18–25.
2. Anonymous. 1975. Holzbrütende Borkenkäfer. *Holz Roh- Werkst.* 33(6):251.
3. Armstrong, F. H., and J. G. Savory. 1959. The influence of fungal decay on the properties of timber. *Holzforschung* 13(3):84–89.
4. Bauch, J., W. Höll, and E. Endeward. 1975. Some aspects of wetwood formation in fir. *Holzforschung* 29(6):198–205.
5. Bavendamm, W., and F. Roch. 1970. Untersuchungen über die natürliche Resistenz von Tropenhölzern gegen Meerwasserschädlinge. *Holz Roh- Werkst.* 28(3):105–117.
6. Beal, F. C., and H. W. Eickner. 1970. Thermal Deg-

radation of Wood Components: A Review of Literature. U.S. For. Prod. Lab. Res. Paper FPL 130.
7. Becker, G. 1958. Holzzerstörende Tiere und Holzschutz im Meerwasser. *Holz Roh- Werkst.* 16(6):204-215.
8. Bletchly, J. D. 1967. Insect and Marine Borer Damage to Timber and Woodwork. For. Prod. Res. Lab., HMSO, London.
9. Borgin, K., G. Tsoumis, and C. Passialis. 1979. Density and shrinkage of old wood. *Wood Sci. Technol.* 13:49-57.
10. Boyce, J. S. 1961. *Forest Pathology*, 3rd ed. New York: McGraw-Hill.
11. Bravery, A. F. 1971. The application of scanning electron microscopy in the study of timber decay. *J. Inst. Wood Sci.* 30(5-6):13-19.
12. Bravery, A. F. 1985. Mould and Its Control. BRE Information Paper IP 11/85.
13. Brill, H., P. Klein, A. Früwald, and J. Bauch. 1980. Alterations of biological and technological characteristics in *Abies alba* Mill. due to fir-dying. In *Natural Variations of Wood Properties*, ed. J. Bauch, pp. 195-205. Hamburg-Reinbek: Mitt. B.F.A.
14. Bues, C. T. 1986. Untersuchung einiger Eigenschaften von Tannen- und Fichtenholz nach 17jähriger Wasserlagerung. *Holz Roh- Werkst.* 44:7-15.
15. Building Research Establishment, U.K. 1972. Marine borers and methods of preserving timber against their attack. Tech. Note No. 59.
16. Cartwright, K. S. G., and W. P. K. Findlay. 1958. *Decay of Timber and its Prevention.* London: HMSO.
17. Chafe, S. C. 1989. Observations on the relationship between wood durability and density. *J. Inst. Wood Sci.* 11(5):182-185.
18. Courtois, H. 1966. Über den Zellwandabbau durch Bakterien im Nadelholz. *Holzforschung* 20(5):148-154.
19. Cowling, E. B. 1961. Comparative Biochemistry of the Decay of Sweetgum Sapwood by White-rot and Brown-rot Fungi. USDA, Tech. Bull No. 1258.
20. Derbyshire, H., and E. R. Miller. 1981. The photodegradation of wood during solar irradiation. *Holz Roh- Werkst.* 39(8):341-350.
21. Desai, R. L. 1968. Photodegradation of cellulosic materials—A review of the literature. *Pulp Paper Mag. Canada* (reprint, pp. 11).
22. Eltringham, S. K. 1971. Marine borers and fungi. In *Marine Borers, Fungi and Fouling Organisms*, pp. 327-335. Paris: O.E.C.D.
23. Essiamah, S., and W. Eschrich. 1985. Changes of starch content in the storage tissue of deciduous trees during winter and spring. *IAWA Bull.* 6(2):97-106.
24. Fengel, D. 1966. Über die Veränderung des Holzes und seiner Komponenten in Temperaturbereich bis 200°C. III. *Holz Roh- Werkst.* 24:529-536.
25. Ferguson, C. W. 1969. A 7104-year annual tree-ring chronology for bristlecone pine, *Pinus aristata*, from the White Mountains, California. *Tree-Ring Bull.* 29(3-4):3-17.
26. French, J. R. J. 1970. Review of control measures against the common furniture beetle, *Anobium punctatum. J. Inst. Wood Sci.* 26(5/2):29-32.
27. Futo, L. P. 1976. Influence of temperature on the photochemical decomposition of wood. *Holz Roh- Werkst.* 34:31-36; 49-54.
28. Galligan, W. L. 1975. Mechanical properties of wood. In *Wood Structures*, pp. 32-54. New York: Am. Soc. Civil Engineers (ASCE).
29. Gambetta, A., and E. Orlandi. 1982. Durabilita Naturale di 100 Legni Indigeni e di Importazione a Funghi, Insetti e Organismi Marini. *Ist. Legno*, 30:47-71, Firenze.
30. Gareth Jones, E. B. 1971. The ecology and rotting ability of marine fungi. In *Marine Borers, Fungi and Fouling Organisms*, pp. 237-258. Paris: O.E.C.D.
31. Giordano, G. 1971. *Tecnologia del Legno*, Vol. 1. Torino: Unione Tipografico-Editrice.
32. Goldstein, I. S. 1973. Degradation and protection of wood from thermal attack. In *Wood Deterioration, etc.*, Vol. I, ed. D. D. Nickolas, pp. 307-340. Syracuse, New York: Syracuse University Press.
33. Greaves, H. 1971. The bacterial factor in wood decay. *Wood Sci. Technol.* 5(1):6-16.
34. Greaves, H. 1973. Selected wood-inhabiting bacteria and their effect on strength properties and weights of *Eucalyptus regnans* F. Müll. and *Pinus radiata* D. Don. sapwoods. *Holzforschung* 27(1):20-26.
35. Haygreen, J. G., and S.-S. Wang. 1966. Some mechanical properties of aspen wet-wood. *For. Prod. J.* 16(9):118-119.
36. Higuchi, T., ed. 1985. *Biosynthesis and Biodegradation of Wood Components.* London, New York: Academic Press.
37. Hoffman, P., R.-D. Peek, J. Puls, and E. Schwab. 1986. Das Holz der Archäologen. *Holz Roh- Werkst.* 44:241-247.
38. Illston, J. M., J. M. Dinwoodie, and A. A. Smith. 1979. *Concrete, Timber and Metals.* New York: Van Nostrand Reinhold.
39. Jackman, P. E. 1981. The fire behaviour of timber and wood based products. *J. Inst. Wood Sci.* 9(1):38-46.
40. Kauman, W. G. 1961. Effect of thermal degradation on shrinkage and collapse of wood from 2 Australian species. *For. Prod. J.* 11(9):445-452.
41. Knight, B. F., and H. J. Heikkenen. 1980. *Principles of Forest Entomology.* New York: McGraw-Hill.
42. Kollmann, F. F. P., and I. B. Sachs. 1967. The effects of elevated temperature on certain wood cells. *Wood Sci. Technol.* 1:14-25.
43. Kollmann, F., and W. A. Côté. 1968. *Principles of Wood Science and Technology. I. Solid Wood.* Berlin: Springer Verlag.
44. Koltzenburg, C. 1975. Zur Entstehung von Verfärbungen in gelagertem Bergahorn holz (*Acer pseudoplatanus* L.). *Holz Roh- Werkst.* 33(11):420-426.
45. Levi, M. P., and R. L. Dietrich. 1976. Utilization of southern pine beetle-killed timber. *For. Prod. J.* 26(4):42-48.

46. Liese, W., ed. 1975. *Biological Transformation of Wood by Microorganisms.* Berlin: Springer Verlag.
47. Liese, W. 1970. Ultrastructural aspects of woody tissue deterioration. *Ann. Rev. Phytopathol.* 8:231–258.
48. Mackay, G. D. M. 1967. Mechanism of Thermal Degradation of Cellulose: A Review of the Literature. Canada Dept. For. Publ. No. 1201.
49. Miller, E. R. 1981. Photodegradation of wood. *BRE News* 54:15.
50. Moulopoulos, C., and G. Tsoumis. 1960. Growth abnormalities of euramerican hybrid poplars cultivated in Macedonia-Greece. *Ann. School Ag. For.* 59:193–237, Aristotelian University (Greek, English summary).
51. Netherlands National Commission for UNESCO. 1979. Conservation of Waterlogged Wood. The Hague.
52. Nickolas, D. D., and J. F. Siau. 1973. Factors affecting the treatability of wood. In *Wood Deterioration, etc.*, Vol. II, ed. D. D. Nickolas, pp. 299–344. Syracuse, New York: Syracuse University Press.
53. Noguchi, M., et al. 1986. Detection of very early stages of decay in western hemlock wood using acoustic emissions. *For. Prod. J.* 36(4):35–36.
54. Passialis, C., and G. Tsoumis. 1984. Characteristics of discolored and wetwood in fir. *IAWA Bull.* n.s. 5(2):111–120.
55. Pechmann v. H., and O. Schaile. 1950. Über die Änderung der dynamischen Festigkeit und der chemischen Zusammensetzung des Holzes durch den Angriff holzzerstörende Pilze. *Forstw. Cbl.* 69(8):441–466.
56. Peek, R., and W. Liese. 1979. Untersuchungen über die Pilzanfähigkeit und das Tränkverhalten nassgelagerten Kiefernholzes. *Forstw. Cbl.* 98:280–288.
57. Peek, R.-D., H. Willeitner, and U. Harm. 1980. Farbindikatoren zur Bestimmung von Pilzbefall im Holz. *Holz Roh- Werkst.* 38:225–229.
58. Raczkowski, J. 1980. Seasonal effects on the atmospheric corrosion of spruce micro-sections. *Holz Roh-Werkst.* 38:231–234.
59. Read, S. J. 1986. Controlling death watch beetles. BRE Information Paper IP 19/86.
60. Schaffer, E. L. 1975. Fire considerations. In *Wood Structures*, pp. 392–410. New York: Am. Soc. Civil Engineers (ASCE).
61. Scheffer, T. C. 1973. Microbiological degradation and its causal organisms. In *Wood Deterioration, etc.*, Vol. I, ed. D. D. Nickolas, pp. 31–106. Syracuse, New York: Syracuse University Press.
62. Schmidt, O., and W. Liese. 1982. Bacterial decomposition of woody cell walls. *Int. J. Wood Preservation* 2(1):13–19.
63. Silva, de D., and W. E. Hillis. 1980. The contribution of silica to the resistance of wood to marine borers. *Holz Roh- Werkst.* 34(3):95–97.
64. Stamm, A. J. 1964. *Wood and Cellulose Science.* New York: The Ronald Press.
65. Trada (Timber Research and Development Association). 1964. *Timber Pests and Their Control.* Bucks, England: High Wycombe.
66. Tsoumis, G. 1956. Une étude sur la résistance des bois aux attaques des xylophages marins en rapport avec la teneur en silice. *Rev. Bois Appl.* 10(12):23–25.
67. Tsoumis, G. 1961. Grading fir logs and lumber in the University sawmill at Pertouli. *Ann. School Ag. For.* 6:103–134, Aristotelian University (Greek, English summary).
68. Tsoumis, G., and E. Voulgaridis. 1978. A study of blue stain in black pine. I. Prevention of attack in the forest. *Geotechnica* B(1):3–11; and 1979. II. Effects on wood properties. *The Forest* 86:9–14 (Greek, English summary).
69. Tsoumis, G. and V. Vassiliou. 1983/84. A study of pine wood from burnt forests. Ann. Dept. For. Nat. Envir., Aristotelian University 26/27: 245–255 (Greek, English summary). Also in Proc. Balkan Conf. "Investigation, Conservation and Utilization of Forest Resources," Sofia 1984.
70. Tsoumis, G. 1991. *Harvesting Forest Products.* (In press).
71. Turner, R. D. 1971. Identification of marine-borer molluscs. In *Marine Borers, Fungi and Fouling Organisms*, pp. 17–64. Paris: O.E.C.D.
72. Wagenführ, R., and C. Scheiber. 1974. *Holzatlas.* Leipzig: VEB Fachbuchverlag.
73. Wächli, O. 1977. Der Temperatureinfluss auf die Holzzerstörung durch Pilze. *Holz Roh- Werkst.* 35(2):51–54.
74. Wächli, O., and P. Tscholl. 1975. Möglichkeiten der Bekämpfung holzzerstörender Insekten ohne Giftanwendung. *Holz Roh- Werkst.* 33(2):49–53.
75. Wangaard, F. F. 1966. Resistance of wood to chemical degradation. *For. Prod. J.* 16(2):53–64.
76. Wang, S., O. Suchsland, and J. H. Hart. 1980. Dynamic test for evaluating decay in wood. *For. Prod. J.* 30(7):35–37.
77. Ward, J. C., and W. Y. Pong. 1980. Wetwood in Trees: A Timber Resource Problem. USDA, Pac. For. Range Exp. Station, Report PNW-112.
78. Ward, J. C., and J. G. Zeikus. 1980. Bacteriological, Chemical and Physical Properties of Wetwood in Living Trees. In *Natural Variations of Wood Properties*, ed. J. Bauch, pp. 133–165. Hamburg-Rinbek: Mitt. BFA.
79. Wayman, M., M. R. Azhar, and Z. Koran. 1971. Morphology and chemistry of two ancient woods. *Wood Fiber* 3(3):153–165.
80. Wazny, J. 1958. Studien über die Einwirkungen von *Merulius lacrymans* (Wulf.) Fr. and *Coniophora cerebella* Pres. auf die mechanischen Eigenschaften befallenen Holzes. *Holz Roh- Werkst.* 16(8):285–288.
81. Wilcox, W. W. 1973. Degradation in relation to wood structure. In *Wood Deterioration, etc.*, Vol. I, ed. D. D. Nickolas, pp. 107–148. Syracuse, New York: Syracuse University Press.
82. Wilcox, W. W. 1978. Review of literature on the effects of early stages of decay on wood strength. *Wood Fiber* 9(4):252–257.
83. Wolf, F., and W. Liese. 1977. Zur Bedeutung von Schimmelpilzen für die Holzqualität. *Holz Roh-Werkst.* 35(2):53–57.

FOOTNOTES

[1] Old waterlogged wood is found in more or less deteriorated condition—plasticized or very brittle—and undergoes heavy shrinkage on drying. The degradation may be abiotic, and involves hydrolysis of wood components. Cellulose and hemicelluloses are lost to a lesser or greater extent, but lignin is more persistent (37). Such wood may be stabilized by chemicals (51), such as polyethylene-glycol (PEG), and this enables the reconstruction of archaeological and other historical findings, such as old sunken ships (ancient Greek, Roman, medieval, etc., see also Chapter 10).

[2] Bacteria, in combination with physicochemical action of inorganic environments (dissolution of extractives, hydrolysis of chemical constituents), may cause, after prolonged storage in water, considerable structural changes of wooden objects (e.g., items of archaeological importance). Although in the water such objects may preserve their form, exposure to the atmosphere may cause drastic changes of shape (warping, excessive shrinkage, collapse) (9). However, if storage is not that long, there seems to be no important deterioration. Fir and spruce wood stored in water for 17 years showed some anatomical and property changes, but these were mainly limited to sapwood. Permeability increased and led to improved drying and absorption, but compressive and bending strength were still within the natural range of green wood (14).

[3] Wetwood has also been attributed to wounding, related to tree age (77), and observed in dying trees (77, 78).

[4] A high correlation between density and durability was reported for eucalypts and some other hardwoods—not softwoods. It may be argued, however, that higher density hardwoods contain greater quantities of extractives (17).

[5] Chemical changes in wood, caused by decay fungi even at the incipient stage, enable detection of attack by staining (57).

[6] In addition to usual laboratory tests of strength, the effects of fungal attack have been studied by electrical vibration of specimens (nondestructive testing). Differences in the frequency of vibration indicate attack at the incipient stage. Also, acoustical emissions (produced by specimens stressed in bending) are indicators of very early stages of decay (53, 76).

[7] Fungi that attack wood of living trees (and sometimes timber) are: *Fomes annosus* Fr. (fir, spruce, pines, and other softwoods), *Fomes pini* Karst. (softwoods), *Fomes pinicola* (Sw.) Cooke (softwoods and hardwoods), *Fomes fomentarius* Fr. (beech, poplar), *Fomes igniarius* Fr. (poplar, willow, birch, elm), *Fomes applanatus* Fr. (beech, poplar, ash, and softwoods), *Polyporus schweinitzii* Fr. (pines, larch), *Armillaria mellea* (Vahl) Quel. (softwoods, in hardwoods also timber), etc. (10, 31).

[8] Insects that attack wood of living trees (and sometimes timber) are: *Xyloterus lineatus* Oliv. (fir, spruce), *Urocerus (Sirex) gigas* L. (fir, pines, larch, seldom poplar and ash), *Cerambyx cerdo* L. (mainly oak, seldom elm and ash), *Platypus cylindricus* F. (oak, beech, chestnut), *Sciapteron tabaniformis* Rott. (poplar), *Aegeria (Trochilium) apiformis* Clerck (poplar, seldom willow, ash, birch, lime), *Cossus cossus* L. (various hardwoods), *Saperda carharias* L. (poplar), etc. (31, 41).

[9] The stages of development (metamorphosis) of termites are: egg–pupa–adult. Pupa resembles the adult form.

[10] Fungi and bacteria (seawater) contribute to attack of wood by marine organisms (61). There are two hypotheses: (a) fungal attack partly decomposes and softens the wood and, therefore, facilitates the entrance of marine organisms; and (b) fungi are indispensable as food to these organisms, which cannot survive with only wood as food (main source of energy is cellulose, lignin cannot be digested) (22). Fungi that attack wood in the sea are *Ascomycetes* and *Fungi Imperfecti*, and they cause soft rot (30).

[11] Species of *Bankia* are similar to *Teredo*, but *Martesia* species are enclosed in shells and are mainly found in tropical waters (8) (Figure 15-13).

[12] Aluminum melts at about 650°C (1200°F), glass at about 750°C (1400°F), and steel glows at 500–600°C (930–1110°F), with resulting loss of about 50% of its capacity to bear loads (39).

III

UTILIZATION

The *utilization* of wood is examined with regard to products of *primary* manufacture (technology of production, properties of products, etc.). The products discussed are: *roundwood products* (poles, piling, posts, etc.), *lumber, veneer, plywood, laminated wood, particleboard, fiberboard,* and *paper.* Separate chapters are devoted to *drying* (air-, kiln-drying), *preservative treatment,* and *adhesion and adhesives. Secondary* products (furniture, etc.), made by further processing the above products, are given a general consideration in the Appendix, together with chemically derived products, wood as a source of energy, exudates and extractives, and bark (including cork) products.

16

Roundwood Products

Roundwood products are poles, piling, masts, posts, and some mine timbers. All keep their cylindrical, tree-trunk form,[1] and they are produced by felling the trees, topping, delimbing, cross-cutting (to reduce length, if needed), and debarking. Poles, piling, masts, and posts are usually rounded ("shaved") by machines to acquire a smooth surface and better appearance, and machined (drilling holes, etc.) and air-dried in preparation for preservative treatment. Such treatment is necessary because roundwood products are used in contact with the ground, and exposed to weather and to destructive organisms.

Poles are produced in varying lengths (up to about 30 m or 100 ft), and basal and top diameters, and they are accordingly separated into classes. In addition to suitable dimensions, poles should be straight (small deviations are permitted), sound (without decay, insect holes, splits, or wounds), with few knots, and little spiral grain and taper (specifications define permissible defects) (4, 5). The woods used include pines, fir, Douglas-fir, larch, cedars, hemlock, chestnut, various Asiatic and tropical species, and others. Preference is given to species with medium density and high strength in relation to weight. Light woods do not possess the necessary strength, and heavy woods are more difficult (and expensive) to handle, transport, and install. Ease of preservative treatment and natural durability are also criteria for selection. Requirements and production of *masts* are similar to those for poles.

Piling is a kind of pole used to reinforce foundations in land structures (buildings) or support structures in water (bridges, docks, and wharves); they are respectively distinguished into foundation and marine piling. The dimensions and processing are similar to poles, and they are made of many of the same species, with the additional requirement that they must have adequate strength to withstand their driving into the ground; common piling woods include certain pines, Douglas-fir, elm, oak, maple, and bald cypress *(Taxodium)*. Piling differs from poles in that a pole is used with the large end set into the ground while a pile is set with the small end in the ground; also, the foundation piling is completely buried in the ground. Piling placed in seawater requires a high natural durability or special preservative treatment against attack by wood-destroying marine borers (see Chapters 15 and 19). The specifications for piling differ from those applied to poles due to differences in carrying loads.

Posts are used for fences and road guard rails. Fence posts are often made of woods that possess a natural durability, such as chestnut and black locust. Otherwise, they have to be treated with preservatives, and such treatment may be done with simple and inexpensive methods by farmers (see Chapter 19); proper treatment is similar to that of poles.

Mine timbers are used in round or sawn form. Wood species are selected for high natural durability and adequate strength; preferred species are chestnut, white oak, maple, hickory, black locust, Mediterranean cypress, certain pines, Douglas-fir, and others. The ability to provide audible warning before failure is a major advantage of wood in mines (3).

There is a certain replacement of wood by other materials (steel, concrete) in the above products, because wood may decay, burn, or be consumed by insects and marine borers.

However, the replacement is limited, because wood has unique advantages, which in the products under consideration are: low initial cost, availability in directly usable form (with little preparatory processing), ease of processing with tools and machines, favorable relationship of strength to weight, satisfactory natural durability of certain species for short-term service, and possibility to considerably increase the durability by preservative treatment.

REFERENCES

1. Bohannan, B. 1975. Wood poles and piles. In *Wood Structures*, pp. 155–165. New York: Am. Soc. Civil Engineers (ASCE).
2. Hunt, G. M., and G. A. Garratt. 1968. *Wood Preservation*, 3rd ed. New York: McGraw-Hill.
3. Jones, L. C. R. 1974. *Use of Timber in Mining*. U.K.: TRADA.
4. Knigge, W., and H. Schulz. 1966. *Grundriss der Forstbenutzung*. Hamburg, Berlin: P. Parey.
5. Panshin, A. J., E. S. Harrar, W. J. Baker, and P. B. Proctor. 1950. *Forest Products*. New York: McGraw-Hill.
6. U.S. Forest Products Laboratory. 1987. *Wood Handbook*. Ag. Handbook No. 72 (revised).

FOOTNOTE

[1] In addition to trunk, round and thin (branch) material is utilized to support agricultural crops (vines, bean stalks), as basket-weaving material (willow), etc.

17

Lumber

Lumber is a solid wood product made by lengthwise sawing of logs. Transverse sawing is secondary and is applied to reduce length or remove defects. Lumber is produced in varying dimensions and used in buildings and other structures and products.

MACHINES

Three basic types of sawing machines are used to produce lumber: frame saw, band saw, and circular saw.[1] These machines differ in constructional details, but their general characteristics are as follows.

A *frame saw* (Figures 17-1 and 17-2) has a number of straight blades (seldom one blade) with teeth on one edge and seldom on both edges; the blades are set at predetermined distances on a frame that reciprocates vertically or, rarely, horizontally. A *band saw* (Figures 17-1 and 17-3) is equipped with an endless, ribbon-like blade, which usually has teeth only on one edge, and revolves on two wheels placed parallel to each other—vertically or, occasionally, horizontally. A *circular saw* (Figures 17-1 and 17-4) is a disk that has teeth on its periphery and rotates on an axle (shaft or arbor); sometimes, the teeth are inserted and can be replaced.[2]

Selection of the proper machine as headsaw depends on respective advantages and disadvantages.

- Frame saws and band saws have relatively thinner blades; therefore, there is less waste of wood in the form of sawdust in comparison to circular saws. In this regard, band saws may be better than frame saws (18). However, modern machines of all three types can give similar kerf losses (41).
- In frame saws, the thickness of lumber is in each case predetermined, whereas in band and circular saws it is possible to select dimensions and turn the log after each saw cut. This possibility may contribute to a better quantitative and qualitative utilization of wood; however, the operators of such band or circular saws should be more knowledgeable and should be alert to take proper decisions quickly.
- Use of a frame saw requires the logs to be classified into diameter classes, because the position of the blades is determined accordingly. This means a greater expense for classification and a larger area for storage of logs.[3]
- Vertical frame saws (i.e., machines with a frame reciprocating vertically) may process logs having a diameter up to about 1.25 m (4 ft), usually 40–80 cm (16–32 in.). In horizontal frame saws (rarely used) the range of diameters is 60 cm to 2.0 m (about 2 to 6 ft) (21). Band saws (vertical, horizontal) are more flexible machines and can handle large logs, 1–2 m (3–6 ft) in diameter. Circular saws cannot saw logs larger than about 40 cm (16 in.), because only about one third of the disk is utilized during sawing (12). There are, however, machines equipped with disks mounted on separate axles (double arbors), which are placed parallel on the same verti-

240 III / UTILIZATION

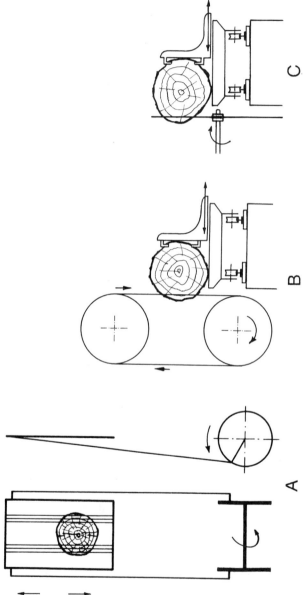

Figure 17-1. Schematic representation of the basic (conventional) types of sawing machines. (A) *Frame saw*: the frame that carries the blades is reciprocating (up–down) by rotation of wheels connected eccentrically to the frame with arms. (B) *Band saw*: an endless blade rotates on wheels; the sawed log may be moved toward the blade, or freed temporarily and turned to be machined from another side. (C) *Circular saw*: sawing is performed by a rotating disk with teeth on its periphery; the log may be moved toward the saw as in the case of the band saw. Note that newer installations of band and circular saws include multiblade equipment; see text and Figure 17-5.

Figure 17-2. A frame saw. (Courtesy of Linck.)

late to wood (species, dimensions, hardness, moisture content, defects), pattern of sawing (number of cuts, lumber thickness), personnel (training, experience, age, number), degree of automation, etc. Frame saws have a lower speed of feeding and sawing, but more blades work simultaneously; each log passes once and not repeatedly, as in the case of band or circular saws, where there are more delays for turning, thickness setting, and return of the carriage. Efficiency is measured in volume of logs processed, or lumber produced, or area of sawn surface per unit of time, and is a result of the combination of machines in a sawmill. Efficiency with respect to labor varies from about 2.5–10 m^3 (90–350 ft^3)/man/day to 35–70 m^3 (1200–2500 ft^3)/man/day (15).

cal plane; with such machines, log diameters may reach 1 m (3 ft) (2).
- Frame saws are more demanding with regard to installation (foundation), due to reciprocating movement of action in comparison to the rotating action of band or circular saws.
- Processing efficiency varies among the three types of headsaws. A comparison is difficult, because of the many factors involved that re-

All machines are served by devices that support and transport the logs. Transportation to headsaws is made on carriages, where the logs are held firmly by suitable hooks ("dogs"), spindles, or sharp chains; in modern installations, sawing small-diameter logs, other supporting and conveying devices are used [e.g., top and bottom chains ("alligator infeed system"), side-feeder chains (Figure 17-6), and conveyor belts and supporting arms (10, 54)

Figure 17-3. A band saw. (Courtesy of Canali.)

Figure 17-4. A portable (mobile) circular saw. (Courtesy of Corinth-American.)

(Figure 17-18)]. Conveyors are also employed for transportation of wood from machine to machine—to sawmill exit (i.e., to the lumber yard). Machines other than headsaws, are resaws, edgers, and trimmers. Resaws are frame, band, or circular saws. Edgers include conventional (single-blade) circular or band saws, or chippers, but other arrangements and combinations also exist. Trimming (transverse sawing) is performed by circular saws.[4]

In addition to sawing machines (and chippers), sawmills are equipped with tools and machines for maintenance and repair work, such as tooth sharpening and blade welding. Energy is as a rule electrical.[5]

SAW BLADES

Saw blades vary in shape, dimensions, tooth morphology, and material of construction (2, 6, 14, 21). Shape is dictated by machine type (frame, band, or circular), and information on dimensions is included in following paragraphs. Main tooth forms are shown in Figure 17-7, which includes other information about saw blades. Each tooth has three angles: a, front angle; b, tooth angle; and c, supplementary angle ($a + b + c = 90°$). Angle a affects the entrance of a tooth in the mass of wood, angle b affects tooth strength, and angle c sawing speed. Angles vary widely depending on machine type, wood species, sawing direction (longitudinal, transverse), and blade material and thickness. In certain tooth forms, angle a is zero or negative (Figure 17-7, 2–3), and for hard woods and thinner blades angle a is smaller and angle b greater.

The cavity between teeth, called "gullet" (Figure 17-7, 1g), should be sufficient to hold the sawdust produced. Its capacity depends on its shape, tooth height, distance between teeth, and blade thickness. The volume of generated sawdust is 3–6 times greater than the volume of wood from which it is produced, and depends on tooth morphology and wood characteristics (density, moisture content). The ca-

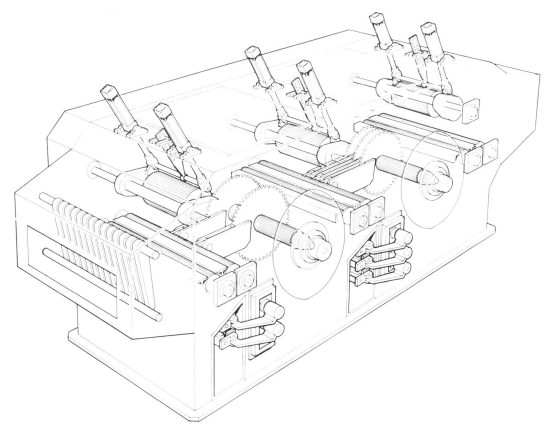

Figure 17-5. A multiple circular saw machine; the system is enclosed for control of noise. (Courtesy of ARI.)

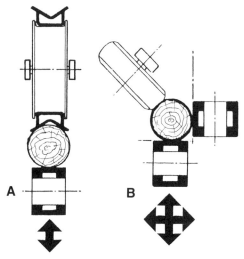

Figure 17-6. Infeed systems for Chip-N-Saws processing small logs. (A) *Alligator* system: top and bottom chains advance the log to chipping heads; the V-shaped links of the top chain center the log. (B) *Side-feeder chains*: the system can be shifted horizontally and vertically to allow varying depth of cut. (Courtesy of Forintek Canada Corp.)

pacity of a gullet should be at least one half the uncompressed volume of sawdust generated from a single sawtooth due to sawdust compaction during sawing (25).

Regarding blade thickness, the following relationships exist: the smaller the free length of blade (the greater the log diameter), the greater its thickness; thickness affects feeding speed (i.e., smaller thickness—lower speed). With frame saws, cuts with fewer blades in a frame and lower quality requirements may be made with thicker blades and a higher feeding speed; within a frame, it is possible to place, simultaneously, blades of different thicknesses (12).

In order to facilitate the movement of a blade in the gap opened by sawing, the teeth are "set," either by alternately bending their tips away from the plane of the blade or by widening the tips; the result is called "spring-setting" or "swage-setting," respectively (Figure 17-7, 4–6). In this manner, the width of cut,

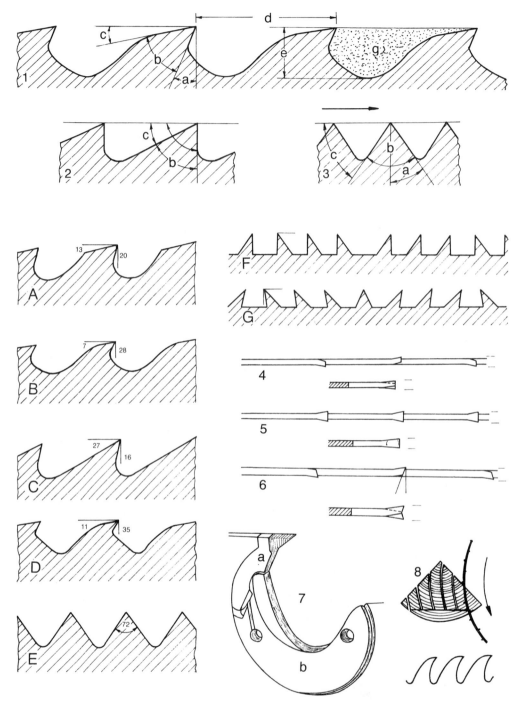

Figure 17-7. Tooth morphology: 1, tooth angles (a, b, c), distance (d), height (e), and gullet (g); 2, angle (a) zero; 3, angle (a) negative; (A–E) tooth types (with indications of angle size); (F, G) teeth of blades of horizontally reciprocating frame saw (F, hardwoods; G, softwoods); 4–6, kerf (4, 6, spring-set; 5, swage-set); 4, 5, longitudinal sawing; 6, transverse sawing; 7, an inserted tooth of circular saw (a, tooth; b, base); 8, sawing barrel staves with a cylindrical saw and (down) its tooth morphology. (After Refs. 6, 14, and 27.)

called "kerf," becomes greater than the thickness of the blade and is equal to thickness plus twice the spring-setting, or is equal to the final width of the tip. Setting is greater in thinner blades and for sawing soft and green woods, and should be proper in each case. Too small a setting will cause overheating due to friction, burning of the sawn wood surface, and tightening or breaking of the blade. Excessive setting will cause unjustified waste of wood, and reduction of the life of the blade.

The upper edges of tooth tips are in one plane, at a 90° angle to the plane of the blade, or they are inclined (Figure 17-7, 4–6); these conditions apply to longitudinal and transverse sawing, respectively.

Specific characteristics of blades in head saws are as follows (2, 6, 12, 14, 21)

In *frame saws*, the blades are usually 1.10–1.50 m (3.5–5 ft) long, but length may range from 80 cm to 2.5 m (2.5–8 ft); width is 12–18 cm (5–7 in.), and thickness 1.2–2.5 mm.[6] Settings are 0.2–0.5 of thickness, or 0.5–0.6 mm for softwoods and 0.4–0.5 mm for hardwoods, and kerfs range from about 3 to 5 mm, and average about 4 mm. Tooth forms are A, B, and C (usually C; see Figure 17-7). In horizontally reciprocating blades, the teeth (form F for hardwoods and form G for softwoods) are placed in two directions, in groups of 3–4, in order to reduce the pressure on those teeth that are placed in one direction, especially when sawing hardwoods (6, 8).

In *band saws*, the length of a blade is related to the diameter of the wheels and their maximum distance, as follows:

$$M = 2\left(a + \frac{\pi d}{2}\right) \quad (17\text{-}1)$$

where

M = length (m, ft)
a = maximum distance of wheels (m, ft)
d = wheel diameter (m, ft)
π = 3.14

The diameter of wheels varies in headsaws from about 1–3 m (3–10 ft). Width of blades is 12–25 cm (5–10 in.). In general, the length of a blade is 6–8 times greater than the wheel diameter; width is 1/10 and thickness 1/1000–1/1250 of wheel diameter. Kerfs range from about 2–5 mm, and typically in band saws they are not as wide as in frame saws (6).

Tooth forms are B or D, usually B. Certain types of band saws have blades on both edges and, therefore, they saw with the carriage both coming and going. Such blades are used with easily sawn woods, mainly softwoods.

In *circular saws*, disks are up to 1.30 m (4 ft) in diameter, and 4.5–6.0 mm thick depending on diameter (1/250 of diameter with good maintenance, otherwise 1/200) (18). Setting is up to 0.7 mm. Kerf is wider than in frame or band saws; it ranges from 4 to 10 mm and averages about 7 mm. Teeth may have forms A to E; A, B for lengthwise, and C, D for transverse sawing; form E is being gradually abandoned (6).

Some circular saws have inserted teeth (Figure 17-7, 7). Such teeth are periodically taken out, resharpened, and replaced; they are made of alloys that can be adapted to the hardness of the processed wood. The disadvantage is that inserted teeth are suited for disks of great thickness (29).

In addition to the above, there are cylindrical saws with teeth on the rim of one end of a hollow cylinder. The diameter of such saws is 20–80 cm (8–32 in.), and teeth have a special morphology (Figure 17-7, 8). Cylindrical saws are specialty equipment for the production of barrel staves.

STORAGE OF LOGS

Logs are usually stored on the ground. Direct contact should be avoided, however, due to risk of fungal attack. Insects may be also involved in open-air storage, as well as the development of checks and splits—and loss by fire. For these reasons, long storage is avoided, and the logs are continuously sprayed or stored in water (pond, river, lake, sea—Figure 17-8). In addition to protection, spraying and especially wet storage contributes to cleaning the logs from

Figure 17-8. Sawmill exterior views (in Greece). (A) A small sawmill; the log storage yard is at the back of the building (sawmill) and the lumber storage yard in front. (B) An integrated wood-using industry—sawmill, veneer, plywood, and particleboard plant; the logs (mainly tropical woods) are stored in seawater. (B. Courtesy of Shelman.)

dirt and small stones attached during harvesting in the forest, and makes their mechanical processing easier. Storage in water also contributes to the removal of substances, such as starch, which favor fungal and insect attack, and reduces the need for ground storage area.

There are disadvantages in water storage: there may be pollution; if the water is stagnant, it favors growth of bacteria[7] and discoloration of wood (24); logs may sink and the water may freeze in the winter; storage in seawater may risk attack by marine borers. The small quantity of salt absorbed by wood is not considered sufficient to adversely affect its quality (19).

Table 17-1. Schemes of Log Grouping for the Frame Saw

A	B	C
	Diameter classes (small end, cm)	
18–25, 26–30	16–20, 21–22	15–30 (every 2 cm)
31–35, 36–40	23–24, 25–26, etc.	31–50 (every 3 cm)
41–47, 48–57	(every 2 cm)	51 and up (every 5 cm)

In sawmills where the headsaw is a frame saw, the logs are separated into groups based on their small-end diameters. Examples of grouping schemes are shown in Table 17-1; the range of diameters included in a group affects the yield of lumber: the greater the range, the greater the reducing effect on the yield.

The logs are stored with or without bark. Softwoods are generally debarked before sawing, but hardwoods may not be debarked, especially if the bark is thin. Debarking increases the service life of saw blades, reveals surface defects (which may affect the pattern of sawing), and facilitates utilization of residues for pulp and paper (50). Long storage with bark invites insect attack. Minimum diameter of sawlogs is 15 cm (6 in.) and sometimes 12.5 cm (5 in.) (54).

METHODS OF SAWING

Sawing to produce lumber can be distinguished as being either *primary* or *secondary*. Primary sawing, also called primary breakdown, is sawing a log by a headsaw; the products are unedged flitches or cants. In secondary sawing, they are resawed, edged, and trimmed to accurate length in order to complete the prismatic shape of lumber. Primary and secondary breakdown may be simultaneous, as in the case of edging by chipping. In other cases, lumber flitches are not edged in the sawmill (e.g., when sawing some valuable furniture woods) (Figure 18-6).

Sawing patterns are influenced by several factors, such as machinery, intended use of the lumber, kind of wood, log diameter, yield, and

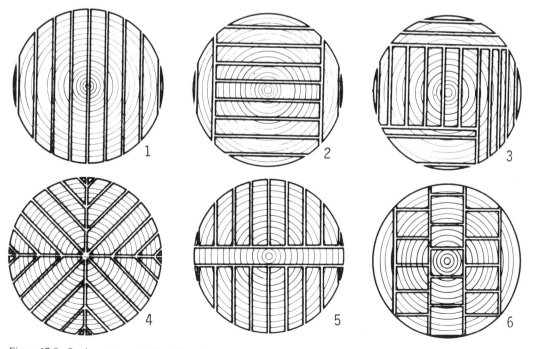

Figure 17-9. Sawing patterns: 1, live (through and through) sawing; 2, cant sawing; 3, sawing for grade; 4, 5, sawing for (radial) grain; 6, sawing dimension lumber.

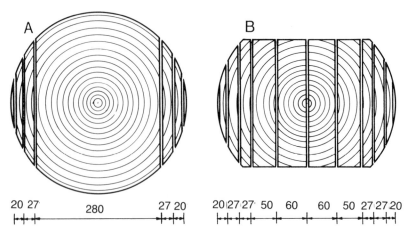

Figure 17-10. A sawing plan for cant sawing in a frame saw (log diameter class 39–40 cm): (A) presawing, (B) sawing the cant. The sawing plan reads: presawing $\frac{1}{280}, \frac{2}{27}, \frac{2}{20}$, and sawing the cant $\frac{2}{60}, \frac{2}{50}, \frac{4}{27}, \frac{2}{20}$. (Numerators are numbers of pieces and denominators show width in mm.)

Figure 17-11. Sawing patterns and taper. (A) Symmetric sawing through the pith; (B) taper sawing (half-taper symmetric); (C) double-taper sawing (full taper asymmetric). (A. Scandinavian practice; B and C. North American practice.) (After Ref. 18.)

cost. Various sawing patterns exist and include "live" sawing, cant sawing, and others. Live sawing is sawing with all sawlines parallel to each other and to the length of a log. Cant sawing is sawing in a pattern where a log is slabbed usually on two parallel sides (sometimes one or three sides), and the resulting cant is reduced to lumber in a secondary stage. Other patterns involve sawing for radial (quarter-sawn or edge-grained) lumber, and sawing "dimension stock" (Figures 17-9 and 17-10). Sawing may be parallel to pith or parallel to bark, and symmetric or asymmetric according to the position of taper (Figure 17-11). Edging is parallel or conical (Figure 17-12); conical (i.e., parallel to bark) edging is sometimes applied to valuable hardwoods intended for solid furniture in order to reduce waste—or the lumber is marketed unedged.

There are various combinations of machines in a sawmill, for example: circular saw–frame saw–edger–trimmer; two frame saws (head saw, resaw)–edger–trimmer; two band saws (headsaw, resaw)–edger–trimmer; a series of double band saws with chipping edgers; double

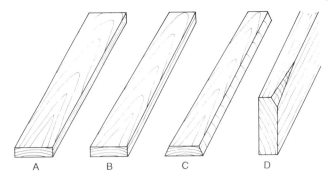

Figure 17-12. Edging of lumber: (A) parallel, (B) conical, (C) unedged, and (D) parallel with wane. (Wane is lack of wood or attachment of bark on any edge).

arbor circular saws (or double band saws, or frame saw)–chippers; multiple band or circular saws–chippers, etc. (8, 13, 32, 33, 39, 40) (Figures 17-13 to 17-18).

The position of sawlines depends on the desired cross-sectional dimensions of the lumber, as such dimensions are accepted in commercial practice and specifications (e.g., see Tables 17-2 and 17-2A).[8] However, the dimensions produced by sawing are not final; there is an oversize allowance to compensate for shrinkage, sawing variation, and planing—if the lumber is marketed planed ("dressed"). Thus, sawing produces the so-called "target" thickness.[9] The

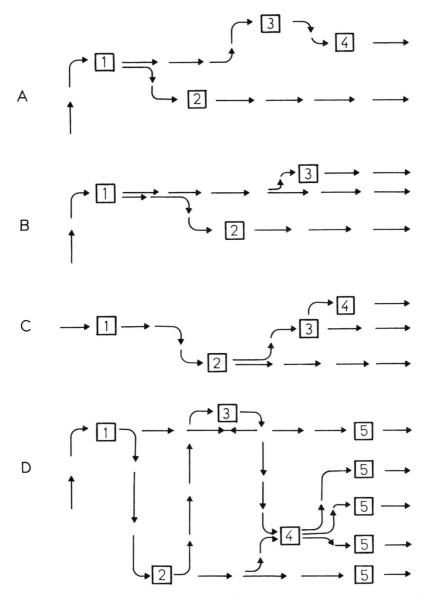

Figure 17-13. Production flow in sawmills: (A) 1, frame headsaw (cant sawing); 2, frame saw (resaw, sawing the cant); 3, circular saw (trimmer); 4, double circular (edger). (B) 1, band headsaw; 2, band saw (edger); 3, double circular (edger). (C) 1, circular headsaw; 2, frame saw (sawing the cant); 3, double circular (edger); 4, circular (trimmer). (D) 1, band headsaw; 2, band saw; 3, band saw (with merry-go-round); 4, circular gang (edger); 5, circular (trimmers). (A–C, European sawmills, small; D, tropical wood sawmill in Ivory Coast.) (A and B, adapted from Ref. 2, and D from Ref. 11.)

250   III / UTILIZATION

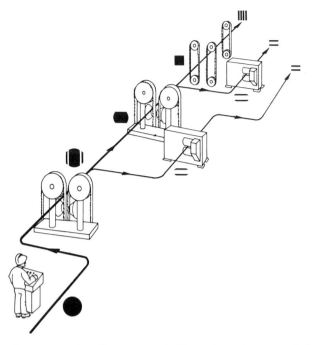

Figure 17-14. Production flow with a series of band saws and chippers that reduce slab and edging residues to chips. (After Ref. 39, reproduced by permission from Springer-Verlag.)

allowance for shrinkage depends on species, grain direction (radial, tangential), and final (air-dry) moisture content. Sawing variation (i.e., deviations from set thickness and width) may be due to such factors as saw-blade vibration, misalignment of blades, feeding speed, and log or cant transport systems; postponement of required maintenance and adjustments, higher speeds of feeding, and thinner blades at low feeding speeds increase sawing variation (for each machine, there is an optimum blade thickness) (42).

On the basis of the above, a "sawing plan" is prepared, the goal of which is to utilize the particular log to the best advantage with regard to yield and/or grade in conjunction with desired dimensions of lumber (Figure 17-10). This sawing plan, different for each diameter or group of diameters, is based on the small end of logs, and may be prepared manually on graph paper or by use of a computer.

Given the various marketable dimensions and kerf, the goal is to "inscribe" target thicknesses and widths on the small end of a log of a certain length, considering also the magnitude of taper (Table 17-3). The log is then placed accordingly to execute the first sawline, or two parallel sawlines in the case of producing a cant. A computer facilitates this work,

Figure 17-15. Schematic representation of a pair of band saws combined with chippers. (After Ref. 12, reproduced by permission from DRW-Verlag Weinbrenner.)

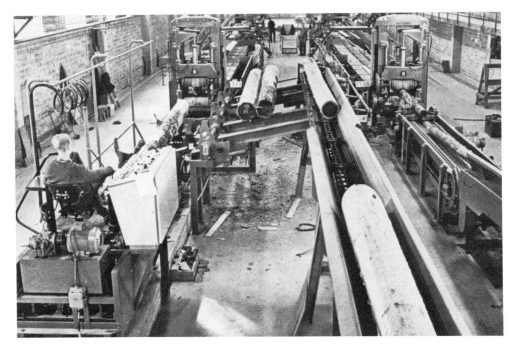

Figure 17-16. Interior of a sawmill with two frame saws and remote control handling of the logs. (B. Thunell, *Schweiz. Z. Forstw.* 122/3:91–104.)

Figure 17-17. Control room and interior of an automated circular saw and chipper sawmill. (Courtesy of ARI.)

Figure 17-18. A small-log sawmill equipped with chipper-canters, profiling units (vertical chipping edgers), and double arbor gang (circular) saws. The sequence of transformation of a log is shown at the bottom. The system is computerized: diameter, length, and taper are electronically measured and used by a computer to calculate the optimum sawing pattern. (Courtesy of Linck.)

and all possible combinations of cross-sectional dimensions can be tried for the best possible result. This goal is also aided by electronic scanning, which provides information on each particular log with regard to existing deviation from straightness, deviation from a circular cross-section, and location of defects, such as knots (54).

A simple explanation of approach in the case of cant sawing is to "inscribe," on the small end of a log, a square or orthogonal parallelogram (Figure 17-19), using the log diameter as the hypotenuse of the Pythagorean theorem ("the square of the hypotenuse of a right-angled triangle is equal to the sum of the squares of the other two sides"). This work is facilitated by use of Figure 17-20, which presents possible selections of side lengths (a, b) within a certain log diameter. In Figure 17-19A, side a determines the width of lumber boards to be produced by sawing the cant (after inversion, see Figure 17-10); and side b determines the number of boards according to desired target thickness; this number may be influenced, to a

Table 17-2. Dimensions of Softwood Lumber

Thickness (mm)		\multicolumn{12}{c}{Width (mm)}											
a	b	75	100	115	125	150	160	175	200	225	250	275	300
16		X	X	X	X	X	X	X	X	X	X	X	X
19	18	X	X	X	X	X	X	X	X	X	X	X	X
22		X	X	X	X	X	X	X	X	X	X	X	X
25	24	X	X	X	X	X	X	X	X	X	X	X	X
32	28	X	X	X	X	X	X	X	X	X	X	X	X
38	44	X	X	X	X	X	X	X	X	X	X	X	X
50	48	X	X	X	X	X	X	X	X	X	X	X	x
63	60–65	X	X	X	X	X	X	X	X	X	X	X	X
75	70–80	X	X	X	X	X	X	X	X	X	X	X	X
100			X	X	X	X	X	X	X	X	X	X	X
125					X	X	X	X	X	X	X	X	
150						X	X	X	X	X	X	X	
175								X	X	X	X	X	
200									X	X	X	X	
250											X	X	X
300													X

[a]Proposed for international standardization (ISO 3179): a, preferred; b, nonpreferred thickness. Preferred lengths vary from 1.5 to 6.3 m at intervals of 0.3 m, or 1.5–6.5 m at intervals of 0.5; the interval 0.5 m is not preferred. The dimensions apply to a moisture content of 20%. For higher moisture contents, the dimensions are 1% greater for every 5% increase of moisture. Similar corrections are made for moisture contents lower than 20%.

Table 17-2A. Dimensions of Hardwood Lumber[a]

Thickness (mm)	\multicolumn{11}{c}{Width (mm)}										
	50	63	75	100	125	150	175	200	225	250	300
19			X	X	X	X	X				
25	X	X	X	X	X	X	X	X	X	X	X
32			X	X	X	X	X	X	X	X	X
38			X	X	X	X	X	X	X	X	X
50				X	X	X	X	X	X	X	X
63						X	X	X	X	X	X
75						X	X	X	X	X	X
100						X	X	X	X	X	X

[a]Based on 15% moisture content.

certain extent, by log taper and wane allowance in lumber. The effect of side b is shown in the following formula (21):

$$b = x(t + o) + (x - 1)k \quad (17\text{-}2)$$

where

x = number of lumber boards
t = board thickness in air-dry condition (cm, in.)
o = oversize of t due to shrinkage (%)[10]
k = kerf, i.e., width of cut made by a saw blade, or blade thickness + 2 × tooth "setting" (mm, in.)

YIELD

The volume of round wood input in a sawmill is transformed to lumber in a proportion that varies from about 30 to 70% (7, 14, 45, 49). The rest is changed to sawdust, slabs, trimmings, or chips.[11] The importance of a high yield of lumber is obvious considering the price

Table 17-3. Nominal, Actual, and Green Target Sizes of Lumber[a]

	Sizes		Allowances			
	Nominal	Actual	Shrinkage	Variation	Planing	Target size
Thickness	1.000 (25.40)	0.750 (19.05)	0.025 (0.64)	0.063 (1.60)	0.078 (1.98)	0.916 (23.27)
	2.000 (50.80)	1.500 (38.10)	0.048 (1.22)	0.063 (1.60)	0.078 (1.98)	1.689 (42.90)
Width	4.000 (101.60)	3.500 (88.90)	0.109 (2.77)	0.063 (1.60)	0.078 (1.98)	3.750 (95.25)
	8.000 (203.20)	7.250 (184.15)	0.224 (5.69)	0.063 (1.60)	0.078 (1.98)	1.614 (193.40)
	12.000 (304.80)	11.250 (285.75)	0.346 (8.79)	0.063 (1.60)	0.078 (1.98)	11.736 (298.09)

Source: Sawmill Improvement Study, Project No. 03-108(3), 1985.
[a] A North American (U.S.) example. Numbers are given in inches (equivalent centimeters in parentheses, 1 in. = 2.5 cm). Note the differences between nominal and actual thicknesses and widths. (The lumber is sold at nominal sizes.) Air-dry actual sizes are at 18% moisture content.
Machines of sawmill: band headsaw–horizontal band resaw–frame (sash gang) saw–multisaw edger.

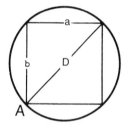 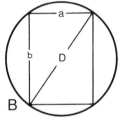

Figure 17-19. Inscription of (A) orthogonal parallelogram and (B) square on the small end of logs in cant sawing (see text).

differential between lumber and residues or chips.

The yield is calculated on the basis of log and lumber volume according to the following relationship:

$$Y\% = \frac{\text{lumber m}^3}{\text{log m}^3} \times 100 \quad (17\text{-}3)$$

The volume of lumber produced from a certain volume of logs is affected by several factors related to wood (log diameter, length, taper, defects, diameter grouping in cant sawing by frame saw), machines (kerf, condition and maintenance of equipment, sawing variation), sawing pattern (lumber dimensions,[12] number of sawlines), depth of planing, and abilities, training, and experience of machine operators.

Larger log diameters give higher yields (Figures 17-21 and 17-22). In sawing with a frame saw, the range of log diameters included in a group is important: more inclusive groups reduce the yield, because a sawing plan is based on a certain diameter. The yield decreases with log length and taper (37) (Figure 17-22). In sawing with a single blade (conventional band or circular saws), the position of the first sawline has a decisive influence on yield (the concept is called "best opening face" or BOF) (16, 17). Other effects on yield are presented in Figure 17-23 in reference to sawing pattern[13] and edging.[14]

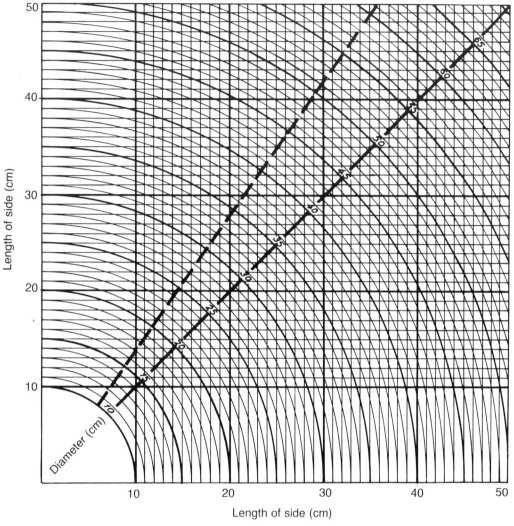

Figure 17-20. Diagram for graphic determination of the length of sides of "inscriptions" (see Figure 17-19). For example, a log 30 cm in small-end diameter, may produce a 21 cm square, or a 20 × 22 cm prism, etc. The diagonal line determines the sides of maximum squares for each particular diameter, and the interrupted line orthogonal parallelograms with 5:7 ratio of sides.

QUALITY OF LUMBER

The quality of lumber depends primarily on the quality of logs, but it may be influenced, within limits, by sawing. This possibility exists especially when the head saw is band or circular—because, with these machines, lumber dimensions (thickness, width) and direction of sawing (by turning the log) may be selected during sawing. Quality may also be influenced by sawing with a frame saw, by properly positioning the log so that defects, such as knots or reaction wood, are distributed to as few lumber boards as possible (Figure 17-24)[15].

In addition to defects, the condition of the surface of lumber constitutes a quality feature. Lumber with rough surfaces makes a bad impression to customers, creates more waste in planing, and may favor fungi and insects by providing "nests" for spores and eggs. Rough surfaces result from unequal setting of teeth, improper tension of blades, high feeding speed, or defective machines.

Some remarks regarding surface quality are

256   III / UTILIZATION

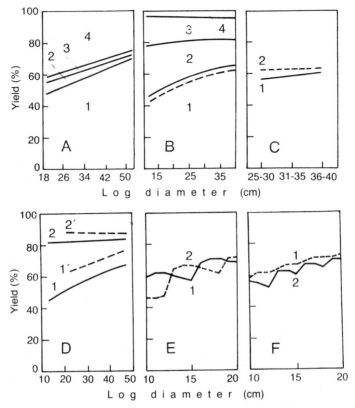

Figure 17-21. Lumber yield. (a) Fir logs (length 4 m, some 3 m, diameter 16-52 cm), frame (cant) sawing variable thickness boards (20-100 cm). Actual sawing 695 logs: 1, lumber; 2, lumber, short (<2 m); 3, batens; 4, slabs, trimmings, sawdust, shrinkage. (B) Theoretical calculation (sawing softwood): 1, lumber (--- approximate actual yield); 2, residues (slabs, trimmings); 3, sawdust; 4, shrinkage. (C) 1, live sawing; 2, cant sawing 180 fir logs. (D) Theoretical (computer) calculation (southern pine, USA): 1, 2, half-taper symmetric sawing (1, lumber; 2, chips), and 1', 2', full-taper asymmetric sawing (1', lumber; 2', chips); the rest is sawdust. (E) Theoretical (computer) calculation, sawing small-diameter oak logs (10-20 cm): 1, sawing through the pith; 2, pith enclosed in the central board. (F) As E above: 1, best opening face; 2, worst opening face. (A, after Ref. 44; B, Ref. 32; C, Ref. 36; D, Ref. 48; and E, F, Ref. 49).

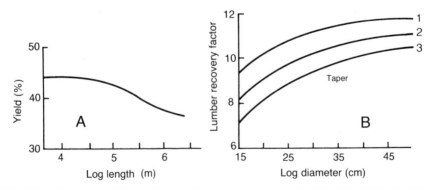

Figure 17-22. Effect of log length (A), log diameter (B), and taper on yield: 1, 2.5 cm (1 in.); 2, 7.5 cm (3 in.); 3, 12.5 cm (5 in.) taper per 5 m (16 ft) log length. The reduction of yield with increasing log length is due to increasing taper. (Data based on computer calculations.) Note that "Lumber recovery factor" is a measure of yield used in the United States; it is an expression of "board feet" of lumber obtained from a given volume of logs (1 bd ft = 12 × 12 × 1 in. = 30 × 30 × 2.5 cm). Calculations are based on nominal dimensions (see text). The maximum value of the lumber recovery factor is 12 or more, and the corresponding yield may reach 100%. (After Ref. 37.)

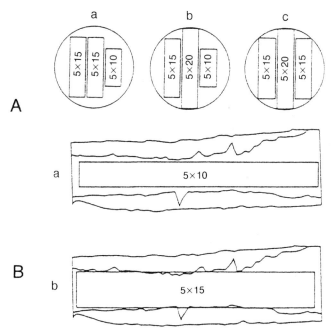

Figure 17-23. Effect of sawing patterns on lumber yield. (A) The yield of pattern (b) is 12.5% higher and that of pattern (c) 25% higher than that of (a). (B) The yield of edging solution (b) is 50% higher than that of (a). Note that logs are 20 cm (8 in.) in diameter and 3.60 m (12 ft) in length. Cross-sectional dimensions of the lumber are in cm; corresponding inches are as follows: 5×10 cm = 2×4 in., 5×15 cm = 2×6 in., 5×20 cm = 2×8 in. (After Ref. 37.)

as follows: quality may be improved by increasing the speed of saw blades; increasing the speed of feeding reduces tooth sharpness by wearing, with the result that the surfaces produced are substantially rougher; the closer the spacing of teeth, the smoother the surface; roughness increases with increased tooth setting; spring-setting produces a better surface than swage-setting; sharpening of all teeth and their alignment are prerequisites to satisfactory

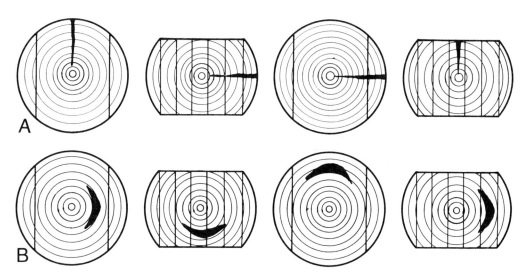

Figure 17-24. Method of sawing to reduce defects in lumber: (A) a split and (B) compression or tension wood. Defects may be confined in one board.

quality; teeth that are not sharpened do not cut, but tear, the wood fibers; tooth angles should be proper in each case; surface quality depends on wood species, and is better in denser, harder, and drier woods (10, 26).

The quality of a lumber board[16] is also affected by the manner of sawing. Quarter-sawn lumber shrinks and swells less in width, warps less, wears more evenly, and in certain species has a more attractive figure due to the presence of ray flecks, which are preferred in oak parquet flooring, for example. On the other hand, flat-sawn lumber shrinks and swells less in thickness, but lumber of large width has a tendency to warp; knots appear smaller in size (round or oval) than the "spike" (longitudinally cut) knots of quarter-sawn lumber; shakes and resin pockets extend through fewer lumber boards; and figure may be more attractive in certain species having a pronounced contrast between earlywood and latewood. Flat-sawn lumber is more easily produced by regular sawing, whereas quarter-sawn lumber is obtained by certain sawing patterns (Figure 17-9), adding to the production cost.

Sawing parallel to bark or to pith may affect the quality of lumber, depending on taper and juvenile wood. Sawing parallel to pith will result in a more uniform distribution of grain deviations due to taper; whereas in sawing through the pith the central lumber boards will have one-sided presence of juvenile wood (see Chapter 6). Lumber including grain deviations and juvenile wood tends to warp.

LUMBER GRADING

Grading[17] is based on certain rules that take into consideration defects, such as knots (intergrown, loose), resin pockets, compression or tension wood, checks, spiral grain, discolorations, and wane (defective dimensions). The rules vary in different species and countries, and the importance of defects is weighed differently in softwoods and in hardwoods.

Softwood grading is based on the number and size of defects ("defect system") (34, 53). Dimensions are also important (see Table 17-4). Further processing to final products is not taken into consideration, although the removal of edge defects could substantially change the grade.

Hardwoods are seldom graded in that manner. Their grading is usually based on the proportion of clear wood that could be produced from a lumber board, in combination with the number and size of useful "units" (34, 47, 53). According to this system (the "cutting system"), a high proportion of clear surface and a small number of clear units denote a better grade. The units are orthogonal, and are produced by cuts that are transverse and/or parallel. They must exceed the minimum possible size[18] that can be marketed, and it is sufficient that one surface is clear (34) (Figure 17-25).

The grades of softwood lumber vary in different species and countries (28). For example, in Germany there are five grades of fir and spruce and two grades of Scots pine (23); in Austria there are also five grades of fir and spruce and four of pine (52) (see Table 17-4); and in Britain there are five grades for softwoods (9). In Scandinavian countries, six grades are recognized for Scots pine and spruce, but usually grades I–IV are not marketed separately—they are offered as one "unsorted" grade (9, 43). The same applies to lumber of Eastern Canada and the U.S.S.R. ("unsorted" grades I–III) (9); whereas lumber produced in northwest North America (U.S., Canada) is classified into seven grades (34, 47).

Hardwood lumber is classified into seven grades in the United States, four in Britain (9), five in France, and four in Germany (2, 3) (three in beech, DIN 68369) or there is no distinction of grades (27).[19]

In general, grading rules are subjective and complicated, so that specialization is required for their practical application. Difficulties also arise because there are no internationally accepted rules. Occasionally, different criteria are applied in different regions of the same country (e.g., the U.S.) (47).[20]

HANDLING THE LUMBER

The lumber is transported outside the sawmill by various conveying mechanisms (rollers, belts, etc.). It is piled in a lumber yard to air-dry or is taken to dry kilns. Exit from the mill

Table 17-4. Grading Specifications for Lumber of Fir, Spruce, and Pine[a,b]

Fir, Spruce

Grade 0

Wood clean, without discolorations and compression wood

Allowed: Per running meter (rm) a tight knot up to 2 × 5 cm or 1 resin pocket up to 5 × 0.5 cm, small checks, small wane, and deflection from level (warping) up to 2 cm per rm

Grade I

Without compression wood

Allowed: Per rm 1 small loose knot (tight knots up to 2 × 5 cm unlimited), light discoloration in patches, small resin pockets, small checks, small wane, and deflection up to 2 cm per rm

Grade II

Without insect galleries

Allowed: Per rm 2 small loose knots (tight knots up to 2 × 5 cm unlimited), slight discoloration in patches, small resin pockets, small checks, small wane

Grade III

Allowed: Few loose knots average in size, sound knots unlimited, discoloration up to 40% of the surface, a small number of resin pockets average in size, checks and wane average in size, limited insect attack

Grade IV

Allowed: All defects as long as the wood is usable

Pine

Grade I

Without radial (diametric) checks
Allowed: Few, small sound knots, limited blue stain, few small resin pockets, few small checks and small wane

Grade II

Allowed: Small, sound, and few loose knots, extensive blue stain, unlimited resin pockets, few straight-end checks (splits) not longer than the width of the piece, small wane

Grades III and IV

As in fir and spruce

[a] An example of appearance grading specifications for European white (silver) fir, European spruce, and Scots pine (Austrian rules, 52). The lumber should be 8 cm (about 3 in.) or more wide and 3(4)–6 m (10–20 ft) long. Other specifications are given in specialized publications. For example, for Swedish grading rules, see Ref. 42; for U.S., Ref. 46.

[b] *Knots*: In fir and spruce, knots up to 0.5 cm in diameter are not taken into consideration. Small knots are those that have a diameter up to 2 cm, average 4 cm, and larger than 4 cm. In pine, knots up to 1 cm are not taken into consideration, small are up to 3 cm, and average larger than 3 cm. In general, the minimum diameter is measured. Maximum diameter should not be larger than 4 times the minimum. Knots that are hard, black, or black in their border are considered sound when intergrown at least over half their circumference. *Resin pockets*: Pockets up to 2 × 0.2 cm are not taken into consideration even if found on wane. Small are 5 × 0.5 cm in size, average 10 × 1 cm, and large when they are larger. *Checks*: Fine checks are not taken into consideration. Small checks should not be longer than the width of the piece, and they should be neither diagonal nor extending through the full thickness of the piece (unless found at the ends). Average checks should not be longer than 1.5 times the width. Large extend through the thickness; they are diagonal and may derive from ring shakes. *Wane*: Wane is bark or lack of wood on the edge of a piece. Small wane is $\frac{1}{4}$ of thickness or up to $\frac{1}{4}$ of the length of a piece. Average up to $\frac{1}{2}$ of width or length. Large should not be wider than $\frac{1}{2}$ the width of the piece. In the worst case, the lumber should be at least "touched" by the saw on all four edges. In general, wane is measured laterally, and the worst situation is taken into consideration.

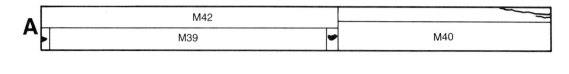
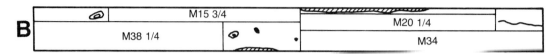
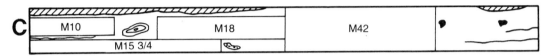

Figure 17-25. Examples of grading hardwood lumber (U.S., board dimensions 3.60 × 0.3 m = 12 ft × 12 in). (A) Grade 1-2 (FAS), (B) grade No. 1 common, and (C) grade No. 2 common (M = units, see text). Note that, in total, seven grades are distinguished: firsts, seconds, selects, No. 1 common, No. 2 common, No. 3A common, and No. 3B common. Firsts and seconds (1-2) are usually separated. In selects, the face is selects and the back surface No. 1 common, but shorter dimensions are permitted. In grades 6 and 7 (Nos. 3A and 3B common), all defects are permitted. Grading is complicated and is based on combinations of defects and dimensions. (Courtesy of U.S. National Hardwood Lumber Association.)

is combined with classification according to species, dimensions, and grade. Lumber to be shipped green should be immersed in a preservative solution for protection against fungi or insects.

REFERENCES

1. AFNOR (Association Francaise de Normalisation). 1947. *Classement d'Aspect des Bois Indigènes (Sciages avivés de feuillus durs)*, NF B53-501.
2. Autorenkollektiv. 1971. *Sägewerkstechnik*. Leipzig: VEB Fachbuchverlag.
3. Autorenkollektiv. 1975. *Werkstoffe aus Holz*. Leipzig: VEB Fachbuchverlag.
4. Boyd, C. W., et al. 1976. Wood for structural and architectural purposes. *Wood Fiber* 8(1):3-84.
5. Cassens, D., and R. Maeglin. 1987. Live-sawing log-grade red oak logs. *For. Prod. J.* 37(10):49-53.
6. Dalois, C. 1977. *Manuel de Sciage et d' Affutage*. France: Centre Technique Forestier Tropical.
7. Dobie, J., and D. M. Wright. 1975. Conversion Factors for the Forest Products Industry in Western Canada. Inform. Rep. VP-X-97. Canad. For. Serv., W. For. Prod. Lab., Vancouver, British Columbia.
8. Esterer, M., and M. Blasy. 1969. Das moderne Sägewerk. *Holz Roh- Werkst.* 27(4):125-147.
9. Findlay, W. P. K. 1975. *Timber Properties and Uses*. London, New York: Granada Publ.
10. Forintek Canada Corp. 1985. Overview of saw technology research at Forintek (Bull. July-August, pp. 8-9); 1986. Infeed systems for Chip-N-Saws (Bull. February, pp. 4-5).
11. Fraser, H. R. 1976. New Ivory Coast mill saws 33,000 m³/yr. *World Wood* 17(9):9-10.
12. Fronius, K. 1965. *Die Arbeiten am Gatter und an anderen Sägewerks-maschinen*. Stuttgart: Holz-Zentr. Verlags.
13. Fronius, K. 1981. Profiling technique. *Holz-Zentralblatt* 63/64:1031-1032; 1041-1042.
14. Giordano, G. 1974. *Tecnologia del Legno*, Vol. 2. Torino: Unione Tipografico-Editrice.
15. Hallock. H. 1974. Automation of decisionmaking in sawmills. In *Economics in Sawmills and Woodworking Industries*, pp. 301-314. Sweden: Proceedings IUFRO Meeting.
16. Hallock, H., and D. W. Lewis. 1971. Increasing Softwood Dimension Yield from Small Logs—Best Opening Face. U.S. For. Prod. Lab. Res. Paper FPL-166.
17. Hallock, H., A. R. Stern, and D. W. Lewis. 1976. Is There a "Best" Sawing Method? U.S. For. Prod. Lab. Res. Paper FPL-280.
18. Hartman, L. A. 1975. North American and Scandinavian sawmilling practices and equipment—main characteristics and differences. In *Nordic and North American Sawmill Techniques*, Proceedings of the International Sawmill Seminar (Jonkoping), pp. 9-26. San Francisco: Miller Freeman.
19. Highley, T. L. 1974. Personal communication.
20. Huber, H. A. 1987. Laser wood cutting. *Proceedings IUFRO World For. Congr. Div. 5*, pp. 165-167. Ljubljana, Yugoslavia.
21. Ingenieurschule f. Holztechnik, Dresden. 1965. *Taschenbuch der Holztechnologie*. Leipzig: VEB Fachbuchverlag.

22. International Organization for Standardization, Switzerland. I.S.O. Recommendation R.738, I.S.O. Standards 3179, 1032.
23. Knigge, W., and H. Schulz. 1966. *Grundriss der Forstbenutzung.* Hamburg: P. Parey.
24. Knigge, W. 1972. Eigenschaftsänderungen des beregneten und wassergelagerten Kiefer-Sturmholzes. *Forstarchiv* 44(3):62–64.
25. Koch, P. 1964. *Wood Machining Processes.* New York: The Ronald Press.
26. Kollmann, F. F. P., and W. A. Côté. 1968. *Principles of Wood Science and Technology. I. Solid Wood.* Berlin: Springer Verlag.
27. König, E. 1970. *Sortierung und Pflege von Rund- und Schnittholz.* Stuttgart: DRW-Verlags.
28. Leight, J. G. 1971. *The Timber Trade—An Introduction to Commercial Aspects.* Oxford, New York: Pergamon Press.
29. Lustrum, S. J. 1972. Circular Sawmills. USDA, Forest Service, State and Private Forestry, SE Area, Atlanta, Georgia.
30. Lutz, J. F., H. H. Haskell, and R. McAlister. 1962. Slicewood—A promising new wood product. *For. Prod. J.* 12(4):218–227.
31. Mills, J. 1976. Survey of portable sawmills. *World Wood* 17(4):31–35,41.
32. Montague, D. E. 1971. Band and Circular Sawmills for Softwoods. For. Prod. Res. Bull. No. 55, HMSO, London.
33. Mugge, H. F. 1970. Das moderne Sägewerk. 2. Fertigungsplanung und Fördetechnik. *Holz Roh- Werkst.* 28(12):453–470.
34. National Hardwood Lumber Association (USA). *An Introduction to the Grading and Measurement of Hardwood Lumber.* Chicago.
35. Peters, C. C., and H. C. Marchall. 1975. Cutting Wood Materials by Laser. U.S. For. Prod. Lab. Report FPL 250.
36. Philippou, J., and J. Barboutis. 1981. Effect of sawing method on lumber yield of *Abies* sp. sawlogs. *Ann. School Agr. For.* Aristotelian University 24:329–346 (Greek, English summary).
37. Steele, P. H. 1984. Factors Determining Lumber Recovery in Sawmilling. U.S. For. Prod. Lab. Gen. Res. Paper FPL 280.
38. Szymani, R., and F. E. Dickinson. 1975. Recent developments in wood machining processes: Novel cutting techniques. *Wood Sci. Technol.* 9(2):113–128.
39. Thunell, B. 1974. Neuzeitliche Bandsägestrassen für die Schnittholzerzeugung in Schweden. *Holz Roh-Werkst.* 32(8):289–294.
40. Thunell, B. 1977. Neuzeitliche Schwedische Sägewerkstechnik mit Kreissägemaschinen. *Holz Roh-Werkst.* 35:461–466.
41. Thunell, B. 1980. Processing Systems. Royal Inst. Tech., Wood Tech. Proc. Rep. 1, Stockholm.
42. Thunell, B. 1982. Technisch-wirtschaftlicher Vergleich zwischen Gatter-, Kreis-, Bandsäge- und Spannersägewerk. Royal Inst. Technology, Rapport 24, Stockholm.
43. Timber Grading Committee of 1958, 1962. Guiding Principles for Grading of Swedish Sawn Timber Redwood and Whitewood. AB Ragnar Lag. Botk. Karlshamn.
44. Tsoumis, G. 1961. Grading fir logs and lumber in the sawmill of the University sawmill at Pertouli. *Ann. School Agr. For.* 6:103–134 Aristotelian University (Greek, English summary).
45. Tsoumis, G. 1973. The sawmill of the University forest at Pertouli. (A technical and economic study for its reoperation). *Ann. School Agr. For.* 15:3–56 Aristotelian University (Greek, English summary).
46. Tsoumis, G. 1991. *Harvesting Forest Products.* (In press).
47. U.S.D.A., Forest Products Laboratory. 1987. *Wood Handbook.* Ag. Handbook No. 72 (revised).
48. U.S.D.A., Forest Service. 1973. Study of Softwood Sawlog Conversion Efficiency and the Timber Supply Problem (I, II). H. C. Mason & Assoc., Inc., Gladstone, Oregon.
49. Vassiliou, V. 1987. Utilization of Small-Dimension Oak Wood for Sawn Products. Dissertation, Dept. For. Natur. Envirn., Aristotelian University Thessaloniki (Greek, English summary).
50. Vypfel, K. 1969. Mechanische Entrindung. *Holzforsch. Holzverwert.* 21(3):52.
51. Warren, W. G. 1973. How to Calculate Target Thickness for Green Lumber. Canad. For. Serv. W. For. Prod. Lab. Inform. Report VP-X-112.
52. Wiener Börsekammer. 1973. *Oesterreichische Holzhandelusancen.* Vienna: Oesterr. Agrarverlag.
53. Willis, W. E. 1970. *Timber—From Forest to Consumer.* London: E. Benn Ltd.
54. Williston, E. M. 1976. *Lumber Manufacturing.* San Francisco: Miller Freeman.

FOOTNOTES

[1] Frame saws are also called "sash gang saws" or simple "gang saws." In North America, the term "gang saw" is applied to multiple circular saws (mounted on a common axle or "arbor"), or multiple band saws; multiple circular saws are also called "rotary gang saws." Historically, frame saws appeared in the thirteenth century, and had a wooden frame with one blade, and later (sixteenth century) more blades; they were powered by water or wind. Circular saws were introduced by the end of the eighteenth century, and band saws appeared first in the beginning of the nineteenth century and attained their present form during the second half of that century (12, 15). Founding of such "sawmills" caused, in some cases, a violent reaction on the part of hand sawers.

[2] In new installations, band and circular sawing machines are equipped with multiple blades of bands or disks (see Figure 17-5), and production of lumber is assisted by incorporated chipping machines, called chippers, which simultaneously reduce slabs (and edgings) to chips. Such arrangements ("Chip-N-Saw" machines) are mainly used with relatively small diameter logs, and dominate modern

sawmills in the United States, Canada, and Scandinavia (13, 18, 42). In this chapter, sawing with band or circular saws refers to single-blade machines unless otherwise specified.

³The requirement for log classification may be overcome if all lumber is of equal thickness; mechanical or hydraulic increase or decrease of the distance of the central blades is also sometimes applied (12). However, lumber usually varies in thickness according to market demands, and such arrangements may not satisfy the maximum yield requirement.

⁴The trend is toward a combination of machines in a sawmill—not purely frame, band, or circular saws (see under "Methods of Sawing").

⁵Occasionally, energy is provided by gasoline-operated engines [e.g., in mobile saws (27, 29), see Figure 17-4], or by burning residues (sawdust, edgings, trimmings, bark) to produce steam which, in turn, is utilized as a source of mechanical or (by transformation) electrical energy. (The latter is rare, where no electricity is available, while the main use of steam is to heat kilns by heat radiation from steam coils.) In older times, as previously mentioned, energy was provided by wind or falling water.

⁶$\frac{1}{16}$ in. = 1.5625 mm; $\frac{1}{32}$ in. = 0.78125 mm.

⁷Bacteria may be beneficial; they attack pit membranes and thus increase the permeability of wood, assisting the penetration of preservatives (see Chapter 15).

⁸Table 17-1 shows a proposed international standardization of lumber dimensions. Lumber sizes differ in different countries, but reference to such information is impractical in this book. For U.S. practice—where specifications are quite complex—see, for example, the *Wood Handbook* (47).

⁹According to a recommendation by the International Standardization Organization (R 738), the following deviations are permitted due to sawing "imperfections": ±1 mm for lumber up to 29 mm thick, ±2 mm for 30–105 mm in thickness or width, and ±3 mm above 105 mm in thickness or width. Length deviations are ±50 mm and −25 mm (22).

¹⁰Oversize due to shrinkage may be calculated on the basis of shrinkage values given in Table 8-1, as an average of radial and tangential shrinkage corresponding to desired level of drying. For example, for spruce lumber to be dried down to 15%, oversize is 2.85% (radial shrinkage 3.6%, tangential 7.8%, average 5.7%, and half of it 2.85%), considering that total shrinkage takes place when wood dries from about 30% (fiber saturation point) to zero (see Chapter 10). The average (2.85) applied to Eq. (17-2), gives the following answer to required side length b (width of cant; Figure 17-19) for production of 6 lumber boards, 2.5 cm (1 in.) thick, and kerf of 2.5 mm (0.1 in.):

$$b = 6(2.5 + 2.5 \times 0.0285) + (6 - 1) \times 0.25$$
$$= 15.42 + 1.25 = 16.7 \text{ cm } (6.7 \text{ in.})$$

¹¹In Canada, it has been estimated that residues are 10% sawdust, 35% residues suitable for chips, and 10% planing residues. In the United States, the following figures have been reported: for softwoods 35% planed ("dressed") lumber, 29% pulp chips, 15% planer shavings and end trim, and 21% fuel (sawdust, bark); for hardwoods 28% planed lumber, 29% pulp chips, 20% planer shavings and end trim, and 23% fuel (sawdust, bark). In both (U.S.) cases, the logs are considered unbarked, the proportion of bark is 10%, and all percentages refer to oven-dry weight (not volume) of wood (4).

¹²There is a difference between nominal and actual thickness. For example, in North America, lumber of 1 in. nominal (marketed) thickness is only $\frac{3}{4}$ in. and that of 2 in. only 1.5 in. thick—in air-dry and planed condition. Thickness (and width) is reduced on account of shrinkage, sawing variation, and planing (51) (see Table 17-3). In Europe, the reduction is less, and the lumber is usually sold nonplaned.

¹³In a study of sawing 180 fir logs, 25–40 cm (10–16 in.) in small-end diameter, the yield was higher in cant sawing as compared to live sawing; respective yields were on the average 62.7% and 59.1%, ranging from 62.3 to 63.2% and 56.5 to 61.1%, with increasing diameter (36) (see Figure 17-21).

¹⁴In order to avoid loss of wood by reduction to sawdust, other methods have been proposed—namely, cutting with knife, water, laser beams, and vibrating blades. Knife-cutting is applied in making veneer (see Chapter 20), but it has also been used to produce thin boards of lumber form (slicewood). Cutting with water has been used experimentally. A fine water filament, 0.1–0.25 cm (0.04–0.1 in.) in diameter, applied under high pressure (up to 300 N/mm² or 43,500 psi), may "saw" wood up to a depth of 30 cm (12 in.); the depth is reduced with increasing speed of feeding the wood—for example, feed 8 m (25 ft)/min depth 1 cm (0.4 in.), and feed 2 m (6 ft)/min depth 2 cm (0.8 in.); "sawing" green wood is easier than dry wood. Use of laser beams is presently rather theoretical, because there are problems due to superficial carbonization of the wood and high cost of equipment. (20, 30, 35, 38).

¹⁵A comparison of cant sawing, live sawing, and grade sawing 66 low-grade red and black oak logs, and grading for maximum value after edging, showed that live sawing produced more value than did cant sawing, and cant sawing more than grade sawing. Live sawing produced more value per unit time than did cant sawing, and cant sawing more than grade sawing (5).

¹⁶The term "board" refers to lumber less than 2 in. (5 cm) in nominal thickness.

¹⁷Appearance grading; see Chapter 11 about grading for structural purposes.

¹⁸In the United States the minimum size of a cutting is 4 in. × 5 ft or 3 in. × 7 ft (10 cm × 1.5 m or 7.5 cm × 2.10 m) for the best grade, and 1.5 in. × 2 ft (2.5 cm × 60 cm) for the worst. The clean surface of a lumber board varies from $\frac{11}{12}$ to $\frac{3}{12}$, respectively (47). Similar rules are applied in France (1), and in Britain—where products, such as railroad ties, are graded according to the manner and

size of defects ("defect system") as in grading softwoods (9).

[19]The wood is required to be healthy. If serious defects are present (decayed knots, fungal attack or discoloration in patches, checks, insect galleries), both the length and width of lumber are reduced by the defective part. Straight checks are not taken into consideration, but lumber with strong deviations is rejected. Hardwood lumber is usually offered unedged in the market (27).

[20]In the United States, grading is complex. Softwood lumber is generally classified into two major categories—construction and remanufacture. Construction includes three categories (stress-graded, nonstress-graded, appearance); and remanufacture includes four categories (factory or shop, industrial clears, molding-ladder-pencil, etc., and lumber for structural laminations). Hardwood lumber includes three basic categories—factory lumber, dimension parts, and finished market products. Factory lumber is graded on the basis of clear material (on one side)—that is, on the basis of clear cuttings or units; there are seven standard hardwood cutting grades. In both cases, in addition to defects, intended use is a grading factor, and grading rules (specifications) exist for different species and regions (West Coast, Southern pine, redwood, etc.). Lumber for structural purposes is mainly softwood (47).

18

Drying

Lumber and other wood products usually contain considerable quantities of moisture immediately after their production. Irregular exit of this moisture will cause defects (checking, warping, etc.); and, if moisture is kept above a certain level, the wood is subject to attack by fungi. For these and other reasons mentioned below, proper drying of wood is necessary.

Why Wood is Dried

Wood should be dried properly (i.e., gradually, uniformly, and to a certain moisture level depending on its intended use). Such drying has important advantages:

1. Shrinkage in use is reduced, and warping and checking are avoided.
2. The wood is protected from attack by stain and decay fungi.
3. The weight of wood is reduced and, therefore, the cost of transportation is lower.
4. Drying results in higher strength—assuming that no defects develop, especially checking. Also, the nail-holding capacity of wood increases.
5. Satisfactory painting, finishing, and (usually) preservative treatment require air- or kiln-dried wood.
6. The high temperatures of kiln-drying kill fungi and insects that may dwell in wood.

Drying Factors

The factors that influence drying are *heat, relative humidity*, and *air circulation*. Heat is needed to evaporate moisture. The higher the air temperature, the faster the rate of moisture exit from the interior of wood to its surface. *Relative humidity* determines the drying capacity of the air. Drier air (lower relative humidity) has a higher drying capacity and can hold more moisture in the form of water vapor. Drying capacity is considerably affected by temperature: the warmer the air, the higher the drying capacity, because a rise of temperature causes a fall in relative humidity. Thus, by controlling relative humidity, it is possible to control the rate of moisture exit and, therefore, the magnitude of stresses that develop in wood due to shrinkage. As a result, defects of wood during drying may be avoided or reduced. *Air circulation* is needed to transport heat to wood, and to remove moisture from its surface. Air circulation affects the drying rate.

Air-Drying and Kiln-Drying

Wood may be dried out-of-doors or in a kiln. Air-drying is usually done in the open air and seldom under shelter; whereas kiln-drying requires special instrumentation to create an artificial climate in a closed space, where temperature, relative humidity, and air circulation can be controlled. In air-drying, the possibility of control is very limited or nonexistent.

In spite of the above, air-drying is not inferior to kiln-drying with regard to product quality (56). However, air-drying requires more time, during which capital is immobilized, and there is a prolonged danger of degradation or loss (e.g., by fire). In addition, air-drying may not be sufficient for certain uses (furniture, flooring, etc.), because the moisture that wood finally attains is dependent on climatic conditions.

Kiln-drying is much faster, and moisture may be reduced or raised to any desired level, independent of local climatic conditions. However, it is more expensive to build a kiln, and the danger of developing defects, or even rendering the wood useless due to improper drying procedure, is much greater in comparison to air-drying.

AIR-DRYING

Air-drying requires a suitable yard and proper piling of lumber.[1]

Selection of Yard

Criteria for selecting a yard are its size and position. The size should be sufficient for current and future needs—storage of lumber, movement of machines (forklifts, trucks, etc.), and construction of buildings and perhaps dry kilns. The yard should be on flat ground, well aerated (not surrounded by hills, trees, or buildings), and drained. If sheltered from winds, the ground remains moist, drying is delayed, and development of fungi is favored. The surface of the yard should be kept clean, without grass or other vegetation that restricts the movement of air under the lumber piles, and without wood residues that favor development of fungi and insects. If the ground is muddy, the yard should be properly surfaced.

Piling

The lumber is placed in piles that are properly constructed and arranged.

Arrangement of Piles. The piles are placed in parallel rows; alleys are left in-between, which serve piling and unpiling the lumber and circulation of the air, while simultaneously providing access for fire-fighting equipment. Alleys separate the lumber into units, and each unit contains a number of piles. The width of alleys varies depending on intended use, manner of piling (manual,[2] mechanical), and kind of piling equipment (forklifts with front or side loading; see Figure 18-1). There are main, secondary, and intermediate alleys. Main alleys serve lumber piling and passage of trucks or fire-fighting equipment; secondary alleys facilitate the movement of piling machines from one alley to another; intermediate alleys help to approach a pile for various reasons (e.g., place or remove drying samples). Small side distances between piles (20–30 cm or 8–12 in.) (29, 36) may also be considered as alleys in the sense that they facilitate air circulation. In some cases, main and secondary alleys are equipped with rails and the lumber is transported on rolling carriages. The width of alleys varies, depending on their use, but width is also a factor that affects air circulation and rate of drying. With regard to orientation, main alleys should have a north–south direction in order to make better use of sun energy (34). The direction of wind flow is also important; arrangement of piles with their length (length of lumber) parallel to the main direction of wind flow may cause end-checking due to faster evaporation of moisture.

Dimensions of Piles. Pile dimensions affect the rate of drying. Width affects the horizontal movement of the air, and height affects its vertical movement. Air enters horizontally in the direction of wind flow, it is cooled by evaporation of wood moisture, becomes heavier, and drops. Evaporating moisture gradually saturates the air, and the lumber stops drying or dries very slowly. In higher piles, the level of saturation is higher. The horizontal movement of air may be obstructed in various ways—that is, by improper (not parallel) placement of piles, low foundations, and underbrush or wood residues on the ground. Horizontal movement is also more difficult in wider piles. The disadvantages of large piles may be overcome by spacing the boards, and leaving flues (usually funnel-like) in the lower part of piles (Figure 18-2). Pile width is affected by forklift capacity (fork length).

The lumber is usually prepiled at the sawmill exit into "packets," about 1–1.20 m (3–4 ft)

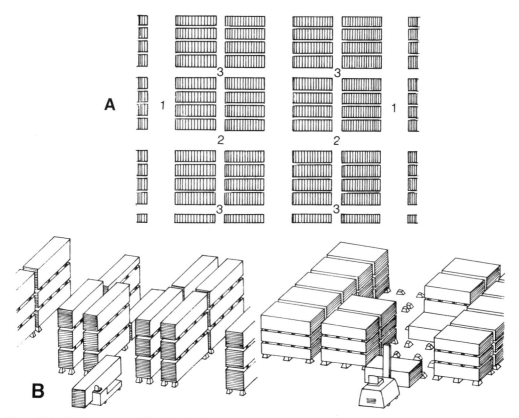

Figure 18-1. (A) Arrangement of lumber piles in an air-drying yard, and (B) side- or front-piling with forklifts: 1, main alleys, 3–4.5 m (10–15 ft); 2, secondary alleys, 2 m (6 ft); and 3, intermediate alleys, 1 m (3 ft). (Adapted from Ref. 19; reproduced by permission from VEB-Fachbuchverlag.)

in height and width, and 3–4 or more such packets are placed on top of one another in the lumber yard (Figures 18-3 and 18-4). The total height of a pile depends on considerations of air movement, stability, and piling and unpiling costs. A 1:3 height-to-width ratio is considered safe with regard to pile stability. Higher piles are recommended for woods that tend to warp in drying (36).

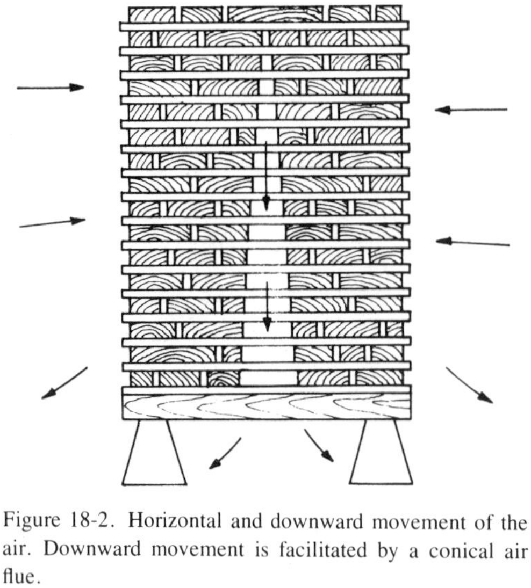

Foundations. Lumber piles are laid down on suitable, strong, and usually movable foundations. These supports are made of concrete or treated timbers, and their height is such that the first row of lumber is about 30–40 cm (1–1.5 ft) above ground. They are placed at distances of 25 cm to 1.50 m (about 1–5 ft), usually 50 cm to 1 m (1.5–3 ft), depending on lumber thickness and wood species (25); shorter distances are recommended for smaller thicknesses and woods tending to warp (e.g., beech and elm). The tops of foundations should be aligned to form an inclined surface, 8–10 cm (3–4 in.) drop per running meter (3 ft), so that excess of water (rain, snow) entering the pile is removed (38, 39).

Figure 18-2. Horizontal and downward movement of the air. Downward movement is facilitated by a conical air flue.

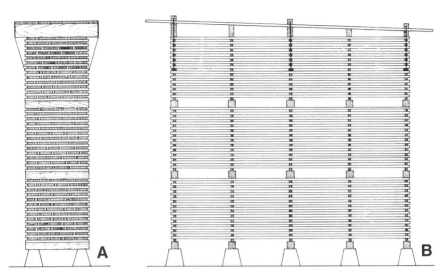

Figure 18-3. A pile made of three small piles (packets). (A) End view, and (B) side view. Also shown: foundations, stickers (in regular vertical arrangement), and roof (supported). (Adapted from U.S. Forest Products Laboratory.)

Figure 18-4. Piling lumber packets with a forklift. (Courtesy of Irion.)

Stickers. The piles are not compact, but incorporate separating stickers between rows of boards. Piling without stickers immediately after sawing, even for short periods and in warm weather, favors development of fungi. Stain fungi may attack within hours or days.

Stickers are prismatic pieces of wood, equal in length to the width of a pile, and rectangular or orthogonal in cross-section, usually 2.5 × 2.5 cm (1 × 1 in.) or 1.5 × 2.5 cm (0.5 × 1 in.).[3] The dimensions should be uniform and the wood air-dry, without defects (knots, compression or tension wood, spiral grain), which may cause warping. Woods susceptible to attack by stain or decay fungi (pine, beech) are not suitable, while oak and other hardwoods containing large amounts of extractives may discolor the drying wood. Metallic stickers (I-beam, etc.) are available.

Cross-section and distance of stickers affect the rate of drying. Softwood lumber will withstand faster drying without waste, and for this reason such lumber may be piled with stickers 2.5 cm (1 in.) thick at any season of the year. In contrast, certain hardwoods, such as oak and beech, develop checks easily when their surfaces dry rapidly, and it is advisable to use 2.5 cm (1 in.) thick stickers in fall or winter piling, and 1–1.5 cm (0.5 in.) thick stickers in spring or summer piling. The same stickers, placed on edge rather than flat, may serve different situations.

The distance between stickers, and their width, are adjusted according to species and thickness of lumber. Fewer stickers are used for softwoods than for hardwoods. In softwoods, the distance between stickers is kept to about 1 m (3 ft); whereas in hardwoods it varies from 50 cm to 1 m (1.5–3 ft), and may be down to 30 cm (1 ft) for relatively thin lumber. The stickers should be placed in vertical rows to support the ends of each board; otherwise, the lumber will warp (see Figure 18-5). Thicker stickers (5–10 cm or 2–4 in. high) are placed between ''packets'' of lumber.

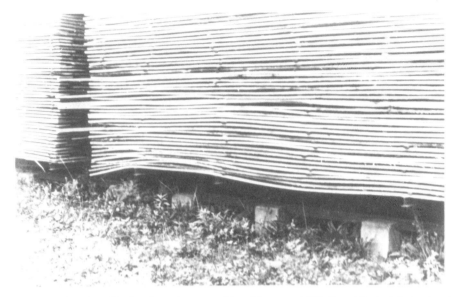

Figure 18-5. Warping of lumber due to improper piling. (Courtesy of U.S. Forest Products Laboratory.)

Figure 18-6. Piling unedged lumber (the original log form is preserved). (Courtesy of Irion.)

Hardwood lumber is sometimes piled in "log form" (Figure 18-6), and sawn wood of short length (parquet flooring, railroad ties, barrel staves, joists) is usually self-piled (i.e., without stickers). Self-piling is unidirectional or cross-wise, and upright (end-racking) or horizontal (Figure 18-7), but the latter two methods offer very limited control of drying.

Grouping, Piling Unequal Lengths, Distance of Boards. Lumber should be piled in groups of similar drying behavior (i.e., separate by species and thickness). Grouping by length, width, growth-ring arrangement (radial, tangential), and lumber grade is also desirable.

Boards of unequal length may cause problems, because protruding ends warp if not supported. To avoid such a defect, a pile is made with its length equal to that of the longest board; shorter boards are placed inward, and empty spaces are left within the pile. All ends are supported, and the pile is prismatic (Figures 18-8 and 18-9).

The boards of each layer may be spaced to facilitate air circulation. The distance is regulated depending on specific conditions (i.e., if slow or fast drying is desired). However, such a procedure is not practical in mechanical piling.

Roof. Each pile is roofed in order to protect the lumber from the direct action of sun, rain, or snow. The roof, made of wood, asbestos-cement, or metal panels, is inclined, longer than the length of the pile (by 30–50 cm or 1–1.5 ft on both ends), and should be supported for protection against strong wind (6) (Figure 18-3). A roof is not needed if the lumber is piled under shelter—in sheds. Sheds have no side walls, but may have partitions with

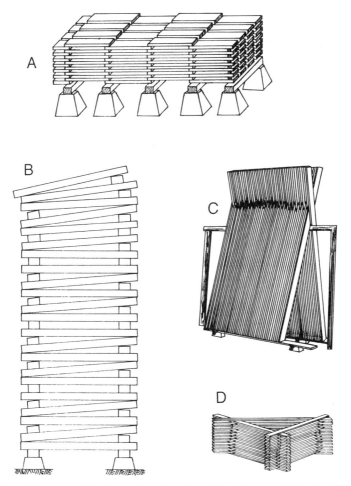

Figure 18-7. Self-piling: (A) parquets, (B) railroad ties, (C) upright piling of lumber, (D) horizontal piling (foundations are also needed in D). (A, C, and D after Ref. 19, reproduced by permission from VEB Fachbuchverlag; B after Ref. 43, reproduced by permission from Springer Verlag.)

openings for air circulation, oriented toward the main direction of wind flow. Sheds are used for piling valuable lumber, or storage of lumber after air-drying; air-dried lumber is piled compact (i.e., without stickers). Drying conditions are milder than in open air and, therefore, drying under shelter may contribute to better quality of lumber with regard to end-checking and discoloration (38, 39).

Protection of Ends. Checking (splitting)[4] of lumber ends is the most common form of degrade during drying. Due to faster evaporation of moisture, shrinkage at the ends is greater, while the rest of the wood is drying and shrinking at a slower pace. Thus, tension stresses develop and the wood may check (see Chapter 10). Checking is associated with the presence of rays which constitute planes of reduced strength.

End-checking may be avoided or reduced by protection from sun or wind. Coatings that slow evaporation of moisture are helpful; they are applied immediately after sawing and before checking starts. Such coatings are paraffin, linseed oil, commercial preparations, or mixtures of ingredients (6); a simple recipe is to mix 100 parts (by weight) of low-quality varnish, 25 parts barium sulfate, and 25 parts magnesium silicate (talc). Coatings are applied cold or hot—by brush, immersion, or spray. Protection by coatings is justified especially for high-value

Figure 18-8. Proper piling (box piling) without protruding ends. The pile is prismatic and its length is equal to that of the longest board. Shorter boards are placed as shown in Figure 18-9. Moisture samples are designated by X. (Courtesy of U.S. Forest Products Laboratory.)

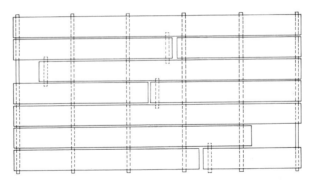

Figure 18-9. Overview of a layer, to avoid protruding ends. The ends of shorter boards are supported separately.

woods and thick lumber (5 cm or 2 in. and up) (38, 39). Protection is also provided by anti-checking irons (S-shaped, etc.), or perforated metallic plates nailed to the ends of thick lumber; their efficiency is lower in comparison to coatings (6).

Rate of Drying

The rate of drying depends on such factors as wood species, thickness, growth-ring arrangement, sapwood or heartwood, manner of piling, conditions in the lumber yard, and climatic factors.

Softwoods and light hardwoods dry faster, but dense hardwoods need more time for their moisture to drop to desired levels.

Thickness is very important; the time required for drying is proportional to the square of thickness (6, 39). This means that lumber 5 cm (2 in.) thick would need four times more time than lumber 2.5 cm (1 in.), assuming the wood is of similar drying behavior. Depending on species and other factors, the time may actually increase 2-4 times or more (39).

Lumber with a tangential growth-ring arrangement on the wide face (flat-sawn) dries faster in comparison to radial, because tangential surfaces expose transversely cut rays, and thus facilitate the exit of moisture.

Sapwood dries faster than heartwood, because anatomical features of the latter (tyloses, pit aspiration, extractives) present obstacles to the exit of moisture.[5]

The effect of piling (dimensions and distance of piles, stickers, orientation of piles, foundations, etc.) and of conditions in the lumber yard (position, ground characteristics) have been previously discussed. Climatic conditions exert a very important influence: drying is much faster during the summer period or dry season (Figure 18-10) and is slowed down under shelter (Figure 18-11).

The rate of drying may cause or reduce defects. Softwoods are less sensitive to faster drying than hardwoods; this property of softwoods is desirable because faster surface drying provides protection from fungal attack (blue stain, molds). Softwoods are also less subject to checking as a result of fast drying.[6]

Control of Moisture

The changing moisture content of wood during air-drying should be known. Thus, measures may be taken to speed or slow the exit of moisture, and when a certain level (below 20%) is reached, drying in piles may discontinue; the lumber is then stored in sheds without stickers. Measurement of the moisture of representative lumber boards is made by use of electric moisture meters, or samples are used and their moisture content is determined by drying and weighing (see Chapter 9). Samples with a known initial moisture content and coated ends are placed in certain positions within each pile, and at intervals they are taken out and weighed. The manner of preparing and handling such samples is described under kiln-drying.

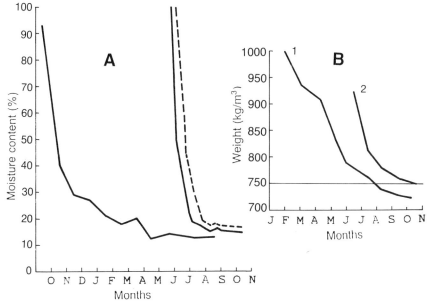

Figure 18-10. Rate of air-drying. (A) Rate of reduction of the moisture of pine *(P. strobus)* lumber piled to air-dry in the autumn (October) and in the summer (July). Continuous line shows lumber 4 × 4 in. (10 × 10 cm) in cross-section, and interrupted line 4 × 6 in. (10 × 15 cm). (B) Rate of reduction of the weight of beech railroad ties piled to air-dry: 1, winter period (February); 2, summer period (July). The horizontal line at the height of 750 kg/m^3 (46.5 lb/ft^3) shows the moisture level suitable for preservative treatment, about 20-25% (100 kg/m^3 = 6.2 lb/ft^3). Note that the diagrams show faster drying during the summer period (A, B), and practically no effect of size during that period (A). (After Refs. 11 and 43.)

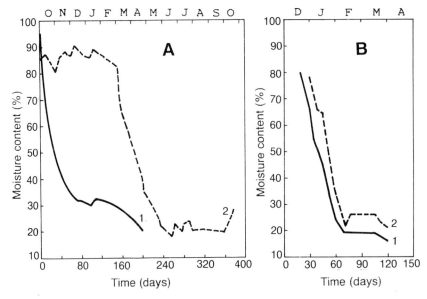

Figure 18-11. Effect of shelter on the rate of air-drying. (A) Drying spruce 5 × 10 cm (2 × 4 in.) in cross-section (Scotland): 1, pile with roof; 2, pile without roof. (B) Drying black pine 2.5 cm (1 in.) thick (Greece): 1, open air, with roof; 2, under shelter (shed). (After Refs. 36 and 54.)

KILN-DRYING

Kiln-drying is usually carried out in steam-heated kilns, equipped with means for temperature and humidity control, air circulation, and removal of moist air. *Conventional* kiln-drying is carried out as follows. (Modifications are discussed under "Other drying methods").

Kiln Types

There are two kiln types—*compartment* and *progressive* (33, 37, 51, 56, 61). In a compartment kiln, the lumber remains stationary during the drying period. Drying conditions (temperature, relative humidity) are changed at intervals, but remain constant for certain periods throughout the kiln. In a progressive kiln, the lumber is progressively moved toward the exit. Drying conditions are not constant throughout the kiln, which is long. At the entrance, the conditions are milder (low temperature, high relative humidity) and progressively become more intense (temperature is raised and relative humidity reduced). Consequently, there always are piles in different drying stages. Periodically, one pile or more is removed, the rest are moved toward the exit, and new piles are introduced in the kiln (Figures 18-12 and 18-13). Progressive kilns are rare, and are mainly used to dry softwoods. The possibility of control is limited, and the final moisture content becomes relatively uniform only if successive loads of wood have about equal initial moisture and similar drying behavior.

Construction—Equipment

Kilns are usually made with bricks, and their floors and roofs are concrete. Metallic (usually aluminum) kilns are also available; they are prefabricated (can be enlarged or reduced in size), mobile, and heat faster (16, 36) (Figure 18-14). In general, kilns should provide thermal insulation; and the door should close tight in order to avoid waste of heat and steam. The inner side of the walls, as well as all metallic surfaces, should be suitably coated for protection against corrosive organic acids; such acids are produced especially when drying beech, oak, chestnut, and other species containing large amounts of extractives.

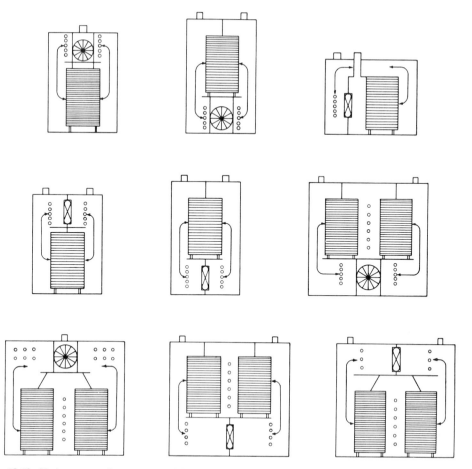

Figure 18-12. Various types of compartment kilns, with single and double piles, and different fan placements. Small circles show cross-sections of steam pipes controlling temperature (relative humidity is controlled by releasing steam through perforated pipes), and arrows show the direction of air circulation. (Courtesy of U.S. Forest Products Laboratory.)

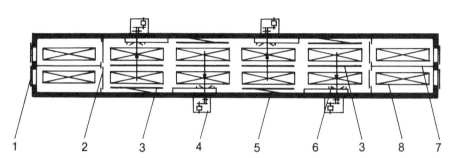

Figure 18-13. Ground plan of a progressive kiln (continuous drier): 1, entrance door; 2, baffle (curtain); 3, heating unit; 4, motor (air circulation); 5 and 6, metallic baffles to direct air flow; 7, separating wall; 8, lumber pile. (After Ref. 21, reproduced by permission from VEB Fachbuchverlag.)

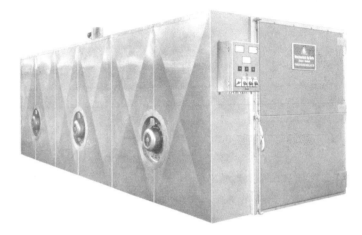

Figure 18-14. Prefabricated dry kiln. (Courtesy of Kiefer.)

Kiln sizes vary according to specific needs, but there is a maximum above which it is difficult to establish uniform conditions. Compartment kilns have a maximum length of 15 m (50 ft), and progressive kilns are usually 25–30 m (80–100 ft) long. A kiln is 2–3 m (6–10 ft) high, and its width depends on the arrangement of piles (single, double rows) and the width of each pile—which should not be greater than 1.80 m (6 ft). A kiln is not full of lumber; about 40% is empty space around the piles for air to circulate. Additional free space is formed by the stickers, and the kiln volume occupied by lumber (lumber volume $\times$ 100/kiln space) is only 30–40% for usual lumber thicknesses of about 15–50 mm (0.5–2 in.). Thus, a kiln 60 m^3 (2100 ft^3) in volume of space (6 $\times$ 4 $\times$ 2.5 m or about 20 $\times$ 13 $\times$ 8 ft in inner dimensions) has a "useful" capacity of about 20 m^3 (700 ft^3) (56).

Heating is usually provided by steam, which is introduced in pipes suitably placed above, below, or between lumber piles. Relative humidity is controlled by the introduction of free steam (rarely by spraying fine drops of water), and the air is circulated by fans placed above or below the piles and sometimes sidewise. The fans are mounted at regular intervals on a shaft placed along the length of the kiln or on short transverse shafts (18, 33, 36, 37). Sidewise mounting on transverse shafts enables the use of large fans in kilns that are low in height (Figure 18-15).

The control of desired conditions is sometimes manual, but modern installations are equipped with automatic mechanisms, which work with compressed air or electricity; they introduce or stop steam, open or close air vents for fresh air to enter or moist air to be removed, and at intervals change the direction of air circulation for the purpose of uniform drying.

The conditions of temperature and relative humidity inside a kiln are monitored by specialized instruments, usually autographic, which are placed outside and provide continuous recording for 1 day or 1 week, or combine recording and control (recorder-controllers) (31). Such instruments (Figure 18-16) measure temperature and relative humidity, and operate by the expansion of mercury or other liquid or gas contained in metallic tubes ("bulbs") or spirals. The latter are placed inside the kiln, near the entrance of air forced by the fans into the lumber piles, and the expansion or contraction of their content is transmitted by tubing to the outside instruments. These "thermometers" are usually placed in pairs—one "dry" and another "wet," the latter with its bulb covered by a "wick" kept continuously wet with water. Occasionally, there are two dry bulbs (one at the entrance and another at the exit of air from a pile) and one wet (at the entrance or

Figure 18-15. An air-circulation fan placed sidewise. Heating units are to the left and right of the fan; a lumber pile is also shown. (Courtesy of Hildebrand.)

exit). Dry bulbs measure temperature, and relative humidity is calculated from tables on the basis of the difference between dry and wet bulb readings. In some kilns, glass (mercury) thermometers are placed inside the kiln, and are read from the outside; however, such practice does not give accurate information. Hair hygrometers are not suitable; their reaction to changing relative humidity is not sensitive enough when exposed to high temperatures (36).

A kiln is additionally provided with equipment to determine the moisture content of wood (balance, oven, desiccator, small saw, electric

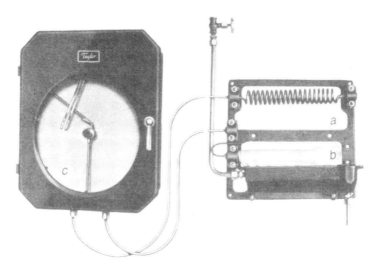

Figure 18-16. A recorder-controller of temperature and relative humidity: a, dry bulb; b, wet bulb; c, revolving recorder. (Courtesy of Forest Products Laboratory, Canada.)

hygrometer), direction and speed of air, and other information (17, 37).

Piling

The general principles of lumber piling, as previously explained, apply also in kiln-drying. Construction of the piles so that board ends do not protrude, proper placement or stickers (unless self-piling is applied), and grouping of lumber in piles of similar drying behavior (separation mainly of species and thickness) are important points to note. Drying conditions are more severe in a kiln in comparison to air-drying; therefore, deviations from proper piling will cause greater degradation. Protruding board ends affect air speed and cause local deviations of its direction, resulting in warping, checking, and nonuniform drying. Air circulation is also affected by lateral spaces between boards; for this reason, each layer is made compact (i.e., without spacing) (48). If the volume of the lumber to be dried is not sufficient, the width of a pile should be reduced—not its height. Warping of lumber boards in the upper layers of a pile may be prevented by placing weights (heavy wooden or concrete beams) on top. The lumber is piled on carriages, which are brought in the kiln on rails (20), but forklifts are also used (Figure 18-17).

The changing moisture content of wood may be followed by use of kiln samples (Figure 18-18) or in other ways. The samples are prepared from representative boards; each sample is 1–1.5 m (3–5 ft) long and is taken at a distance of at least 50 cm (1.5 ft) from one end of the board. The width is equal to the width of the lumber; in lumber of varying width, the sample is 20–25 cm (8–10 in.) wide (36, 56). The bark is removed. In case of wide variation of moisture, samples are taken both from wet and drier material. The ends of each sample are coated with asphaltic paint, or some synthetic resin preparation resistant to water and high temperatures—then, the sample is weighed without delay. Two small specimens taken at both ends of the kiln sample (Figure 18-19A and B) are also weighed and placed in the oven to determine their moisture content. Usually, six kiln samples are inserted in each pile (Figure 18-20) in positions of fast and slow drying, and in such a manner that they can be easily

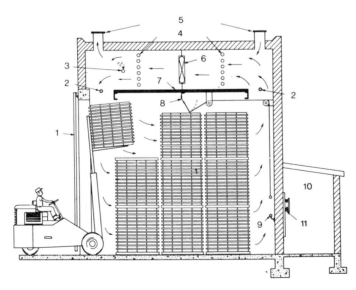

Figure 18-17. Mechanical piling of lumber packets in a dry kiln: 1, door; 2, dry bulbs; 3, steaming pipes; 4, heating pipes; 5, ventilators (automatic); 6, fan; 7, false roof; 8, baffle; 9, wet bulb; 10, control office; 11, recorder-controller. (Courtesy of U.S. Forest Products Laboratory.)

DRYING 277

Figure 18-18. Placement of a moisture sample in a lumber pile. Kiln instruments are shown on the left and the special door on the right side of the photograph. (Courtesy of G. F. Wells, Ltd.)

removed for periodic weighing, and returned to their place. Fewer samples (3–4) may be used if there is no great moisture content variation in the lumber and considerable experience exists with the species to be dried.

Drying Procedure

Drying Schedules. The conditions of temperature and relative humidity applied in drying a load of lumber are specified by a *drying*

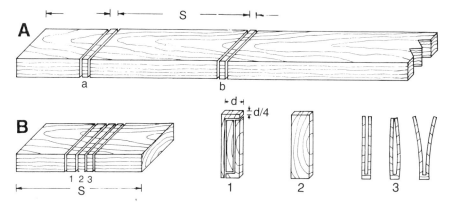

Figure 18-19. Monitoring moisture content and casehardening. (A) An end piece, about 50 cm (1.5 ft) long, is removed from a representative board, then a moisture content sample (*a*), 1–1.5 cm (0.5 in.) wide, is cut, as well as another kiln sample (*S*), 1–1.5 m (3–5 ft) long, and another moisture sample (b). Samples *a* and *b* are immediately weighed and placed in an oven to determine their moisture content—which is also the moisture content of sample *S* (average of *a* and *b*). Sample *S* is covered on both ends with a hydrophobic substance (e.g., immersed in wax), weighed immediately, and placed in the pile to be dried. (B) Shortly before the end of drying, small samples (1, 2, 3) are cut from sample *S* to determine moisture content distribution (1), average moisture content (2), and check for casehardening (3). Sample 1 is separated into shell and core; core thickness is about $\frac{1}{4}$ of total.

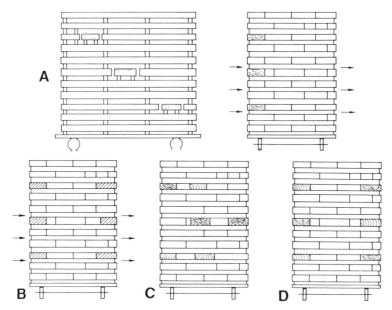

Figure 18-20. Placement of kiln samples in side view (A, left), and A (right), B-D, end views (different suggestions— C, overhead fans; D, side fans). (After Refs. 7, 20, and 36.)

schedule. Drying schedules have been empirically developed (by trial and error) in various Forest Products Laboratories (England, France, U.S.A., etc.) (20, 36, 37) with the intent of drying lumber in such a way as to minimize both drying time and drying defects. Examples of schedules are shown in Tables 18-1, 18-2, and 18-2A.

All schedules are indicative and, within limits, may be adapted to specific conditions according to experience or purpose of drying. For example, there are special schedules for faster drying, more careful drying to avoid adverse effects on strength[7] (drying wood for ladders, sporting goods, etc.), or drying pines without surface exudation of resin (36, 37).

A schedule is applied by setting certain initial conditions of temperature and relative humidity, according to species and initial moisture content of the wood, and making changes at successively lower moisture content levels. This requires monitoring of moisture content, which is done by periodic weighing of kiln samples. Use of samples is the most certain way of drying by application of schedules, but this is a cumbersome and sometimes impractical procedure; one has to enter a working kiln at intervals to remove and place back the samples. However, it is often necessary to do it when drying high-value lumber susceptible to drying defects (45).

The initial moisture of each sample is calculated as an average of the moisture of the two small specimens cut out from its ends (Figure 18-19A and B). With the progress of drying, the samples are periodically weighed at intervals, depending on the rate of moisture loss. After each weighing, the current moisture content of a sample is determined from the relationship

$$\text{CMC } (\%) = \frac{\text{CW} - \text{CDW}}{\text{CDW}} \times 100 \quad (18\text{-}1)$$

where

CMC = current moisture content
CW = current weight
CDW = calculated dry weight

The "calculated (oven-) dry weight" (M_o) is found from the relationship

$$Y(\%) = \frac{M_x - M_o}{M_o} \times 100 \quad (18\text{-}2)$$

$$M_o = \frac{100 \, M_x}{Y\% + 100} \quad (18\text{-}3)$$

Table 18-1. Examples of Kiln-Drying Schedules[a]

Wood mc %	Temperature Dry	Temperature Wet	Relative humidity (%)
Schedule 1 (°C)			
Green	35	30.5	70
60	35	28.5	60
40	40	31	50
30	45	32.5	40
20	50	35	35
15	60	40.5	30
Schedule T4-C2 (°F)			
Green	110	106	
40	110	105	
35	110	102	
30	120	106	
25	130	100	
20	140	90	
15	180	130	
Schedule 10 (°C)			
Green	70	65	80
50	75	67	70
30	80	68.5	60
20	90	69	40
Schedule T6-C4 (°F)			
Green	120	113	
40	120	110	
35	120	105	
30	130	105	
25	140	100	
20	150	100	
15	180	130	

[a]Schedule 1, Black locust, maple, elm, Antiaris, Kosipo, Sapele, Sipo; schedule 10, fir, spruce, pine (*P. radiata*), basswood, Abura; schedule T4-C2, white oak; schedule T6-C4, mahogany. 1, 10, Lumber of random width, 25–40 mm (1–1.5 in.) thick, and air speed 1–1.5 m (3–5 ft)/s; for larger thicknesses and lower or higher speeds, relative humidity is increased by 5–10% at each level. T4-C2, T6-C4, Lumber 4 × 4 in. (10 × 10 cm), 5 × 4 in. (12.5 × 10 cm), 6 × 4 in. (15 × 10 cm). Note that air speed varies from 1 to 2.4 m (3–8 ft)/s, depending on predrying (air-drying, low speed), difficulty of drying (decreases with difficulty), and distance of air travel (increases with distance). Attainable speeds are affected by the size, speed, and placement of fans.
Source: Refs. 36 and 37.

where M_x is the initial weight of the sample, and Y is the initial moisture content (both known).

Monitoring moisture by weighing kiln samples is sometimes eliminated by the application of "time schedules"; changes of drying conditions are made not according to moisture but rather at certain time intervals. Time schedules are based on long experience, after repeated drying of certain species and thicknesses of lumber; they are not as precise as moisture-content schedules, but are often used for drying softwoods and certain hardwoods that dry easily. Even in such cases, however, it is helpful to use kiln samples near the end of drying.

Kiln-drying by application of schedules is called *semi-automatic* when the drying conditions are set by the kiln operator but maintained by instruments. There are kilns, however, that are entirely *automatic*. After setting certain starting conditions, automatic instruments, which may be controlled by a computer (14, 16, 26, 32), maintain heating of the kiln and wood, control temperature, relative humidity, and air circulation according to changing moisture content, and switch-off the kiln when the moisture of the lumber reaches the desired final level. Moisture is continuously monitored by electrodes connected to certain points (up to 8) (16) in a lumber pile, or even by "noncontact" sensors. The procedure is based on changes of electrical properties of wood with changing moisture content. Aside from the expense involved and some technical difficulties, which are a subject of current research, such automation is not applicable to woods that are difficult to dry—and electrical measurements are of no practical value above the fiber saturation point, where critical changes in drying conditions are sometimes necessary (45, 51).[8]

Stages of Drying. There are three stages in kiln-drying: *preparatory*, actual *drying*, and *equalization* of moisture content (see Table 18-2).

In the *preparatory* stage, the intention is to preheat the wood. Its temperature should be raised to the first level provided by the schedule (on the average, 40–65°C or 100–150°F), stepwise but fast; during this time, a difference of 3–4°C (5–7°F) is maintained between dry and wet bulb temperature. In drying green lumber, the relative humidity may be raised up to 100%, and the preheating time is reduced. If the lumber has been air-dried to a moisture content of 25–30%, the relative humidity is 70–75% or at least two steps higher than that corresponding to the wetter sample (37, 56).

Table 18-2. Examples of Kiln-Drying Schedules (°C) (Drying Stages)[a]

m.c %	A						B						C					
	1	2	3	4	5	6	1	2	3	4	5	6	1	2	3	4	5	6
>50	55	53.5	1.5	92	19	—	60	56.5	3.5	85	14	—	—	—	—	—	—	—
50–40	55	53.5	1.5	92	19	—	60	55.5	4.5	80	13	—	—	—	—	—	—	—
40–30	55	53.5	1.5	92	19	—	60	54.5	5.5	76	12	2.7	—	—	—	—	—	—
30–27	60	58	2	90	18	—	68	62	6	75	11	2.7	—	—	—	—	—	—
27–24	60	58	2	90	18	—	68	61	7	71	10	2.7	—	—	—	—	—	—
24–21	60	58	2	90	18	—	68	59	9	64	9	2.7	—	—	—	—	—	—
21–18	—	—	—	—	—	—	70	59	11	58	7.5	2.7	—	—	—	—	—	—
18–15	—	—	—	—	—	—	70	56	14	50	6.5	2.7	65	62	3	87	15	1.1
15–12	—	—	—	—	—	—	70	50.5	19.5	37	5.5	2.7	65	59.5	5.5	76	12	1.1
12–8	—	—	—	—	—	—	70	48.5	21.5	33	4.5	2.7	65	54	11	56	8	1.1
>50	65	63	2	—	18	—	70	64	6	75	11	—	—	—	—	—	—	—
50–40	65	63	2	—	18	—	70	63	7	71	10	—	—	—	—	—	—	—
40–30	65	63	2	—	18	—	70	61	9	64	9	4	—	—	—	—	—	—
30–27	70	68	2	—	17	—	80	70.5	9.5	65	8	4	—	—	—	—	—	—
27–24	70	68	2	—	17	—	80	68	12	58	7	4	—	—	—	—	—	—
24–21	70	68	2	—	17	—	80	65	15	50	6	4	—	—	—	—	—	—
21–18	—	—	—	—	—	—	85	66	19	43	5	4	—	—	—	—	—	—
18–15	—	—	—	—	—	—	85	61	24	34	4	4	80	77.5	2.5	93	15	1.1
15–12	—	—	—	—	—	—	85	59	26	31	3.5	4	80	75.5	4.5	85	12	1.1
12–8	—	—	—	—	—	—	85	55	30	23	3	4	80	70.5	9.5	65	8	1.1

[a]The schedules apply to lumber less than 34 mm (about 1.5 in.) thick and the following species. Upper schedule: fir, spruce, Scots pine; lower schedule: beech, plane (sycamore), basswood, Aleppo pine, mahogany (Khaya), Afrormosia, Afzelia, Limba, Meranti, Padauk. (A–C) Drying stages: (A) preparatory, (B) drying, (C) equalization. 1, Dry bulb (°C); 2, wet bulb (°C); 3, difference, wet-dry (°C); 4, relative humidity (%); 5, equilibrium moisture content (%); 6, moisture quotient (see text).
Source: Ref. 20; reproduced by permission.

Table 18-2A. Examples of Kiln-Drying Schedules (°F) (Drying Stages)[a]

m.c %	A						B						C					
	1	2	3	4	5	6	1	2	3	4	5	6	1	2	3	4	5	6
>50	131.0	128.3	2.7	92	19	—	140.0	133.7	6.3	85	14	—	—	—	—	—	—	—
50–40	131.0	128.3	2.7	92	19	—	140.0	131.9	8.1	80	13	—	—	—	—	—	—	—
40–30	131.0	128.3	2.7	92	19	—	140.0	130.1	9.9	76	12	2.7	—	—	—	—	—	—
30–27	140.0	136.4	3.6	90	18	—	154.4	143.6	10.8	75	11	2.7	—	—	—	—	—	—
27–24	140.0	136.4	3.6	90	18	—	154.4	141.8	12.6	71	10	2.7	—	—	—	—	—	—
24–21	140.0	136.4	3.6	90	18	—	154.4	138.2	16.2	64	9	2.7	—	—	—	—	—	—
21–18	—	—	—	—	—	—	158.0	138.2	19.8	58	7.5	2.7	—	—	—	—	—	—
18–15	—	—	—	—	—	—	158.0	132.8	25.2	50	6.5	2.7	149.0	143.6	5.4	87	15	1.1
15–12	—	—	—	—	—	—	158.0	122.9	35.1	37	5.5	2.7	149.0	139.1	9.9	76	12	1.1
12–8	—	—	—	—	—	—	158.0	119.3	38.7	33	4.5	2.7	149.0	129.2	19.8	56	8	1.1
>50	149.0	145.4	3.6	—	18	—	158.0	147.2	10.8	75	11	—	—	—	—	—	—	—
50–40	149.0	145.4	3.6	—	18	—	158.0	145.4	12.6	71	10	—	—	—	—	—	—	—
40–30	149.0	145.4	3.6	—	18	—	158.0	141.8	16.2	64	9	4	—	—	—	—	—	—
30–27	158.0	154.4	3.6	—	17	—	176.0	158.9	17.1	65	8	4	—	—	—	—	—	—
27–24	158.0	154.4	3.6	—	17	—	176.0	154.4	21.6	58	7	4	—	—	—	—	—	—
24–21	158.0	154.4	3.6	—	17	—	176.0	149.0	27.0	50	6	4	—	—	—	—	—	—
21–18	—	—	—	—	—	—	185.0	150.8	34.2	43	5	4	—	—	—	—	—	—
18–15	—	—	—	—	—	—	185.0	141.8	43.2	34	4	4	176.0	171.5	4.5	93	—	1.1
15–12	—	—	—	—	—	—	185.0	138.2	46.8	31	3.5	4	176.0	167.9	8.1	85	12	1.1
12–8	—	—	—	—	—	—	185.0	131.0	54.0	23	3	4	176.0	158.9	17.1	65	8	1.1

[a]See footnote of Table 18-2.

Actual *drying* takes place in the second stage. The procedure may be understood by an example of drying maple lumber. According to the schedule applicable to this species (Table 18-1, Schedule 1), and if the moisture content of the lumber is high (e.g., 85%), the dry bulb is set at 35°C and the wet bulb at 30.5°C. Drying starts under these conditions, and is continued until moisture falls to 60%. When this is attained (as determined by the average moisture content of the samples), the settings of the temperature controller (Figure 18-16) are changed to 35°C and 28.5°C. When moisture falls to 40%, temperatures are set at 40 and 31°C, and so on, until wood reaches the desired final moisture (e.g., 15%). Corresponding manipulations in °F, and drying oak lumber, are shown in Schedule T4-C2 (Table 18-1).

If the initial moisture content is, for example, 30%, because the lumber has been air-dried, drying does not start with the indications corresponding to 30%, but rather one step higher. Application of the schedule starts after a few hours.

The third stage is applied shortly before the end of drying, with the purpose of *equalizing* the moisture content of the lumber. The procedure starts when the moisture content of the drier sample is about 2% lower than the desired final moisture content (e.g., 8% if the final is 10%). Conditions in the kiln are changed a little so that the equilibrium moisture content corresponds to the moisture content of the drier sample (8% in our example). For this reason, the temperature is raised a little and the relative humidity is set correspondingly. Drying is continued until the moisture content of the wetter sample falls to the desired level (10%).

The lumber should not be removed from the kiln immediately after the end of the drying procedure. It should be left there until the temperature of the kiln falls to within about 20°C (68°F) of the outside temperature; otherwise, and depending on the sensitivity of species, there is the danger of developing defects, such as surface checking and reverse casehardening (36).

Moisture Quotient. In kiln-drying, the rate of drying is more important than in air-drying with respect to degradation of the wood. Control of this rate may be done by calculating the "moisture quotient" (Trocknungsgefälle) (16, 17, 25, 59), which is an expression of the relationship between current average moisture content of the lumber (as determined from the samples) and equilibrium moisture content corresponding to the set conditions of temperature and relative humidity. For example, if the current average moisture content is 40% and the equilibrium moisture content 10% (corresponding to 70°C or 160°F temperature and 70% relative humidity, see Fig. 9-6), the quotient is 4.

Moisture quotient affects the rate of drying more than temperature. A low quotient increases the time of drying and makes it uneconomical; whereas a high quotient reduces the time but leads to high moisture content differences in the mass of wood and may cause defects (checking, etc.). When the moisture content of the drying lumber is below the fiber saturation point, the quotient should be kept as constant as possible. Thus, if the lumber is valuable, thicker than 2.5 cm (1 in.), has high shrinkage, and is prone to checking, the quotient recommended is 1.8–2.5 (average 2.0) for softwoods and 1.3–1.8 (average 1.5) for hardwoods. In other cases (with lumber less than 2.5 cm thick, and species not sensitive to fast drying), corresponding quotient values are 3.0–4.0 (softwoods) and 2.0–3.0 (hardwoods) (17, 25).

Above the fiber saturation point, the concept of moisture quotient is not valid, because in this region there is no equilibrium moisture content.

Correction of Defects. Periodic inspection of samples and drying lumber may reveal defects, such as checking, warping, casehardening, honeycombing, collapse, or molding.[9] In most cases, these defects may be corrected by changes of the drying conditions—unless they result from abnormal wood features (spiral grain, compression or tension wood, knots, etc.); in that case, the possibility of intervention does not exist or is very limited.

Surface *checks* may close by increasing the moisture content of surface layers, and this may be achieved by raising the relative humidity to

85–90%, with a small simultaneous increase of temperature by 5–10°C (10–20°F). These conditions are held 0.5–1 h, depending on lumber thickness, and discontinued when the checks close. If they are relatively deep, the relative humidity should not be raised above 80–85%, but the conditions are held longer.

Warping due to the above-mentioned abnormalities cannot be corrected, but may also be caused by incorrect piling (stickers not in vertical rows, protruding lumber ends). Placement of weights on top of the piles may help to avoid or reduce this defect. Also, it has been observed that high temperatures favor warping (48), and this should be taken into consideration in choosing or adapting drying schedules.

The possible presence of *casehardening* may be examined during drying, but usually this is done shortly before the end—after the equalization of moisture. The defect is recognized by suitable sectioning of a kiln sample; a cross-section about 1 cm (0.5 in.) thick is taken at a distance of 20–25 cm (about 1 ft) from one end, and is separated by sawing in two or four strips of equal thickness (or shaped like a fork). The strips may number three, or six in thicker boards. If there is casehardening, the strips or arms of the fork bend toward one another, but the extent of casehardening, as expressed by the extent of bending, can be judged after drying to a uniform moisture content under room conditions (after 12–24 h). At the same time, it is possible to study the moisture distribution by separating another section into "shell" and "core" (Figure 18-19) and determining their moisture content.

Casehardening will cause warping after resawing the lumber. The defect may be corrected by changes of drying conditions—raising relative humidity to 90% and temperature by about 10–15°C (20–30°F), unless the latter is already high (e.g., near 80°C or 175°F) (36, 48). Another procedure is to keep the temperature constant and to raise the relative humidity to a level that achieves a 4% increase in equilibrium moisture content. These conditions are held for 2–6 h for softwood lumber (48) and 16–24 h for hardwood lumber, 2.5 cm (1 in.) thick, and up to 48 h for lumber 5 cm (2 in.) thick (56). If the defect is corrected, repetition of the above test finally shows that the strips or arms of the fork remain straight, although at the beginning they may bend outward. However, if the duration of the above change of conditions is longer than necessary, reverse casehardening may occur.

A similar treatment is applied for disguising *honeycombing*, but *collapse* is handled by more drastic changes—temperature 100°C (212°F) or at least 85°C (185°F) and relative humidity 100% for 4–8 h. Collapse occurs in certain eucalypts, oaks, and other species when very wet heartwood is kiln-dried at relatively high temperatures; it may be avoided by predrying (air-drying).

Molding fungi may develop under low-temperature drying conditions, 40–45°C (100–110°F). The treatment involves an increase of temperature to 65–70°C (150–160°F) and relative humidity to 95–100%—if wood moisture is lower than 30%, and 75–80% if higher. These conditions are maintained for 0.5–1 h. Another procedure is to raise the temperature to 70°C (160°F) and the relative humidity to 100% for 3 h. After that, the conditions are returned to those specified by the schedule; however, the reduction of temperature should be gradual and the relative humidity kept high to avoid checking. The difference between dry and wet bulb is kept at 3–4°C (5–7°F) or less, if the lumber is green (48).

Insects in wood are killed by a combination of temperature at 50–60°C (120–140°F) or higher and relative humidity 60–100%. The duration depends on lumber thickness (increases with thickness) and kiln conditions (better results are obtained with higher temperatures in combination with high relative humidities); for *Lyctus* only 0.5 h is recommended at 100% relative humidity. In wood products (furniture, etc.), both temperature and relative humidity are not high—temperature no higher than 50–55°C (120–130°F) and relative humidity no higher than 60% (36, 48). Sterilized wood may be reinfested.

Duration of Drying

The time needed to kiln-dry is much shorter than air-drying. Comparative data are given in

Table 18-3. Times for Air- and Kiln-Drying Lumber (2.5 cm, 1 in. thick)[a]

Species	Air-drying Green-20%	Kiln-drying 20-60%	Kiln-drying Green-6%
	Days		
Douglas fir	10-180	—	2-7
Fir	—	—	3-5
Pine	15-200	2-3	3-10
Spruce	20-150	—	3-7
Ash	60-200	4-7	10-15
Beech	70-200	5-8	12-15
Birch	40-200	5-8	3-15
Black locust	—	5-8	12-16
Chestnut	—	4-8	8-12
Elm	50-180	4-8	10-17
Lime (basswood)	40-150	3-5	6-10
Oak	70-300	5-12	16-40
Plane (sycamore)	30-150	4-7	6-12
Poplar	50-150	3-5	6-10
Walnut	70-200	5-8	10-16
Willow	30-150	5-8	12-16

Source: Refs. 37 and 39.
[a] Similar results are reported elsewhere (16): pine 80-200 days, spruce 90-200, ash 60-200, beech 70-220, birch 60-220, chestnut 60-180, mahogany 70-160, maple 60-200, walnut 60-200. In Greece (54) air-drying of 2.5 cm (1 in.), 5 cm (2 in.), and 8 cm (about 3 in.) thick lumber of fir, black pine, and beech gave the following results: fir 8-17 days, pine 30-100, beech 130-200, depending on season of the year, location, species, and thickness of lumber (53). See Figures 18-11, and 18-24.

Table 18-3. Considerable reduction of kiln-drying time is achieved by predrying (air-drying) as shown in Table 18-3. The data are indicative, and variations exist depending on factors that affect the rate of drying. Certain factors (softwood-hardwood, sapwood-heartwood, thickness, etc.) have been discussed under air-drying. Others include drying characteristics of a wood species, range of moisture content change (initial-final), speed of air circulation, quality of drying lumber, and accepted drying defects.

Final Moisture Content

The final moisture content depends on the purpose of drying or the intended use of the wood (17, 36). For example, if the purpose of drying is to protect the wood from fungal attack during storage or shipment, it is sufficient to bring the moisture content to a level of about 20%. For building components exposed out-of-doors (window frames, etc.), the recommended level of moisture is 12-15%, which represents an average of the expected variation. Indoor uses require lower moisture contents; for example, wood to be used for furniture, flooring, or interior paneling should be dried down to 6-8%—that is, to an equilibrium moisture content roughly corresponding to a temperature of 20-25°C (70-80°F) and a relative humidity of 30-40% (see Figure 9-6). A higher moisture content will cause defects due to shrinkage (gaps in flooring, etc.). The cost of drying should also be taken into consideration; the lower the moisture content, the higher the cost; therefore, low moisture contents, without justification, should be avoided.

Storage of Lumber

The hygroscopicity of wood is not substantially affected by air- or kiln-drying. Dry wood may again absorb moisture. Therefore, if not immediately used,[10] the lumber should be stored under conditions that protect it from absorbing high quantities of moisture. In practice, this is done by piling without stickers and under a shelter, where temperature is sometimes controlled. Depending on relative humidity, the

Table 18-4. Time (Days) and Method of Drying Hardwoods

Method	Red oak		Yellow poplar	
	10 × 10 cm (4 × 4 in.)	10 × 20 cm (4 × 8 in.)	10 × 10 cm (4 × 4 in.)	10 × 20 cm (4 × 8 in.)
1. Air-drying[a] (green to 25%)	120–180	270	60	90–120
2. Low-temperature drying[b]				
predrying	29–32	77	13	—
dehumidification	12–42	64	—	15
3. Conventional kiln-drying[c]	25–30	60–75	3–5	10–12

[a]Forced air-drying or "fan air-drying" (lumber in a shed with fans to control air circulation) considerably reduces the drying time. For low-density, easy-to-dry hardwoods (in the eastern U.S.), up to 1 month is needed for 10 × 10 cm (4 × 4 in.) lumber and 1–2 months for 10 × 20 cm (4 × 8 in.).
[b]Closed chamber, drying temperature below 55°C (130°F): predryers, temperature 30–40°C (80–100°F), relative humidity 50–80%, lumber green to 25% mc; dehumidification, drying down to 3–8% mc from initial 42–70% mc; solar dryers belong here, but are not suitable for high production.
[c]High temperature drying is also applied; drying is extremely fast (see pp. 285–286).
Source: Data derived from Ref. 60.

temperature should be maintained at levels that would keep wood moisture to an approximate desired equilibrium.

OTHER DRYING METHODS

Other methods include: drying with solar energy, dehumidification (condensation) kiln-drying, high-temperature drying, continuous rising temperature drying, chemical drying, vapor drying, boiling in oils, drying with solvents or high-frequency electricity, and others. Some of these methods have no practical importance, mainly for reasons of high cost.

Drying with Solar Energy

This method is a variation of air-drying. Drying is carried out in "solar dryers" or "solar kilns" (3, 10, 20, 35). There are two basic types: greenhouse and solar collector. In the latter type, heat captured by the collector is transferred inside the kiln by air or a closed circuit of water. Solar drying is faster than air-drying, and slower in comparison to conventional kiln-drying (Figure 18-21), but the cost of installation of solar dryers is low, especially for the greenhouse type, and their operation is simpler than dry kilns. The moisture content of wood can be reduced to lower levels (down to 7%) (7, 40) in comparison to air-drying; this is important especially in regions of high latitude, and in the fall or winter time (42, 62). Solar drying may be combined with kiln or dehumidification drying (40); the wood dries faster and the effect of environmental conditions is reduced.

Dehumidification Kiln-Drying

In this method (1, 40, 60), the moisture evaporated from the wood is not removed from the kiln chamber in vapor form, as in conventional dry kilns, but is condensed and removed as a liquid. During condensation, the moisture gives its latent heat to a cooler. This heat is recaptured when the circulating air passes above the condensation pipes, and is used for additional evaporation of moisture from the wood. The dryer is tightly closed, but new loads may be introduced, while ready (dry) ones are removed. Temperature is low, usually under 40°C or 100°F (with improved cooling chemicals it may go up to 80°C or 175°F), and relative humidity 30–50%. The initial cost is low in comparison to conventional kilns, and the operation of such dryers is simpler, although the method also has certain disadvantages. Drying is generally slower in comparison to conventional kiln-drying, especially at low moisture contents (under 12%) (40).[11] Although total energy consumption is reduced, electrical energy is required, which is often more expensive than alternate sources of energy. (Noteworthy, however, is a reported 67% reduction of energy costs for drying softwoods and 43% for hardwoods, in comparison to conventional kiln-drying) (40). The control of

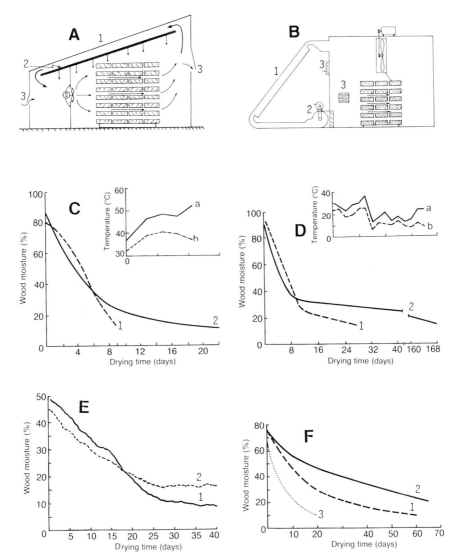

Figure 18-21. Drying with solar energy. (A) Solar dryer of greenhouse type (1, transparent cover; 2, metallic surface; 3, vents) and (B) with solar collector (1, collector; 2, motor, air circulator; 3, vents). (C–F) Comparison of drying rates. (C, D) Yellow poplar lumber 2.5 × 10 cm (1 × 4 in.) and 2.5 × 15 cm (1 × 5 in.) in cross-section (C, July–August; D, November–December–May), (E) spruce 4 × 9 cm (1.5 × 3.5 in.), (F) oak 10 × 10 cm (4 × 4 in.): 1, solar drying; 2, air-drying; 3, kiln-drying; a, fluctuation of maximum temperature of solar dryer; b, maximum air temperature. (A, courtesy of JAKRAP; B–F, after Refs. 9, 35, and 42.)

moisture is not very accurate, but the results are satisfactory in practice, depending on the purpose of drying. Dehumidification drying is usually applied to hardwoods (16).

High-Temperature Drying

The difference between this method and conventional kiln-drying is that high temperatures are used. The temperature of the dry bulb is always higher than 100°C (212°F), while the wet bulb may be below 100°C (drying with superheated humid air) or 100°C (drying with superheated steam; i.e., only steam and no air).

Drying with high temperatures is very fast (Figure 18-22). For example, pine lumber, 2.5 cm (1 in.) thick, was dried from 95% to 9% moisture content in 11–12 h (56), spruce lum-

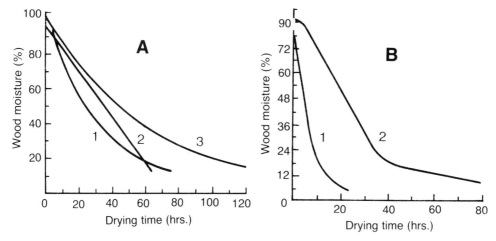

Figure 18-22. High-temperature drying and other methods. (A) Rate of drying western hemlock (*Tsuga*) lumber 5 × 20 cm (2 × 8 in.) in cross-section: 1, high temperature; 2, conventional kiln-drying; 3, constant rising temperature (CRT). (B) Rate of drying pine lumber (*P. resinosa*) 5 × 10 cm (2 × 4 in.) in cross-section: 1, high temperature; 2, conventional kiln-drying. (After Refs. 2 and 47.)

ber 5 cm (2 in.) thick from 40–90% to 8–12% in 16 h, and birch 2.5 cm (1 in.) thick from 15% to 4% in 20 h (6). The productivity of such kilns was estimated to be 5–10 times greater in comparison to conventional dry kilns. The wood is reported to become less hygroscopic (it holds less moisture content at equilibrium under the same conditions) (40, 47), and has a higher dimensional stability (36).

However, there are important disadvantages:

- Special kilns are needed, with high thermal insulation and resistant coating of their walls. Kilns are usually metallic, prefabricated, with aluminum-covered walls, and have a relatively low volume capacity, 1–8 m^3 (35–280 ft^3). Common kilns wear fast (6, 24, 56).
- The method cannot be used with woods that have a high initial moisture, and their anatomical structure does not allow for its fast exit. It has been used successfully to dry various softwoods, but hardwoods should be predried to a moisture content of 20–25%, and care is still needed because certain species (oak, walnut, etc.) tend to collapse (56).

High-temperature drying may cause discoloration of the wood (the color becomes darker), exit of resin on the surface, and loosening of knots. Exposure to high temperatures at high moisture contents may also result in reduction of some strength properties (modulus of rupture, modulus of elasticity, toughness), due to thermal degradation of the wood (19, 45). Such effects depend, however, on species and time of exposure, and some references report no change or even improvement of some properties (23, 24, 47).

Although the time is reduced, the cost of drying may not be lower, because high-temperature kilns are expensive, and more thermal and electrical energy are required. Also, a greater fan power is needed for a faster circulation of air or steam (56).[12]

Drying with Continuous Rising Temperature

Instead of stepwise rising of temperature in conventional kiln-drying, this method advocates a continuous increase, and control of the rate of heat transmission from the kiln air to the wood, so that a constant rate of drying is achieved (Figure 18-22A). Drying starts with a low initial temperature of 60°C or 140°F (typical) and the difference between dry and wet bulb is kept constant; final dry bulb temperature is 100°C or 212°F (typical). The speed of drying is faster even in comparison to high-temperature drying, and the method (CRT) is credited with reduction of time, energy, and drying defects; it is not a widely known method (2, 45).

Chemical Drying

This method is based on the property of certain chemicals to hold moisture and reduce the shrinkage of wood (see Chapter 10). It is possible to dry lumber or other wood products of large thickness, because the higher moisture held in the surface layers and the reduction of shrinkage permit use of lower relative humidities, which would normally cause surface checking.

Common salt (NaCl) and urea are effective (6), but cause oxidation of metals; corrosion inhibitors have to be incorporated, or all metallic surfaces in the kiln should be cleaned with water after each drying run. Saws and other machines processing wood dried in this manner also need to be cleaned. In addition, chemicals increase the cost of drying, and dried wood tends to "sweat" when the relative humidity of the air is high (higher than 80%). For all these reasons, the method is not used except in special cases.

The chemicals are applied by immersing the wood in saturated solutions, brushing, or dry sprinkling on the surface of the wood. The lumber is then piled for a few days without stickers. Green wood favors the entrance of dry chemical (powder) inside its mass, while the effectiveness of solutions depends on the transverse permeability of wood. This method is also known as "salt seasoning" (36). Among the chemicals used is polyethylene glycol (see Chapter 10).

Vapor Drying

Instead of exposure to air, drying may be accomplished by use of organic vapors (e.g., xylene). The wood is placed in a hermetically closed cylinder (similar to the ones used for preservative treatment with pressure; see Chapter 19), and is exposed to the high temperature of the boiling liquid, 100–200°C (212–390°F). The moisture evaporated from the wood, mixed with vapors, is taken outside the cylinder and condensed, but the two liquids do not mix because their density is different. The water is measured and discarded, whereas the chemical is introduced back into the system and reused. Finally, a vacuum is applied, which removes the chemical absorbed by the wood.

With this method, drying is very fast, 3–20 times faster in comparison to conventional kiln-drying. However, the cost is high, because more electrical energy is needed (higher horse power of motors, due to the higher density of the vapors in comparison to air) (6), as is more heat to produce steam. The method has been successfully employed in practice to dry railroad ties and poles prior to their preservative treatment (24), but some doubts have been expressed with regard to quality of lumber (fine checks may appear) (36).

Drying by Boiling in Oils

The heat needed to evaporate moisture is provided by immersing the wood in hot oil with a boiling point higher than that of water. The method is usually applied in combination with preservative treatment of wood that has a high initial moisture. Raising the temperature of the preservative (usually creosote) is combined with the application of vacuum in a treating cylinder (see Chapter 19); in this manner, drying temperatures are lower than 100°C (212°F).

A more general application of this method to dry lumber is not justified, because special treating installations are needed, and the wood absorbs a certain quantity of oil (according to a study, about 4%), with the result that its color is changed and it becomes more flammable (24). Also, defects may appear (checking, casehardening, and sometimes honeycombing), because high temperatures are used and it is not possible to control the rate of moisture exit from the wood.

Drying with Solvents

This method is similar to the one used to determine the moisture content of wood by distillation (see Chapter 9). The wood is placed in an air-tight chamber (distillator) and is sprayed for a few hours with hot acetone (about 90°C or 200°F). After that, the liquid (a mixture of acetone, wood moisture, extractives) is removed, while air is introduced and circulated to complete the drying and to remove the remnants of the solvent. After drying, the solvent is reclaimed by distillation (experimentally up to 90%) and is reused (6).

This method has been initiated to remove resin from softwood knots, but was applied experimentally in the United States to dry lumber (pine, oak, redwood) by use of other solvents (e.g., methanol) with good results. The advantages are relatively fast drying (oak sapwood, 2.5 cm or 1 in. thick, was dried to a moisture content of 10% in 30 h) and collection of extractives, but the main disadvantage is high cost (6).

Drying with High-Frequency Electricity

Theoretically, drying wood by placing it in a field of high-frequency electric current presents considerable interest. The wood is heated fast (by 5–20°C/min or 10–35°F/min) (24) and uniformly; therefore, drying proceeds quickly. The method is suitable for automation; the material (lumber or other wood products) is transported on a conveyor and, passing through an electric field, it is dried in a continuous flow. Pine lumber, 25 mm (1 in.) thick, was dried in less than 3 h (conventional kiln-drying took 3 days); shoe-lasts, about 10 cm (4 in.) thick were dried from 40% to 8% moisture content in 2 h (6, 56). However, refractory species (where the anatomical structure does not facilitate the exit of evaporated moisture) may develop defects (checking) and strength may be reduced (24). The high cost for equipment and energy are also limiting factors for use of this method except for high-value products (45).

Other Methods

Other methods employ vacuum, centrifuging, and ultraviolet radiation. The main attraction of vacuum-drying is that it allows vaporization and removal of moisture at temperatures considerably below 100°C (the boiling point of water) almost as rapidly as with high-temperature drying above 100°C (212°F). A combination of high-frequency and vacuum-drying has also been successfully tried (15, 52). However, such methods are considered uneconomical (24, 46). In centrifuging, the lumber is placed in a revolving cage, where temperature and relative humidity are controlled; drying is said to be fast and economical, without defects, but the method is rather theoretical without known practical application (6). Drying by ultraviolet radiation (i.e., exposure of separate pieces of wood to ultraviolet light) has technical limitations for practical application and is considered uneconomical (50). Microwave drying has also been tried (2).

STEAMING

Steam is used in kiln drying to control relative humidity in the drying chamber, for the purpose of preventing or correcting wood defects (checking, casehardening, etc.). However, steaming is also applied for other reasons, such as change of the natural color of certain woods, or preparation for veneer production and steam bending.

Steaming to change the natural color is applied to beech, walnut, maple, and some other species. European beech is often steamed; the wood attains a pinkish-brown color, color irregularities caused by "red heart" (see Chapter 7) may be masked, and wood-working properties are considered improved. As a result, steamed beech is more easily marketed, at a higher price, and used in making certain types of furniture and other products.

Steaming may be carried out in conventional dry kilns, but this is rare, because steaming is wearing on equipment. Usually, special steaming chambers are used, made of concrete or bricks, with an internal coating resistant to acids (acetic and formic acids are produced during steaming) (24, 56), but prefabricated metallic units also exist with a coating of stainless aluminum alloy (18) (Figure 18-23). The lumber is piled green and without stickers. Steaming is carried out under atmospheric pressure or a little higher (1.1–1.7 atm), with a gradual increase of temperature from about 70°C (160°F) to 100°C (212°F) (25), or 60–

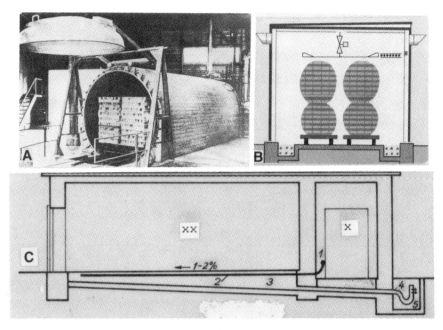

Figure 18-23. Steaming chambers. (A) A prefabricated unit (aluminum lined). (B) A combined dry kiln and steaming chamber. (C) Longitudinal section of a brick-concrete construction: xx, steaming chamber; x, control room; 1, main steam supply; 2, piping to chamber; 3, ditch to discharge condensates; 4, condensate trap; 5, drainage canal. (A and B courtesy of Hildebrand.)

90°C (140–190°F) (7, 56), depending on wood species, and is continued for 24–48 h or more, depending on lumber thickness (18 h for each cm of thickness) (7, 56). After steaming, the lumber is not abruptly exposed to the environment, but is gradually cooled in the chamber, and subsequently piled with stickers for air-drying.

Steam acts by dissolving or altering (oxidizing, decomposing) extractives and other chemical constituents of wood (24, 56). Aside from improving the color,[13] it is believed in woodworking practice that steaming improves machinability, gives smoother surfaces, and protects wood from warping. Research results are as follows: steamed wood exhibits a small increase of shrinkage (12, 55), (which should be attributed to the removal of extractives), and a small reduction of hygroscopicity (emc) (28). Some references report a better dimensional stability (the wood is "un-nerved") (7, 56), which may be due to release of growth stresses (27). Removal of extractives and greater shrinkage affect density; dry density shows a small increase (55, 57), and basic density a small reduction (28). Most mechanical properties (bending, compression, hardness) remain practically unchanged after steaming, but toughness is reduced. Screw-holding capacity is lower (55, 57). Machining tools wear faster (12). Moisture content is reduced a little after steaming green wood, but remains fairly high; the moisture of green beech lumber, after steaming at 60–80°C (140–175°F), was reduced by about 20–25%, but a moisture content of 50–60% was retained (18). In air-drying that follows, steamed and nonsteamed lumber did not show important differences (55, 57) (Figure 18-24). In kiln-drying, some workers observed a faster drying of steamed lumber (in oak lumber,[14] drying time was reduced up to 50%), but others dispute the existence of differences if steaming is done at temperatures lower than 100°C (212°F) (56). Walnut steamed for 3 days at 100°C was kiln-dried initially faster, but later there was no substantial difference between steamed and nonsteamed wood (9).[15]

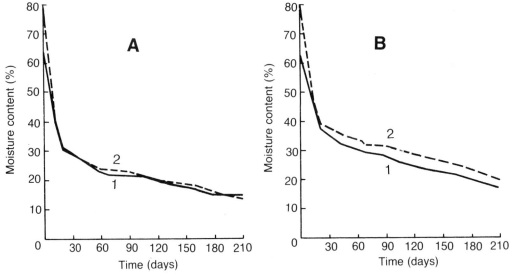

Figure 18-24. Rate of air-drying of steamed (———) and nonsteamed (-------) beech lumber, (A) 5 cm (2 in.) thick and (B) 8 cm (3 in.) thick. Drying in a shed from November to June. (After Refs. 55 and 57.)

REFERENCES

1. Alvarez Noves, H. and J. I. Fernandez-Golfin Seco. 1990. Solar drying of sawn lumber in Spain. *Holz Roh-Werkst.* 48:173–178.
2. Arganbright, D. G. 1979. Developments in applied drying technology, 1971–1977. *For. Prod. J.* 29(12):14–20.
3. Bois, P. J. 1977. Constructing and Operating a Small Solar-heated Lumber Dryer. U.S. For. Prod. Lab. Tech. Rep. No. 7.
4. Boone, R. S. 1986. High-temperature kiln-drying red maple lumber—Some options. *For. Prod. J.* 36(9):19–25.
5. Brauner, A. B., and E. M. Conway. 1964. Steaming walnut for color. *For. Prod. J.* 14(11):525–527.
6. Brown, W. H. 1965. *An Introduction to the Seasoning of Timber.* Oxford, New York: Pergamon Press.
7. Centre Technique du Bois. 1977. Conseils Pratiques pour le Séchage des Bois. Paris.
8. Chech, M. Y., and D. R. Huffman. 1974. High-temperature Drying of Mixed Spruce, Jack pine and Balsam Fir. Dept. Environ. Canada, Publ. No. 1337, Ottawa.
9. Chen, P. Y. S., and E. C. Workman. 1980. Effect of steaming on some physical and chemical properties of black walnut heartwood. *Wood Fiber* 11(4):218–227.
10. Chen, P. Y. S. 1981. Design and tests of a 500 BF solar kiln. *For. Prod. J.* 31(3):33–38.
11. Forest Products Laboratories, Canada. 1951. Canadian Woods. Ottawa.
12. Forest Products Research Laboratory, England. 1955. The Steaming of Home Grown Beech. Leaflet No. 16.
13. Forest Products Research Laboratory, England. 1967. The Steam Bending Properties of Various Timbers. Leaflet No. 45.
14. Forrer, J. B. 1984. An electronic system for monitoring gradients of drying wood. *For. Prod. J.* 7/8:34–38.
15. Harris, R. A., and A. C. Lee. 1985. Properties of white pine lumber dried by radiofrequency/vacuum process and conventional kiln process. *Wood Fiber Sci.* 17(4):549–552.
16. Hustede, K. 1979. *Schnittholztrocknung.* Stuttgart: Deutsche Verlags-Anstalt.
17. Janik, W. 1965. *Handbuch der Holztrocknung.* Leipzig: VEB Fachbuchverlag.
18. Hildebrand, R., ed. 1970. *Kiln Drying of Sawn Timber.* Oberboihingen, Wuertt: Hildebrand Maschinenbau.
19. Ingenieurschule f. Holztechnik, Dresden. 1965. *Taschenbuch der Holtechnologie.* Leipzig: VEB Fachbuchverlag.
20. Joly, P., and F. More-Chevalier. 1980. *Théorie, Pratique & Economie du Séchage des Bois.* Dourdan, France: H. Vial.
21. Kalnin's, A., E. Mikits, K. Upmanis, and J. Staprans. 1972. Kontinuierlich arbeitende Trockenkammer für Schnittholz. *Holztechnologie* 13(4):199–201.
22. Klamke, F. A. and P. N. Peralta. 1990. Laser incising for lumber drying. *For. Prod. J.* 40(4):48–54.
23. Koch, P., and W. L. Wellford. 1977. Some mechanical properties of small specimens cut from 1.79 inch-thick southern pine dried for 6 hours at 300°F or for 5 days at 180°F—A comparison. *Wood Fiber* 8(4):235–240.
24. Kollmann, F. F. P., and W. A. Côté. 1968. *Principles of Wood Science and Technology. I. Solid Wood.* Berlin: Springer Verlag.
25. König, E. 1972. *Holz-Lexikon.* Stuttgart: DRW-Verlag.
26. Kordes, W. 1980, 1981. Regelungstechnik für die

Schnittholztrocknung heute. *Holz Roh Werkst.* 38:419–422, 39:11–15.
27. Kubinsky, E. 1971. Der Einfluss des Dampfens auf die Holzeigenschaften. *Holzforsch. Holzverwert.* 23(1):1–11.
28. Kubinsky, E. 1971. Influence of steaming on the properties of *Quercus rubra* wood. *Holzforschung* 25(3):78–83.
29. Langendorf, G., and H. Eichler. 1973. *Holzvergütung.* Leipzig: VEB Fachbuchverlag.
30. McCollum, M. P. 1986. Effect of high-temperature drying on the grade yield and shrinkage of southern pine lumber. *For. Prod. J.* 36(3):51–53.
31. McIntyre, S. 1975. Operation and Maintenance of Pneumatic Record-Controllers for Dry Kilns. W. For. Prod. Lab. Canada, Inform. Report VP-X-92.
32. McLauchlan, T. A. 1975. Continuous Moisture Meters for Kiln Dried Dimension Lumber. W. For. Prod. Lab. Canada, Inform. Report VP-X-147.
33. Panshin, A. J. 1949. Lumber Dry Kilns and Their Operation. Michigan State College, Ag. Expt. Station Sp. Bull. 359.
34. Peck, E. C. 1956. Air Drying of Lumber. U.S. For. Prod. Lab. Rep. No. 1657.
35. Peck, E. C. 1962. Drying 4/4 red oak by solar heat. *For. Prod. J.* 12(3):103–107.
36. Pratt, G. H. 1974. *Timber Drying Manual.* Dept. Environ., BRE, HMSO, London.
37. Rassmussen, E. 1961. *Dry Kiln Operator's Manual.* U.S.D.A., Ag. Handbook No. 188.
38. Rietz, R. 1970. Air Drying Lumber in a Forklift Yard. U.S. For. Prod. Lab., Note FPL-0209.
39. Rietz, R. C., and R. H. Page. 1971. *Air Drying of Lumber.* U.S.D.A., Ag. Handbook No. 402.
40. Rosen, H. N. 1981. Drying processes for the year 2000. *Proc. 17th IUFRO World Congress, Div. 5*, pp. 183–196, Kyoto.
41. Schmidt, K. 1986. Untersuchungen über die Anwendbarkeit der Schallemissionens-analyse (SEA) zur Steurung der Holztrocknung. Proceedings 18th IUFRO World Congress, Div. 5, pp. 102–112. Ljubljana, Yugoslavia.
42. Schneider, A., F. Engelhardt, and L. Wagner. 1979. Vergleichende Untersuchungen über die Freilufttrocknung und Solartrocknung von Schnittholz unter mitteleuropäischen Wetterverhältnissen. *Holz Roh Werkst.* 37(11):427–433.
43. Schulz, G. 1969. Freilufttrocknung von Buchenschwellen. *Holz Roh Werkst.* 27(9):326–333.
44. Siau, J. F. 1984. *Transport Processes in Wood.* Berlin, New York: Springer Verlag.
45. Simpson, W. T. 1983/84. Drying wood: A review. In *Drying Technology*, Vols. 2(2) and 2(3), 2(2):235–264, 2(3):353–368. New York: Marcel Dekker.
46. Simpson, W. T. 1987. Vacuum drying red oak. *For. Prod. J.* 37(1):35–38.
47. Smith, W. B., and J. F. Siau. 1979. High temperature drying of red pine. *J. Inst. Wood Sci.* 8(3):129–133.
48. Stevens, W. C., and G. H. Pratt. 1952. *Kiln Operator's Handbook.* London: H.M.S.O.
49. Stevens, W. C., and N. Turner. 1948. *Solid and Laminated Wood Bending.* London: H.M.S.O.
50. Timber Research and Development Association (TRADA). 1969. Timber Drying. U.K.
51. Timber Research and Development Association (TRADA). Contemporary Drying (undated).
52. Trofatter, G., R. A. Harris, J. Schroeder, and M. A. Taras. 1986. Comparison of moisture content variation in red oak lumber dried by a radio-frequency/vacuum process and a conventional kiln. *For. Prod. J.* 36(5):25–28.
53. Tsoumis, G. 1955. Moisture content of wood under the climatic conditions of Greece. *Technica Chronica* 371/372:158–163 (Greek, English summary).
54. Tsoumis, G., and E. Voulgaridis. 1980. Experimental air-drying of pine, fir and beech lumber. *Geotechnica* 1:26–31 (Greek, English summary).
55. Tsoumis, G., and E. Voulgaridis. 1984. Properties and drying behavior of steamed beech wood. *Proc. Balkan Sci. Conf.* 3:50–58. Sofia.
56. Villière, A. 1966. *Séchage des Bois.* Paris: Dunod.
57. Voulgaridis, E., and G. Tsoumis. 1982. Effects of steaming on properties of beech wood. *Ann. Dept. For. Nat. Environ.* 25:165–181 Aristotelian University (Greek, English summary).
58. Wassipaul, F., M. Vanek, and A. Mayrhofer. 1986. Klima und Schallemissionen bei der Holztrocknung. *Holzforsch. Holzverwert.* 38(4):73–79.
59. Weinbrenner, R., ed. 1965. *Holztrocknung.* Stuttgart: DRW-Verlag-GmbH.
60. Wengert, E. M., and F. M. Lamb. 1988. Matching a drying system to a mill's requirements. *World Wood* 29(5):31–33.
61. Williston, E. M. 1976. *Lumber Manufacturing.* San Francisco: Miller Freeman.
62. Yang, K. C. 1980. Solar kiln performance at a high latitude, 48° N. *For. Prod. J.* 30(3):37–40.

FOOTNOTES

[1]The discussion that follows applies to lumber unless otherwise specified.

[2]Manual handling of lumber is a rare situation in lumber yards; forklifts are practically in general use.

[3]Other sticker dimensions are 1×2, 1.5×2, 2×2, 2×2.5, 2.5×3, 3×3, 3×4, 4×4 cm, and in English units 0.5×1, 1.5×1.5 in., etc.

[4]Checking has the meaning of external separation of wood tissue as a result of drying stresses. Checks occurring on the ends of boards often develop into splits.

[5]Pit aspiration (see Chapters 3 and 4) reduces the rate of drying by reducing the permeability of wood; aspiration occurs mainly in earlywood and when moisture approaches the fiber saturation point; the effects on drying increase with increasing temperature (44).

[6]The rate of air-drying may be accelerated by installation of large fans, 1.5–2 m (5–6 ft) in diameter between the

piles ("fan air-drying"). This arrangement is not effective during cold and humid weather, but may be combined with raising the temperature to 20–40°C (70–95°F) under shelter. Drying time is very much reduced in comparison to conventional air-drying (20, 24, 56). Laser incising has also been tried to accelerate drying (in conventional kiln-drying) and was effective in some species, but led to reduction of bending strength (MOR), and does not appear to be economical (22).

[7]Exposure of wood to temperatures higher than 60°C (140°F) will cause small, permanent reduction of strength, but such reduction is insignificant for most uses, because the temperatures of kiln-drying are relatively low and their action is of short duration (37).

[8]Changes of moisture content may also be monitored by acoustic emissions caused by microcracks developing in wood due to drying stresses. Such emissions can be used as warning signals that defects are about to occur, thus allowing control of drying rate. The changing speed of sound may also provide a monitoring system (41, 58).

[9]Except for molding, these defects are caused by differential shrinkage. It may be noted here that shrinkage in a kiln may be different from normal; it tends to be greater at higher temperatures and smaller with a faster drying rate (51).

[10]Sawing and other machining of wood should not be done immediately after kiln-drying, but after cooling to the temperature of the environment.

[11]Table 18-4, derived from another bibliographical source (60), does not clearly support this contention.

[12]Experimental high-temperature drying (dry bulb thermometer 150°C or 300°F, and maximum possible difference between dry and wet bulb) has been applied by continuous passage of lumber boards through a kiln. Lumber 25–40 mm (1–1.5 in.) thick was dried in 3–4 h. To avoid warping, the drying boards were supported by revolving rolls (2, 40).

[13]Steaming beech has little effect on its color if the wood is air-dry, or there may be a change to an undesirable gray (16). In steaming walnut, it has been observed that higher temperature, higher moisture content, and longer steaming time, contribute to faster and more uniform change of color (5).

[14]Prolonged steaming of oak at temperatures higher than 60°C (140°F) may cause collapse (56).

[15]Steaming preparatory to bending is not applied to green wood, but a minimum moisture of about 15% (exterior layers) is required. The increased plasticity of wood is attributed to heating (25). Ash, elm, maple, oak, plane (sycamore), and hickory have good bending properties (13, 49). Steaming for veneer production is discussed in Chapter 20.

19

Preservative Treatment

Wood may be protected from the action of destructive agents—such as fungi, insects, marine organisms, and fire—by preservative treatment (i.e., introduction of suitable chemicals into its mass). In this manner, the durability of wood is considerably increased[1].

The destruction or degradation of wood in the form of logs or products has much greater economic consequences than the attack of standing trees in a forest. Harvesting and processing increase the value of wood, and replacement of products is usually expensive. Other problems may also arise, such as the interruption of electric power or telecommunication during the replacement of poles.

Increasing the natural durability of various wood products is especially important for products exposed to environmental conditions that favor biodegradation, such as poles, piling, railroad ties, posts, and mine timbers. The consequences of a quick degrade are greater in countries that are poor in local production of wood; however, the problem has wider implications due to a worldwide deficiency of wood, which is expected to intensify in the future (see Introduction).

PRESERVATIVES

Chemicals suitable as preservatives should be toxic to attacking organisms, or render wood less flammable. In addition, it is desirable that they should be able to enter more or less easily into the mass of wood, not leach or evaporate, not be hazardous to humans or animals, not be flammable or odorous, not decompose or discolor wood or oxidizing metals, not be expensive, and be chemically stable (exterior factors or time should not affect their action) (17, 21). None of the available preservatives meet all these requirements, but their deficiencies in this regard vary depending on intended use. Chemicals that have a wider use satisfy a greater number of these requirements.

The suitability of a certain chemical with regard to toxicity may be evaluated in a laboratory, or by out-of-door experiments, but actual service data are more realistic. Laboratory evaluation is based on testing the resistance of treated and nontreated specimens of wood. The specimens are placed on a nutrient (culture medium), such as agar or soil, inside small glass containers (Petri, Erlenmayer, Kollé). The containers are then placed in a chamber under controlled conditions of temperature, and the loss of mass of wood is determined (21). A laboratory evaluation may also be based on the quantity of preservative retained in a specimen of wood after repeated washing with water, or exposure to simulated changes of environmental conditions. However, such results are only indicative and comparative. Out-of-door experiments, where wood specimens of relatively large dimensions are exposed to a natural environment (Figure 19-1) are of similar value. The properties of fire retardants are judged in laboratories with specimens or entire wooden structures (e.g., doors).

Wood preservatives against fungi, insects, and marine organisms are divided into three categories—*water-borne*, *oils*, and *organic solvent* (oil-borne). Fire retardants belong to the

Figure 19-1. Samples of wood, in the form of small posts or stakes, placed in the ground to test suitability of preservatives. (Courtesy of *Reader's Digest*.)

water-borne type, but they are separately discussed because their manner of action is different.

Water-Borne

Water-borne preservatives consist of salt solutions of various inorganic chemicals, such as copper, chromium, arsenic, fluorine, mercury, and others. Such salts, usually in mixture, are dissolved in water and provide toxic solutions (8).

Water-borne preservatives are available in powder form or liquid concentrates. They enter easily into the mass of wood, are not flammable, usually are not odorous, are less costly than other preservatives, and treated wood can be painted after drying. Disadvantages of these preservatives are that the wood should be dried after treatment, they do not offer protection against mechanical wear, and they leach when wood is used under damp conditions or exposed out-of-doors (8, 17).

However, water-borne preservatives do exist that present a high resistance to leaching by fixation (i.e., formation of insoluble compounds in the wood). Copper-chromium-arsenic (CCA) belongs to this category, and has the largest usage, dominating the market of water-borne preservatives; it is used for treating poles and other exposed structures, including wood in water, such as piling. CCA is the subject of criticism, however, due to its arsenic content. In some countries, it is banned from uses where treated wood comes in contact with humans or animals; however, in other countries fixation is considered adequate and CCA safe even for treating wood for playgrounds, inhabitable places, food containers, and greenhouses. Cautious handling is necessary by workers in treating plants, while burning of treated wood is dangerous and inhaling of sawdust is suspected as hazardous (16). Aside from health effects, a progressive structural weakening of CCA-treated timber has been reported, particularly in poles (eucalypt and pine), which failed in service by peripheral separation of growth rings between earlywood and latewood, and by radial splits (4).

Other water-borne preservatives are ammoniacal copper arsenite (ACA), copper-chromium-boron (CCB), copper-chromium-phosphorous (CCP), fluorine-chromium-arsenic-phenol (FCAP or FCA), copper-chlorophenol, and others. Commerical preparations are offered under various trade names, such as Celcure, Tanalith, Boliden, Chemonite, Wolmanit, Basilit, and others (8).

Oils

Creosote is the main representative in this category. This preservative is very effective for treatment of railroad ties, poles, and pilings. Creosote is mainly produced from tar obtained from dry distillation of bituminous coal. Tar is distilled under increasing temperature of 200–400°C (390–750°F) and gives oils of different boiling points; the portion boiling above 355°C (670°F) is called "residue" (36). Creosote is available in different formulations: if the proportion of the higher boiling point components is low, it is called "low residue" or "light oil," and inversely "high residue" or "heavy oil"; heavy oils ensure long-lasting protection, but light components facilitate penetration in the wood.

Creosote has high toxicity against fungi, insects, and marine borers; it is insoluble in water, has low volatility, does not usually oxidize metals, has a high electrical resistance, and protects wood against weathering. It is easy to determine the depth of creosote's penetration, and different formulations can be applied to different practical needs. The disadvantages are that treated wood becomes more flammable (flammability is reduced with time), and it is difficult or impossible to paint. Also, creosote may "bleed" (i.e., exude on the surface of treated wood), has a disagreeable odor, and presents health hazards; for the latter reason, there are restrictions to its application.

Aside from bituminous coal, creosote may be produced from lignites, but this has a lower toxicity and is mainly used as a solvent to pentachlorophenol; other sources of creosote are peat, shale, or wood.

Organic Solvent

The main representative of this type of preservative is pentachlorophenol; it is produced by reacting chlorine with phenol, and is available in crystalline form. Pentachlorophenol is prepared with a variety of solvents, including petroleum, creosote, and volatile solvents (15).

The advantages of this preservative are its chemical stability, low solubility in water, low volatility, and high toxicity to fungi and insects (by the addition of insecticides). With choice of a suitable solvent, the wood is kept clean, may be painted and glued (43), and acquires water repellency. Pentachlorophenol does not oxidize metals and is not flammable—but wood is more flammable than untreated wood until the solvent is evaporated. This preservative presents health hazards and should not be used in inhabitable places; there are restrictions to its use similar to those of creosote (42). Pentachlorophenol in heavy oil possesses similar advantages and disadvantages to creosote. Water-soluble sodium salts of chlorophenols are used for preventing blue stain of wood. Trade names or organic solvent preservatives are Xylamon, Lindane, Rentokil, Dieldrin, and others (36)[2].

Fire Retardants

Fire retardants are not toxic; they are applied to reduce the flammability of wood and act by (a) creating a barrier to the spread of flame or (b) generating noncombustible gases. Chemicals of the first type speed the formation of a charred wood layer or form a massive layer of foam, while those of the second type provide fire-retardant action by melting, evaporation, or thermal decomposition (26).

Fire retardants contain silica and other chemical elements; the most popular compounds are ammonium phosphate, ammonium sulfate, zinc chloride, boric acid, and borates (36). Commercial preparations have such trade names as Minalith, Pyresote, Non-Com, and others (21).

Fire retardants are applied as coatings or by impregnation of wood. They are water-borne and therefore leachable; their effectiveness is reduced if treated wood is exposed to high humidities or comes in contact to water (26). Under such conditions of wood use, they are corrosive to metals, and may interfere with gluing or painting. Exterior-type fire retardants contain a water-soluble organic compound which polymerizes during kiln-drying of wood, and gives a leach-resistant, noncorrosive product

(43). Fire retardants may be effective against fungi or insects by mixing toxic chemicals (e.g., fluorine salts).

PREPARATION OF THE WOOD

The preparation of wood refers to its moisture, machining, and whether or not bark is removed.

With few exceptions, it is necessary that the wood is air-dry; its moisture content should be reduced to a maximum of 20–25% (i.e., below the fiber saturation point). Presence of water in cell cavities or excessive water in cell walls resists the entrance of preservatives[3].

Drilling holes, framing, and other machining applied to poles, railroads ties, or other products, as a necessary preparation to installation, should be carried out before any preservative treatment; otherwise, treated wood will be subsequently removed and untreated wood exposed. Another type of machining, called "incising," is applied to aid treatment by facilitating the axial movement of a preservative inside wood (11, 19, 21, 26, 33). Incisions are short slits made at intervals by knives (sometimes needles) through the surface of wood (mainly poles and ties, sometimes sawn wood); they are about 1 cm (0.4 in.) long, 0.5 cm (0.2 in.) wide, and 0.6–1.5 cm (0.25–0.6 in.) deep. In poles, incisions are made only in their base and to a length of about 90 cm (36 in.) of which 2/3 will be in the ground and 1/3 above it; ties are incised on all lateral surfaces (Figure 19-2). Incising is applied prior to air-drying to woods that are difficult to treat, such as spruce, hemlock (*Tsuga*), Douglas-fir, white oak, and others (35)[4].

The bark, both outer and inner, is impermeable to preservatives and, as a rule, is removed; exception is made in applying a preservative by hydrostatic pressure, in which case the bark is kept intact during treatment.

Finally, poles and other round-wood products (piling, highway marking posts) are "shaved" to remove irregularities and make their surface smooth.

METHODS OF TREATMENT

Introduction of preservatives into wood may be done by two categories of methods: (i) without pressure, and (ii) with pressure.

Treatment Without Pressure

This group includes brushing, spraying, immersion, hot and cold bath, diffusion and others.

Brushing. The preservative is applied by means of a brush. The method is simple and

Figure 19-2. Incising railroad ties; inserted is an incised pole base. (Photograph courtesy of U.S. Railroad Tie Association; insertion from Ref. 21.

economical but of limited effectiveness. Abundant preservative should be placed on the surface of the wood and inside checks, and this should be repeated from time to time, depending on conditions of exposure.

Spraying. This treatment requires an apparatus to spray the preservative with some pressure—such as a hand pump used to spray agricultural crops. The method is faster but its effectiveness is low, similar to brushing. Dispersion of preservative in the air may be hazardous to the health of workers; in addition, a considerable quantity of material (up to 50%) is lost.

Both brushing and spraying are mainly used for application of insecticides (35). Spraying is also suitable for fire-retardant coatings and for application of preservatives in a forest (e.g., to protect logs against blue stain).

Immersion. The wood is immersed in a preservative for a period of seconds, minutes, days, or weeks, depending on the desired degree of protection. Short-term immersion is applied in some sawmills, where the lumber is immersed immediately after sawing and while sitting on a conveyor that transports it to the lumber yard. Wood products may be immersed in total (e.g., lumber) or partly (e.g., the base of posts). The degree of protection depends on the type of preservative, wood characteristics (structure, density, moisture content, proportion of heartwood, roughness of surface), and duration of immersion. Immersion in creosote (also brushing) is more effective if the preservative is hot (8).

Hot and Cold Bath. This is a variation of immersion with the difference that the wood is first immersed in hot and subsequently in cold preservative. The treatment can be applied in three ways:

- Use two open tanks of which the first contains hot and the other cold preservative. The wood is quickly transferred from the first to the second tank.
- Empty the first tank and replace hot with cold preservative.
- Interrupt heating and allow the preservative to cool gradually.

The mechanism of impregnation is as follows. When wood is placed in a hot preservative, moisture evaporates and the air contained in cell cavities expands. In the following cold bath, the air contracts and penetration is aided by creation of a partial vacuum.

Creosote and organic solvent preservatives are mainly used, but water-borne preservatives are also suitable (except CCA, ACA, and other fixing formulations introduced by pressure in closed cylinders). With creosote, the temperature of the hot bath is about 90°C (190°F). For water-borne preservatives, the high temperature is adjusted to avoid losses from evaporation, and water is added at intervals in order to maintain the desired concentration of the solution. Preservatives that produce a sediment when heated are not suitable. The cold bath is usually not heated—except in cold weather (21).

In comparison to previous methods, the hot and cold bath ensures a faster and deeper penetration. The time that wood remains in each bath varies from 1 to 12 h or more. Small-dimension products and permeable species may be totally impregnated, but this is costly and usually pointless. It has been observed that in that case, a second hot bath of short duration (about 0.5 h) will contribute to the exit of surplus preservative and will reduce "bleeding" in service (8, 21).

Hot and cold bath treatment is applied by use of special installations (tanks, steam piping to heat the preservative, crane to transfer the wood from hot to cold bath; Figure 19-3), but may also be done with common metallic barrels—if the preservative does not react with metal (e.g., some copper compounds react with steel). The latter procedure is used by farmers to treat the base of fence posts[5].

A variation of this method involves application of a hot bath in water and cold bath in a preservative. This treatment is advantageous when the preservatives used may ignite if heated over a free flame (21).

Diffusion. In contrast to other methods, preservative treatment by diffusion is applied to green wood (e.g., poles or constructional timber). A water-borne preservative (boron salts, fluorine compounds) in paste form is brushed

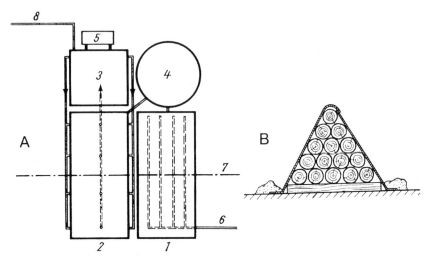

Figure 19-3. Treatment by (A) hot and cold bath and (B) diffusion. (A) Ground plan of an installation for hot and cold bath treatment: 1, hot bath; 2, cold bath; 3, cooling the preservative; 4, storage of preservative; 5, pump; 6, heat piping; 7, crane; 8, entrance-exit of cold water. (B) Treatment by diffusion. After the preservative is applied on their surface, the poles are stacked and covered. (After Ref. 25, reproduced by permission from VEB-Fachbuchverlag.)

on the surface of the wood, and is diffused from there into its mass, mainly in sapwood. The mechanism of treatment is, therefore, different in comparison to other methods; the preservative does not move through the capillary structure of wood—the vehicle of transportation is the water contained in wood. Diffusion is, by definition, equalization of concentration of a solution, and involves movement from a place of high concentration to a place of zero concentration until equalization takes place.

The method is mainly applied to species that are difficult to treat by other methods, such as spruce, hemlock, and Douglas-fir. After application of the paste, the wood is closely stacked, covered with waterproof paper or a plastic sheet, and left in that condition for some time (with poles it may take up to 3 months) (36). The method may be applied in the forest immediately after the trees are felled and converted to poles or posts.

Treatment by diffusion is also applied to standing poles partially attacked by decay at their base. In that case, the soil is dug out around the pole to a depth of about 50 cm (20 in.), and the decayed wood is removed or charred (21); preservative paste is then applied to the exposed surface of the pole up to a height of about 50 cm above ground—this is covered by waterproof paper or plastic, and the soil is placed back. A similar treatment is applied to poles by use of bandages containing the preservative (water-borne or creosote). The method is called "Osmose" process, but the term is also used in connection with the previously discussed diffusion process.

The disadvantages of treating wood with leachable preservatives can be minimized by "double diffusion"; this involves successive soaking of wood in chemicals forming compounds that are highly resistant to leaching (e.g., copper sulfate and sodium chromate (19, 24, 35).

Injection and Other Methods. Standing poles or railroad ties showing signs of decay are sometimes treated by injection of a preservative. This is done by use of an applicator equipped with a hollow needle ("Cobra" process); a viscous preservative is injected in certain positions from which it is diffused (36, 41). A similar treatment applies the preservative by injection through plastic nozzles inserted into previously opened holes (36).

A variation of diffusion into green wood[6] involves spraying such wood with usually hot preservative, then closely stacking and covering it (36). Fused borate rods placed in wood are sometimes used to treat products, such as window frames. Other methods include disinfection of the soil around the base of a pole (an ineffective method) and superficial charring of wood.

Treatment with Pressure

This category includes treatment by (a) hydrostatic pressure and (b) pressure in closed cylinders. The latter is the main methodology.

Hydrostatic Pressure. This method, called the "Boucherie" process (or "sap displacement" method), is traditionally applied to green poles with bark. The tank containing the preservative is placed on scaffolding, at a height of 10–15 m (30–50 ft), and the poles on the ground with their butt a little higher than their top (Figure 19-4). A disk, about 5 cm (2 in.) thick, is cut off from the butt of each pole to produce a clean surface, where a tight metallic cap is attached and connected by piping to the tank containing the preservative. Diffusion is involved in advancing the preservative inside the wood, but impregnation is assisted by hydrostatic pressure (1–1.5 atm) and is completed in 8–14 days, depending on wood species. The preservative (colored copper sulfate, CCB, FCAP) passes through the sapwood and comes out from the top end. Treated poles are stacked, and debarking is done later (26).

Modifications of this process include elimination of the scaffolding, use of open tanks or closed cylinders, caps with inflated cuffs, application of vacuum to the capped end while the other end is submerged in preservative, and debarking (36, 39). Thus, the preservative passes through sapwood at a much greater speed (20–30 times greater by application of vacuum) (24), and the traditional Boucherie process is gradually abandoned. Experimental investigations seek replacing the sap of living trees by preservative, prior to felling (27).

Pressure in Closed Cylinders. Impregnation by pressure in closed cylinders is the most effective method of protecting wood that is used under conditions that favor attack by fungi, insects, and marine borers. Such processes are mainly applied with creosote or other preservatives, which cannot be introduced, in adequate quantity and depth, by previous methods; some water-borne preservatives (CCA, ACA, etc.) are introduced in this manner.

The treatment is carried out in specially equipped plants. The main equipment is a closed cylinder or cylinders, up to 3 m (10 ft) in diameter and 60 m (200 ft) long, which is able to withstand pressures up to 8 atm or more. Other equipment includes storage and measuring tanks, pressure and vacuum pumps, a steam-producing facility, incising and pole-

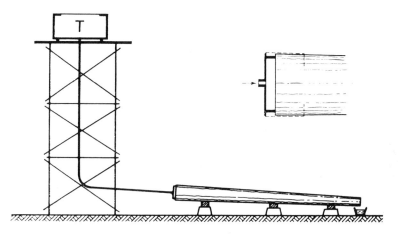

Figure 19-4. Treatment of a pole by hydrostatic pressure: (T) tank with preservative. (The attachment covering the base of the pole is shown enlarged.) (After Ref. 25, reproduced by permission from VEB-Fachbuchverlag.)

Figure 19-5. Installation to treat poles by pressure in a closed cylinder. (Courtesy of Scholz & Co.)

shaving machines, tram cars, and measuring instruments (21, 26) (Figures 19-5 and 19-6).

Preservatives are introduced into wood by two main methods: (a) *full-cell* and (b) *empty-cell*. In full-cell treatment, the aim is to introduce as much preservative as possible to coat the cell walls and more or less fill the cell cavities; whereas in empty-cell treatment some of the preservative is drained and little remains in the cavities (36, 43). The usual process is empty-cell, because it is less expensive without being inferior in effectiveness. Full-cell treatment is used in special cases (e.g., in treating piling with creosote for use in places of high danger of marine borer attack); it is also used with rapid-fixing water-borne preservatives (CCA, ACA, etc.).

Full-Cell Treatment. The treatment is applied in five stages: vacuum—fill the cylinder with preservative—pressure—release of pressure—vacuum (Figure 19-7). Specifically, after the wood is placed in the cylinder and its door is hermetically closed, a vacuum of 600 mm (24 in.) Hg (Torr) is applied and maintained for

Figure 19-6. Treatment of lumber with fire retardants in a closed cylinder. (Courtesy of Koppers Co., Inc.)

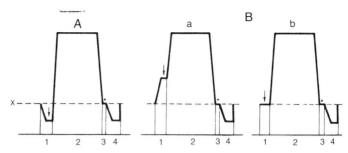

Figure 19-7. Diagrams showing the application of pressure processes. (A) *Full-cell* process: 1, initial vacuum (entrance of preservative while the vacuum is maintained); 2, pressure; 3, release of pressure-exit of preservative; 4, final vacuum. (B) *Empty-cell* process—(a) *Rueping*: 1, initial pressure (entrance of preservative while the pressure is maintained); 2, pressure; 3, release of pressure-exit of preservative; 4, final vacuum. (b) *Lowry*: 1, entrance of preservative at atmospheric pressure; 3, release of pressure-exit of preservative. Note that arrows indicate entrance of preservative (filling the cylinder), and dots (·) its removal.

at least 0.5 h (15 min to 1 h or more, depending on species). Subsequently, without introducing air, the cylinder is filled with preservative, and more is pumped in until the pressure is raised to the desired height (7–8 atm). This pressure is applied until the wood retains a certain quantity of preservative (according to specifications) or to the point of refusal. During the pressure period, water-borne preservatives are generally used at room temperature, because they may decompose if heated (35). Creosote is heated to 60–95°C (140–200°F). Finally, the pressure is released, while at the same time preservative is removed from the cylinder, and typically a low vacuum (400 Torr for 5–10 min) is applied to prevent bleeding of preservative from the surface of treated wood (21, 22, 26).

A variation of this method is applied by the gradual raising of pressure to improve the entrance of preservative (22, 26), but the effectiveness of such handling is doubted (21). Another variation applies double vacuum (*Vac-Vac* process). After an initial vacuum and filling the cylinder, the system is brought back to atmospheric pressure, and this helps the entrance of preservative through peripheral and transverse wood surfaces. When penetration stops, the preservative is removed and a new vacuum (higher than the initial) is applied; this is kept until a predetermined quantity of preservative enters the wood. The vacuum is released and the system is brought back to atmospheric pressure, under which the preservative penetrates deeper. The method is effective especially with lumber intended to be used for joinery (windows, etc.), where low retention and deep penetration of preservative are required (19).

Empty-Cell Treatment. The treatment is applied in two variations, known as "Rueping" and "Lowry" processes. In both, the initial vacuum (characteristic of the full-cell treatment) is missing, but they differ with regard to application of an initial pressure. The Rueping process includes the following stages: initial pressure—fill the cylinder with preservative—raise pressure—release pressure—vacuum. The Lowry process is different in that the initial pressure is missing (i.e., treatment starts at atmospheric pressure) (Figure 19-7). In both cases, main preservatives are creosote and pentachlophenol, but water-borne types may also be employed. The initial pressure is 1.5–4 atm[7] and is maintained for 5–10 min. Subsequently, and while this pressure is maintained, hot preservative (usually creosote) is introduced, the pressure is raised up to 7–8 atm, maintained for about 1 h or more, and released simultaneously with the removal of the preservative. Finally, a 600 Torr vacuum is applied for 10 min. This vacuum and the initial pressure, as applied in the Rueping process, are helpful for the excess preservative to exit from the wood. Thus, good penetration is attained with less preservative, and therefore empty-cell treatment is less costly.

Woods that are difficult to impregnate (e.g,

beech) are treated by the "double Rueping process" or a combination of Rueping (with creosote) and full-cell treatment (with a water-soluble preservative, usually sodium fluoride), or other modifications (20). Double Rueping includes the following stages: pressure (0.5–4 atm) for about 15 min—fill the cylinder while pressure is maintained—raise pressure (7–8 atm) and maintain it usually for 60 min—release pressure and remove the preservative—vacuum (at least 600 Torr) for 30 min—pressure (2.5–4 atm) for 15 min—fill the cylinder while pressure is maintained—raise pressure (7–8 atm for 180 min)—release pressure and remove the preservative—vacuum (at least 600 Torr) for 30 min (22)[8].

The full-cell and empty-cell processes are also applied with pentachlorophenol in liquified petroleum gas. After treatment, the solvent remaining in the wood is evaporated under reduced pressure, removed from the cylinder, recompressed, cooled, and reliquified (*Cellon* process) (19, 21). Impregnation becomes easier, but the solvent, which is gas under atmospheric pressure and temperature, must be stored and used under pressure in order to keep it in liquid form and to avoid losses from evaporation. Also, the solvent is flammable; therefore, special caution is needed to prevent dangers from explosion and fire.

A variation of the empty-cell process (MSU) is reported to give full-cell results with chromated copper arsenate (CCA). The method advocates: initial pressure—fill cylinder and raise temperature—maximum pressure and hold—reduce pressure, remove preservative—fixation period (increase temperature that helps fixing the preservative components in the wood)—release pressure, allow "kickout" of preservative—final vacuum. The "kickout" preservative can be segregated, treated, and returned to the working tank, and in this manner both cost-reduction and antipollution requirements are met. This method could be applied to other preservatives and procedures (5).

CRITERIA OF EFFECTIVENESS

The effectiveness of preservation can only be judged on the basis of the service life of treated wooden structures (Figure 19-8), but this is not an immediate and practical solution of the problem. For this reason, the quantity of preservative retained in the wood, and its depth of penetration, are taken as criteria of effectiveness.

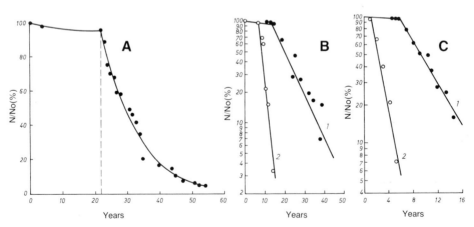

Figure 19-8. Service life of railroad ties and posts. (A) Ties (pine, spruce, larch, poplar) treated with creosote by the Rueping process. (B) Ties of pine (*P. banksiana*, as in A): 1, treated with creosote by the Lowry process; 2, untreated. (C) Posts: 1, treated with a water-soluble preservative (borax-boric acid); 2, untreated. Note that N/No (%) is proportion of ties or posts in service (N) in relation to their original number (No) in percent of the original. In all cases there is an initial period during which all (or almost all) structures are in service (in case A this period is 20 years or more). After that, their number is drastically reduced. Treatment shows such reduction (A–C) and prolongation of total service time (B, C). Data from Canada. (After Ref. 12; reproduced by permission from the Society of Wood Science and Technology, USA.)

Table 19-1. Treatment Specifications for Poles[a]

Species	Retention kg/m^3 (lb/ft^3)	Depth of penetration
Fir	128(8)	2 cm (~1 in.)
Pine	128(8)	6 cm (~2 in.)
Beech	112(7)	85% of sapwood
Oak, red	96(6)	65% of cross-section
Oak, white	96(6)	100% of sapwood

[a]Requirements of the State Utility Company of Greece for imported and locally produced poles. Other sources vary in requirements. German practice: poles (pine) 80–90 kg/m^3 (5–5.5 lb/ft^3), and railroad ties, pine 63 kg/m^3 (4 lb/ft^3), oak 45 kg/m^3 (3 lb/ft^3), and beech 160 kg/m^3 (10 lb/ft^3) (44).

The quantity retained (retention) is expressed in kg/m^3 (lb/ft^3, Tables 19-1 and 19-2). This is calculated from the volume of preservative consumed, conversion of volume to weight (on the basis of density and temperature of the preservative), and reference of such weight to the volume of the wood. In the case of water-borne preservatives, the calculations are based on dry weight of preservative.

The depth of penetration is observed on transverse sections or preferably by taking borings. If a preservative is not colored (e.g., sodium fluoride, zinc chloride), use of a suitable reagent will produce intense coloration. A uniform depth is important.

A combination of quantity and depth is a desired condition. The quantity may be determined by chemical analysis of depth samples. The required levels of these criteria are given by specifications, which vary according to species and intended use of the wood. Tables 19-1 and 19-2 show such specifications.

FACTORS AFFECTING TREATMENT

The quantity of preservative retained in the wood and its depth of penetration are affected by method of treatment, type of preservative, and wood characteristics. The factors pertaining to wood refer to its preparation, especially with respect to moisture content, and to such characteristics as anatomical structure, sap-

Table 19-2. Recommended Preservatives and Retentions[a]

Product and use	Preservative (lb/ft^3)		
	CCA	Creosote	Penta
Poles, utility			
Normal service conditions	0.60	7.5	0.38
Severe decay and termite areas (and structural poles)	0.60	9	0.45
Piling			
Soil or freshwater use and foundations	0.80	12	0.60
In salt water, severe hazard[b]	2.5(8) and 1.5	NR	NR
moderate hazard	NR	20	NR
Ties, railroad	—	8	—
Posts			
Guardrail and sign, round	0.50	10	0.50
sawn four sides	0.60	12	0.60
Fence round, half-round, quarter-round	0.40	8	0.40
sawn four sides	0.50	10	0.50
Lumber and timber			
Above ground	0.25	8	0.40
Soil or freshwater, nonstructural	0.40	10	0.50
structural (foundations, bridges)	0.60	12	0.60
In salt water	2.5	25	NR
Plywood			
Above ground	0.25	8	0.40
Soil or freshwater contact	0.40	10	0.50

[a]Minimum net retentions (1 lb/ft^3 = 16 kg/m^3).
[b]Severe hazard *Limnoria tripunctata*, moderate hazard *Pholads*. The retentions are based on two assay zones, 0–0.50 in. (0–1.2 cm) and 0.50–2 in. (1.2–5 cm). NR = not recommended.
Source: Society of American Wood Preservers, Inc.

wood or heartwood, density, species, and grain (direction of impregnation).

The *structure* of wood is a basic factor, because the entrance and movement of a preservative depends mainly on its cellular characteristics (cell cavities, pits, etc.), which form passages. This is important mainly for oils and oil-borne preservatives, because water-borne preservatives pass through the cell walls by diffusion, although much more slowly.

Differences in structure between softwoods and hardwoods give a different importance to characteristics of anatomical structure. In softwoods, the main cellular elements (i.e., tracheids) have a relatively short length and closed ends (see Chapter 3); therefore, the movement of preservatives is slower and the condition of the pits is of decisive importance. Pit aspiration by translocation (Figure 19-9) or deposition of extractives on their membranes, renders impregnation difficult or impossible. Pit aspiration occurs both in heartwood and in sapwood after drying, especially in the region of earlywood; latewood, which has fewer and smaller pits, is treated better (21, 31, 34). The extent and degree of aspiration may differ. The difficulty of treating spruce is attributed to pit aspiration, but the small permeability of rays is also important, because they have a small proportion of ray tracheids, which are considered to be the main radial passage. In Douglas-fir, deposition of extractives has also been observed on pit membranes of the sapwood (29).

Aspiration may be restricted by special drying methods (e.g., use of solvents; see Chapter 18) and freeze-drying; however, these methods are expensive for use in preparation for preservative treatment. It should be noted that pit membranes of sapwood may be returned to their normal position if wood is kept in water for some time. This may have practical importance for treatment with water-borne preservatives and pressure—but not CCA, ACA, and others, which are not used with wet wood (31).

In hardwoods, the vessels create passages of long length due to the absence or perforation of transverse walls in their members. For this reason, and also because their structure is different (pit membranes without openings), the role of pits is smaller. Movement is also affected by tyloses, which may plug the vessels. Tyloses are formed mainly in heartwood, but in some species they are also present in sapwood. For example, tyloses are the reason that treatment of white oaks is difficult in comparison to red

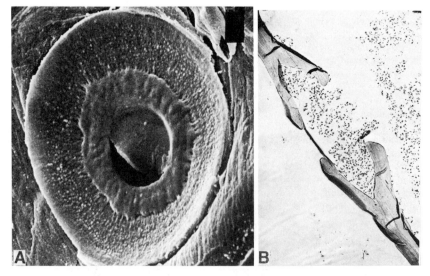

Figure 19-9. Displacement of pit membranes (pit aspiration) hinders penetration. (A) A torus lowered on the aperature (pine, SEM 1800×). (B) Fine particles are not allowed to pass (2200×). (A, promotional photograph by Zeiss; B, courtesy of W. A. Côté).

oaks (see Keys in Appendix II). For the same reason, it is difficult to treat the "red heart" of beech (Chapter 7).

Other structural characteristics are of lesser importance with respect to impregnation. Resin canals exercise a favorable effect in pines, but in other softwoods (spruce, larch, Douglas-fir) their presence does not contribute to an easier treatment in comparison to species without such canals. In heartwood, resin canals are plugged with tylosoids (see Chapter 3). Parenchyma cells (axial and ray) do not seem to exercise an important role, nor do the fibers of hardwoods. In general, sapwood is treated more easily than heartwood. This makes treatment of poles and posts easier and more uniform than treatment of square material.

Density is an index of empty spaces in the mass of wood and, therefore, it may be assumed that it affects impregnation. However, there is no general relationship between ease of impregnation and density, because the basic factor is anatomical structure. A relationship seems to exist within groups of wood (e.g., certain light ring-porous hardwoods are easily impregnated, but in softwoods the situation may be opposite); for example, pitch pine is treated easily, although its density is higher than other species (29).

The effect of *species* is dictated by the previous factors. Certain species are easy and others difficult to treat (Table 15-1). However, the ability of a wood to absorb preservative does not guarantee its durability; this property is influenced by other factors, such as toxicity, resistance to leaching, and distribution of a preservative in the mass of wood (at both macroscopic and microscopic levels), and exposure conditions of the treated wood (9).

The *grain* of wood is very important with regard to entrance of preservatives. Impregnation is much easier in the axial direction due to the cellular arrangement. The ratio of axial-to-side permeability (as expressed by the depth of impregnation) is, on average, 15:1 for oils and oil-borne preservatives and 20:1 for water-borne preservatives (21).

Artificial improvement of permeability, and therefore improvement of preservative treatment, is possible by enzymes and microorganisms (bacteria, molding fungi), which are able to cause perforation of pit membranes (and erosion of the walls mainly of nonlignified parenchyma cells; see also Chapter 15) (23, 27, 28, 31). In practice, such a result may be achieved for spruce poles, for example, by storing them in water (3). It has been observed that, by short-time enzymatic action, it is possible to treat all sapwood in spruce. Spraying with water-containing bacteria may also reduce surface "bleeding" of creosote, because the permeability of wood is increased and, by application of a final vacuum, more creosote is removed (3). On the other hand, bacteria may decompose (detoxify) creosote and grow in treated wood (28, 38).

EFFECT ON THE PROPERTIES AND USES OF WOOD

The introduction of preservatives in the mass of wood affects not only its durability but also other properties of this material. The changes induced depend on the properties of preservatives and the conditions of treatment. Some effects have been previously discussed, others relate to painting, flammability, odor, strength, electrical resistance, gluing, weight of wood, and oxidation of metallic connectors.

Painting of wood treated with creosote is problematic. The difficulty is less pronounced some time after treatment, but the results are not satisfactory. Also, the wood may "bleed." However, the main products treated with creosote (poles, piling, railroad ties) are not painted; therefore, the practical importance of this situation is nonexistent or limited. With pentachlorophenol, the type of solvent is important, because the preservative itself has no adverse effects; heavy oils create problems similar to those of creosote, whereas the behavior of wood treated with light solvents is influenced by the method of treatment. Treatment by simple methods (e.g., short-time immersion) causes little difficulty, but pressure treatment may cause exudation of preservative or formation of crystals on the surface of wood (17). The Cellon method of treatment does not interfere with painting, because pentachlorophenol is deposited in solid form. Also, wood treated with water-borne preservatives is painted without difficulty after drying, while some preservatives of this category impart color to light-colored woods (8, 21) (a type of CCA is called "Greensalt").

Flammability is affected after treatment with creosote; the wood becomes more flammable, but after some time, when the more volatile (and more flammable) compounds are evaporated, there is no practical problem. Wood attacked by decay fungi is also more flammable. Certain water-borne preservatives (e.g., zinc chloride) reduce flammability, but high doses reduce the strength of wood as well (21). Preservatives containing copper (or copper and chromium) may cause self-ignition; for example, poles may burn after a creeping fire (as when burning the residual stalks of agricultural crops, such as wheat or corn) if they are not soaked with water.

The *odor* imparted by creosote, pentachlorophenol, and other preservatives is reduced with time, but has no practical importance when the wood is used out-of-doors.

The *strength* of wood treated under pressure in closed cylinders may be reduced as a result of high temperatures and pressures that may be employed. Pressure is more important, especially when heating is prolonged, but short duration allows higher pressures. The combined effect of temperature and pressure depends mainly on wood species and type of preservative. Light woods are affected less. Under the same conditions of temperature and pressure, woods treated with water-borne preservatives have a greater tendency to collapse (21). In railroad ties, creosote (or creosote in petroleum) acts as a "greasing" or "lubricating" agent, reducing damages from friction and strokes. The effect of pentachlorophenol depends on the solvent; if the preservative is deposited as solid (in the form of crystals) in the mass of wood, its properties are not affected. Water-borne preservatives containing chromium increase hardness but decrease the toughness of wood (17). In general, the reduction of strength that may be normally caused by preservative treatment is slight (7).

Electrical resistance is not substantially affected by creosote, but water-borne preservatives may reduce it, unless by-products are formed that are insoluble in water (see Chapter 14, Figure 14-3).

Gluing of wood treated with water-borne preservatives is easier in comparison to oils and oil-borne preservatives. The latter reduce the polarity of wood surfaces and interfere with wetting and bonding. In comparison to untreated wood, the strength of joints is somewhat reduced, but is considered satisfactory if gluing is performed under proper conditions (21). When possible, it is preferable to treat after gluing, but joints may form barriers to the passage of preservatives, and glued structures may be difficult to handle depending on their size and shape[9].

The *weight* of wood is affected in proportion to the weight of retained preservative, but changes of weight may also be due to changes of the moisture content of the wood.

Finally, *oxidation* is a problem only with water-borne preservatives when the moisture content of wood is kept high.

OTHER EFFECTS

Preservatives are toxic chemicals; thus, poisoning, skin irritation, and other adverse effects may be caused by contact or inhalation on the part of workers if proper precautions are not taken (2, 26, 37). Such health hazards have led to restrictions on use or prohibition of certain preservatives (42). A note was previously made about CCA. Environmental pollution may also occur by discharging preservation residues, but such pollution by wood preservation plants is small in comparison to other industries, because preservatives are stored and reused (27, 40).

SUBSTITUTION OF WOOD

Wood may deteriorate (decay, burn, etc.), and for this reason it is sometimes replaced by metals, plastics, or cement. In the area of treated products, the substitution refers to reinforced concrete poles, and reinforced concrete or steel railroad ties[10]. Such products may be more expensive than those made of wood, and they do not always last longer. Steel ties are subject to oxidation and may present problems of connection and insulation, while the service life of concrete ties is shorter due to repeated stressing^2; their stiffness is also higher in comparison to wood, and this relates to travel comfort. Reinforced concrete poles are also more prone to break under impact. In spite of substitution, on a worldwide basis, more poles and ties are wooden (32).

REFERENCES

1. Adolf, F. P. 1976. Untersuchungen zum Tränkverfahren von enzymatisch vorbehandeltem Holz. *Holz Roh Werkst.* 34(5):163–166.
2. Baker, M. J. 1979. Developments in the protection of wood and wood-based products. *J. Inst. Wood Sci.* 8(4):161–166.
3. Banks, W. B. 1970. Prevention of Creosote Bleeding from Treated Pine Poles. Timberlab Paper No. 34, For. Prod. Res. Lab., England.
4. Bariska, M., A. Pizzi and W. E. Conradie. 1988. Structural weakening of CCA-treated timber. *Holzforschung* 42(5):339–345.
5. Barnes, H. M. 1985. Trends in the wood-treating industry: State-of-the-art report. *For. Prod. J.* 35(1):13–22.
6. Bodig, J. 1984. Personal communication.
7. Bodig, J., and B. Jayne. 1982. *Mechanics of Wood and Wood Composites.* New York: Van Nostrand Reinhold.
8. British Wood Preserving Association—Timber Development Association (undated). Timber Preservation.
9. Building Research Establishment. 1979. The Resistance of Timbers to Impregnation with Wood Preservatives. Princes Risborough Lab., England.
10. Déon, G. 1978. *Manuel de Préservation des Bois en Climat Tropical.* France: Centre Technique Forestier Tropical.
11. Forintek Canada Corp. 1985/1986. Incising technology update (and Wood poles or concrete poles). *The Forintek Review* (December 1985/January 1986).
12. Gillespie, T., H. P. Sedziak, and J. Krzyzewski. 1969. The analysis of service life experiments on wood treated with preservatives. *Wood Sci.* 2(2):73–76.
13. Gjovik, L. R., and H. L. Davidson. 1975. Service Records on Treated and Untreated Fenceposts. U.S. For. Prod. Lab. Res. Note FPL-068.
14. Graham, R. D. 1973. History of wood preservation. In *Wood Deterioration and Its Prevention by Preservative Treatment,* Vol. I, ed. D. D. Nickolas pp. 1–30. Syracuse, New York: Syracuse University Press.
15. Graham, R. D. 1980. Converting 1979 wood preservation problems into opportunities for 1999. *For. Prod. J.* 30(2):17–20.
16. Greaves, H. 1985. CCA-Treated Timber: Facts, Figures and Comments on Health and Safety in Use. CSIRO, Div. Chem. and Wood Tech., Forest Products Newsletter n.s. No. 1, Australia.
17. Hartford, W. H. 1975. Chemical and physical properties of wood preservatives and wood preservative systems. In *Wood Deterioration and Its Prevention by Preservative Treatment,* Vol. II, ed. D. D. Nickolas, pp. 1–120. Syracuse, New York: Syracuse University Press.
18. Helsing, G., and R. D. Graham. 1976. Saw kerfs reduce checking and prevent internal decay in pressure-tested Douglas-fir poles. *Holzforschung* 30(6):184–186.
19. Henry, W. T. 1973. Treating processes and equipment. In *Wood Deterioration and Its Prevention by Preservative Treatment,* Vol. II. ed. D. D. Nickolas, pp. 279–298. Syracuse, New York: Syracuse University Press.
20. Hösli, J. P. 1980. Untersuchung über die Imprägnierbarkeit von Buchenholz mit Steinkohlenteeröl. *Holz Roh- Werkst.* 38(3):89–94.
21. Hunt, G. M., and G. A. Garratt. 1968. *Wood Preservation,* 3rd ed. New York: McGraw-Hill.
22. Ingenieurschule f. Holztechnik, Dresden. 1965. Taschenbuch der Holztechnologie. Leipzig: VEB Fachbuchverlag.
23. Johnson, R. B. 1979. Permeability changes induced in three western conifers by selective bacterial inoculation. *Wood Fiber* 11(1):10–21.
24. Kollmann, F. F. P., E. W. Kuenzi, and A. J. Stamm. 1975. *Principles of Wood Science and Technology. II. Wood Based Materials.* Berlin: Springer Verlag.
25. Langendorf, G. 1961. *Handbuch für den Holzschutz.* Leipzig: VEB Fachbuchverlag.
26. Langendorf, G. and H. Eichler. 1973. *Holz-Vergütung.* Leipzig: VEB Fachbuchverlag.
27. Levi, M. P. 1973. Control methods. In *Wood Deterioration and Its Prevention by Preservative Treatment,* Vol. I, ed. D. D. Nickolas, pp. 183–216. Syracuse, New York: Syracuse University Press.
28. Liese, W., ed. 1975. *Biological Transformation of Wood by Microorganisms.* Berlin: Springer Verlag.
29. Liese, W., and J. Bauch. 1967. On anatomical causes of the refractory behaviour of spruce and Douglas-fir. *J. Inst. Wood. Sci.* 19:3–14.
30. Morrell, J. J., and M. E. Corden. 1986. Controlling wood deterioration with fumigants—A review. *For. Prod. J.* 36(10):26–34.
31. Nickolas, D. D., and J. F. Siau. 1973. Factors influencing the treatability of wood. In *Wood Deterioration and Its Prevention by Preservative Treatment,* Vol. II., ed. D. D. Nickolas, pp. 299–343. Syracuse, New York: Syracuse University Press.
32. Öllmann, H. 1979. Holzschwellenaufkommen und Bedarf weltweit. *Forstarchiv* 50(7/8):165–168.
33. Perrin, P. W. 1978. Review of incizing and its effects on strength and preservative treatment of wood. *For. Prod. J.* 28(9):27–33.
34. Petty, A. 1970. The relation of wood structure to preservative treatment. In *The Wood We Grow* (Soc. For.

Britain), pp. 29–35. England: Oxford University Press.
35. Purslow, D. F. 1974. *Methods of Applying Wood Preservatives*. London: BRE, H.M.S.O.
36. Richardson, B. A. 1978. *Wood Preservation*. London: The Construction Press.
37. Ruddick, J. N. R. 1980. Wood Preservation and Its Application to Hardwoods of Western Canada. Forintek Canada Corp., Symposium Proceedings, Sp. Bull. No. SP-2.
38. Scheffer, T. C. 1973. Microbiological degradation and the causal organisms. In *Wood Deterioration and Its Prevention by Preservative Treatment*, Vol. I. ed. D. D. Nickolas, pp. 31–106. Syracuse, New York: Syracuse University Press.
39. Stalker, I. N., and C. S. McClymont. 1976. The effectiveness of the suction sap-displacement method with some hardwoods and softwoods. *J. Inst. Wood Sci.* 7(3):5–9.
40. Thomson, W. S. 1973. Pollution control. In *Wood Deterioration and Its Prevention by Preservative Treatment*, Vol. II, ed. D. D. Nickolas, pp. 345–396. Syracuse, New York: Syracuse University Press.
41. Tsoumis, G. 1956. Preservative treatment of railroad ties, poles and other wood structures by the method COBRA. *Technica Chronica* 107/108:28–31 (Greek).
42. U.S.D.A. 1981. The Biologic and Economic Assessment of Pentachlorophenol, Inorganic Arsenicals, Creosote. Vol. I, Wood Preservatives. Tech. Bull. 1658-II.
43. Wilkinson, J. G. 1979. *Industrial Timber Preservation*. London: Ass. Business Press.
44. Willeitner, H. 1986. Holzschutz. In *Holz-Taschenbuch*, eds. R. von Halasz, and C. Scheer, pp. 51–79. Berlin: Ernst & Son.

FOOTNOTES

[1] Pole durability was reported to increase from 6–12 to 30–70 years, piling from 1 year or less to 10–12 years, railroad ties from 2–3 to 35 years (36) (oak from 12–15 to 25 years, pine from 5–7 to 15–18 years, beech from 2–3 to 30–35 years), posts from 10 or less to 50 or more years (13), and mine timbers from 2–3 to 35 or more years (36).

[2] Another category of chemicals that may be used to arrest and prevent attacks of wood by decay fungi or marine borers are fumigants, such as NaMDC (Sodium N-methyl dithiocarbamate), MIT (methylisothiocyanate), chloropicrin, and trademark preparations. They are considered effective in treating poles, piling, timbers, export logs, chips, and other wood products. Nearly all fumigants are applied as liquids, whose vapors diffuse through wood; however, they are hazardous to health, and elaborate safety precautions should be taken. There are also problems of environmental pollution, including disposal of fumigant-treated wood. Efforts are made to develop solid fumigants (capsules, pellets) (30).

[3] Exception is made in introducing certain water-borne preservatives by diffusion or hygrostatic pressure ("Boucherie" process), where wood to be treated is in "green" condition (see further in this chapter under "Methods of Treatment"). A third case is the so-called "Boulton" process, where green wood (poles and piling) is placed in a closed cylinder and flooded with hot preservative, usually creosote; as a result, the moisture content of wood is reduced, and subsequent preservative treatment by application of pressure is facilitated (35).

[4] In addition to aiding preservative treatment, incising has been observed to facilitate drying and reduction of checking, especially in hardwoods, but may cause a reduction of strength (up to about 50% in loblolly pine—after laser incising, *For. Prod. J.* 40/4:48–54, 1990). It has been observed that poles may be protected from large checks and susequent interior decay by making a longitudinal-radial sawcut from their base to a height of 1.5 m (5 ft) from the ground or 2/3 of height (18).

[5] Sale of creosote, pentachlorophenol, and inorganic arsenicals is now banned in many countries for health reasons, except for licenced applicators, such as treating plants (42).

[6] The concept of diffusion applies also to treatment of air-dry wood (e.g., by brushing, spraying, immersion, or contact to preservative). Vehicle of transport is the preserving liquid or moisture absorbed by exposed wood (contact with wet ground, atmospheric moisture). In such cases, diffusion is acting in combination with capillary movement.

[7] The initial pressure is lower with woods that are easy to treat and higher in the opposite case. Low pressure is also applied to species with a low transverse compressive strength, and to steamed wood in order to avoid the incidence of collapse or checking (21).

[8] Experience is very helpful in setting the conditions of treatment (i.e., magnitude and duration of pressure and vacuum).

[9] Experimental beams, about 15 cm × 8 cm in cross-section and 3 m long (6 in. × 3 in. × 10 ft) made by gluing 2.5 mm (0.1 in.) thick Douglas-fir veneers with phenol-resorcinol resin, were thoroughly treated with creosote and pressure, in contrast to similar beams of solid Douglas-fir, where penetration was limited to a very small depth. The successful treatment of glued beams was attributed to checks ("loose" faces) of the veneers (6).

[10] Wood products that are or may be treated are many. Aside from poles, piling, and railroad ties, they include mine timbers, naval constructions, agricultural tools, lumber (for windows, exterior doors, etc.), fence and roadmarking posts, bridge timbers, and others.

20

Veneer

Veneer is a term applied to thin sheets of wood, usually 0.5–1.0 mm and sometimes up to about 10 mm (0.4 in.) in thickness (9, 18). Veneers are sometimes used as such in certain products (matches and match boxes, novelties), but usually they are glued into plywood or laminated wood products.

Use of veneered wood originated long ago. In ancient Egypt, small veneer pieces of valuable woods (e.g., ebony) were used for decorative purposes in combination with other materials, such as precious stones, metals, and animal horns. In the tombs of the Pharaohs, archaeologists have found very artistic furniture made in this manner. Wood was similarly used in ancient Greece and Rome (20).[1]

The manner of veneer production in those times is not accurately known. An Egyptian mural is interpreted as showing production by knife-cutting (see Chapter 21, Figure 21-1); other methods were probably splitting, axe-cutting, and sawing, but the available hand tools did not allow for the production of large pieces. Hand production was practiced for about 3000 years, up to the beginning of the nineteenth century, when mechanical saws were first used, and later machines with cutting knives.

Today, the production of veneer is also made by cutting (peeling, slicing) and (seldom) by sawing; however, machines are available that work fast and accurately, and produce large quantities of varying dimensions and controlled quality. Development of the veneer industry is closely related to improvements in adhesives and gluing methods, limited availability of valuable wood species, and certain advantageous properties of products made with glued veneer in comparison to solid wood.

WOOD AS RAW MATERIAL FOR VENEER

Source

Technically, all species may be used for production of veneer. In practice, about 50 species are preferred (see Table 20-1) for various reasons, such as availability, ease of processing, aesthetics, and wood structure and properties (physical, mechanical, gluability). Softwood and hardwood species with a uniform texture, without large differences between earlywood and latewood, are preferred, but this is not a limiting factor; for example, southern yellow pine[2] and Douglas-fir, which have an abrupt transition between earlywood and latewood, are major species of veneer for softwood plywood in the United States, and black pine, with a similar structure, is used in Europe. Most suitable hardwoods are diffuse-porous species, and ring-porous (e.g., oak) with narrow growth rings where latewood is almost missing. Tropical woods are extensively used, and have the advantage that most of them do not form pronounced growth rings (8).

There are two categories of veneer woods—*decorative* and *utility* (Figure 20-1). The distinction is not clear, because woods of the second category may be used to produce decorative veneers by special processing methods or suitable utilization of growth abnormalities.

Table 20-1. Veneer Woods

1. Temperate
   a. Softwoods: cedar, Douglas-fir, fir, hemlock, pine, redwood, spruce
   b. Hardwoods: ash, beech, birch, cherry, elm, eucalypt, hickory, maple, poplar, sycamore, walnut, yellow poplar
2. Tropical
   a. African: Afara, Afzelia, Aiélé, Amazakoué, Antiaris, Avodiré, Beté, Bubinga, Difou, Ebony, Idigbo, Iroko, Kosipo, Limba, Mahogany, Makoré, Niangon, Obeche, Okoumé, Opepe, Padauk, Sapele, Satinwood (East African), Tiama, Walnut (African), Zebrano.
   b. American: Cativo, Rosewood (Brazilian, Honduras), Primavera, Satinwood, Mahogany.
   c. Asian: Meranti (Lauan), Ramin, Rosewood (Indian), Seraya, Teak

Decorative species, due to color and grain, have a greater value, because they are used to surface furniture, interior plywood paneling, and related products. In contrast, utility (nondecorative) species are used for partitions of furniture and cabinets, and other structural uses, or in products coated with paint or other materials.

Veneer logs vary widely in length and diameter; they are usually up to about 3 m (10 ft) long with minimum diameter 15 cm (6 in.), maximum length about 7 m or 23 ft, and maximum diameter 2 m or 6 ft. The logs are usually brought to factories in long lengths, which are later reduced according to intended use and machine capacity.

Storage

Veneer logs, especially those intended for decorative veneer, are more valuable than saw logs, and for this reason better care is needed for their protection against deterioration. Protective measures include storage in water (navigated waters, artificial ponds), or on land (cold decking) with continuous water spraying (sprinkling). Certain valuable species, such as tropical woods transported long distances, are given protection by end-coating or metallic (S-shaped) bars, and they are stored under shelter. Prolonged storage is avoided.

Preparation

The preparation of veneer logs includes cross-cutting, sorting by grade and size, debarking, and heating to facilitate their processing.

Cross-cutting is done by use of large chain saws or circular saws. Cross-cutting in the factory, rather than in the forest, contributes to reduction of wood waste, and end-checking during storage is avoided.

Debarking is sometimes applied to long logs by ring debarkers or hydraulic power (water under high pressure); "blocks" or "bolts" (final-size logs after cross-cutting) are debarked by "lathe" debarkers, which remove the bark by a scraping action after a heat treatment (6, 7).

Heating, in hot water or steam, softens the wood and facilitates cutting.[3] Additional effects of heating are prolonged life of knives,

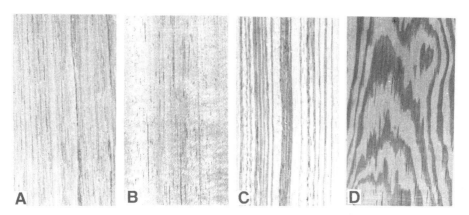

Figure 20-1. Decorative and utility veneer: 1, teak; 2, mahogany (Khaya); 3, Zebrano; 4, pine.

increased rate of production, equalization of heartwood–sapwood moisture, improvement of wood color (certain susceptible, light-colored woods may develop local discolorations), killing fungi and insects that may inhabit wood, and improvement of veneer quality. Improvement of quality refers to uniform thickness, smooth surfaces, and less checking.

Heating in water is a milder treatment but more expensive in comparison to steaming; more water and more time and energy are needed to heat the logs (Figure 20-2). The technique is mainly applied to decorative and rare species, and to logs with abnormal growth (stem forkings or "crotches," burls, etc.) that produce veneer with pronounced figure.

The temperature of the water is usually not higher than 80–90°C (180–200°F) (5, 19). Species of very high density, difficult to cut, are exposed to lower temperatures. Low temperatures (35–40°C or about 90–100°F) (24) are also applied to woods containing abnormal growth. The duration of heating depends on wood species or density (Figures 20-2 and 20-3), log diameter, moisture content (time after felling in the forest), water temperature,

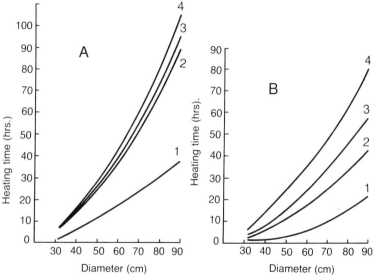

Figure 20-2. Dependence of heating time on log diameter and (oven-dry) density of wood: (A) hot water, (B) steaming (steam temperature 100°C or 212°F). Note that numbers 1, 2, 3, and 4, in both A and B, indicate wood densities of 0.4, 0.5, 0.6, and 0.7 g/cm³, respectively; in A, they also indicate respective water temperatures of 49°C (120°F), 66°C (150°F), 77°C (170°F), and 93°C (200°F). (After Ref. 13.)

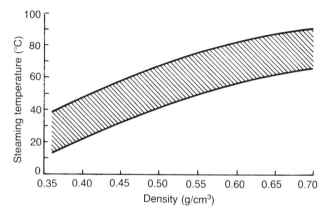

Figure 20-3. Optimal temperatures for softening wood by steaming in relation to (oven-dry) density. (After Ref. 23.)

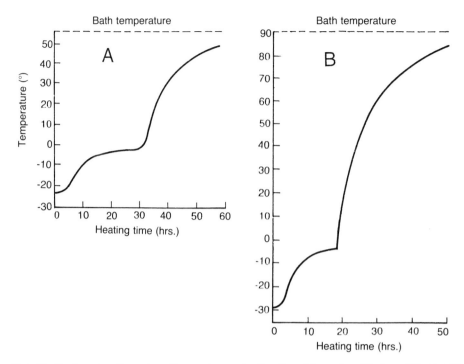

Figure 20-4. Heating in water—frozen logs. Temperature vs. heating time at the center of the logs: (A) aspen (distance from center 6 in. or 15 cm) and (B) red oak, 8.25 in. (20.5 cm). (Adapted from Ref. 25).

and whether logs are frozen (Figure 20-4). It is important for the desired temperature to reach the interior of wood up to the depth of processing.

Steaming is applied in two ways: (a) by direct introduction of steam into the steaming chamber (*direct steaming*), and (b) by evaporation of water from the bottom of the steaming chamber (*indirect steaming*). In direct steaming, the steaming chamber is simpler, but the temperature is raised too rapidly; this may cause degradation (checking), especially if steam is acting directly on a part of the log, because of stresses developing due to thermal expansion of the wood. Indirect steaming is preferred, because the temperature is raised slowly; therefore, the treatment is milder and the danger of degradation is reduced. Indirect steaming requires a more complex installation, but the possibility of reusing the condensed steam is an economic advantage (5, 19).

Heating installations include *vats*, where a batch of logs is placed and heated by water spray, steam, or a combination (steam and water); *tunnels*, where a system of chain conveyors continuously transports logs through a space heated as in the previous case; or hot-water *tubs*, where the logs are placed in hot water (6, 7).

Heating in water or steaming are not always needed. Soft woods (e.g., poplar) in green condition, are easily cut without preparation.

METHODS OF PRODUCTION

There are three methods of producing veneer: (a) rotary cutting (peeling), (b) slicing, and (c) sawing. Rotary cutting produces a continuous sheet, but in slicing and sawing the sheets are separate.

Rotary Cutting

This is the usually method; about 90% of all veneer (practically all softwood veneer) is produced by rotary cutting. The main use of such veneer is to make plywood. Small quantities are utilized in veneer form or used for decorative purposes, but decorative veneer is usually produced by the other methods.

Rotary cutting is made by revolving prepared veneer logs in a lathe and against a knife, which peels a continuous sheet of wood; the length of the knife is equal to the length of the log. The length of a log varies depending on the intended use of the veneer and is determined by the length of the lathe (Figure 20-5). Large machines may process shorter logs after suitable shortening of the distance between the spindles that hold the revolving log, or by use of "back rolls" (see below).

The speed of a lathe varies from about 50 to 300 spindle revolutions per minute. Productivity, as measured by the speed of veneer output (m/min), depends on log diameter and lathe speed. For a given number of spindle revolutions, the smaller the log diameter, the slower the average speed of veneer output. Speeds range from about 30 to 350 m/min. (Pine logs with an average diameter of about 30 cm or 12 in. can be peeled in 20 s) (6). Veneer produced at low speed has a coarse surface and nonuniform thickness, and knives wear faster (19). Veneer may be peeled to about 0.5 in. (1.2 cm) for laminated wood products.

A lathe is equipped with a "pressure bar," which is placed parallel to the knife and is of equal length (Figures 20-6 and 20-7). The bar compresses the wood shortly before the knife enters, in order to avoid formation of checks in the veneer. Preparation of wood by heating has the same purpose, but a properly made and positioned pressure bar is the decisive factor. The bar may be fixed or equipped with rollers (28) (Figure 20-7). Knife and bar move automatically toward the rotating log.

The production of good quality veneer requires proper setting of the lathe. The angle at which the knife is mounted is very important. Angles vary from about 15–23°; greater angles are used for softwoods with a pronounced density difference within growth rings, and hard, intergrown knots, and smaller angles (16–20°) for hardwoods with uniform structure (19). An average angle of 21° is considered suitable to serve different wood species at least in starting trials (22). The knife tip is set at or very near the axis of the lathe spindle or center of the peeled log. This distance, as well as the vertical distance between pressure bar and knife, is adjusted by trial in order to avoid defects and to obtain an optimum surface smoothness of the veneer. Too much pressure exerted by the bar will crush the wood, and too little pressure will

Figure 20-5. Veneer production by rotary cutting (peeling). Interior of a factory: logs are shown transported to the machine (lathe). (Courtesy of Cremona.)

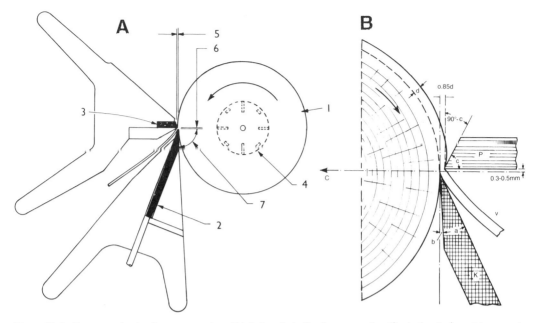

Figure 20-6. Veneer production by rotary cutting. (A) 1, log; 2, knife; 3, pressure bar (firm); 4, attachment to support the log (chuck); 5 and 6, distances, knife edge—pressure bar; 7, cutting angle; 8, veneer. (B) Detail of cutting: K, knife; P, pressure bar; a, knife angle; b, angle knife-to-wood; c, angle of pressure bar; d, veneer thickness; C, center of log; v, veneer. (A, courtesy of U.S. Forest Products Laboratory; B, adapted from Refs. 5 and 19).

produce badly checked veneer. The thicker the peeled veneer, the deeper are the checks. The vertical distance between knife and bar is set at 80–95% of peel thickness, depending on wood species (6). A lathe should not be used to peel many different species and log lengths, or to produce a wide range of veneer thicknesses, in order to avoid the need for frequent change of settings.

Veneer produced by peeling has two sides—outer or *tight* and inner or *loose*; the latter has fine or large checks, called "lathe checks," but these can be regulated by proper preparation of the wood and positioning of the pressure bar (12) (Figure 20-8). Lathe checks are taken into consideration in making plywood, where loose sides are placed inward in surface layers.

Peeling is continued to a minimum diameter, according to the condition of the wood and the design of the lathe. If the heartwood is soft, with decay and checks, the log ("bolt" or "block") is supported by "chucks" of a larger

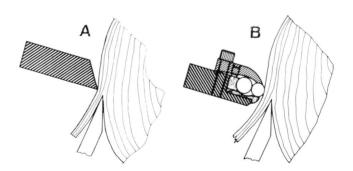

Figure 20-7. Pressure (nose) bars: (A) fixed, (B) roller (double). (After Ref. 28.)

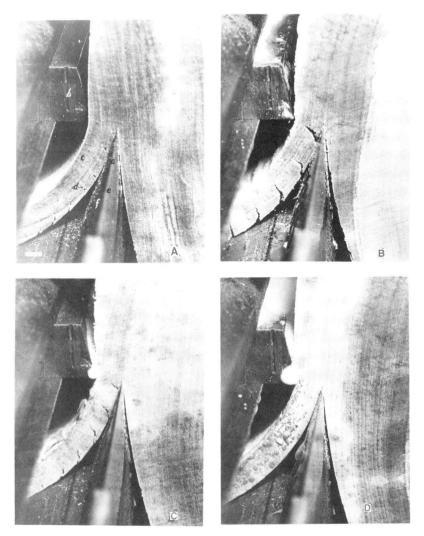

Figure 20-8. Effects of pressure bar on veneer quality. (A) Conventional production—small checks in the "loose" side: a, knife; b, pressure bar; c, veneer; d, checks; e, face of knife; f, back of knife. (B) The bar does not exercise proper pressure; checks and surface roughness are pronounced. (C) The bar is placed at a distance to knife edge 93% of veneer thickness; veneer is relatively smooth and checks less pronounced. (D) The distance is 86% of veneer thickness; there are no checks, the veneer is tight and smooth, but the wood overcompressed. Note that these results are based on experiments of rotary cutting birch veneer, 8 mm or $\frac{5}{16}$ in. thick. (Courtesy of U.S. Forest Products Laboratory).

diameter; therefore, the core that remains also has a larger diameter. Normally, a minimum diameter is about 10 cm (4 in.), but long cores may be reduced in length and transferred to smaller machines for further peeling, where their diameter may be reduced to about half (20); "backup rolls" (cylindrical supports revolving opposite the knife at midpoint) also permit peeling to smaller cores without shortening the bolt (6).

Considerable progress has been made regarding automatic handling of bolts (i.e., placement, centering, support, and core ejection). In some cases, this is done by use of computers. Each bolt is electronically scanned as to diameter variation along its length and shape (taper or cross-sectional deviation from circular), and the optimum spindle center (which may be off-pith of the bolt) is computed for maximum yield of usable veneer.

Figure 20-9. A veneer rewinding machine. (Courtesy of Babcock.)

Peeling produces a continuous veneer ribbon—after some initial, irregular pieces (called "fishtails") are removed—until the bolt is rounded to a cylinder. This ribbon is then transported by conveyor to one or more green veneer clippers per lathe, or rewound (reeled) into rolls or led directly to veneer dryers (Figures 20-9 and 20-19). A clipper is a guillotine-type knife, employed to remove defects (large knots, decayed or discolored parts, pitch pockets, splits, or voids, etc.), and to produce individual sheets of acceptable (widest possible) width for the predetermined plywood panel dimensions. Clipping is manual (by a worker activating a knife) or semiautomatic (the worker marks the clipping position with a special marking pencil and the knife is automatically activated by its presence); however, in some cases, it is computerized and automated by signals originating from defects or width dimensions (6). Green veneer clippers are eliminated in the case of veneer transported directly to dryers; clipping is done after drying. However, this system (continuous drying and subsequent clipping) is adaptable only to good grade bolts, and requires special arrangements to catch up with the fast production of veneer (more tray storage levels, additional clippers).

Clipped veneer is sorted: sapwood and heartwood sheets are stacked separately, because they are dried with different schedules unless they are present in small amounts. Improvement of grade is done by patching (i.e., removal of defects and insertion of plugs of sound wood); this is done by patching or plugging machines.

Variations of rotary cutting are sometimes applied by eccentric positioning of the log in the machine or placement of half or parts of a log (Figure 20-10).[4] This is done for the purpose of improving the figure of veneer; in common rotary cutting, with a centered log, the veneer has a tangential and usually nondecorative figure. In these variations, the speed of production is lower (19).

New developments are a "spindless" machine, an "incising" roller, and powered backup rolls. In a spindless machine, the log is placed between three revolving rolls and against a stationary knife (Figure 20-12). Both lower

VENEER 317

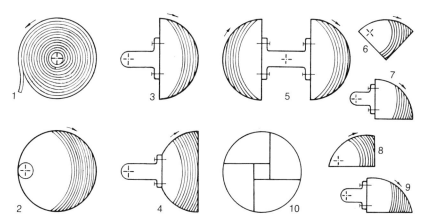

Figure 20-10. Rotary cutting methods: 1, continuous cutting; 2, eccentric placement; 3-5, half logs; 6-9, log sections (method of section preparation is shown in 10). (Adapted from U.S. Forest Products Laboratory, and Ref. 19).

rolls move gradually upward and toward the upper roll, which revolves in a steady position. The method is suitable for small-diameter logs; recovery of veneer is higher (core size is smaller) than with a conventional lathe (4). The incizing ("tenderizing") roller arrangement is shown in Figure 20-13.[5]

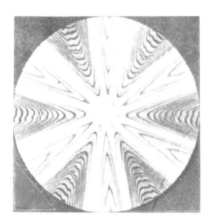

Figure 20-11. An attractive figure (table-top) produced by gluing sectors of elm veneer. (After Ref. 18, reproduced by permission from DRW-Verlag Weinbrenner.)

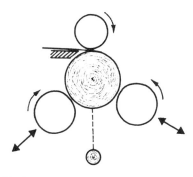

Figure 20-12. Principle of a "spindless" rotary cutting machine. (Courtesy of Durand-Raute Industries).

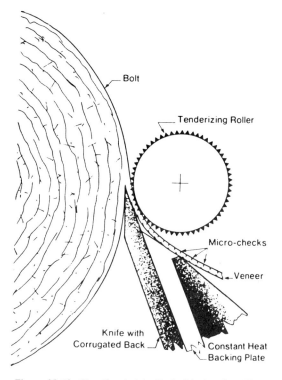

Figure 20-13. The "tenderizing" (incising) roller. The knife is steam-heated and the roller introduces a pattern of holes on the tight side of the veneer. This treatment results in reduced drying time, reduced splitting and breaking, and—according to laboratory tests—virtual elimination of "press blows" during subsequent manufacture of plywood. (Courtesy of Forintek Canada Corp.)

Slicing

This method is exclusively used to produce decorative veneer. The wood is mounted on the slicing machine, not in round form but after sawing into "flitches" (Figure 20-14). The flitches are then prepared by heating in water or steaming, and are subsequently sliced lengthwise into veneer strips by a machine, where the wood or a knife move rhythmically in a vertical or horizontal direction (Figure 20-15). Slicing machines are also equipped

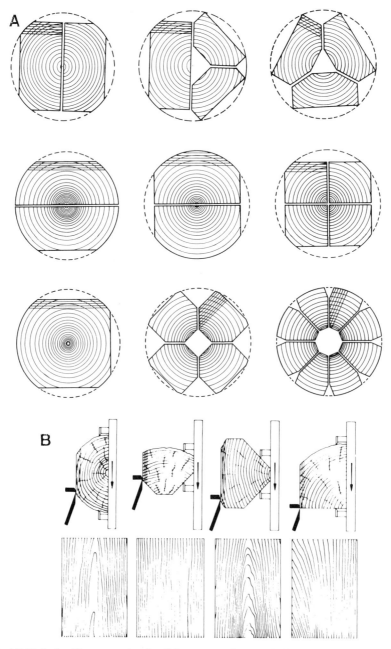

Figure 20-14. (A) Methods of log preparation for slicing veneer. (Same methods may be applied in veneer production by sawing.) (B) Figure produced in relation to log preparation. (B after Ref. 15, reproduced by permission from Springer-Verlag.)

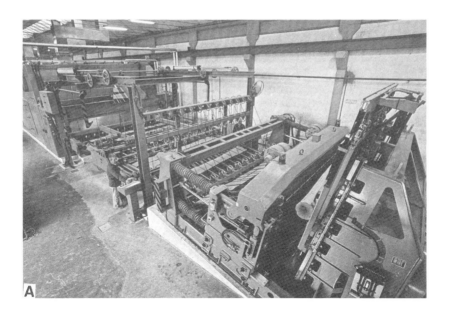

Figure 20-15. Veneer slicing machines: (A) vertical (with automatic transfer of veneer), (B) horizontal (veneer handled by hand—an older arrangement). (Courtesy of A, JEM and B, Cremona.)

with a pressure bar and, as in rotary cutting, the settings of bar and knife are decisive with regard to veneer quality (Figures 20-16 and 20-17).

After each cut, the knife or the wood move closer by a distance equal to veneer thickness. Veneer thicknesses vary from about 0.5 to 5 mm. (Some horizontal slicers produce very thin decorative sheets, 0.2 mm in thickness.) Slicing continues until the flitch is reduced to a 1.5–2 cm (about $\frac{1}{2}$–1 in.) residue. Some slicers are equipped with a vacuum table to hold the flitch—the bottom side of which is made smooth by planing. Thus, not only mounting the wood by clamps is eliminated, but the remaining flitch residue is very thin (17, 22). Another innovation is that the sliced veneer, both in horizontal and vertical machines, is removed automatically by means of a vacuum conveyor and not by hand.

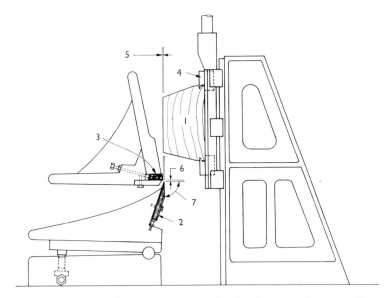

Figure 20-16. Veneer production by vertical slicing: 1, wood; 2, knife; 3, pressure bar; 4, attachment to support the flitch; 5 and 6, distance knife edge-pressure bar; 7, angle of cutting. (Courtesy of U.S. Forest Products Laboratory.)

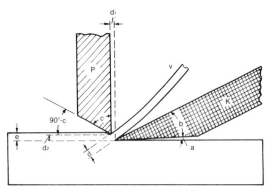

Figure 20-17. Arrangement for horizontal reciprocation: K, knife; P, pressure bar; a, angle knife–wood; b, knife angle; c, angle of pressure bar; d, actual distance knife-pressure bar (d_1, vertical—d_2, horizontal distance); e, veneer thickness; v, veneer. (Adapted from Refs. 5 and 19.)

Sawing

The production of veneer by sawing is an older method, but it is seldom used because a considerable part of the wood is transformed to sawdust. The veneer is of high quality (there are no "loose" and "tight" faces) and finds specialized uses (e.g., musical instruments). The logs are preshaped as for slicing, but no heat treatment is applied. Flitches are sawed with special frame saws or circular saws. Minimum veneer thickness is about 1 mm.

DRYING

In the past, veneer was dried in kilns similar to those used to dry lumber, and sometimes it was air-dried. Presently, veneer and plywood industries are equipped with specialized dryers. They are prefabricated metallic chambers, 8–30 m (25–100 ft) or more in length and up to 5 m (15 ft) wide, where temperature, air circulation, and speed of veneer transport are controlled. Veneer sheets, or a continuous ribbon, are introduced through one end of the dryer, move slowly, and come out dry from the other end.

Heat is generated by natural gas or steam. Steam is the oldest and most frequent source of heat, and it is often generated from boilers operating on wood residues. Temperatures vary (60–180°C or 140–350°F, and sometimes higher) (6, 9, 20), according to species, initial moisture content, and veneer thickness, and drying time is respectively affected. Air is circulated by large fans, placed parallel or transverse to the direction of veneer movement (11, 19) (Figure 20-18). The veneer is transported by various propeller mechanisms that include drums (rollers) and endless belts (wire mesh).

Air flow may be longitudinal (along the

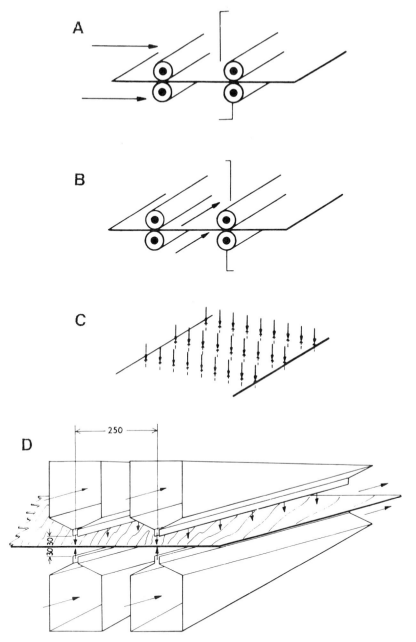

Figure 20-18. Air circulation in veneer drying: (A) parallel to veneer travel; (B) transverse to veneer travel; (C, D) transverse to veneer surfaces (through the veneer); (D) jet dryer (numbers in mm). (After Refs. 10 and 11, reproduced by permission from DRW-Verlag Weinbrenner.)

length of the dryer), or transverse (cross-circulation). In the first type, the direction of air flow is opposite to the direction of veneer movement; the temperature is raised from entrance to exit (it falls at the entrance due to evaporation of moisture), resulting in slower drying at the beginning. Transverse movement of the air favors fast drying at the beginning and a reduced total drying time; moisture is evaporated at a high speed from the surface layers, and there is fast diffusion from the interior outward (5, 11).

Figure 20-19. Continuous drying of rewound veneer after peeling (dryer on the left with entering veneer sheet, and veneer rolls on the deck at right). (Courtesy of JEM.)

Propelling the veneer by *drums* (pairs revolving in opposite directions and placed at short distances from one another) contributes to flattening and better appearance of the veneer (the smaller the drum distance, the better); however, this arrangement is not suitable for fragile sheets and small thicknesses (< 1 mm), and also when the direction of veneer movement is transverse to the grain of the wood.

Endless *belts* (perforated and revolving in opposite directions) are inferior with regard to flattening, but they help a safe passage of sheets that are fragile, variable in length and width, and traveling parallel or transverse to the wood grain (Figure 20-19).

A combination of intermittent *drums* and *plates* is seldom used. Propelling is served by drums, while hot plates open and close rhythmically. The sheets are forwarded when the plates are open, forwarding stops and the plates close, the plates open and forwarding continues, and so on. Such "progressive plate dryers" have the advantage of flattening the veneer, and their installations require a small space, 4–7 m (about 15–25 ft) in length (15), but productivity is relatively small.

Related to the above is *platen (press) drying*, where veneers are dried between the plates of a hot press (3). Hardwood veneers and fragile sheets obtained from "burls" or other growth abnormalities, and tending to become wavy, are sometimes dried in this manner. Platen drying is 2–5 times faster than drum (roller) drying, reduces shrinkage, saves energy (steam, electricity), and dried veneers are generally flatter. Disadvantages relate to complexity of equipment, handling problems, inactivation of veneer surfaces (reduced wettability) when drying at temperatures higher than about 150°C (300°F), patterning of veneer surfaces by the mesh used to ventilate the press plates, and iron stain problems (16). A variation of the method includes a carousel configuration [i.e., a system of revolving multistoried (8–14) hot presses, moving in sequence as in a carousel to load and unload veneer (3)]. Continuous press-drying is under development (21).

In the so-called *jet dryers*, air circulation is

Table 20-2. Veneer Drying Times[a]

Species	Thickness (mm)	Moisture content (%)		Temperature °C (°F)	Time (min)	Type of dryer
		Initial	Final			
Beech	1.5	50	5–7	80 (175)	22	Belt
Oak	0.8	40	5–7	100 (212)	4	Belt
Walnut	0.8	65	8	140 (285)	2.5	Drum
Oak[b]	0.6	60–80	10–12	180 (350)	1	Jet

[a]Based on data from Ref. 19.
[b]Also beech, birch, cherry, Makoré, maple, walnut (0.8–1.5 mm).

transverse to the veneer surfaces (Figure 20-18D). The sheets are propelled by drums or belts, but in–between there are "boxes" with small openings through which hot air is forced at a high speed and "impinges" (hits) the veneer; hence, these dryers are also called "impingement" dryers. Jet dryers are in common use, because for a comparative expense the speed of drying is 2–3 times higher in comparison to drum or belt dryers (Table 20.2), the factory space required is reduced, and drying quality is improved (10, 19). The system is also suitable for drying a continuous ribbon produced by peeling or rewound (16).

Another type of dryer consists of large, *perforated* and slow-revolving *drums* (Figure 20-20). The sheets are held on the surface of the drums by vacuum, and are slowly transferred from one drum to another. Such dryers are suited only for thin veneers (5, 10).

Dryers are made of prefabricated sections (up to 50), and it is possible to make them longer if necessary. Veneer sheets may move simultaneously at different levels (up to 5).

Decorative veneers are introduced in the dryer in sequence of cut, in order to utilize the figure of wood to the best advantage. This is desirable in making furniture, wall paneling, and other products.

Waving or checking of the ends of veneer sheets sometimes occurs, especially in drying large sheets and certain tropical species. Such defects may be avoided or reduced by short overlapping of the ends of consecutive sheets in the dryer, or attaching perforated tapes (paper or plastic). Another defect is discoloration, which may be caused by high temperatures or contact of sheets to worn metallic parts (5). Collapse may also occur by exposing veneers to high temperatures early in drying, when their moisture content is above the fiber saturation point (16).

The drying rate of veneer is influenced by such factors as thickness, density, initial (green) moisture content, and sapwood or heartwood. This indicates that in many cases segregation of species and/or sapwood from heartwood is desirable to minimize variations in final moisture content. If such variations exist, the most common method of control is to hold the dried veneer in piles for an equalizing period before subsequent processing. Equations are given in the literature to calculate the effect of thickness and temperature on the

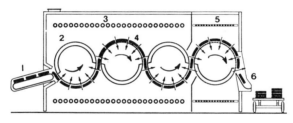

Figure 20-20. A drum dryer: 1, entrance; 2, drum; 3, heating pipes; 4, veneer; 5, chamber for reduction of temperature; 6, exit. (After Refs. 5 and 10)

drying rate, but there is no general agreement as to the form of such relationships (16).

YIELD

The yield of veneer (i.e., the proportion of original volume transformed to sheets) depends on the method of production, log characteristics, and shrinkage of the wood. The yield is generally low; the volume of sheets may be less than 50% of the original round wood volume (6, 26). Log diameter is especially important: large diameters give a higher yield.

Loss of wood volume comes from several factors, including log oversize (if any), taper, deviation from circular cross-section, method of rotary cutting, core diameter, method of preshaping logs into flitches, trimming the sheets to dimensions and shapes suitable to final use, sawdust in case of sawing, and shrinkage of the wood. In addition, the volume may be reduced due to removal of defects, such as knots, checks, discolorations, and parts attacked by fungi or insects. These residues are burned to produce energy, or they may be utilized as raw material for making various products (particleboard, fiberboard, paper) after chipping or defibering.

Although the yield is relatively low in volume, a certain volume of veneer is much more valuable than lumber of the same volume and species. Veneer sheets cover a much greater surface (Table 20-3, Figure 20-21), and usually their volume is not considerably reduced in making final products—in contrast to lumber. The surface that may be covered by veneer

Table 20-3. Comparative Surface Areas of Lumber and Veneer[a]

Lumber:					
Thickness	19	25	32	50	mm
Surface	52	40	31	20	m^2/m^3
Veneer:					
Thickness	0.5	1.0	3.0	5.0	mm
Surface	2000	1000	333	200	m^2/m^3

[a]This is the upper, exposed surface, as in flooring, paneling, etc. The total surface area (including back and sides) is double or more, especially in lumber; total surface is an important factor in drying.

Figure 20-21. Whole logs transformed to veneer by slicing. (Courtesy of Hess-Ehrenreich, reproduced by permission from DRW-Verlag Weinbrenner.)

produced by slicing can be easily calculated from its thickness and the original volume of a flitch—minus the residue. In the case of sawing, the width of kerf is decisive. In rotary cutting, the length of veneer produced is found from the following approximate relationship:

$$M = \frac{\pi(d_1 + d_2)(d_1 - d_2)}{4a} \quad (20\text{-}1)$$

where

M = length of ribbon (m)
π = 3.14
d_1 = initial log diameter (m)
d_2 = diameter of remaining core (m)
a = veneer thickness (m)

Calculation of the surface is based on log length, which determines the width of the veneer in the sense of a peeled ribbon. The length of ribbon may be calculated in feet also by use of the above formula, setting d_1, d_2, and a in inches and dividing the result by 12.

REFERENCES

1. American Walnut Association (undated). Fine Hardwoods Selectorama. Indianapolis.
2. American National Standard for Hardwood and Decorative Plywood. ANSI/HPMA HP 1983.
3. Anonymous. 1980. Researchers develop a new technology for platen drying wood veneer. *For. Prod. J.* 30(11):12–13.
4. Anonymous. 1988. Schwachholzverarbeitung und Wertmaximierung. *Holz Roh- Werkst.* 46:113–115.
5. Autorenkollektiv. 1975. *Werkstoffe aus Holz*. Leipzig: VEB Fachbuchverlag.
6. Baldwin, R. F. 1980. *Plywood Manufacturing Practices*, 2nd ed. San Francisco: Miller Freeman.
7. Baxter, D. G. 1977. *Softwood Plywood Manufacture and Practices in the Pacific Northwest*. Tacoma, Washington: Plywood Research Foundation.
8. Chudnoff, M. 1984. *Tropical Timbers of the World*. U.S.D.A., Forest Service, Ag. Handbook 607.
9. F.A.O./U.N. 1966. Plywood and Other Wood-Based Panels. Rome.
10. Fecht, P. 1965. Furnier-Durchlauftrockner mit Düsenbelüftung. In *Holztrocknung*, ed. R. Weinbrenner. pp. 151–169. DRW-Verlag.
11. Fessel, F. 1965. Trocknung in Dampf-Luft-Gemischen. In *Holztrocknung*, ed. R. Weinbrenner. pp. 69–108. DRW-Verlag.
12. Fleischer, H. O. 1949. Experiments in rotary veneer cutting. *For. Prod. Res. Soc. Ann. Meeting* 3:137–155.
13. Fleischer, H. O. 1965. Heating Rates for Logs, Bolts and Flitches to be Cut into Veneer. U.S. For. Prod. Lab. Report No. 2149.
14. Forintek Canada Corporation. 1988. A piece of the puzzle. *Forintek Review* July-August: 4–5.
15. Fuchs, F.-R. 1981. Moderne Messerfurnierherstellung. *Holz Roh-Werkst.* 39:179–192.
16. Hartley, J. 1986. Hardwood veneer drying—A review. *Proceedings World For. Congr. Div.* 5, pp. 77–88. Ljubljana, Yugoslavia.
17. Ingenieurschule f. Holztechnik, Dresden. 1965. *Taschenbuch der Holztechnologie*. Leipzig: VEB Fachbuchverlag.
18. König, E. 1972. *Holz-Lexikon*, Band I, II (2. Auflage). Stuttgart: DRW-Verlag.
19. Kollmann, F., ed. 1962. *Furniere, Lagenhölzer und Tischlerplatten*. Berlin, New York: Springer Verlag.
20. Kollmann, F. F. P., E. W. Kuenzi, and A. J. Stamm. 1975. *Principles of Wood Science and Technology. II. Wood Based Materials*. Berlin, New York: Springer Verlag.
21. Loehnertz, S. P. 1988. A continuous press dryer for veneer. *For. Prod. J.* 38(9):61–63.
22. Lutz, J. F. 1974. Techniques for Peeling, Slicing and Drying Veneer. U.S. For. Prod. Lab. Res. Paper 228.
23. Mörath, E. 1949. Das Dämpfen und Kochen in der Furnier- und Sperrholzindustrie. *Holztechnik* 29(7):129–134.
24. Perry, T. D. 1948. *Modern Plywood*. London, New York: Pitman.
25. Steinhagen, S. P., H. W. Lee, and S. P. Loehnertz. 1987. LOGHEAT: A computer program for determining log heating times for frozen logs. *For. Prod. J.* 37(11/12):60–64.
26. U.S. Forest Products Laboratory. 1962. The Manufacture of Veneer. Report No. 285.
27. U.S. Forest Products Laboratory. 1973. Veneer Species of the World. Madison, Wisconsin.
28. Walser, D. C. 1978. New developments in veneer peeling. In *Modern Plywood Techniques*, Vol. 6. San Francisco: Miller Freeman (reprint, pp. 12).

FOOTNOTES

[1]Reference to cutting wood into thin layers for veneering is made by Pliny the Elder (23–79 A.D.) in his *Natural History* (Book XVI).

[2]A group of species including longleaf (*Pinus palustris*), shortleaf (*P. echinata*), loblolly (*P. taeda*), and slash pine (*P. elliottii*).

[3]Softening is attributed to physical changes of lignin and pectic substances. Cutting is also facilitated by the resulting high moisture content, which has a "lubricating" ef-

fect, reducing the friction between wood and cutting knife (19).

⁴This method (stay-log) reduces bending of the residual wood in comparison to the core of rotary cutting, and the wood is better utilized (19). A method of producing veneer by a machine resembling a pencil sharpener should be included among the variations (18). The method is applied to small-diameter logs, and produces a continuous sheet, which is cut into sectors. Circular or other shapes of surfaces, showing attractive figures, are produced by sidewise gluing of the sectors (Figure 20-11).

⁵Experiments have shown that incised veneer dried faster (as much as 15%), press blows were virtually eliminated (in plywood pressed at high moisture contents, up to 10%), bond quality was improved, and preservative treatment of glued products was easier (14).

21

Adhesion and Adhesives

Gluing of wood has been practiced since ancient times, as already mentioned in the preceding chapter. Natural substances and preparations thereof were used at first, including tar, resins, gums, bee's wax, animal hides, and bones. A natural glue made from animal hides seems to have been known in ancient Egypt (Figure 21-1).

Such adhesives were used for thousands of years, up to the seventeenth century, and "animal glue" (made of animal hides and bones) was in common use up to some years ago. The art of their production was known only to craftsmen working with wood; it was kept secret and transmitted from generation to generation. The first "factory" to produce animal glue was founded in Holland in 1690; later, similar plants were established in England (1700) and the United States (1808).

The use of adhesives increased after the invention of wood-working machines (circa 1800). However, their use on an industrial scale started much later, reportedly in 1875, by gluing veneer to produce plywood. Since that time, the use of adhesives was greatly extended and, gradually, with the progress of Chemistry, new substances have entered the market.

In addition to enlarging the array of products, bonding wood with adhesives contributes indirectly to the conservation of forest resources, because it offers a possibility for more complete utilization by transforming residues to useful products.

MECHANISM OF ADHESION

For a long time, adhesives were considered mysterious products. The reasons for adhesion were examined after 1920, in relation to requirements for strong joints in certain constructions (e.g., wooden airplanes).

The current view on the subject is that the adhesion is due to mechanical anchoring, molecular (physical) attractive forces, and the development of chemical bonds between wood and adhesive (Figures 21-2 and 21-3).

According to the mechanical view (theory of *mechanical adhesion*), the adhesive spread on a wood surface enters exposed pores (cell cavities), where it is solidified and anchored. Water-soluble adhesives are not limited to cell cavities, but can also enter cell walls (37) (between microfibrils); therefore, their anchoring positions are greater in number. This view does not explain adhesion of nonporous materials (glass, plastics, metals) and tropical woods of high density which are glued with nonwater-soluble adhesives. It has also been observed that the hardened adhesive shrinks in the cell cavities and does not always keep contact with cell walls; and, further, that the plugging of pores with wax does not obstruct adhesion.

The main reason for adhesion is considered to be the development of molecular attraction (van der Waal forces, hydrogen bonds) between wood and adhesive (theory of *specific adhesion*) (9, 15). This view is supported by

Figure 21-1. Schematic representation of an Egyptian mural (circa 1500 B.C.). (A) A workman is gluing a sheet of wood (*1*) to another (*2*), knife (*3*), piece of wood (*4*), geometric instruments (*5*, *6*), and ready furniture (*7*). (B) A workman is pressing small sacks of sand (*8*, *9*), probably hot, to the construction being glued (*10*); vessel containing glue (*11*), probably animal, is placed on fire (*12*); pieces of glue (*13*), and probably a dish with leftovers (*14*). (C) A workman is spreading glue with a brush (*15*) on veneer placed on a support (*16*). (Note that according to a different interpretation, the workman is cutting a sheet to desired shape with a knife, *15*). (After Refs. 27 and 38.)

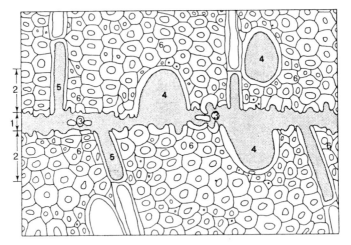

Figure 21-2. Schematic representation of the mechanism of mechanical adhesion. The adhesive enters exposed cell cavities of the surfaces to be glued: 1, main glue line; 2, outer (extended) glue line; 3, bubbles; 4, vessels (pores); 5, parenchyma (filled with glue); 6, fibers. (After Ref. 7.)

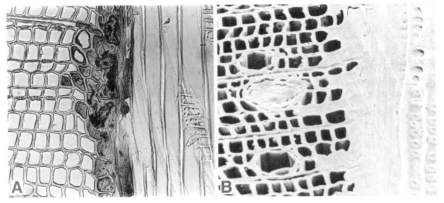

Figure 21-3. Glue lines in spruce plywood glued with phenolic resin. (A) Light microscope photograph; adhesive has penetrated into tracheids mostly ruptured during veneering (110×). (B) Scanning electron micrograph (SEM, 170×); adhesive has filled tracheids, a resin canal, and rays. (Photographs courtesy of N. P. Kutscha, reproduced by permission form the Society of Wood Science and Technology, USA.)

the above-mentioned observations, and may also be based on the possibility of gluing certain materials without an adhesive (e.g., quartz sheets). This could be done with other materials if their surfaces were absolutely (in an atomic level) clean. (It should also be noted that in the case of quartz, adhesion is not possible after the surfaces are exposed to the air, even for a very short time) (8, 9). Because such clean surfaces cannot be practically produced, molecular forces cannot be activated without the intervention of an adhesive. If a thin layer of water or oil is placed on the plane that separates the material or materials to be glued, these liquids act as adhesives and may produce strong joints until they are evaporated or absorbed. However, substances employed as adhesives should produce durable joints, resistant to the conditions of use. Such substances should be characterized by high cohesion and, considering that molecular attractive forces are affected by the polarity (electric attraction) between the adhesive and the material to be glued, the two materials should be polar; strong joints cannot be produced between polar and nonpolar materials (8, 9). Wood is polar (the property is due to free hydroxyls of cellulose molecules); therefore, substances suitable as adhesives should also be polar.[1]

The formation of primary chemical bonds between wood and adhesive is also possible, especially with adhesives containing formaldehyde (37). The relative importance of secondary forces and primary chemical bonds is a matter of discord and depends on the type of adhesive (9, 37).[2]

FACTORS OF ADHESION

In addition to the type of adhesive, the factors affecting adhesion (quality of joints) are the condition of the wood surface, its wettability by the adhesive, and wood moisture content, among others.

Condition of the Wood Surface

According to the theory of specific adhesion, in order to produce molecular attractive forces, it is necessary to have a perfect contact between the surfaces to be glued. This means that the surfaces should be smooth and clean. Perfectly smooth surfaces cannot be produced in practice even with materials such as metals or glass; hence, this is more difficult with wood due to its anatomical structure (cell morphology, early- and latewood, etc.). The role of establishing contact is undertaken by the adhesive, which bridges surface irregularities, but the development of attractive forces between adhesive and wood requires the wood surface to be clean. Atmospheric dust, atmospheric moisture or organic (e.g., oil) vapors, theoretically even in a monomolecular layer, and visible dirt, obstruct the formation of strong joints; for this reason, the surface to be glued should be prepared by planing shortly before gluing. Sawn surfaces are usually rough; they expose torn fibers, obstruct a good contact, contribute to local development of stresses, and consume more adhesive.[3] Sanding also has an adverse effect, except in cases of surface modification produced by machining wood with dull tools, or overheating veneer during drying or plywood during hot-pressing; in such cases, light sanding can improve bonding (5, 8, 9).

Wettability

The wettability of wood by an adhesive is measured by the angle of contact to the surface of the wood (Figure 21-4). This angle should approach zero in order to produce the strongest joints. Wettability is affected by various factors, which relate to adhesive (surface tension, temperature, viscosity) and to wood (density, porosity, extractives) (37). Woods of lower density (higher porosity) are better wetted; whereas extractives in excessive amounts, or nonpolar extractives such as terpenes and fatty acids, have an adverse effect. Wettability is also affected by the surface cleanliness of wood and machining conditions. For example, tools that are not well-sharpened cause overheating or compaction; the wood is "burned" or its surface "hardens." Drying veneer at high temperatures, higher than about 160°C (320°F), results in the reduction of wettability (i.e., surface inactivation) (16).

Figure 21-4. Measurement of the angle of contact between adhesive and wood (generally, between a liquid and the surface of a solid). Wettability is higher at smaller angles.

Moisture Content

High moisture content has an adverse effect on attractive forces and contributes to "fluidity" of the adhesive; this results in excessive absorption by the wood and formation of weak joints. Moisture changes may also produce stresses due to shrinkage and swelling, and cause breakages of the continuity of a joint. Inclusion of air bubbles due to moisture evaporation in the plane of adhesion, when high temperatures are applied, has the same effect. Very low moisture content may also cause problems because it affects wettability, obstructs the entrance of adhesive, and causes premature hardening (24).

Moisture is contained in the wood, but is also added by water-dispersed adhesives, or removed by application of high temperatures. Depending on type of adhesive and gluing procedure (hot or cold pressure), the moisture content of wood may range from about 2 to 12% (hot pressure 2-8%, cold pressure 8-12%) (8, 21, 24, 30). A general rule is to have the wood at 6-8% moisture content for interior applications (such as furniture, doors or woodwork) and 12-14% for exterior uses (such as exposed lumber-type beams); the moisture content of wood prior to gluing should approximate that expected during use of the product (5).

In addition to the quantity of moisture, its distribution is important; it should be as uniform as possible, both in a piece of wood and between pieces, in order to avoid development of stresses due to shrinkage or swelling and, therefore, weakening of joints. These stresses differ depending on species (stresses are higher in denser woods)[4] and growth-ring arrangement (tangential-radial).

Other Factors

In addition to the above requirements, prerequisites to successful adhesion are:

- Good quality adhesive,[5] suitably prepared. Long or unsuitable storage[6] of adhesives result in reduced adhesive properties or obsolescence (Figure 21-5). In every case, it is necessary to comply with the instructions of the manufacturer.
- Uniform and controlled spreading in predetermined quantity.[7] More adhesive than required (i.e., a great thickness of glue line), in addition to cost, will give joints of low quality, due to voids which may result from contraction of the adhesive or inclusion of air, and also because the strength of joints will be affected by the properties (mechanical and other) of the adhesive. Thus, the strength of joints increases when the thickness of the adhesive is reduced (Figure 21-6). Theoretically, a monomolecular layer is sufficient; however, in practice, the least thickness, after the adhesive is hardened, is about 0.1 mm (18).

Control of the time that intervenes between spreading and assembly. This "assembly" time is distinguished into "open assembly"

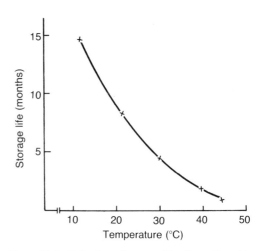

Figure 21-5. Reduction of storage life of urea-formaldehyde in relation to the temperature of storage space. (Courtesy of Borden, Inc.)

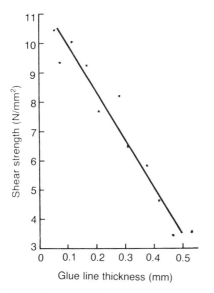

Figure 21-6. Relationship between strength of joint and glue line thickness. (Adapted from Ref. 8.)

and "closed assembly." Open-assembly time is the time between spreading and contact of the surfaces to be glued, and "closed" between contact and application of pressure. During open assembly, some solvent of the adhesive is evaporated; whereas during closed assembly the adhesive may harden prematurely, resulting in production of weak joints, or complete failure due to absorption of the solvent by wood or chemical reaction of adhesive components. For these reasons, it is necessary to control both "open" and "closed" assembly times according to experience, testing, or instructions of the manufacturer.

- Sufficient and uniform pressure exerted for sufficient time. Pressure is needed to ensure the closest possible contact between the surfaces to be glued, and for the adhesive to form a thin, continuous layer, uniform in thickness, without damage to the strength of wood. The height and duration of pressure depends on type and viscocity of the adhesive, species (density) of wood, type of joint, and intended use of the glued product. In certain types of adhesives, heat is also needed. The level of temperature depends mainly on type and other factors related to the adhesive (Figure 21-7).
- Final regulation of the moisture content of the glued product because moisture is added or lost during adhesion.

ADHESIVES

Adhesives are *natural* or *synthetic*.

Natural

Natural adhesives include substances of *plant* origin (starch and soya glue) and *animal* origin

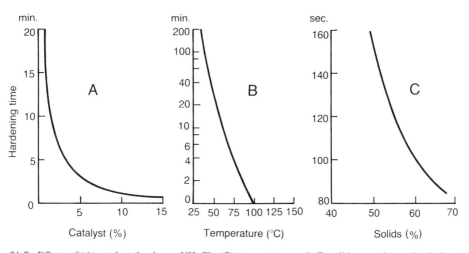

Figure 21-7. Effects of (A) catalyst (hardener, NH_4Cl), (B) temperature, and (C) solid proportion on hardening (curing time) of urea-formaldehyde. (Adapted from Ref. 2.)

(glue from hides or bones, casein, and blood albumin) (15, 27).

Up to about 1930, the use of natural glues was extensive (they were the only ones available), but afterward they were replaced, almost entirely, by synthetic adhesives. This replacement has taken place because natural glues lack durability under adverse conditions; whereas synthetic resins are easy to use, are superior in resistance to moisture, microorganisms, high temperatures, and chemicals, and are comparatively cost-effective. Among natural glues, only casein has important applications in the wood-using industry (8). Therefore, the following discussion of natural glues is primarily of historical interest.

Plant Origin

Starch Glue: The raw material is starch from various sources—mainly corn and Cassawa roots (a plant of tropical America, *Manihot utilissima*), rarely wheat, potatoes, or bananas. The glue is prepared by mixing starch with water and then heating. Instead of heating, sodium hydroxide may be added, although a high proportion may cause discoloration of light-colored woods. Starch glue is viscous and hardens by evaporation or diffusion of moisture in the wood; therefore, redrying of the glued product may be needed. The disadvantages of this glue is that the joints are not resistant to moisture, fungi, and insects, although improvement is possible by the addition of a synthetic resin.

Soya Glue: This glue is prepared from a protein contained in the seeds of soya, an asiatic plant, *Glycine soja*, the seeds of which are utilized to produce oil, flour, and other products. In the trade, it is offered in the form of a powder containing other chemical substances, such as calcium or sodium hydroxide, or these chemicals are added during preparation by simple mixing with water. Soya glue is chemically similar to casein glue. It hardens partly by loss of water and partly by chemical reaction. It is alkaline, like casein, but the joints are comparatively less resistant to moisture, and are susceptible to attack by fungi and insects. Its properties may be improved by the addition of a synthetic resin.

Animal Origin

Hide or Bone Glue: This glue is known as the main animal glue. The adhesive capacity is due to collagen—a protein found in hides, bones, and muscles of animals, especially cattle. Animal glue is used in the form of a warm aqueous solution at a temperature of about 40–60°C (100–140°F). To ensure uniform heating, the containing vessel is heated in a warm water bath. Hardening takes place by cooling and loss of moisture. The joints are not resistant to water or high relative humidities, but may be improved by the addition of chemicals, such as oxalic acid and paraformaldehyde. Another disadvantage is that the glue may be attacked by fungi and insects (15, 27).

Casein Glue: Casein is a protein contained in milk. It is prepared by oxidation, which separates a sediment from whey. Whey may be used as animal food, whereas the sediment (raw casein) is washed, pressed, dried, and reduced to powder (21).

The glue is prepared by mixing such powder with water, although liquid casein is also available. The addition of calcium hydroxide increases the resistance of joints to moisture (insoluble calcium caseinate is formed in the glue line), whereas sodium hydroxide prolongs the "working life" (i.e., the period during which the glue is suitable for use). Other chemicals are also used to improve the joints, or as antifoam additives. Casein glue may be used at low temperatures, even out-of-doors in the winter (as low as to 0°C or 32°F), and keeps its adhesive capacity as powder (many months), after preparation (several hours), and after spreading (up to 1.5 h). The glue provides strong joints, but is not suitable for products exposed to the weather or to environments of high relative humidity. Although joints may be attacked by fungi and insects, improvement is possible by the addition of toxic chemicals. Machining of glued wood may wear cutting tools. Casein glue is alkaline—it facilitates gluing resinous woods and surfaces that have a greasy appearance, but may cause discoloration of light-colored woods and skin irritation to workers (8).

Blood Albumin Glue: Albumins contained in the blood of animals, mainly cattle, possess adhesive properties. Glue prepared from this source was extensively used during the first World War for building wooden airplanes, as the joints were resistant to moisture. Gluing required high temperatures (above 70°C or 160°F) and, therefore, hot-pressing (27).

Synthetic

Synthetic adhesives, called *synthetic resins* because their appearance is similar to natural resins, are products of the chemical industry. The raw materials necessary for their production are petroleum, coal, or natural gases. Experimental preparation started in 1872, and these adhesives appeared on the market in the 1930s (21). In the following years, the use of synthetic resins increased continuously at the expense of natural glues.

The first synthetic resin was phenol-formaldehyde, which originally (1929) appeared in the form of thin sheets (Tegofilm), and later (1935) as a liquid. Urea-formaldehyde became available almost simultaneously (1931), melamine-formaldehyde by the end of the same decade, and resorcinol-formaldehyde a little later (1943) (21). Other synthetic resins are products of the same period (e.g., polyvinyl emulsions, 1929) (23), but newer ones also exist.

Synthetic resins are divided into *thermosetting* and *thermoplastic*. Under the influence of heat, thermosetting resins at first soften and later harden permanently. In contrast, thermoplastic resins stay soft as long as their heating is continued, and harden when cooled. Thermosetting resins harden by chemical reaction, which is accelerated by heat or a catalyst; whereas in thermoplastic resins, hardening is a physical process and the result of evaporation of solvent or a decrease in temperature. Thermoplastic resins create joints of lower strength and diminished resistance to moisture, heat, and solvents; they also show creep under long-term loading.

Thermosetting. Thermosetting resins are produced by a controlled reaction of their constituents (e.g., phenol and formaldehyde), which is interrupted before completion. The intermediate product is condensed to a viscous liquid, dried, and reduced to powder, or used to impregnate paper sheets. Continuation and completion of the reaction take place during adhesion by application of heat, catalysts, or both. Because the reaction is slowly continued during storage, the resins are stored at low temperatures to prolong their working life.

Catalysts or hardeners are acids, paraformaldehyde, ammonium salts, or other chemicals, which are added in liquid or powder form; they are used to control the time between spreading and pressure, as well as the temperature required for adhesion.

Synthetic resins, both thermosetting and thermoplastic, are mixed with various substances for the purpose of reducing the cost or improving their properties. Such additives are called "extenders" or "fillers," although the distinction is not clear. The main purpose of extenders is the reduction of cost, but some (e.g., wheat or other grain flour) contribute to adhesion because they contain starch. Fillers are chemically neutral (e.g., walnut four); they are added to improve the working properties of an adhesive (e.g., viscosity) or the properties of glue line (e.g., brittleness). Extenders are added in proportions of 100% or higher, whereas fillers are usually up to about 25% (22, 27). Large amounts of additives reduce the cost, but lower the quality of joints. Improvement of the quality of urea joints may be made by the addition of stronger resins (melamine or resorcinol, 10–20%); natural glues are also sometimes improved by the addition of thermosetting resins. Proper use of such resins, according to the manufacturer instructions, will provide strong joints resistant to moisture, fungi, and insects.

Thermosetting resins include the following types.[8]

Phenol-Formaldehyde: Phenol-formaldehyde is available as a liquid (viscous, dark red in color) or a powder, or in the form of impregnated paper sheets (film). Powder is preferred for long storage life, and may used to produce liquid resin by mixing with water. As previously mentioned, phenol-formaldehyde first appeared in the form of sheets. Sheets are clean to work with, and better for fragile, figured veneers, although their cost is higher.

Phenol-formaldehyde is available in formulations that require high temperatures for hardening, 115–150°C (240–300°F), and others that may be used at room temperature, about 20°C (70°F), while special formulations are suitable for a wide range of temperatures, 10–90°C (50–200°F) (18).[9]

The quality of joints glued with phenol-formaldehyde is very good. Proper adhesion gives high strength and durability under the most difficult conditions of use. The joints are resistant to cold and boiling water, they are not attacked by fungi, insects, and chemicals, and are resistant to high temperatures that cause carbonization of wood. Resins that harden at room temperature, due to the addition of strong acids as catalysts, provide joints of lower but better quality than those produced by urea-formaldehyde.

Phenol-formaldehyde has certain disadvantages: the joints are dark-colored, light-colored veneers may discolor, and greater care is needed in comparison to other types of synthetic resin. In addition, workers may experience skin irritation if proper protection is not taken, and certain formulations produce a disagreeable odor even after hardening.

Urea-Formaldehyde: Urea-formaldehyde is available in liquid or powder form, and seldom as impregnated sheets. Resins of this type harden both at room and high temperatures, 95–130°C (200–260°F). Special formulations require intermediate temperatures, up to 70°C (160°F) and some are applied by spreading the resin on one wood surface and the catalyst on the other; hardening takes place during their contact and, depending on the catalyst, this may be achieved within a very short time. Resins hardening at low temperatures utilize strong acids as catalysts, but such acids may attack the wood (5). Figure 21-7 shows the effect of various factors on hardening of urea-formaldehyde.

Urea-formaldehyde is usually extended to reduce the cost. The proportion of additives (wheat flour, etc.) may reach 100–200%. Additives degrade the quality of joints, but the desired quality depends on the intended use of the glued product. Formulations without additives give strong joints, resistant to moisture at room temperature, but which are susceptible to high temperatures—degrading at temperatures of 65°C (150°F) or higher, especially when the relative humidity is high. Urea-formaldehyde is not suitable for exterior use, but its performance may be improved by adding (10–20%) melamine-formaldehyde or resorcinol-formaldehyde ("reinforced" urea-formaldehyde) (18). Very high proportions of extenders reduce moisture resistance.

Urea-formaldehyde is a synthetic resin of common use. By proper selection of catalysts and additives, it is possible to adapt it to varying uses. It produces joints that are colorless or have a light brown color, and are resistant to microorganisms. The resin may be used with high-frequency current (for fast adhesion), and its cost is lower in comparison to other thermosetting adhesives, but has a moderate blunting effect on cutting tools. Other disadvantages are that urea-formaldehyde has relatively poor resistance to long periods of wetting, and it emits formaldehyde vapors, which may be hazardous to health. The emission is affected by the molar ratio of formaldehyde to urea. This ratio varies from 1.2:1 to 2.0:1. Resins with a low ratio have a lower free formaldehyde content and longer "pot life," but they are slower to cure and produce joints of lower strength, stiffness, and durability (resistance to water). This problem exists mainly with particleboard (12) (Chapter 24).

Melamine-Formaldehyde: Synthetic resins of this type are usually offered in the market as water-soluble powder, because, in liquid form, their preservation in storage is difficult; impregnated sheets are also available (23). Hardening temperatures vary from about 50 to 100°C (120–212°F) (8). The joints are colorless, and are resistant to moisture (also in boiling water) and microorganisms. However, the cost of melamine-formaldehyde is high; for this reason, it is seldom used pure—only when the color of phenol-formaldehyde is undesirable. It is usually employed to improve urea-formaldehyde. Melamine-formaldehyde has disadvantages including brittleness of joints, blunting of tools that machine glued products, and difficulty of cleaning mixing equipment. Improvement of the first two is possible by the addition of polyvinyl acetate (up to 30%) (20), and the equipment may be cleaned with chemical solutions if water is not effective (8).

Resorcinol-Formaldehyde: Resorcinol-formaldehyde is usually available as a liquid. It is dark-colored, hardens under a large variation of temperature, 5–100°C (40–212°F), and may be used with a relatively high moisture content of wood (up to 18%). The joints are strong, similar to those produced by phenol-formaldehyde, although resorcinol is also expensive. There are formulations of phenol-resorcinol which preserve most of the advantageous properties of resorcinol, and they are less expensive (35).

Epoxy resins: Epoxy resins are seldom used for gluing wood, because they are expensive, and the joints are not better than those produced by phenol-formaldehyde or resorcinol-formaldehyde. They are more suitable for gluing metals or other materials, or woods with a high moisture content (18). The joints have high strength (also to long-term loads), and they are resistant to water and other solvents and microorganisms; however, their quality may vary depending on hardening temperature (at low temperatures, moisture resistance is lower) and added catalysts or additives (26). Epoxy resins are available in the form of powder, particles, paste, or liquid (20). They harden at room or high temperatures, up to about 200°C (400°F), and exhibit a small shrinkage. A light pressure is sufficient for adhesion (8).

Isocyanate Adhesives: Isocyanate adhesives form strong bonds at high moisture contents, up to 20% (32); they cure at room temperature (with a catalyst) or at elevated temperatures, and no formaldehyde vapors are emitted. Important disadvantages also exist. They are expensive, may liberate vapors that are hazardous to health before curing (29), and may create manufacturing problems, such as bonding to metals (e.g., caul plates of presses). These adhesives are considered the most important alternate to traditional thermosetting synthetic resins (phenol-, urea-) for the composite panel industry (19). Tannins may be added to reduce cost (29).

Thermoplastic. This group mainly includes emulsions of *polyvinyl acetate*. These adhesives are available in the form of a viscous milky liquid (a mixture of polyvinyl acetate and usually water), ready for use; they harden at room temperature (by evaporation or diffusion of water in the mass of wood), and produce colorless joints (35). These properties, in combination with ease of use (e.g., application by brush), cleanliness, speed of adhesion, and almost unlimited storage life (when the solvent is not allowed to evaporate), have contributed to their extended use, especially in furniture. Their thermoplastic property is of no practical importance when the glued products are not exposed to high temperatures, higher than 55°C (130°F), and the lower rigidity of the joints is an advantage for certain uses, such as dowel joints. However, their creep under long-term loading and their lack of resistance to moisture constitute disadvantages. Improvement of these properties is possible by the addition of thermosetting resins (urea- or phenol-formaldehyde) and preparation of new types of vinylic copolymers (35).

Thermoplastic adhesives are also produced from *cellulose* (cellulose ethers and esters), but they are expensive and present difficulties in use. For this reason, they are of limited practical importance (1).

Elastomers. Elastomers or elastomeric adhesives are based on natural and synthetic rubber materials. They include polymers, such as nitriles, butylic gum, and neoprene (synthetic rubber). Gluing is fast with little pressure or simple contact ("contact adhesives"). Some are available in solid form (particles, thread), which melt by application of heat and harden by cooling ("hot melts") (15). Elastomers have specialized uses, such as gluing wood to metals or plastics, and edge-gluing of veneers (see Chapter 20), but some (e.g., neoprene) are used in on-site building construction, e.g., attaching plywood subfloors to joists (31). These adhesives are more effective in transferring stress than nails and "gap-filling"—therefore, they can be spread thick.[10]

Newer Developments

Worth noting are some other efforts to develop new adhesives. Such efforts relate to the utilization of products with a natural base, or waste and other materials. Considerations of increased cost, and projected long-term problems of availability of raw materials for synthetic resins (e.g., petroleum) also support these efforts.

Tannins contained in considerable quantities in the wood or bark of certain species of forest tress (see Appendix) are utilized for the production of "tannin-formaldehyde" (a thermosetting resin related to phenol-formaldehyde). Such usage of tannins derives from their phenolic nature (4, 29, 36). Tannin resins, fortified with small amounts of phenol-, urea-, or resorcinol-formaldehyde, give joints of good strength and resistance to moisture (3, 10, 12, 29) and are used to glue plywood, laminated wood, particleboard, and other products; unfortified tannin resins are also used in making particleboard.

Another source is *pulping waste*—namely, sulfite spent liquor (see Chapter 26). Its potential as an adhesive is due to its high lignin content (11, 12, 13, 14, 25). This source has been experimentally investigated with good results. The main drawback is the required longer pressing time in comparison to synthetic resins and, therefore, higher costs; other disadvantages are higher pressing temperatures, problems of corrosion, and dark coloration. Improvement of lignin-base adhesives is possible by combination with phenol- or urea-formaldehyde resins. Efforts have also been made to utilize lignin from Kraft pulping.

Finally, note should be made of the possibility of gluing by application of *nitric acid*, followed by a mixture of lignosulfate ammonium–furfural–alcohol–maleic acid, or nitric acid or hydrogen oxide and dialcohols or diamines. The method has been applied experimentally in making particleboard. Such a possibility of "gluing without glue" is explained as follows: oxidation of the components of wood results in the formation of groups of organic acids in cellulose and lignin, and action of the other chemicals between the oxidized surfaces leads to chemical adhesion. However, this method—as well as the other systems of chemical adhesion that have been proposed—is far from practical application for various reasons, such as cost of chemicals, danger from their use, lack of "gap-filling capability," and variable effectiveness in different cases (17, 28, 37).

STRENGTH OF JOINTS

The strength of glued joints is measured by various tests (26). A common test for solid wood utilizes a shear-like specimen, as shown in Figure 21-8A and B, and measures shear strength and "proportion of wood failure"; the latter is the proportion (percentage) of total surface area of the broken joint covered by wood as opposed to exposed adhesive. The accepted level

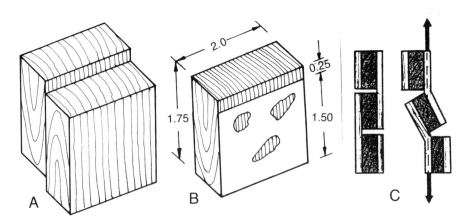

Figure 21-8. (A) A test specimen for determination of the shear strength of a glued joint of solid wood. Dimensions are given in inches (1 in. = 2.54 cm), according to American specification ANSI/ASTM D 2559-76. (B) The plane of failure shows adhering wood (striated areas) of the sheared surface and a low "percentage of wood failure." (C) Evaluation of glue-bond quality of a 3-ply plywood: (a) original specimen, (b) after testing (diagrammatic presentation—not in scale); this type of failure shows a strong bond (not failure along glue lines). (C, after Ref. 6.)

of wood failure depends on intended use (exterior, interior) of the glued product. For example, according to U.S. specifications (ANSI/ASTM D 2559), and for structural laminated wood products (Chapter 23) used under exterior exposure conditions (wet use), the average wood failure should not be less than 75%.

Another test for solid wood (British Standard BS 1204) makes use of veneer-thin specimens; their ends are superimposed and glued, and the glued specimen is tested in axial tension.

A specimen for plywood (Chapter 22) is shown in Figure 21-8C, and is also tested in axial tension. A simple test for plywood is the so-called "knife test," where the adhesion is evaluated by forcible separation at or near the glue line by a knife or knife-like tool (30). Other mechanical tests for bond quality also exist.

The performance (residual strength) of joints is tested after exposure tests—single exposure (i.e., soaking and heating or boiling in water) or cyclic exposure (i.e., exposure to alternating temperature and/or moisture levels) (12).

REFERENCES

1. Anonymous. 1974. Adhesives. *The Encyclopaedia Britannica* 1:88–90.
2. Autorenkollektiv. 1975. *Werkstoffe aus Holz*. Leipzig: VEB Fachbuchverlag.
3. Ayla, C., and N. Parameswaran. 1980. Macro-and microtechnological studies of beechwood panels bonded with *Pinus brutia* bark tannin. *Holz Roh- Werkst*. 38:449–459.
4. Ayla, C., and G. Weissmann. 1981. Verwendung der Polyphenole aus der Rinde von *Pinus brutia* Ten. zur Herstellung von Holzleimen. *Holz Roh- Werkst*. 39:91–95.
5. Blomquist, R. F., et al., eds. 1983. Adhesive Bonding of Wood and other Structural Materials. EMMSE Project, Materials Education Council, The Pennsylvania State University.
6. Bodig, J. and B. A. Jayne. 1982. *Mechanics of Wood and Wood Composites*. New York: Van Nostrand Reinhold.
7. Bosshard, H. H. 1975. *Holzkunde*, Vol. 3. Basel: Birkhaeuser Verlag.
8. Chugg, W. A. 1964. *Glulam: The Theory and Practice of the Manufacture of Glued Laminated Timber Structures*. London: E. Benn Ltd.
9. Collett, B. M. 1972. A review of the surface and interfacial adhesion in wood science and related fields. *Wood Technol*. 6:1–42.
10. Coppens, H. A., M. A. E. Santana, and F. J. Pastore. 1980. Tannin formaldehyde adhesive for exterior-grade plywood and particleboard manufacture. *For. Prod. J*. 30(4):38–42.
11. Cupta, R. C., S. P. Singh, and S. Jolly. 1978. Phenol-lignin-formaldehyde adhesives for plywood. *Holzforsch. Holzverwert*. 30(6):109–112.
12. Dinwoodie, J. M. 1983. Properties and performance of adhesives. In *Wood Adhesives: Chemistry and Technology*, ed. A. Pizzi, pp. 1–57. New York, Basel: Marcel Dekker.
13. Dolenko, A. J., and M. R. Clarke. 1978. Resin binders from Kraft lignin. *For. Prod. J*. 28(8):41–46.
14. Drechsel, E. R., C. E. Shuler, and N. P. Kutscha. 1978. Spent sulfite liquor binder for eastern spruce flakeboard. *For. Prod. J*. 28(5):36–48.
15. Gillespie, R. H., D. Countryman, and R. F. Blomquist. 1978. *Adhesives in Building Construction*. U.S.D.A. Ag. Handbook 516.
16. Hartley, J. 1986. Hardwood veneer drying—A review. *Proceedings IUFRO World Forestry Congress, Div. 5*, pp. 77–88, Ljubljana, Yugoslavia.
17. Johns, W. E., H. D. Layton, T. Nguyen, and J. K. Woo. 1978. The nonconventional bonding of white fir flakeboard using nitric acid. *Holzforschung* 32(5):162–166.
18. Knight, R. A. G. (R. J. Newall). 1971. Requirements and Properties of Adhesives for Wood. For. Prod. Res. Bull. No. 20 (5th ed.), U.K.
19. Koch, G. S., F. Klareich, and B. Extrum. 1987. *Adhesives for the Composition Panel Industry*. Park Ridge, New Jersey: Noyes Data Corp.
20. Koenig, E. 1972. *Holz-Lexikon*, Bond I, II (2. Auflage). Stuttgart: DRW Verlag.
21. Kollmann, F. ed. 1962. *Furniere, Lagenhölzer und Tischlerplatten*. Berlin: Springer Verlag.
22. Kollmann, F. F. P., E. W. Kuenzi, and A. J. Stamm. 1975. *Principles of Wood Science and Technology. II. Wood Based Materials*. Berlin: Springer Verlag.
23. Marian, J. E. 1967. Brief history of wood gluing. *Wood Sci. Technol*. 1:183–186.
24. Moult, R. H. 1977. The bonding of glued-laminated timbers. In *Wood Technology: Chemical Aspects*, ed. I. S. Goldstein, pp. 283–293, Washington D. C.: Am. Chem. Soc.
25. Nimz, H. H. 1983. Lignin-based wood adhesives. In *Wood Adhesives: Chemistry and Technology*, ed. A. Pizzi, pp. 247–288. New York, Basel: Marcel Dekker.
26. Parker, R. S. R., and P. Taylor. 1966. *Adhesion and Adhesives*. Oxford, New York: Pergamon Press.
27. Perry, T. D. 1944. *Modern Wood Adhesives*. New York: Pitman.
28. Philippou, J. L. 1981. Applicability of oxidative systems to initiate grafting on and bonding of wood. *J. Wood Chem. Technol*. 1(2):199–227.
29. Pizzi, A. 1983. Tannin-based adhesives. In *Wood Adhesives: Chemistry and Technology*, ed. A. Pizzi, pp. 177–246, New York, Basel: Marcel Dekker.
30. Sellers, T. 1985. *Plywood and Adhesive Technology*. New York, Basel: Marcel Dekker.
31. Steiner, P. R., and J. F. Manville. 1978. Elastomeric

construction adhesives: Influence of solvent retention on strength. *Wood Fiber* 10(3):229–234.
32. Steiner, P. R., S. Chow, and S. Vadja. 1980. Interaction of polyisocyanate adhesives with wood. *For. Prod. J.* 30(7):21–27.
33. Suchsland, K. 1957. Einfluss der Oberflächenrauhigkeit auf die Festigkeit einer Leimverbindung am Beispiel der Holzverleimung. *Holz Roh- Werkst.* 15(9):385–390.
34. U.S.D.A., Forest Products Laboratory. 1960. Proceedings of the Symposium of Adhesives for the Wood Industry. Report No. 2183.
35. U.S.D.A., Forest Products Laboratory. 1968. Selection and Properties of Woodworking Glues. Res. Note FPL-0138.
36. Voulgaridis, E., A. Grigoriou, and C. Passialis. 1985. Investigations on bark extractives of *Pinus halepensis* Mill. *Holz Roh- Werkst.* 43:269–272.
37. Wellons, J. D. 1977. Adhesion to wood substrates. In *Wood Technology: Chemical Aspects*, ed. I. S. Goldstein, pp. 150–168. Washington D.C.: Am. Chem. Soc.
38. Wilkinson, G. 1878. *The Manners and Customs of the Ancient Egyptians* (cited by T. D. Perry). London: J. Murray.

FOOTNOTES

[1] Polarity in adhesives is due to their molecules containing groups (radicals) of –OH, –NH, –CO–NH, etc. Presence of atoms of O and H (and N) contributes to the formation of hydrogen bridges, which bond wood and adhesive together.

[2] Another interpretation of adhesion recognizes the following mechanisms (5):

1. *Mechanical interlocking:* as explained in text.
2. *Interdiffusion theory:* the molecules of a liquid adhesive may dissolve and diffuse in the substrate (wood or other); however, the compatability of adhesive and substrate molecules is limited, and the interdiffused layer is very thin (0.5–1.0 mm).
3. *Adsorption and surface reaction:* physical attraction of molecules (adsorption) is due to the development of van der Waals forces, but chemical reaction (chemical bonding) may also be involved.

[3] A study has shown, however, that the shear strength of joints increased with surface roughness—up to a certain point, beyond which the strength decreased (33).

[4] High density woods are difficult to bond because their cellular structure (thick walls, narrow pit openings, small lumina) hinders the penetration of adhesives (5).

[5] Adhesive quality relates to such factors as pH, viscosity, spreading characteristics, and rate of strength development; pH affects the rate of hardening (polymerization), and when low (high acidity) can cause corrosion of metals (machine parts, etc.). In turn, pH is affected by extractives contained in the wood.

[6] The "storage life" (storage time) of an adhesive is divided into "shelf life" (time between manufacture and starting preparation for use) and "pot life" (time between preparation and spreading). These time periods vary according to type of adhesive, form (liquid, particle, powder), and environmental temperature. Shelf life is at least 3 months (liquid urea) up to an almost unlimited maximum (dry animal glue); and pot life varies for synthetic resins from less than half an hour to a few days (18). Prolonged storage of prepared resin, or storage in an environment where temperature is high, results in excessive increase of viscosity and consequent adverse effects on bonding.

[7] The quantity of adhesive, called "spread," is measured in grams per square meter, or pounds per thousand square feet of joint; the adhesive is placed either on one of the surfaces to be glued (*single spreading*) or both (*double spreading*).

[8] Resins having the same name may have different properties due to different formulation, and for this reason it is better to refer to "type" (e.g., phenol-formaldehyde type, urea-formaldehyde type, etc.).

[9] Resins that harden at low temperatures, 10–30°C (50–85°F), are very sensitive to changes of temperature; raising a temperature by 7°C (12°F) may reduce the hardening time by about half (18). Such effects have an obvious significance with respect to storage life.

[10] Elastomers are considered a separate group (1), or they are included with the thermoplastic adhesives (15).

22

Plywood

Plywood is a panel product made by gluing a number of veneer sheets together or gluing veneer sheets to a lumber-strip core; it is characteristic of plywood that the grain direction of successive layers is at right angles—but the central layer (core) of all-veneer plywood is often made by gluing two sheets with parallel grain. The term ''plywood'' mainly refers to an all-veneer construction; lumber-strip core plywood (''core plywood,'' ''face-glued blockboard'') is made in comparatively small quantities. In general, the number of plywood layers is odd (usually 3 or 5, sometimes 7 or 9), but may be even (4 or more) when two central veneer sheets are glued parallel (Figure 22-1).

SELECTION AND PREPARATION OF VENEER

Veneer sheets are selected according to the intended use of plywood. In decorative plywood (furniture, wall paneling), face veneers are of higher grade and value, and they are selected for their figure and color; whereas core and back layers are of lower-grade veneer of the same or other species. In plywood for constructional purposes (sheathing, concrete forms), the main criterion is strength and not the appearance of the product.

Decorative veneers are mainly produced from hardwoods (oak, walnut, birch, elm, tropical woods, such as teak, mahogany, etc.), and they are usually made by slicing. Utility veneers (see Chapter 20) are made from both softwoods (pine, Douglas-fir, spruce) and hardwoods (poplar, beech, maple, tropical woods, etc.), always by rotary cutting. Usual thickness is 0.6–0.8 mm for decorative veneers and 1.5–3.0 mm for utility veneers.

The veneers should have a smooth surface, uniform thickness, and proper moisture content. Most veneers are dried to less than 5% moisture content; overdried veneer, 2% or less in moisture content, is brittle and tears easily when handled (10, 18, 22, 28).

After drying (and clipping in the case of drying an entire veneer ribbon), the veneers are graded and stored to cool before gluing. Narrow veneers are edge-jointed (usually by a rotary planing head), then edge-glued to produce the required panel width. Edge-gluing machines use hot-melt adhesives in a thread-like form reinforced with nylon or fiberglass, or perforated paper bands to simply keep the edges in contact; direct edge-gluing is also performed by machines that spread glue, bring the edges together, and bond by application of heat.

PREPARATION OF LUMBER-STRIP CORES

Lumber-strip cores are made of wood free of serious defects, relatively low in density, and without large differences between radial and tangential shrinkage. Species in common use are spruce, fir, pine, poplar, birch, various tropical woods, such as Okoumé, Gaboon, Obeche, and others (6, 14).

Strips are variable in width, thickness, and length. Width varies from less than 1 cm (about 0.5 in.) to 12 cm (5 in.), and thickness is 1–2 cm (about 0.5–1.0 in.). Automated machines

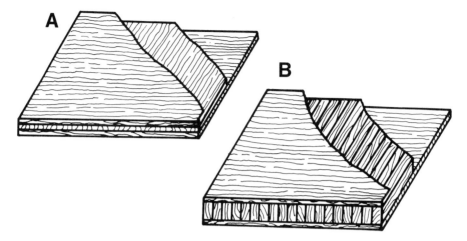

Figure 22-1. (A) All-veneer three-layer plywood. (B) Three-layer core plywood.

gluing cores may utilize very short pieces. Strips are usually produced by sawing (with frame or multiple circular saws), but thin strips (6–8 mm) may be made of rotary-cut veneer.[1]

The strips are usually narrow. Wide strips with a tangential growth-ring arrangement (flat-sawn) have a natural tendency to warp and, if used, they are grooved in order to reduce stresses induced by shrinkage or swelling; the grooves are longitudinal, of small depth (1–2 cm or 0.5–1 in.), and they are made at intervals of 3–4 cm (1–1.5 in.) (14). Ideally, the strips should be alternatively reversed in growth-ring arrangement (Figure 22-2/4, 6), but the procedure is not practical, and in industrial production they are positioned at random. Moisture content is low, as in the case of veneer.

Cores are made by machine and seldom by hand. Machines are continuously fed with strips from one end, glue applied at the entrance is cured by heat, and a continuous layer reaches the exit, where it is sectioned at predetermined panel lengths. Production is on the order of 60–100 m^2/h (600–1000 ft^2/h). In handmade cores, after spreading the glue, the strips are arranged side by side to produce the required panel width, and pressure is applied by clamping devices. In the case of continuous production, each layer (core) is placed separately and clamped in a clamp carrier, which rotates slowly. During rotation, the glue hardens, and when the frame comes back to the operator, he removes the glued core and places another in its position.[2]

After trimming to final dimensions (length, width) by sawing, the cores are planed or sanded to produce smooth surfaces in preparation for veneering. This should be done after conditioning the cores (i.e., equalization of wood moisture to environmental conditions). Premature machining would cause small surface depressions at the joints after drying (sunken joints), due to evaporation of moisture added by the adhesive and the resulting shrinkage. Traces of such depressions would show on the veneered surface of the panel.

Cores may be hollow, and lumber strips are placed parallel or crossed (Figure 22-2/7–9). Cores are also made by use of veneer strips, planer shavings, or other materials (see Figure 22-2/10–12, remarks at the end of this chapter, and Figure 22-10).

ADHESIVES

Plywoods are glued with thermosetting resins: phenol-formaldehyde is used for exterior-type plywood (intended for outside or "marine" use) and urea-formaldehyde for interior-type plywood; interior plywood with limited water resistance may also be produced by use of reinforced urea resins, and sometimes by natural polyphenols (tannins) mixed with synthetic resins.

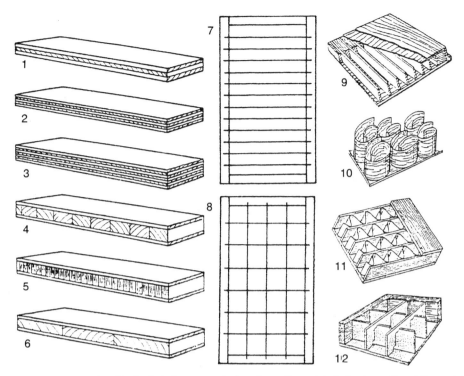

Figure 22-2. Plywood (1-6, solid; 7-12, hollow core): 1, three-layer (three-ply); 2, five-layer; 3, multilayer; 4, blockboard; 5, laminboard; 6, battenboard; 7, parallel strips; 8, crossed strips; 9, detail of a parallel strip construction with face veneers; 10, core of planer shavings; 11, strips of undulating veneer; 12, crossed hardboard. (After Refs. 4, 6, and 24.)

Adhesives are prepared shortly before use by the addition of water, fillers, extenders, and catalysts (see Chapter 21). Resin solids vary from 22 to 30% for exterior plywood and from 12 to 18% for interior use (sometimes as high as 30% with urea-formaldehyde). Additives for phenolic resins include furafil (a ground residue of corn cobs from production of furfural) (1), nut shell flour, wheat or soya flour, bark powder, talc (magnesium silicate), caustic soda, and others. Urea-formaldehyde is prepared by adding wheat flour and ammonium chloride as a catalyst (22).

Adhesives are applied by roller spreader (Figure 22-3), spray, curtain coater, or extruder glue applicator. Rollers are hard (rubber or

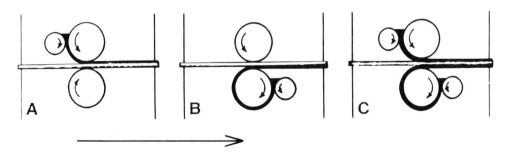

Figure 22-3. Mechanisms of glue application by roller spreaders. (A) Three rollers (spreading the upper surface). (B) Three rollers (spreading the lower surface). (C) Four rollers (spreading both surfaces) (A, B, single spreading; C, double spreading). (After Ref. 9.)

neoprene, corrugated) or soft (sponge); curtain coating is a system where a thin sheet (curtain) of adhesive is passed onto veneer as it is conveyed under a glue reservoir; and extrusion refers to the adhesive being extruded and placed on the veneer in the form of parallel lines or "rods" (Figure 23-9). Accurate weight and even distribution are prerequisites to a high quality bond. Spreads (amount of prepared adhesive per unit area of wood surface) vary from about 100 g/m^2 to 500 g/m^2 (20–100 lb/1000 ft^2), depending on several factors pertaining to wood, adhesive, and manner of application (22, 32).[3] More glue than required produces bonds that are weaker than the wood being bonded and adds to the cost of production. Proper use of adhesives is assisted by following the instructions of the manufacturer, but glue-bond tests are needed for quality control of the product.

Application of glue is followed by assembly of the panels. Assembly (or layup) is manual or semi-automatic (a worker is replaced by a mechanical sheet carrier), but automatic systems are also in use.

PRESSURE

Pressure is applied with hot or cold pressure plates (sometimes called "platens").

Hot Pressure

Plywood is usually produced by hot-pressing in large, hydraulic presses.[4] Steam is the main heating medium, but hot water and hot oil are also used. The presses are multistoried, with 5–25 (up to 70) openings ("daylights") and large plate dimensions—up to 1.5 × 3 m (5–10 ft) for all-veneer panels and 1.8 × 5.0 m (6–15 ft) for core plywood (4, 14). Depending on panel thickness and height of openings, one or sometimes two, panels are loaded in each opening. There are two methods of loading a hot press—manual and automatic. Automatic loaders (and unloaders) are used in most factories (Figures 22-4 and 22-5).

Pressure is applied by progressive upward closing of the openings of a press, or all openings may close simultaneously. The amount of pressure depends on wood species (density) and

Figure 22-4. A multistoried press. (Courtesy of Fjellmann.)

Figure 22-5. Glue spreading (+) and veneer laying in preparation for pressing and production of plywood. (Courtesy of JEM.)

varies from 0.7 to 1.75 N/mm² (100–250 psi) [i.e., 0.7–1.05 N/mm² (100–150 psi) for low-density species (spruce, poplar), 1.05–1.4 N/mm² (150–200 psi) for medium-density species (walnut, mahogany), and 1.4–1.75 N/mm² (200–250 psi) for high-density species (oak, birch, beech) (28)]. The pressure should be neither so high as to crush the wood nor too low (the adhesive is not uniformly distributed), but it should bring the glued surfaces to adequate contact.

The temperature depends on the requirements of the adhesive, and it may be raised to about 180°C (350°F); however, in most cases it varies between 120 and 150°C (250–300°F)—urea formaldehyde 120°C (250°F) and phenol-formaldehyde 150°C (300°F) (22). The required temperature should be attained at the most remote (from each press plate) plane of adhesion; this distance affects the duration (time) of application of pressure and temperature. Mathematical formulas are available in specialized literature for accurate calculations (17). In practice, the time required varies from about 2 to 15 min or more, and depends on panel thickness, type of adhesive, and degree of curing (7). With respect to the latter, and for reasons of economy, the adhesive is not fully cured in the press, but the panels are removed when their adhesion is sufficiently strong to withstand further processing. Completion of curing ("postcuring") is accomplished in storage.

Cold Pressure

Cold pressure is used with natural glues or synthetic resins cured at room temperature. Pressure is applied with hydraulic or mechanical (screw) presses, which usually have one opening. Sometimes, instead of a press, clamps or a combination of clamps and a press are used.

The wood components to be glued are assembled in panels, and pressure is applied in "stacks" (bundles or bales) made of several panels by the following methods. (a) A stack is placed in the press, and after the desired pressure is applied, clamps are fastened at intervals of 25–30 cm (about 1 ft) along the length of the panels to maintain the pressure. The stack is then taken out of the press and left clamped until the adhesive hardens. (b) The stack is briefly pressed in the press until an initial curing of the adhesive is achieved; it is taken out of the press and left without pressure until the adhesive cures. (c) Pressure is rarely applied only by clamps equipped with pressure-measuring dials (i.e., torque wrenches). Depending on the type of adhesive, curing under cold pressure may require 4–12 h or overnight (6, 7).

Cold pressures are relatively high, 1–2.5 N/mm^2 (150–350 psi), depending on wood species (6); in addition to gluing requirements, a high pressure is needed to avoid irregularities of panel thickness due to manner of pressing—not in individual panels but in stacks. Mechanical presses and clamps do not provide accurate and uniform pressure, and repeated tightening may be needed to follow thickness reduction due to curing of the adhesive.

CONDITIONING, TRIMMING, SANDING

After their exit from the press, hot panels are stacked in bundles to cool before trimming, sanding, and other finishing operations. In most factories, the panels are automatically fed to trimming (sizing) saws,[5] then stacked for further processing. Sanding follows by machines equipped with sanding drums or belts, which sand both surfaces in one pass. (Nonuniform removal of wood from the two surfaces of a panel will destroy the balanced construction and cause warping.) The panels are then inspected and graded, and may be patched to remove natural defects (not removed during veneer patching) and manufacturing defects (defective or incorrectly placed veneer plugs, gaps at edge-glued joints, chipped areas, handling damage) (1). Finally, they are finish-sanded, marked (with designations of species, thickness, grade, and sometimes use), packaged in units about 75–80 cm (2.5 ft) high, strapped with steel or plastic bands, and moved to storage or loaded for shipment (Figure 22-6). In storage rooms, 5–6 stacks are placed on top of one another.

PROPERTIES OF PLYWOOD

Shrinkage

The anisotropy of shrinkage and swelling, which characterizes wood, is largely reduced in plywood. The high shrinkage and swelling of wood in tangential and radial directions is considerably reduced due to positioning of successive layers at a grain angle of 90°. Some results of research in the United States concerning the shrinkage of plywood panels, about 2.5–12.5 mm (0.1–0.5 in.) high, made of various species (birch, poplar, basswood, chestnut, elm,

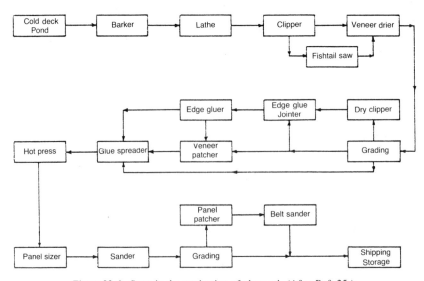

Figure 22-6. Steps in the production of plywood. (After Ref. 25.)

maple, walnut, spruce, mahogany, etc.) are as follows: shrinkage from green to oven-dry condition, parallel to the grain of the surface veneers was, on the average, 0.45% (0.2-1%) and perpendicular to the grain 0.67% (0.3-1.2%) (3, 7). In thickness, the shrinkage of plywood is not substantially different from that of solid wood—under the assumption of same growth-ring arrangement (tangential–radial) and same moisture content change. An investigation of panels made of three veneers, equal in thickness, and hardwood species (including many tropical woods) showed that shrinkage was 10–25 times lower in length or width in comparison to thickness (11, 28). In general, the shrinkage of plywood depends on number and thickness of layers, species of wood or woods, and magnitude of moisture content change. (Accurate determination is possible but complex.) Panels made of five veneers present a higher dimensional stability (smaller difference of shrinkage or swelling in length and width) in comparison to three-layered panels of the same thickness (7).[6]

Irregular shrinkage or swelling may cause warping of plywood panels. This may also result from wood defects (e.g., compression or tension wood), and lack of symmetry of the construction. Symmetry or "balanced construction" is sought by gluing an odd number of layers—but the same effect is achieved with an even number of layers, when the middle layer is made by parallel gluing two veneer sheets. If the middle layer (core) is considered as the "neutral plane," the arrangement of layers should be such that each veneer, of a certain thickness and species, glued on one surface should have a corresponding veneer of the same thickness and same or similar species (with regard to shrinkage) glued on the opposite surface. Aside from thickness and species, the layers glued on the two sides of the neutral plane, and at an equal distance from it, should have the same grain orientation and the same moisture at the time of gluing. In other words, theoretically, the part of plywood found above the neutral plane should be a "mirror image" of the part below that plane (Figure 22-7). In practice, ideal symmetry is seldom realized, especially with regard to grain coincidence,

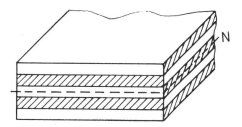

Figure 22-7. The concept of symmetry in a five-layer plywood panel (N = neutral plane).

and, sometimes, for reasons of economy, different layers are made of different species (26). The importance of symmetry is smaller in thicker plywoods. For example, in core plywood panels, made with careful selection of intermediate layers (cross bands), surface veneers may be dissimilar without adverse effects. The thickness of the middle layer is also related to the flatness of a panel. A three-layer, all-veneer panel with double-sheet middle layer is better in this respect. In three-layer core constructions, the middle layer should have 50–70% of the total thickness (27). With regard to grain direction, research has shown that deviations as small as 5° between corresponding layers may cause warping.

Balanced constructions may also warp, however, after a nonuniform gain or loss of moisture. The influence of moisture changes is greater in plywoods made of denser woods due to their higher shrinkage and swelling (26).

Shrinkage and swelling may cause surface checking of plywood due to tension or compression stresses resulting from moisture changes. The magnitude of stresses depends on wood characteristics, such as shrinkage and swelling, density, and growth-ring arrangement (radially cut veneer sheets shrink and swell less in comparison to tangential). Veneer thickness is also important: thin sheets are less prone to checking than thicker ones. Due to the importance of thickness, relatively thin veneers are used on surfaces and cross-bands. In panels exposed to large moisture changes, the maximum thickness of surface veneers, made of wood relatively high in density, should not exceed 3 mm.

Surface checks may also be caused by improper positioning of veneers (i.e., placement of the "loose" side on the surface).

Shrinkage or swelling of a lumber core may result in showing–through traces of core elements. This may be avoided by proper drying of the wood and use of five layers with veneer cross-bands at least 1–1.5 mm thick.

Mechanical Properties

Placement of successive layers with their grain at right angles tends to reduce the difference in mechanical properties of plywood panels, in length and width, in comparison to solid wood (26).[7] Table 22-1 shows a comparison of solid wood and three-ply, equal-veneer-thickness plywood, parallel ($\parallel$) and perpendicular ($\perp$) to the grain (in the case of plywood parallel and perpendicular to the grain of surface veneers) (27); the differentiation of modulus of rupture and modulus of elasticity in static bending is noteworthy.

In plywood, modulus of rupture in the parallel direction is shown to be reduced by 18% and modulus of elasticity by 4%, but transverse values of these properties are about double in comparison to solid wood. A greater number of layers results in greater uniformity of strength along the length and width of panels (Figure 22-8) and better distribution of stresses that develop due to loading; the proportion of grain (thickness of layers) in different directions is also an important factor. Plywood presents a good resistance to dynamic loads due to stress distribution over a large area by the crosswise glued consecutive layers (20).

Plywood has a high resistance to splitting. It is practically impossible to split plywood. This contributes to the avoidance of end-checking,

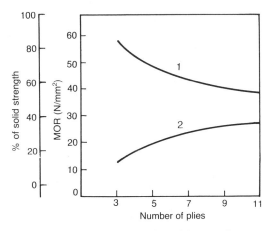

Figure 22-8. Effect of the number of layers (plies) on modulus of rupture: 1, parallel to face layers; 2, perpendicular to face layers (Yellow poplar, all-veneer plywood). (After Ref. 30.)

but is also important with regard to holding nails and other connectors.

Plywood is advantageous in comparison to solid wood because the elements of wood (veneer, lumber-strip core) are glued after drying (dry wood has a higher strength), and defects (e.g., knots) have a much lower influence on strength due to better distribution in plywood.[8] By proper combination of species, number and thickness of layers, and grain direction, it is possible to produce plywood with the desired mechanical properties.

The variation in combinations of species, layers, and other factors (type of adhesive, veneer quality, etc.) in plywood manufacture makes difficult a presentation of mechanical property values. However, equations are available in specialized literature (26, 29) which, on the basis of the properties of the wood species composing the layers, allow such calculations in certain cases.

As in solid wood, moisture has a reducing effect on the mechanical properties of plywood up to the fiber-saturation point. Above this point, strength is practically constant (3).

Other Properties

Other properties also differ between plywood and solid wood, due to differences of behavior

Table 22-1. Comparison of Mechanical Properties of Solid Wood and Plywood[a]

	Solid Wood		Plywood	
	$\parallel$	$\perp$	$\parallel$	$\perp$
Modulus of rupture (%)	100	8	82	17
Modulus of elasticity (%)	100	4	96	7

[a]A generalized presentation (27). (See Fig. 22-8)

between axial and transverse directions in these two products. For example, thermal expansion in length and width of plywood is different because the axial expansion of solid wood is lower in comparison to transverse. In the direction of thickness, no substantial differences exist with regard to thermal, electrical, or acoustical properties in comparison to solid wood. Plywood is used with good results for interior paneling and siding (exterior wall covering).

Plywood and solid wood show some small difference in hygroscopicity due to the incorporation of other materials in plywood (adhesives, catalysts), exposure to high temperatures during pressing, and other factors involved in making this product (16).

Plywood panels are easily sawn to various shapes, may be bent more easily than solid wood, present a favorable relationship between surface and weight, cover large surfaces, and in many cases have a better aesthetic appearance in comparison to solid wood.

TYPES AND GRADES OF PLYWOOD

Different types and grades of plywood exist, depending on the quality of joints (type of adhesive) and the quality (grade) of veneers. Specifications differ in different countries. For example, in the United States, two types of softwood plywood (interior and exterior use) and four types of hardwood plywood (technical 1, 2, and 3) are recognized. According to specifications, the quality of plywood panels is judged on the basis of the surface veneers, and five grades are recognized in softwoods (N, A, B, C, D) and four in hardwoods (1, 2, 3, 4). Aside from veneer defects (knots or their patching, checks, resin pockets, insect galleries, discolorations), specifications refer to other plywood characteristics, such as smooth surface, number of layers, and dimensions (2, 11, 19, 26, 28).

In Germany (DIN 68705), plywoods are distinguished into interior and exterior use (and a special type resistant to termites), and three grades (I, II, III) are recognized on the basis of defects (14).

CURVED PLYWOOD

In addition to flat panels, plywood is made in curved forms (simple or compound), with a curvature that depends on intended use. Curved plywoods are usually produced by simultaneous bending and gluing, and sometimes by bending flat panels.

The first method is preferred because the curvature is better preserved. In this method, after spreading glue, the veneers are properly arranged (successive layers at a 90° angle) and placed between "male" and "female" forms having the desired curvature. Application of hot or cold pressure follows in a conventional press, but curvatures are also produced by use of "bags" and their inflation (with steam, water, or air), or application of vacuum (Figure 22-9).

In order to bend flat plywood panels, a prior "plastification" is required, which (as in the case of solid wood[9]) is usually performed by moisture or heat or both. The magnitude of possible bending depends on various factors, such as wood moisture content, direction of bending in relation to the grain of the surface veneers, thickness and number of layers, veneer quality, and bending technique. Pertinent information is given in specialized literature (26). For example, a hardwood panel, 0.16 in. (4 mm) thick with a 10% moisture content, may be bent, parallel to the grain of the surface veneers, up to a radius (breaking radius) of about 3 in. (7.5 cm). Under the same conditions, the radii are about 7 and 12 in. (17 and 30 cm) for panels 0.32 in. (8 mm) and 0.48 in. (12 mm) thick.[10] In general, the radii are greater for softwood than hardwood plywood (i.e., hardwood plywood may be bent less than softwood plywood of the same thickness). Also, greater bending (smaller radius) is possible if plywood is saturated with hot water, and less bending if a panel is bent perpendicular to the grain direction of the surface layers.

PLYWOODS OF SPECIAL CONSTRUCTION

Plywood panels may be surfaced with metals, plastics, or other materials, or their veneers

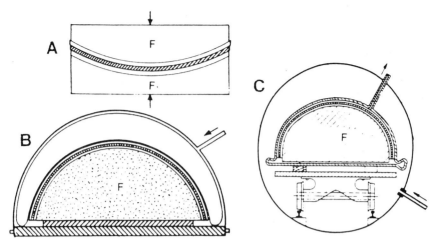

Figure 22-9. Production of curved (molded) plywood. (A) In a press between "male" and "female" form (F); (B, C) by use of form and bag; (B) application of "fluid pressure" (inflation of bag); (C) application of vacuum (deflation of bag) and placement in a heated chamber to control temperature. (After Refs. 15, 21 and 23.)

may be impregnated to achieve a superficial hardness, or resistance to microorganisms, fire, or other destructive agents. For example, plywood is overlaid with reinforced plastics or sheets of aluminum, and used in exterior paneling of railroad cars, refrigerator trucks, airplanes, etc. In other cases, the surface of plywood panels is roughened by eroding the softer earlywood by sand-blasting, or it is grooved to simulate lumber boards, or coated with resin-impregnated paper with photographic reproduction of decorative figures and veneer colors. Plywood is also combined with a middle layer of foamed plastic or honeycombed paperboard in "sandwich" constructions (Figure 22-10), which have good sound- and thermal-insulating characteristics as well as other advantages (e.g., less cost and less weight). Other con-

Figure 22-10. A "sandwich" construction made of plywood faces and a core of honeycombed paperboard. (Courtesy of U.S. Forest Products Laboratory.)

structions have cores made of a wood-strip frame, particleboard, fiberboard, or other panel materials.[11]

REFERENCES

1. Baldwin, R. F. 1980. *Plywood Manufacturing Practices*, 2nd ed. San Francisco: Miller Freeman.
2. Baxter, D. G. 1977. *Softwood Plywood Manufacture and Practices in the Pacific Northwest*. Tacoma, Washington: Plywood Research Foundation.
3. Biblis, E. J. 1979. Effect of moisture on softwood plywood, particleboard and composite wood panels. In *Symposium on Wood Moisture Content—Temperature and Humidity Relationships*, pp. 85-91, U.S. For. Prod. Lab.
4. Blankenstein, C. 1962. *Holztechnisches Taschenbuch*. München: Hanser Verlag.
5. Bowyer, J. L. 1979. Faceglued blockboard from low-grade northern hardwoods. *Wood Fiber* 11(3):184-196.
6. Clark, W. 1965. *Veneering and Wood Bending in the Furniture Industry*. Oxford, New York: Pergamon Press.
7. F.A.O./U.N. 1966. *Plywood and Other Wood-Based Panels*. Rome.
8. F.A.O./U.N. 1976. Proceedings of World Consultation on Wood-Based Panels, Brussels. San Francisco: Miller Freeman.
9. Gillespie, R. H., D. Countryman, and R. F. Blomquist. 1978. *Adhesives in Building Construction*. U.S.D.A., Ag. Handbook 516.
10. Hancock, W. V. 1977. Improvements in veneer yields through better peeling techniques. In *Modern Plywood Techniques*, Vol. 5. San Francisco: Miller Freeman (reprint, p. 12).
11. Hardwood Plywood Institute. 1962. *Hardwood Plywood Manual*. Arlington, Virginia.
12. Illston, J. M., J. M. Dinwoodie, and A. A. Smith. 1979. *Concrete, Timber and Metals*. New York: Van Nostrand Reinhold.
13. International Organization for Standardization, Switzerland. I.S.O. Standards 1096, 1097, 1098, 1954, 2074, 2426, 2427, 2428, 2429, 2430, 2477.
14. Kollmann, F., ed. 1962. *Furniere, Lagenhölzer und Tischlerplatten*. Berlin: Springer Verlag.
15. Kollmann, F. F. P., and W. A. Côté. 1968. *Principles of Wood Science and Technology. I. Solid Wood*. Berlin: Springer Verlag.
16. Kollmann, F. F. P., E. W. Kuenzi, and A. J. Stamm. 1975. *Principles of Wood Science and Technology. II. Wood Based Materials*. Berlin: Springer Verlag.
17. Maclean, J. D. 1955. The Rate of Temperature Change in Wood Panels Heated Between Hot Plates. U.S. For. Prod. Lab. Report 1299.
18. Moult, R. H. 1977. The bonding of glued-laminated timbers. In *Wood Technology: Chemical Aspects*, ed. I. S. Goldstein, pp. 283-293. Washington, D.C.: Am. Chem. Soc.
19. National Bureau of Standards. 1983. U.S. Product Standard PS-183. Construction and Industrial Plywood.
20. Perkins, N. S. 1962. *Plywood: Properties, Design and Construction. Douglas-fir*. Tacoma, Washington: Plywood Association.
21. Perry, T. D. 1948. *Modern Plywood*. New York, London: Pitman.
22. Sellers, T. 1985. *Plywood and Adhesive Technology*. New York, Basel: Marcel Dekker.
23. Stevens, W. C., and N. Turner. 1948/1970. *Solid and Laminated Wood Bending*. London: H.M.S.O.
24. Timber Research and Development Association (TRADA). 1966. *Plywood: Its Manufacture and Uses* (revised) U.K.
25. U.S. Department of Labor. 1968. Technological Changes in Plywood Occupations.
26. U.S. Forest Products Laboratory. 1955. *Wood Handbook*. U.S.D.A., Ag. Handbook No. 72.
27. U.S. Forest Products Laboratory. 1962. Properties of Ordinary Wood Compared with Plywood. Tech. Note 131.
28. U.S. Forest Products Laboratory. 1964. Manufacture and General Characteristics of Flat Plywood. Res. Note FPL-064.
29. U.S. Forest Products Laboratory. 1987. Wood Handbook. U.S.D.A., Ag. Handbook No. 72. (revised)
30. Wangaard, F. F. 1958. Wood. In *Engineering Materials Handbook*, ed. C. L. Mantel, Section 29, pp. 29-1 to 29-37. New York: McGraw-Hill.
31. Watkins, E. 1980. *Principles of Plywood Production*. New York: Reichhold Chemicals.
32. Wood, A. B. 1963. *Plywoods of the World*. Edinburgh, London: Johnston & Bacon, Ltd.

FOOTNOTES

[1]Three types of cores (core plywood) are recognized according to the width of strips: laminboard, up to 7 mm (0.3 in.), blockboard, 7-30 mm (0.3-1.2 in.), and battenboard, greater than 30 mm (1.2 in.) (13).

[2]"Faceglued blockboard" is a product in which individual core pieces are not glued to one another, but the panel is held together only by gluing face veneers in three-layer to five-layer constructions (5).

[3]The amount of adhesive depends on such factors as wood species, veneer or core thickness, type of glue, resin solids, moisture content of wood, quality of surface (smooth, rough), wood and atmospheric temperature, and single spreading (spray, curtain, extruder) or double spreading (rollers); more adhesive is needed with greater thickness and roughness, higher temperature, and lower moisture content. Recommendations vary: reported spreads for U.S. southern pine plywood are 180-230 g/m^2 or about 40-50 lb/1000 ft^2 (single spreading) and 370-470 g/m^2 or about 80-100 lb/1000 ft^2 (double spreading) for veneer thick-

nesses of 2.5–4.2 mm (0.1–1.7 in.) (22). Another source reports 120–170 g/m^2 (25–35 $lb/1000\ ft^2$) for urea-formaldehyde and 150–250 g/m^2 (30–50 $lb/1000\ ft^2$) for casein (32).

[4]In some factories, prepressure at room temperature (0.1–1 N/mm^2 or 15–150 psi) is applied usually to a whole hot-press load in order to consolidate the assembled layers of individual panels. If good consolidation occurs, prepressure contributes to a better quality plywood but the cost is higher (7, 31).

[5]Plywood is available in a wide range of thicknesses from 3 mm (0.1 in.) for all-veneer plywood to about 30 mm (1.2 in.) for lumber-core plywood; the latter is produced in thicknesses of about 13 mm ($\frac{1}{2}$ in.) and up. Surface dimensions (length, width) also vary, and press-plate dimensions vary accordingly. Proposed international standards (ISO 2777), referring to all panels (plywood, particleboard, fiberboard), recommend the following dimensions or their combination: length 80, 210, 240, 270, 330 cm and width 60, 90, 120 cm. In the United States, all-veneer plywood is manufactured in 4 × 8, 4 × 10, 5 × 8 and 5 × 10 ft, and core plywood up to 6 × 16 ft (1 ft = 30.5 cm, 1 in. = 2.5 cm).

[6]Anisotropy, both in dimensions and strength, is reported to be reduced from 5:1 in three-layer, equal-veneer-thickness plywood to 1.5:1 in a nine-layer construction (in solid wood, it is about 40:1); however, the increase of layers makes the product more expensive. For most uses, three-layer plywood is considered to satisfy both properties and cost (12) (also see note 7).

[7]Certain plywood properties may be equalized; in all-veneer plywood, where the core layer is 50% of the total thickness, tension strength and shear through thickness (i.e., perpendicular to thickness) are equal along the length and width of a panel.

[8]Defects seldom have a greater than 10% influence on any mechanical property of plywood, whereas knots may render solid wood practically useless for constructions where strength is needed.

[9]Wood (and plywood) may also be plasticized by chemical means, such as action of ammonia. Desirable shapes can be obtained by placement in liquid ammonia or exposure to ammonia vapors under pressure. After evaporation of the ammonia, the wood hardens and keeps the new form (29).

[10]These figures were derived from experiments using good quality plywood; a "factor of safety" is applied in practice (26).

[11]These various composites are not actually plywood, because they do not have the basic characteristic of crossed wood grain of alternate layers. Exception is made in the case of composite plywood (COM-PLY) made of oriented strand particleboard as core and exterior layers placed at "right" angles to the particle orientation in the core. The product performs structural functions similar to plywood satisfactorily.

23

Laminated Wood

Laminated wood (laminated timber, glued laminated timber, glulam) is produced by gluing two or more layers or "lamellae" of wood with their grain parallel. This is the main difference between this product and plywood, where the grain of two successive layers usually forms a right angle. In addition, laminated wood is not made in panel form, as is plywood, but is produced in a variety of shapes and sizes according to intended use. In general, this product is long in length, and the grain of wood is parallel to length. Lamellae may differ in species, number, shape, and dimensions, and vary from lumber to thin veneer. The shape of the product is straight to curved, and—if curved—bending is, as a rule, simultaneous to production.

Parallel gluing of wood lamellae is applied in making furniture, sporting goods (tennis rackets, some skis), and other items (rifle stocks, shoe lasts, novelties) (1), but the main products—and main subject of this chapter—are laminated timbers (i.e., load-carrying members—beams, arch-shaped supports, parts of bridges, naval and other structures—see Figure 23-1). Sometimes, a combination of laminated wood and plywood or particleboard is made as, for example, in producing box-shaped or I-beams (Figure 23-2).

The industrial production of laminated wood is closely related to the improvement of adhesives, but the idea of using wood lamellae is old. Production of laminated timbers started in the sixteenth century, when arch-shaped and long-span constructions were made, but the wood was connected with iron clamps (18). Gluing was first performed by use of casein in Germany (1906), and later in Switzerland and Scandinavian countries, but production of laminated timbers in large scale started in the United States a few years before the second World War (1935) in conjunction with the appearance of synthetic resins—although casein was also applied for interior-use products. During that period and later, many building constructions were made, such as gymnasiums, churches, theaters, airplane hangers, factories, and warehouses, as well as constructions for the needs of war, such as mine sweepers, propellers for helicopters and airplanes, boat keels, and aircraft-carrier decking (Figure 23-3). Today, laminated wood products are manufactured in many countries.

ADVANTAGES

Laminated wood products present the following advantages:

1. Production of various sizes and shapes, which in most cases cannot be made from the finite dimensions of a tree. This advantage offers great possibilities for architectural design.
2. Improved utilization of wood by reduction of waste due to the possibility of utilizing relatively small dimension wood (Figure 23-4).
3. Improved strength, because it is easier to dry wood of smaller thickness without degrade, in comparison to solid wood of large dimensions; in addition, it is possible to reinforce positions where greater strength is re-

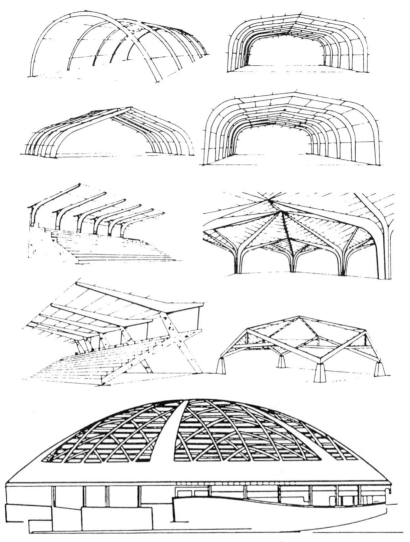

Figure 23-1. Schematic representation of various types of laminated wood structures. Bottom: The world's largest wood laminated structure in Tacoma, Washington (diameter 160 m or 530 ft, height 47.5 m or 157 ft). (Upper sketches, courtesy of Centre Technique du Bois; bottom, after Ref. 6.)

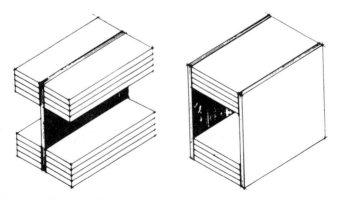

Figure 23-2. Combinations of plywood and laminated wood to produce beams of various cross-sections. (After Ref. 18.)

LAMINATED WOOD 353

Figure 23-3. A recreation building with laminated wood arches. (Courtesy of Timber Engineering Co.).

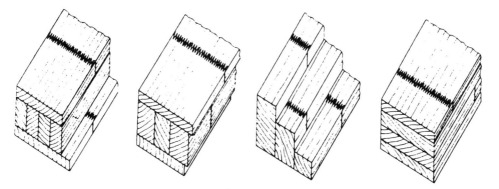

Figure 23-4. Examples of utilization of small-dimension wood for production of window members. (Note application of "finger-jointing," see Figures 23-8 and 23-9). (After Ref. 19, reproduced by permission from Springer-Verlag.)

quired, by selection and placement of proper species or grades of wood in those positions, and by proper design. Improved strength is also accomplished by removal or redistribution of defects, such as knots. Laminated wood exhibits a favorable relation between strength and weight, in comparison to similar structures made from other materials, because wood has such a property.
4. Improved durability, due to the possibility of better preservative treatment of lamellae for protection against fungi and insects, or by placement of durable species on exposed surfaces (Figure 23-5).
5. Laminated timbers of large thickness are resistant to fire (Figures 23-6 and 23-7, see also Chapter 15).

SELECTION AND PREPARATION OF WOOD

The factors that are important in the production of laminated wood are: species, quality, dimensions, moisture, mechanical preparation, and, in certain cases, preservative treatment of wood.

Species vary depending on their availability and intended use of the product. For example, ash and hickory are preferred for sporting goods, and spruce, fir, Douglas-fir, pine, oak, elm, mahogany, teak, and other species for laminated timbers. Theoretically, any species may be used if it can be adapted to the production process and the requirements of the product. A combination of species is also possible (4, 5, 18, 25). In that case, species are selected

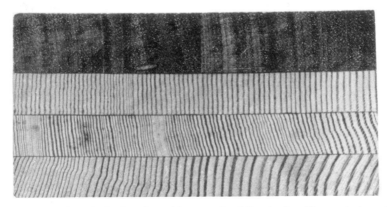

Figure 23-5. Transverse section of a laminated wood beam made of Douglas-fir with a tropical wood on the upper (exposed) surface; part of a ship decking.

Figure 23-6. A laminated wood beam after fire. (Courtesy of Centre Technique du Bois.)

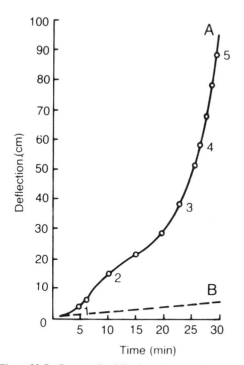

Figure 23-7. Comparative behavior of (A) steel beam and (B) laminated wood beam under the action of fire: 1, 402°C (755°F); 2, 525°C (978°F); 3, 715°C (1320°F); 4, 724°C (1335°F); 5, 783°C (1442°F). (Courtesy of U.S. Lumber Manufacturers Association)

that have a similar behavior with regard to shrinkage and swelling. However, even for a given species, this property is important due to differences originating from growth-ring arrangement. A combination of flat-sawn (tangential) and quarter-sawn (radial) wood pieces is not advisable if the construction will be exposed to moisture changes.

The *quality* (grade) of wood, as determined by defects (e.g., knots, grain deviations) and growth characteristics (e.g., width of growth rings) affects the strength of laminated structures, but the possibility of positioning grades according to the magnitude of developed stresses, reduces the requirement for very high-quality wood. It should be noted, however, that certain defects (e.g., knots, resin pockets) have an adverse effect on gluing, and others, such as blue stain, are objectionable for reasons of appearance. Also, production of members bent to a small radius requires straight-grained wood, without defects. Heartwood is preferable in constructions exposed to the weather, but preservative treatment can overcome this durability requirement (4, 12).

With regard to *dimensions*, lamellar thickness is the main factor of importance. Laminated timbers are usually made by gluing lumber, and thickness may be up to 5 cm (2 in.).[1] Limitations in maximum thickness are posed by the need of drying without defects, and bending to the desired curvature, if curved elements are produced. Also, thickness affects processing costs for production of lamellae, waste of wood, and quantity of adhesive. It is necessary that thickness be uniform; otherwise, difficulties will arise in pressing and gluing.

With regard to the relation of thickness and bending, it should be taken into consideration that wood is bent cold—not prepared by steaming[2] or other treatment. Different species behave differently in this respect (cold bending) but, in general and under the same conditions, hardwoods can be bent more than softwoods (4, 25) (see also Chapter 22).[3]

The other dimensions of lamellae may be given by gluing. The desired width is obtained by side-wise gluing (for laminated timbers minimum width is 4 cm—1.5 in. Length-wise gluing is performed after proper shaping of the ends; gluing transverse surfaces does not give strong joints, because such "butt" joints cause concentration of stresses between adjoining lamellae.

Laminated wood is also produced by gluing veneer sheets. A product, "Laminated-Veneer-Lumber" (LVL), is suitable for furniture (table tops, sofa frames, chairs, kitchen cabinets, shelving, etc.) (11). A similar product, "Press-Lam" (7), or "Micro-Lam" (14), is made by continuous laminating rotary-cut veneer,[4] and possible uses are as structural members, such as high-strength beams (including I-beams), building joists, and trusses, etc. "Loose" faces in veneer (Chapter 20) favor preservative treatment (20).

Moisture content is important for reasons discussed in previous chapters (Chapters 9 and 10). For laminated timbers glued without application of high temperatures (a common procedure), moisture should not exceed 15% (8–15%) (10, 17). Between adjacent lamellae, moisture content differences should not be higher than 3%, and between all lamellae of a construction not higher than 5%. Within a lamella, the distribution of moisture should be practically uniform. These precautions are necessary to avoid excessive stresses due to shrinkage or swelling, which may result in breakage of joints.

The wood is preferably kiln-dried and, until used, it is stored, if possible, under controlled conditions of temperature and relative humidity, so that its moisture content is kept at the desired level. Both warm wood (immediately after exit from a kiln) and cold wood (stored during wintertime in a nonheated space) may affect gluing.

Machining wood includes planing, removal of defects, and shaping of ends for length-wise joining of lamellae.

If the lumber has not been planed in the sawmill, planing is done shortly (2–3 days) before gluing. The purpose of planing is to produce smooth surfaces and lamellae of uniform thickness, as well as to detect defects. If a long time has lapsed, an additional light planing may be required before gluing.

Parts containing serious defects (knots, pitch pockets, discolorations) are removed by cross-cutting. Patching, sometimes applied on veneer and plywood (see Chapters 20, 22), is seldom done with lumber lamellae.

Preparation for end-joining of lamellae is performed by machining in various ways and usually by "finger-jointing" (Figures 23-8 and 23-9). The mechanical effectiveness of "finger" or "scarf" joints is influenced by the form and slope of end-shapes (8, 18); slopes of 1:8 to 1:12 are considered most effective (12).

Members of constructions exposed to conditions that favor attack by fungi, insects, or other organisms, or for protection from fire, should be treated with preservatives (13, 16). The treatment is performed before or after gluing. All types of preservatives (creosote, pentachlorophenol, water-borne) may be used, but pentachlorophenol is preferred; the solvent is different according to intended use of the laminated member (architectural, bridge, etc.). Treated wood may be glued with synthetic resins of phenol or resorcinol type, but the surfaces should be cleaned with steam, or wiped-off in the case of oily preservatives. Gluing of treated wood requires a higher temperature and pressure. Wood treated with salt-type, water-borne preservatives is relatively more difficult to glue; higher temperature and longer pressure time is also needed. However, only those parts exposed to the environment should be treated, and thin members may be protected by surface coatings. Treatment after gluing is possible but not practical for large, especially curved, members.

PRODUCTION TECHNIQUE

After preparation of the wood, production of laminated wood includes end-joining, spreading the adhesive, assembling the lamellae, pressure (by control of temperature, if needed), and final processing.

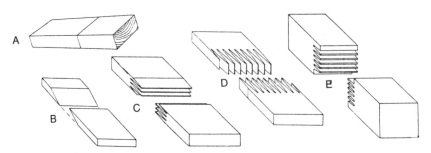

Figure 23-8. Types of end joints: A. butt joint, B. plain scarf, C-E. finger joints (C. structural vertical; D. structural horizontal, see Figure 23-9; E. nonstructural. (Courtesy of U.S. Forest Products Laboratory)

LAMINATED WOOD 357

Figure 23-9. (A) Manner of producing finger-joint components, and (B) components interpositioned. (A, courtesy of American Institute of Timber Constructions.)

End-joining is done by application of glue on the prepared joints and their assembly; high-frequency curing is applied in some cases.

In general, adhesives are synthetic resins (resorcinol, phenol-resorcinol) for exposed structures, but casein is also used for interior applications (9, 13, 15, 17). The glue is spread on the lamellae by roller spreaders or extruder applicators (Figures 22-5 and 23-10). Rollers may apply a double spreading of glue (i.e., both faces of a lamella are spread). With extruder applicators, the glue is pumped through orifices of a pipe, and applied on one surface of each lamella, in the form of ribbons placed at parallel distances of about 5 mm (0.2 in.); the system has advantages, such as less waste of

Figure 23-10. Glue extruder. (Courtesy of American Institute of Timber Constructions.)

glue, relative cleanliness, and longer assembly time—which is desirable in making large laminated structures.

Spreading is followed by assembly of the lamellae. The manner of assembly depends on the laminated product and the form of its parts, and is related to the method of pressure application. If the product is made with veneers, which can be processed by presses used in the production of plywood, such presses are used. In some cases, there are presses exercising a continuous pressure, with plates having the form of endless steel links (as in army tank tracks) or metallic bands (2, 24). These presses are continuously fed and produce a continuous member (e.g., an endless beam) which, after its exit, is sectioned into desired lengths. However, laminated members, especially curved, are usually assembled by placing the lamellae, according to a predetermined sequence and on the basis of a prototype plan of real size, in special steel forms arranged on the ground or on supports. Pressure is applied by clamps (Figure 23-11). The forms are different for straight and curved members, and the clamps are placed at short distances of about 25–50 cm (1–2 ft), which in every case is estimated on the basis of form dimensions and thickness of lamellae (4).

Assembly should be done at a certain speed, because in each case the time available between spreading the adhesive and applying the pressure (assembly time) is fixed. Also, "open assembly" time (between spreading the adhesive and contact of the surfaces to be glued) and "closed assembly" time (between contact and application of pressure) should be controlled. These times, as well as the time (duration) of pressure, differ according to adhesive, wood, and temperature, and they are decided in each case on the basis of experience, experimentation, literature, or instructions of the firm producing the adhesive, as in the case of plywood. Pressure varies by criteria similar to those discussed in plywood (Chapter 22). For laminated members, pressures recommended (with clamps) are 0.7 N/mm^2 (100 psi) for softwoods and 1.0 N/mm^2 (150 psi) for hardwoods (4).

If *heat* is needed, this is done by use of a cover under which heating elements are placed, or the work is performed inside a heated chamber. The time of heating depends on thickness and number of lamellae, and thermal diffusivity of wood. Heating may be also applied by high-frequency electricity, which can raise the temperature very fast to the desired level (2, 4, 15). (High-frequency heats the glue line with-

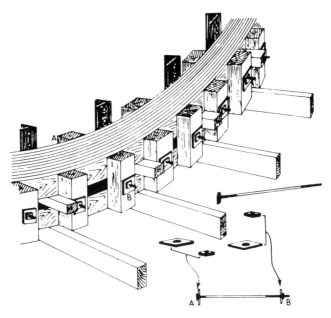

Figure 23-11. Method of production of a curved member by use of clamps. (After Ref. 4, reproduced by permission.)

Figure 23-12. A general view of the interior of a laminated timber factory. (After Ref. 18.)

out appreciably heating the wood.) Relative humidity is seldom controlled. In rare cases, laminated timbers are made on location of use, but usually they are produced in factories (Figure 23-12).

After releasing the pressure, curved members show a tendency of slight change of curvature or "springback." The magnitude of such movement may be calculated in advance; therefore, it can be taken into consideration during production. Thinner lamellae and higher temperatures of gluing reduce springback to a minimum (4).

Finally, the product is made ready for use by processing to final size and appearance. In laminated timbers, such processing includes sawing-off excess dimensions,[5] planing, attachment of metallic accessories to enable connection of members, and protective coating of ends or the whole construction. Before transporting to the site of installation, laminated timbers are protected by waterproof paper or plastic cover.

REFERENCES

1. Capron, J. H. 1963. *Wood Laminating*. Bloomington, Illinois: McKnight.
2. Carruthers, J. F. S. 1965. The Risborough continuous laminating machine. *Wood* (October, reprint).
3. Centre Technique du Bois. 1973. Les Structures en Bois Lamellé-Collé. No. 93, Paris.
4. Chugg, W. A. 1964. *Glulam: The Theory and Practice of the Manufacture of Glued Laminated Timber Structures*. London: E. Benn Ltd.
5. Curry, W. T. 1955. Two-species laminated timber construction. *Wood* 20:122–123.
6. Edlund, B. 1984. Developments in the design and construction of wood structures. *12th Congress, Intern. Assn. Bridge and Structural Engineering Proceedings*, pp. 151–162. Vancouver, B.C.
7. FPL Press-Lam Research Team. 1972. FPL Press-Lam process: Fast efficient conversion of logs into structural products. *For. Prod. J.* 22(11):11–18.
8. Freas, A. D. 1956. Factors Affecting Strength and Design Principles of Glued Laminated Constructions. U.S. For. Prod. Lab. Report No. 2061.
9. Freas, A. D., and M. L. Selbo. 1953. Fabrication and Design of Glued Laminated Wood Structural Members. U.S.D.A. Agr. Bull. 1069.

10. Hann, R. A., et al. 1970. Moisture Content of Laminated Timbers. U.S.D.A. FS Res. Paper FPL 149.
11. Hoover, W., et al. 1987. Markets for hardwood Laminated-Veneer-Lumber. *For. Prod. J.* 37(10):57–62.
12. Ingenieurschule f. Holztechnik, Dresden. 1965. *Taschenbuch der Holztechnologie.* Leipzig: VEB Fachbuchverlag.
13. Koch, P. 1972. Utilization of the Southern Pines (II). U.S.D.A., Ag. Handbook No. 420.
14. Kunesch, R. H. 1978. Micro-Lam: Structural laminated veneer lumber. *For. Prod. J.* 28(7):41–44.
15. Marian, J. E. 1967. Wood, reconstituted wood and glued laminated structures. In *Adhesion and Adhesives,* Vol. 2, eds. R. Houwink, and G. Salomon, pp. 166–280. Amsterdam: Elsevier.
16. McKean, H. B., et al. 1953. Laminating and Steam-Bending of Treated and Untreated Oak for Ship Timbers. Leaflet No. 11, Timber Engin. Co., Washington, D.C.
17. Moult, R. H. 1977. The bonding of glued-laminated timbers. In *Wood Technology: Chemical Aspects,* ed. I.S. Goldstein, pp. 283–293. Washington, D.C.: Am. Chem. Soc.
18. Papaioannou, D. G. 1969. Laminated wood construction members. *Technica Chronica* 5:1–27 (Greek).
19. Ruegge, K. 1976. Stand der Technologie beim Keilzinken von Holz. *Holz Roh- Werkst.* 34:403–411.
20. Schaffer, E. L. 1972. FPL examines new structural material: Press-Lam. So. Lumberman (December 15, reprint).
21. Stevens, W. C., and N. Turner. 1948/1970. *Solid and Laminated Wood Bending.* London: HMSO.
22. Stevens, W. C., and N. Turner. 1957. Springback of laminated bends. *Wood* 22:44–48.
23. Suchsland, O. 1980. Theoretical analysis of yield and strength potential of two-ply lumber. *For. Prod. J.* 30(3):41–47.
24. Taylor, R. B. 1964. A Continuous Process for Manufacture of Laminated Lumber. Spring meeting ASME, Chattanooga, Tennessee.
25. U.S. Forest Products Laboratory. 1987. *Wood Handbook.* USDA, Ag. Handbook No. 72 (revised).

FOOTNOTES

[1] Short pieces of lumber have been used to produce two-layer glued lumber (23), and making glued railroad ties has also been suggested; although technically possible, the cost of such products is high.

[2] Bending solid wood, to produce curved members, is usually done after steaming; in this manner, the bending properties of various woods are considerably improved. This procedure cannot be applied here, because high moisture content hinders gluing. It could be applied by "pre-bending," and gluing after drying (21). It is also possible to bend straight laminated members after steaming.

[3] An empirical relationship, applicable to cold-bending of most species, is $S/R = 0.02$, where $S =$ thickness of layer and $R =$ least radius of curvature. For example, a layer (lamella) 2.5 cm (1 in.) thick may be bent to a radius smaller than about 1.20 m (4 ft). According to another version (21): least radius (for clear, air-dry, straight-grained wood) = thickness × 50.

[4] Production is as follows: thick veneer, 6–12 mm (0.25–0.5 in.), as it comes out of a lathe, is press-dried (5–15 min), and parallel laminated after spreading with a synthetic resin that hardens (2–4 min) by the residual heat of the wood; the thick (and wide) laminated sheets are then sawn and edge-dressed into members of desired shape.

[5] The ends of certain members, which are intentionally made longer, are used to prepare quality-control specimens. Quality control entails tests of strength (shear at glue line) and delamination under alternating conditions of moistening–drying and pressure–vacuum.

24

Particleboard

Particleboard is a panel product made by gluing particles together (i.e., small pieces of wood or other ligno-cellulosic material); wood is the main source. The particleboard industry is relatively new (started in the 1940s in Germany) (11, 27, 29, 30); it is a dynamic industry, established and grown rapidly in many countries (Figure 24-1). Such growth is due to: (a) the possibility of utilizing small-dimension wood including residues from other wood-using industries; (b) availability of synthetic resins that facilitate mass production by fast curing; and (c) suitability of the product for a variety of uses (furniture, building construction, etc.).

Particleboard is produced in thicknesses that range from 2 mm to 4 cm (about 0.1–1.5 in.), and preferably in the density range of 0.50–0.80 g/cm^3 (lower and higher densities are rare). On the basis of production methodology, particleboard is distinguished into "flat-pressed" and "extruded": in the former, particle-grain orientation is parallel and in the latter perpendicular to panel surfaces. Flat-pressed boards are most common; they are single-layer or multilayer (usually three-layer, sometimes five-layer), but graded boards are also produced, and in a relatively newer development the particles are specifically oriented. Extruded boards are single-layer (Figures 24-2 and 24-11).

RAW MATERIALS

Wood, adhesives, and additives (wax to reduce hygroscopicity and sometimes fungicides, insecticides, and fire-retardants) are the materials used for making particleboard (29). In addition to wood, boards are sometimes made by use of flax residues, and the possibilities exist to utilize other agricultural residues, such as straw and cotton stalks (20, 52). Adhesives are usually synthetic resins (see Chapter 21).

Wood is utilized in various forms, such as residues from other wood-using industries (lumber, veneer, plywood, including planer shavings and sawdust) (8), small-diameter roundwood obtained from thinnings, or plantations and coppice forests, harvesting residues, and defective (e.g., crooked) logs (53, 56). Such material may be delivered to the factory in the form of chips.

Softwoods and medium-density hardwoods (0.40–0.60 g/cm^3) are preferred, but heavier hardwoods are not excluded. Pine, spruce, fir, Douglas-fir, poplar, beech, oak, and tropical woods are among the species used (14, 29, 42, 56). Wood density affects energy consumption, wear of knives, and board density. Board density is a major factor, considering that wood will be compressed to produce boards having a density 5–20% higher than that of the wood, in order to obtain maximum bond and sufficient strength and durability according to intended use (14, 39). Addition of resin will also contribute to density; and, if the resulting boards are heavy, above a certain density level specified by standards, they may not be commercially acceptable. For these reasons, relatively light woods are preferred or mixed with heavier woods. In addition to wood density, extractives are important, because they affect gluability (rate of resin curing) and the color of the prod-

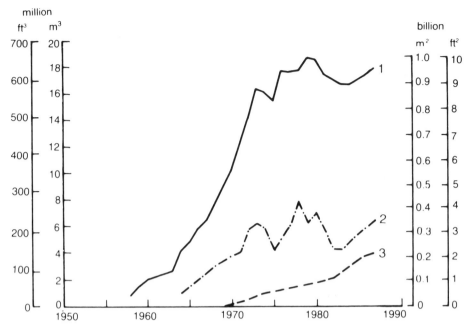

Figure 24-1. Development of yearly production of particleboard: 1, Europe (EEC and EFTA countries); 2, U.S.A.; 3, MDF fiberboard. *Note:* Surface areas (m², ft²) calculated on the basis of a 19 mm (3/4 in.) panel thickness. World production data are as follows (m³): 1965, 9200; 1970, 19,100; 1975, 30,200; 1980, 41,300; 1984, 43,500 (1 m³ = 35.3 ft³). (Courtesy of H.-J. Deppe, EEC, and FAO Yearbooks.)

uct (42). Availability may impose mixture of species (29).

Moisture content of the material is an important factor, because it affects particle dimensions and cost of drying; moisture should be near the fiber-saturation point or a little higher (30–40%). Dry wood generates a high proportion of fine particles (dust).

Bark may be used, but wears knives due to inclusion of foreign matter (soil, dirt, stones, etc.) attached during harvesting in the forest; in addition, bark produces a high proportion of dust, and requires a greater quantity of adhesive. Bark affects the appearance of panels (dark spots show on the surface), and above a certain proportion has an adverse effect on strength and dimensional stability (40, 51). Removal of the bark may be avoided if its thickness is small and if the particles are placed in the interior of the panel.

The production of particleboard permits efficient utilization of wood. The product yield is higher (75–90% or more) (29, 42) in comparison to lumber or plywood (on the average, about 50%).

PRODUCTION AND MORPHOLOGY OF PARTICLES

After an initial developmental period during which particleboard was produced by breaking residues and other wood of low value into irregular particles, research has shown that the dimensions of the particles constitute a decisive factor affecting the properties of the product. Particle length, width, and thickness are all important. For example, modulus of rupture and modulus of elasticity in static bending were found reduced with increasing thickness, but increased with increasing particle length (39). Width has no substantial influence on properties, but may affect other variables, such as flow of particles during production, uniformity of board texture, and surface smoothness. For this reason, the width is generally kept small, except in certain special boards.

In spite of the importance of particle dimensions, accurate control is not always practical for reasons of cost, because such control requires wood of relatively good quality, proper moisture content, and preferably of round form.

PARTICLEBOARD

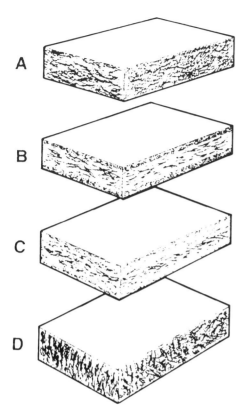

Figure 24-2. Types of particleboard: (A) single layer, (B) three-layer, (C) graded, (D) extruded. Single-layer boards are practically homogeneous, whereas three-layer boards possess three distinct layers of different particle size in core and surface layers; in graded boards there is a gradual, symmetrical reduction of particle size from center to surface. Extruded boards are also single-layer but, in contrast to the others, have their particle grain largely at right angles to the surfaces. (After Ref. 1).

Thus, cost in combination with technological developments, efforts for a more complete wood utilization, and considerations that the extent of dimensional control of particles depends on the intended use of particleboard—have all led to an increasing use of residues from other wood-using industries and workshops.

Particles are produced by cutting, breaking, or friction, and by use of machines which include: chippers, cutter mills, flakers, impact mills, hammermills, and attrition mills (12, 29, 30, 42).

Chippers produce large chips, such as those used in pulping wood to make paper (see Chapter 26). There are *disk* and *drum* chippers (Figure 24-3), and they utilize round wood or large-size residues (slabs, edgings). The chips are not used as such, but only after reduction to small size. Short material is processed in drum-type cutter mills, which can chip manufacturing rejects of lumber, furniture, etc. Drum-type chippers are equipped with screens to separate oversize chips.

Flakers produce particles of small thickness called "flakes." There are three types of machines: drum, disk, and ring flakers. Drum and disk flakers process round and split wood of large dimensions (predetermined or random-size length), whereas wood of small diameter and sawmill residues are first reduced to chips (as above) and then flaked in a ring flaker.

Drum flakers (Figure 24-3B) have a revolving cylindrical cutterhead with projecting knives; the amount of projection regulates flake thickness. Some knives are scoring (i.e., they are not continuous but their edges are at intervals or have notches in order to produce flakes of desired length). The wood is fed with its long axis parallel to drum length.

Disk flakers have several knives set in the disk face, and some are scoring knives, which cut slight grooves at distances equal to flake length. The machine operates on the same general principle as a disk chipper, but the material is fed edge-wise (instead of end-wise), against the cutting disk. Disk flakers produce flakes of high quality, but they are not widely used for reasons of cost; flaking is relatively slow.

Ring flakers (Figures 24-3C and 24-4) have knives placed in a ring, which is either stationary or rotating, counter or in the same direction to an interior rotating impeller (rotor). Flakes are cut off in thickness regulated by the clearance between knife edges and impeller plates.

Impact mills, such as "ring refiners," are utilized to "grind" particles or residues (planer shavings, sawdust, etc.) to produce fine, surface particles. The wood is processed by impact, at a high speed, between rotating impeller blades and a "grinding path," and the fine particles are discharged through a screen.

Hammermills (Figure 24-3D) use the beating action of rotating hammers or steel strips to reduce the size of material by breaking and splitting. The particles exit through a screen of dif-

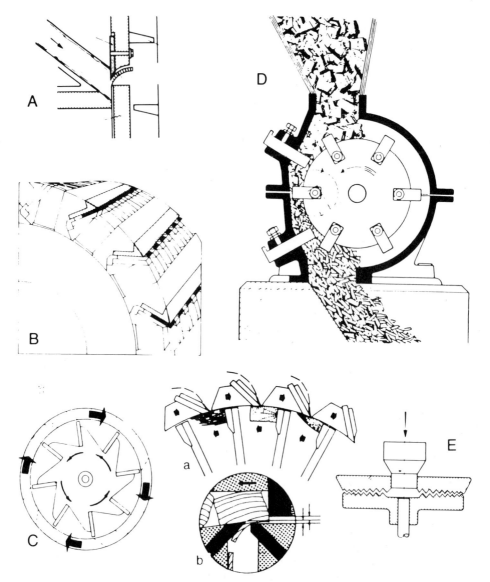

Figure 24-3. Machines for production of particles: (A) disk-knife chipper (inclined feeding); (B) drum flaker (part of a cylindrical head with interrupted and continuous knives); (C) ring flaker (a, b—details of flake production from chips); (D) hammermill; (E) attrition mill (dentate). (Courtesy of Hombak, Pallmann, Rauma Repola, and Voith.)

ferent shape and size openings, but little control of particle geometry is possible.

Finally, *attrition mills* (Figure 24-3E) produce bundles of fibers for particleboard or fiberboard by attrition and shear of wood between plates or disks, one or both of which are rotating.

With regard to morphology, the particles are mainly separated into *flakes*, *slivers*, *fines*, and *fiber bundles*.

Flakes are the most common form (Figure 24-5 and 24-6). Their dimensions vary: thickness is 0.2–0.5 mm, length 10–50 mm (0.4–2 in.), and width 2–25 mm (0.1–1 in.). The ratio of length to thickness is 60–120:1 or higher for better utilization of the natural strength of wood along the grain (23, 35). This ratio is up to 200:1 in surface flakes (42).

Large, square flake-type particles about 5 × 5 cm (2 × 2 in.) to 7 × 7 cm (2.5 × 2.5 in.)

Figure 24-4. Ring flaker (Courtesy of Pallmann.).

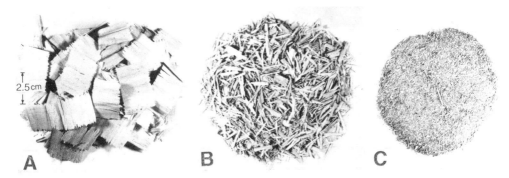

Figure 24-5. Material for production of a three-layer particleboard. (A) "Flat" flakes produced from roundwood (poplar, pine, little beech) with a drum flaker; this material, reduced to (C) in a ring flaker (ring refiner), is mainly utilized for the surface layers. Material for the middle layer (B) is mainly produced from residues (tropical wood veneer or lumber, poplar residues from the production of cores for plywood, etc.); these are first machined in a chipper and subsequently in a ring flaker; a small proportion comes from the drum flaker after reduction in a ring flaker. (Procedure at Balkan Export Industries, Thessaloniki.)

in length and width, and 0.6–0.8 mm thick are called *wafers*; and particles similar to narrow wafers and sometimes a little longer are called *strands* (see Figure 24-30). Both wafers and strands are made from round wood.

Planer shavings are similar to flakes; they have advantages (small thickness, split ends which favor forming and gluing) and disadvantages (nonuniform thickness at the two ends, reduced strength, and twisting form which leads to folding during forming and pressing). Planer shavings are extensively used for particleboard.

Slivers (or splinters) are produced by breaking residues in a hammermill; they have the form of coarse splinters up to 5 mm thick and up to about 1.5 cm (0.5 in.) in length (Figure 24-6B). These particles are usually combined with flakes, because alone they do not produce good quality particleboard.

Particles in the form of *fines* are produced by impact mills, or they are sawdust or sanding dust (Figure 24-6C). Such particles are used to obtain smooth surfaces.

In addition to the above, particles may have the form of *fiber bundles*. Such material is mainly used in fiberboard (see Chapter 25), although sometimes fibers are also included in particleboard to improve surface smoothness.

Finally, wood may be used in the form of

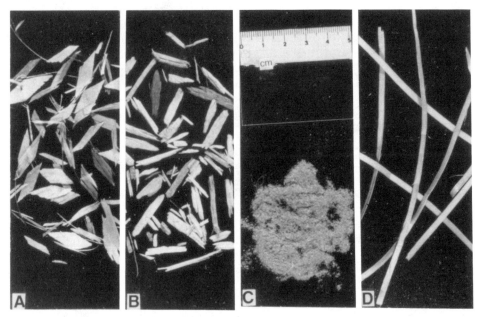

Figure 24-6. Types of particles: (A) flakes, (B) slivers, (C) fines, (D) excelsior. (The scale applies to all.)

excelsior [i.e., long, up to 25 cm (10 in.), and thin ribbons (Figure 24-6D).] In this case, the consolidating material is not an adhesive but cement or gypsum; cement- and gypsum-particleboards are also made by use of wood particles or fibers.

CLASSIFICATION OF PARTICLES

All particles do not have uniform or desired dimensions. There may be particles of large dimensions, which are removed from the line of production or brought back for reduction, and small particles, which are also removed or separated for further use as fines. In certain cases, the particles are separated into two or more dimension groups.

This work is done mechanically by oscillating sieves (23), or by air-classification equipment, where separation is based on particle surface-to-weight ratio (Figure 24-7). A combination of these two methods is also applied (fluid-bed classification). Classification may be done after drying. This solution is preferable, because during drying and storage many particles break. Sometimes, drying and classification are combined (15). During classification, metallic and other foreign materials are removed.

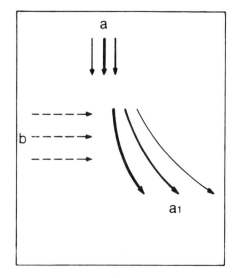

Figure 24-7. Particle classification by air current: a, particles; b, air current; a_1, classification by weight. The lighter particles, shown in a thin line, are thrown further than the heavier particles. (After Ref. 29).

STORAGE AND TRANSPORT

Storage of particles is done with a double purpose—to have a supply of material in case machines of particle production are temporarily out of order due to damage (material supply storage); and to control the flow of material from one production phase to another (material production storage). If the material is supplied in the form of chips or particles, it is stored outside in large open piles or under shelter. Storage of green or dry particles between phases of production is made in silos or bins. Exit from such storage is assisted by various mechanisms, such as rotating screws or shafts, a moving belt at the bottom of the silo, and other.

Transport of particles between different production phases is accomplished by various types of mechanical conveyors (belts, etc.) or by air. The system should transport the particles with minimum damage.

DRYING

Particle moisture is one of the most important factors in particleboard manufacturing. A high initial moisture increases the cost of drying. The amount of moisture after drying usually is 3–6%, depending on resin type and amount (29, 42). A higher moisture may cause steam pockets during curing in a hot press. Drying is performed by exposing the particles to hot air, and the rate and degree of drying is regulated by temperature and time. Temperatures vary with material of different moisture content, and between and within dryers: high temperatures, up to 500°C (900°F) are employed (in jet dryers) at the entrance of particles, and low temperatures, down to 100°C (212°F), at their exit.

Dryers are *rotating* (rotary) and *fixed* (23, 29, 33, 39, 42). Examples are shown in Figures 24-8 and 24-9.

Rotating dryers are "single-pass" and "triple-pass." Both are made of heated drums equipped with interior "flights" or "wings." The particles are briefly held on these flights; and, by rotation of the drum or drums, they are gradually transported to the exit. Three-pass dryers are made of three drums, and both temperature and air speed differ in the respective three compartments; in some types, the interior drums have no flights but are dentate (Figure 24-8A). Figure 24-8B shows the longitudinal view of a drying system, and Figure 24-8C a dryer heated by a rotating bundle of coils.

Fixed dryers are shown in Figure 24-9, and are a "jet" dryer and a "flash tube" dryer, respectively. In the "jet" dryer, the wet particles are fed into a fixed drum and form a "bed" along its length, occupying one quarter to one third of its depth. Hot air is introduced through a bottom slot extending along the length of the drum. The air blows through the particles, and causes a spiral (spinning) circulation (see Figure 24-8C); forward movement of the particles is assisted by raker arms. The dry particles are blown by a fan to a cyclone and discharged at the bottom of it. "Flash" dryers are made of long and large diameter tubes, in a single or double arrangement; in this type (mostly used for drying "fibers"), lengths range from 45 to 70 m (150–230 ft) and diameters from 1 to 1.40 m (3–4 ft). Some fixed dryers are vertical.

BLENDING (MIXING) PARTICLES AND ADHESIVE

Adhesion of particles is usually done by use of urea-formaldehyde for interior-type particleboard (furniture, floor underlayment in housing, etc.), and phenol-formaldehyde for exposed and structural particleboard.[1] The adhesive is usually applied in an aqueous solution, containing 35–60% water (40–65% solids) (35, 42). Use of powder presents problems with respect to uniformity of distribution on the particles. The quantity of adhesive is usually fixed by weight and is about 6–7% (6–7 g of solids per 100 g of dry wood). More adhesive (8–10%) is sometimes added in surface layers of three-layer or five-layer boards. Wax (aqueous solution with 50% solids in a proportion of 0.75–1.0%) (26) is also added to reduce hygroscopicity and, therefore, improve dimensional stability, and other additives include fungicides, insecticides, and fire retardants.

The adhesive is usually added by spraying[2] in combination with stirring the particles, or sometimes by a system of drums (Figure 24-10), and mixing is performed by two methods—*discontinuous* (batch) and *continuous*.

In discontinuous mixing, a certain quantity of particles is placed in a mixer to fill no more than one third of its capacity. The particles are

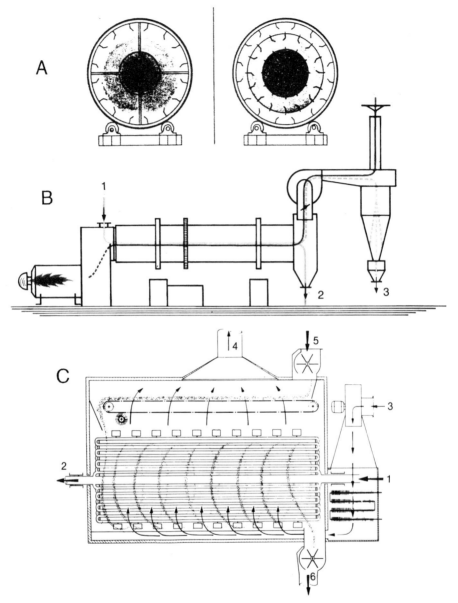

Figure 24-8. Rotating dryers. (A) A single-pass (left) and a triple-pass dryer (right). (B) Longitudinal view of a dryer: 1, entrance and 2, exit of particles; 3, exit of dust. (C) Longitudinal view of a dryer where heat is provided by a rotating bundle of pipes: 1, entrance of hot water; 2, exit of hot water; 3, entrance of air; 4, exit of air; 5, entrance of particles; 6, exit of particles. (Courtesy of A, Manufacturing-Engineering-Construction, M-E-C Co.; B, Overhoff-Altmayer, and C. Schilde)

stirred by rotation of the mixer or by revolving arms. Adhesive is added by spray jets. The mixed particles are removed, after 3–10 min, mechanically or by air. Accurate proportioning of particles and adhesive is an advantage, but the method is slow.

The continuous method is more common. The particles are mechanically and continu-ously pushed inside a cylinder or tube-like tank. Each time, a predetermined quantity of parti-cles is introduced; after this is done, the supply is automatically interrupted (29, 35). The ad-hesive, also in a predetermined quantity, is sprayed and stirred, or applied by revolving drums, or added in other ways (14, 23). In this manner, a mixture of particles and resin flows

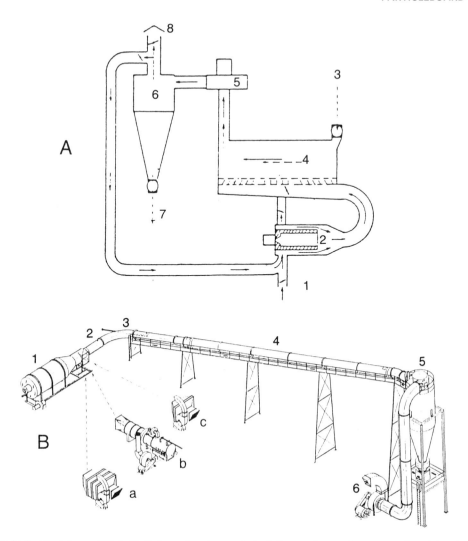

Figure 24-9. Fixed dryers. (A) A "jet" dryer: 1, entrance of air; 2, heat exchanger; 3, entrance of particles; 4, dryer drum; 5, fan (blower); 6, cyclone; 7, exit of particles; 8, exhaust. (B) A "flash tube" dryer: 1, furnace; 2, particle (or fiber) injection; 3, accelerator duct (tube); 4, main duct, 5, negative cyclone; 6, fan. Alternate or supplementary heat sources: a, steam coils; b, solid fuel burner, wood suspension; c, thermal oil. (Courtesy of A, Büttner and B, M-E-C Co.)

continuously toward the exit of the mixer in controlled speed.

Irrespective of the method, it is important that the distribution of the adhesive is uniform; otherwise, adhesion may be in parts defective or nonexisting. Lack of uniformity may be due to malfunction of the mixer, resin viscosity and temperature, and differences in moisture and dimension of the particles.

After addition of the adhesive, the moisture content of the particles is raised to about 8–14% (1, 14, 42).

FORMING

Forming the mat (i.e., spreading the resin coated particles for a preliminary shaping of boards) is the next step in particleboard production. This step is preparatory to application of pressure.

The type of particleboard depends on forming procedure and mat-forming materials. Single-layer boards are made by using the same type of furnish (particles, resin, and additives) throughout; however, in layered boards, parti-

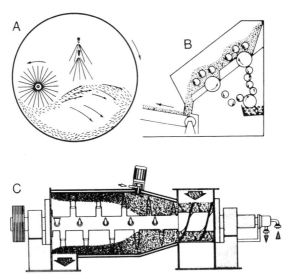

Figure 24-10. Blending (mixing) of particles and adhesive by (A) spraying, (B) drums, and (C) revolving shaft with arms. (Courtesy of A, and C, Dais; B, Fahrni)

cles and amount of resin differ in surface layers and the core. The surface layers may be made from smaller or thinner particles. Also, more adhesive is added in these layers with the result that the boards acquire smoother and denser (harder) surfaces. In five-layered boards, three sizes of particles are used (the smaller and thinner on the surfaces), and in graded boards there is a gradual increase of dimensions from the two surfaces to the core. In all cases, the particles tend to be laid down with a preferential orientation in the direction of board formation (machine direction), and their length tends to be parallel to the plane of the board; perpendicular arrangement is rare, only in extruded particleboard (see Figure 24-2 and 24-11). A new development, called "oriented strand board" (OSB) has a combined, three-layer ply-

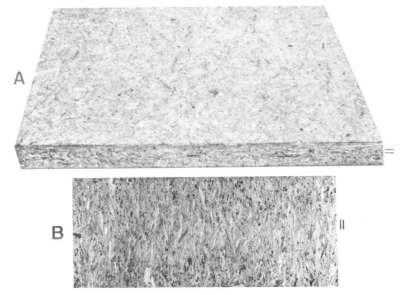

Figure 24-11. (A) Flat-pressed and (B) extruded particleboard with, respectively, parallel and perpendicular orientation of the particles in relation to panel surfaces.

wood-type arrangement (see under "Newer Developments").

Forming is a decisive stage of production, because lack of uniformity of particle distribution within a board or along its length and width will cause respective differences in density. Such differences affect mechanical and other properties, and may cause warping.

There are two forming methods—*discontinuous* (batch) and *continuous*. In discontinuous forming, the supply of particles is interrupted between successive cauls, whereas in continuous forming the supply is continuous. Discontinuous forming produces separate boards. Separate boards may also result from a continuous supply of particles, but this method can produce a continuous, long board, which is sectioned after coming out of the press.

Forming is done by particles of a certain

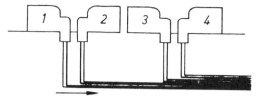

Figure 24-12. Schematic representation of forming a mat for three-layer boards by four particle-spreading heads (1–4). Head 1 is spreading the lower layer, 2 and 3 the middle layer, and 4 the upper layer. (After Ref. 40, reproduced by permission from VEB-Fachbuchverlag.)

quantity (volume or weight) falling from one or more forming heads (dropping points) on metallic (steel or aluminum) cauls or a belt. Single-layer boards may be formed in a single process from one forming head, but the distribution of particles is more uniform if forming is done in stages by more than one forming head (Figures 24-12 and 24-15). Various systems are

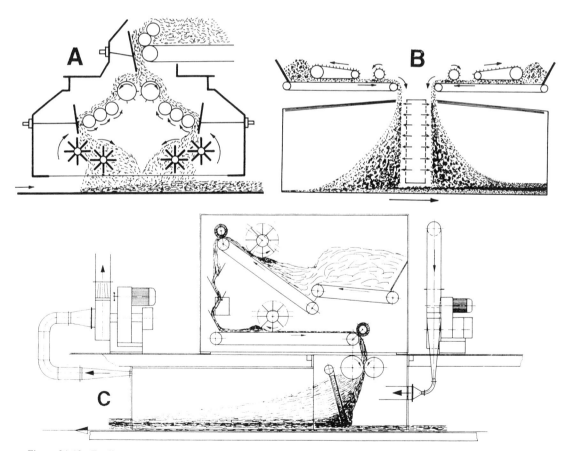

Figure 24-13. Feeding systems. (A) Revolving forming heads; the horizontal arrow shows the direction of movement of the band on which the mat is laid. (B) Forming by air; feeding is double and the particles are classified according to size by air. (C) A system of air-forming after a mechanical (roller) preseparation of particles. (Courtesy of A. Schenk, B. Baehre, and C. Würtex)

applied, such as (a) reciprocating movement of the caul under the forming head, (b) temporary stop of the caul or belt and reciprocating movement of the forming head, and (c) forming by three or four forming heads, each one of which forms part of the mat or a different layer. All three systems are used in making single-layer boards, but three- or other multilayered boards are made by use of system (c).

Forming is accomplished by rollers, air, or vacuum, or a combination (Figure 24-13). Vacuum is applied to lay material in fiber form (42). Continuous forming is often combined with continuous pressing to form a continuous board. There are various systems, as explained later in this chapter.

PRESSING

Pressure is applied by hot presses, usually with many openings (daylights), such as those used in making plywood (Figures 24-14 and 24-15). Single-opening, steel-track belt, and cylinder-type pressing is also applied in production of continuous boards (Figure 24-16).

Usually, and prior to final pressing, the mat of particles is prepressed. The main purpose of

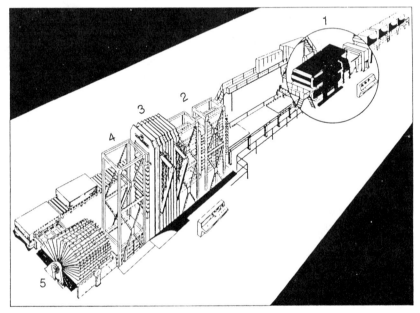

Figure 24-14. Particleboard production line: 1, prepress; 2, loader; 3, multi-opening press; 4, unloader; 5, cooler. (Courtesy of Dieffenbacher)

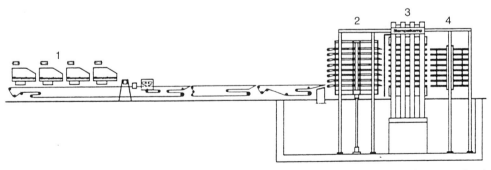

Figure 24-15. Mat forming and pressing. The mat is formed on a belt by four forming heads (*1*), and transported to the loader (*2*), and subsequently to the press (*3*) and the unloader (*4*). (Courtesy of Siempelkamp)

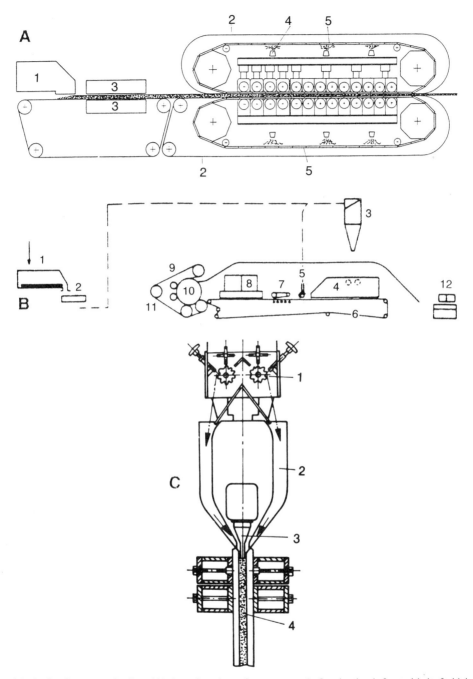

Figure 24-16. Continuous production. (A) A steel-track continuous press: 1, forming head; 2, steel belt; 3, high-frequency and contact heating with prepressing; 4, heater; 5, steel tracks. (Range of board thickness 6–22 mm.) (B) Production of Mende particleboard: 1, dose-regulating bin; 2, adhesive blending; 3, cyclone; 4, forming station; 5, scalping roller; 6, forming belt; 7, prepress; 8, high-frequency heating; 9, steel press belt; 10, heating drum; 11, pressure rollers; 12, saw. (C) Kreibaum press: 1, entrance of particles (flow control); 2, falling shutes; 3, piston; 4, formed board. (Courtesy of A, Bison-Baehre & Greten; B, and C, after Refs. 3 and 50, reproduced by permission from Springer-Verlag and VEB-Fachbuchverlag.)

prepressing is consolidation of the mat, reduction of mat thickness, and avoidance of resin-curing problems (i.e., precuring the faces and under-curing the core); this may happen in hot pressing the original mat thickness, particularly in making thick boards, and would result in large variation of density between faces and core, and lower internal bond. Prepressure also contributes to reduction of the height of the openings of the hot press and, therefore, allows for placement of more openings within a certain height of press and its faster closing. Prepresses are single-opening platen presses, or continuous presses equipped with belts or metal-tracks. Prepressure is cold, or sometimes the mat is preheated but not to the extent of resin curing; the pressure varies between 1 and 2 N/mm² (150–300 psi), depending on species (density) of wood, particle dimensions, and moisture content (29).

A little higher moisture content of the mat surfaces, during hot pressing, can improve the quality of boards; this may be accomplished by spraying water, 100–150 g/m² (20–30 lb/1000 ft²), or use of outer layer particles with a little higher (about 1%) moisture content (25, 26, 42). The boards produced have smoother and denser (harder) surfaces (because the compressibility of the particles is improved) and higher bending strength. In addition, the steam produced by evaporation of such moisture contributes to faster transfer of heat from the hot-press platens to the mat and, therefore, to faster curing of the core.

In the hot press, pressure is applied to stops that control board thickness, or the platens stop at the desired board thickness without stops. The pressure that develops depends on wood density, particle dimensions, and desired board density; it is higher with denser woods and particles of larger dimensions, and for medium-density particleboard (0.40–0.80 g/cm³) varies between 1.5 and 3.5 N/mm² (200–500 psi). Press closing time is also an important factor, because it affects resin curing in the faces and the core, and contributes to density and other property variation; the pressure is higher with faster closing. Temperature at pressure is 160–220°C (320–430°F) for phenolic resin and 140–200°C (280–400°F) for urea-formaldehyde (33, 40).

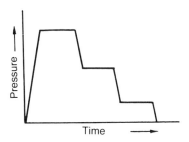

Figure 24-17. Pressure-time diagram for production of particleboard. (Variations in the manner of raising pressure to maximum and its duration will affect the properties of boards, such as surface density, bending strength, and internal bond) (Adapted from Refs. 21, and 29).

Figure 24-17 shows a pressure-time diagram. In an initial phase, the press is closed until the platens contact the mat. Subsequently, the pressure is increased up to the desired level, and kept constant for the time prescribed for each particular case. After that, it is reduced to a low of 0.1–0.2 N/mm² (15–30 psi) to facilitate a gradual escape of steam from the board, and release of the pressure of steam on the particles. This low pressure is held for a short time (it plays a safety role for steam that may be enclosed, and prevents expansion of the board until polymerization of the adhesive is completed), and finally the pressure is released and the press opened (33, 42).

Pressure time increases with board thickness, level of pressure, temperature, and moisture content of particles, and is reduced with wood density and size of particles. The type of resin is also important. In general, pressure time is about 10–12 min for a board of about 2 cm (1 in.) thick, which corresponds to 0.5–0.6 min/mm of thickness. (Some literature gives much lower values, down to 0.15 min/mm.)

Presses operate automatically. The prepressed mat is successively (board by board) placed in a press loader, and when this is filled, it moves into the press, unloads the mat, and returns to its position. After pressing, the boards are placed in an unloader (Figures 24-14 and 24-15). In the loader, and later in the press, the mat is lying on cauls, or it is transferred from cauls or a belt directly to the press, and pressed without cauls. The press gradually closes upward (the lower openings close first), but presses also exist where all openings close simultaneously. Presses are heated by steam, hot water, or hot oil, and sometimes by high-frequency electricity.

Continuous pressing is shown in Figure 24-16. Figure 24-16B shows the so-called "Mende" system (Bison), that produces thin boards, 2–6 mm thick, in a manner similar to the production of paper (see Chapter 26). The board is fairly flexible so that, during the production process, it can pass around a heated drum. Such particleboard competes with plywood and fiberboard for certain uses.

Continuous supply of particles and simultaneous pressing is also applied in the Kreibaum process (extruded board). The particles are successively pushed between two parallel, heated press platens, arranged in a vertical or horizontal position. A continuous board is produced, and sectioned to desired lengths as it comes out of the press. Thickness is regulated by the distance of the press platens, and it is possible to produce very thick and perforated boards by placing heated pipes between the platens, parallel to the direction of production (Figures 24-16C and 24-18). This is a simple and economical system (with low demand on raw material quality, and low proportion of added resin—3.5–5%), but has not found a wide application (1, 14, 42). This is attributed to the slow rate of production and the properties of the product (see "Properties of Particleboard").

CONDITIONING, SANDING, TRIMMING

After removal from the press, the boards are very hot. Immediate storage in high stacks and under conditions that retard loss of heat may have adverse effects, such as discoloration, detachment of surface particles during sanding, and reduction of strength, especially with use of urea-formaldehyde resin (15, 21). For this reason, the boards are piled for a few days in small stacks or "packets," but this is time-consuming. In modern factories, coolers are used, where the boards are placed separately on rotating mechanical devices (Figure 24-14) or transported on edge through a chamber of controlled air temperature. During this time, and in storage, a post-curing of the resin takes place. During cooling, a more uniform distribution of moisture is also achieved; nonuniform moisture content may create stresses with resultant warping of boards.

After conditioning, the boards are trimmed to give predetermined dimensions of length and width. Specifications differ in different countries.[3] Trimming is performed with circular saw systems. The residues are reused, or burned to produce steam.

Sanding follows to improve the appearance of boards and serve further processing, such as overlaying with decorative sheets of veneer. Sanding is performed with drum or wide-belt sanders, and it should be done carefully, because nonuniform reduction of thickness may destroy the symmetry of layered boards and lead to warping. Extruded boards are seldom sanded (14) (Figure 24-19).

Finally, the boards are marked to designate type and sometimes the commercial standard, then strapped into packets, and stored or shipped.

Additional processing may be applied, in same or other plants, such as veneer overlaying (common procedure with extruded boards), overlaying decorative paper sheets impregnated with synthetic resins, printing wood grain on board surfaces, and coating with varnish. Particleboard is also produced with surface designs made by use of proper press platens.

PROPERTIES OF PARTICLEBOARD

The properties of particleboard are a subject of extensive research, and there are many publications relative to this matter. The effort has contributed to important improvements of properties in comparison to formerly produced particleboard.

Figure 24-18. Extruded and perforated particleboards.

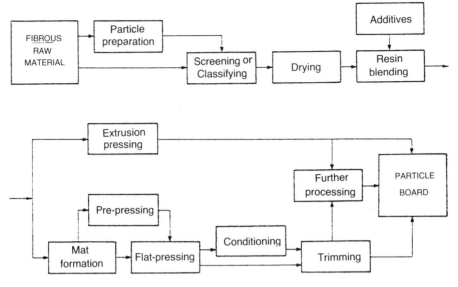

Figure 24-19. Steps in the production of particleboard. (Courtesy of FAO.)

The destruction of the natural structure of wood by transformation to particles, changes its anisotropy with regard to properties. The transformation is more drastic than in plywood, and many variables (particle morphology, arrangement in the board, method of board production, thickness, wood species, type and amount of adhesive, additives, etc.) contribute to the production of boards with different properties.

Density

Particleboard is produced in low density (0.25–0.40 g/cm^3), medium density (0.40–0.80 g/cm^3), and high density (0.80–1.20 g/cm^3). Boards of high density are heavy, and difficult to handle. Board density is basically affected by the density of wood, as previously explained under "Raw Materials," and, in turn, it affects all other physical and mechanical properties of the product.

The density of a board is seldom uniform throughout its thickness; variations constitute the so-called "density profile," which should be nearly symmetrical and not skewed (Figure 24-20). An extreme variation of density between faces and core should be avoided, however, because it indicates an inferior board with low internal bond in the core.

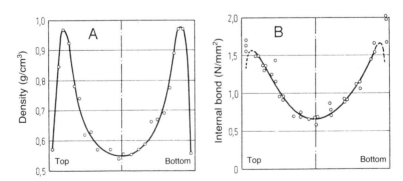

Figure 24-20. Variation of (A) density and (B) internal bond within thickness of a three-layer, 19 mm (3/4 in.) thick particleboard. (After Ref. 45, reproduced by permission from Springer-Verlag.)

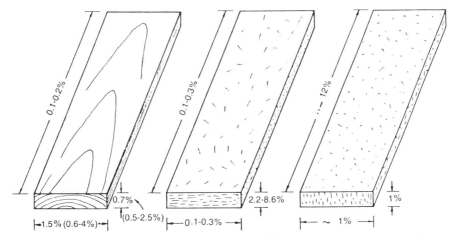

Figure 24-21. The concepts of "length", "width," and "thickness" in lumber and particleboard, and the respective changes of dimensions (swelling) for a relative humidity range of 60–90%. (A) Lumber (flat-sawn); values refer to spruce wood and (in parentheses) to various species. (B) Flat-pressed particleboard. (C) Extruded particleboard.

Shrinkage and Swelling

Under the influence of moisture changes, flat-pressed particleboard exhibits a good dimensional stability (low shrinkage and swelling) in length and width. In contrast, the change of thickness is significant. The following data are characteristic: A moisture content change from 6 to 18% caused a 4–9% increase of thickness (length and width increased only by 0.4%), and at 21–24% moisture content, thickness increased by about 12% (29). An increase of relative humidity from 60 to 90% caused a 2.2–8.6% increase of thickness, whereas length and width increased only by 0.1–0.3%.

The behavior of extruded particleboard is different, displaying a greater variability in length. In the case reported above, where relative humidity increased from 60 to 90%, the length of such boards increased by 12%, whereas width and thickness increased by about 1% (Figure 24-21). Veneer overlay on both surfaces reduced length-wise swelling to less than 1% (1).

In all boards, as in solid wood, shrinkage and swelling is higher when made with heavier woods (23, 42) (Figure 24-22).

Density differences along the length and width of a board will result in differential shrinkage and swelling, and lead to warping. Density variation in thickness usually exists normally, as previously mentioned. Board properties are adversely affected by a nonsymmetrical density profile, and a board will also warp with moisture changes.

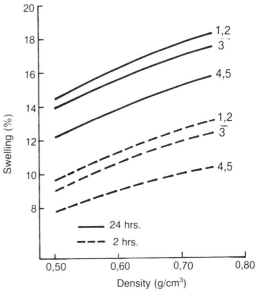

Figure 24-22. Effect of density on thickness swelling of particleboard made of different species; 1, spruce; 2, poplar; 3, pine; 4, oak; 5, beech. (After Ref. 28.)

Other factors which influence shrinkage and swelling in thickness are geometry of particles and amount of resin (Figure 24-23A). The change of dimensions is smaller in boards made with thinner particles and more adhesive (6).[4]

Thickness swelling was greatly reduced by acetylation of the particles; however, in some

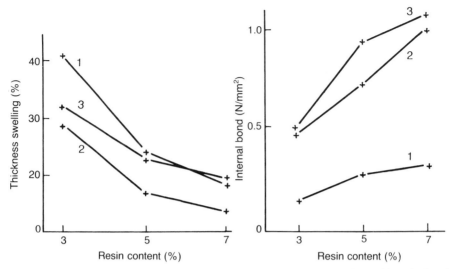

Figure 24-23. Effects of resin content. (A) Thickness swelling, (B) internal bond strength: 1, white oak; 2, red oak; 3, sweetgum (*Liquidambar styraciflua*). (After Ref. 55)

cases, other properties (internal bond, moduli of rupture, and elasticity) were also reduced (60).

Mechanical Properties

Mechanical properties are affected by many factors, such as density, quantity of adhesive, particle dimensions and orientation, and moisture content. Main mechanical properties, in relation to the uses of particleboard, are static bending (moduli of rupture and elasticity), internal bond (transverse tension), and toughness (9, 44).

The *density* of a board is an important index of strength. In general, boards of higher density have a higher strength, but the relationship is not linear (it depends on type of board and manner of loading). Wood density is also important. For a given board density, woods of lower density give boards of higher strength (Figure 24-24). This is so, because a given weight contains more particles of a lighter wood, and during pressure such particles contact each other better. The quantity of *adhesive*, when increased within certain limits (about 6–10%), improves the mechanical properties (Figure 24-23B), but the composition of an adhesive is also important—for example, the U/F mole ratio in urea-formaldehyde resins (Figure 24-25). Important particle *dimensions* are length and width, as previously explained. The influence of particle *orientation* is generally expressed by differences between

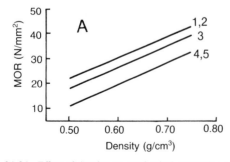
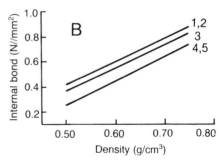

Figure 24-24. Effect of density on mechanical properties (A, modulus of rupture; B, internal bond) of particleboard made of different species: 1, spruce; 2, poplar; 3, pine; 4, oak; 5, beech. (After Ref. 28.)

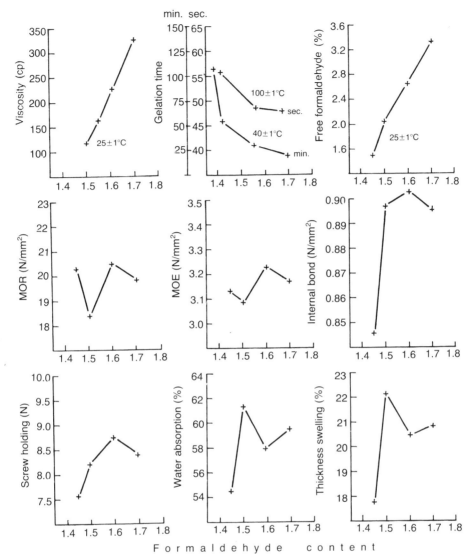

Figure 24-25. Effect of the mole ratio of formaldehyde to urea (1.4:1, etc.) on properties of resin and particleboard. Release of formaldehyde vapors and viscosity increase, gelation time decreases, and best ratio for properties is 1.6:1. (Results from a study of particleboard made from residues of Philippine woods.) (After Ref. 34).

flat-pressed and extruded boards; extruded boards have a lower strength in bending and lower toughness, but higher internal bond. *Moisture* contributes to reduction of strength. A change of moisture content from 5 to 15% reduced static bending by 25–50%. Creep is higher in particleboard (and fiberboard) in comparison to solid wood, and increases (up to 10 times) at high moisture contents (10). The effect of moisture is greater in particleboard glued with urea-formaldehyde than phenol-formaldehyde (46) (Figure 24-26). At the same moisture content, the strength properties decrease with increasing *temperature*.

Particleboard is mechanically inferior in comparison to solid wood and plywood, but its properties are considerably improved by overlays of veneer or other materials; a preferred orientation of particles also improves the properties. However, the practical importance of mechanical and other properties depends on intended use. For example, the principal prop-

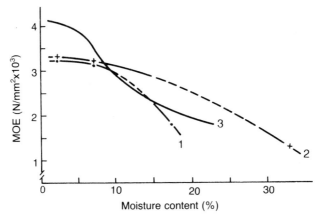

Figure 24-26. Effect of moisture on modulus of elasticity for (1) urea- and (2) phenol-formaldehyde-bonded particleboard compared to hardboard (3). (All measurements at 20°C or 68°F). (After Ref. 62; reproduced by permission from the Forest Products Research Society, USA.)

erties of structural particleboard are high bending strength (modulus of rupture) and high modulus of elasticity, whereas boards for furniture primarily need a smooth surface to accomodate overlays, and mechanical demands are secondary.

Other Properties

Other important properties of particleboard are *springback*, *nail* and *screw holding capacity*, *thermal conductivity*, *sound insulation*, *hygroscopicity*, *quality of surface*, and *machining* behavior. All these properties are affected by board density.

After removal from the press, flat-pressed boards exhibit an increase of dimensions, especially thickness. This change is partly due to capture of moisture from air and, therefore, it is a swelling phenomenon, but a board does not return to the former dimensions after the moisture is lost (Figure 24-27). The phenomenon is called "springback," and is attributed to release of stresses from the compressed particles. Repetition of adsorption–desorption cycles has an additive effect (i.e., the initial dimensions increase after each treatment). Springback is added to swelling, and it may cause an increase of thickness up to 20% or more (42).

The capacity to hold nails or screws is higher in single-layer, and lower in three-layer boards with a middle layer of low density. It is also

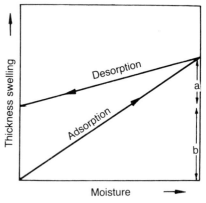

Figure 24-27. Schematic representation of the phenomenon of springback. Dry board after its exit from the press adsorbs moisture and swells. After desorption does not return to its original dimension but retains a permanent swelling (a, recoverable change; b, springback). (After Ref. 42.)

higher perpendicular to the edges. In addition to density, this property is related to internal bond strength (29).[5]

Thermal conductivity and sound insulation are considerably affected by density (Figure 24-28). Water adsorption also varies with density (it is reduced when density increases) (58), and is generally high. Within 24 h, the moisture content of particleboard may increase up to 100% or more (21). Hydrophobic additives (wax) have a reducing effect on hygroscopicity. Particleboards exhibit a lower equilibrium moisture content in comparison to solid wood (Figure 24-29), but this property is influenced

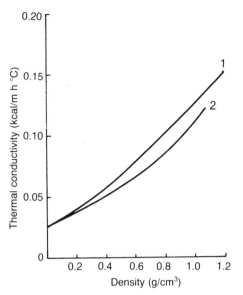

Figure 24-28. Relationships of thermal conductivity and density of (1) particleboard and (2) fiberboard. (After Ref. 32.)

Surface quality is mainly affected by the size of particles. Fine particles on the surface contribute to production of smooth boards, which is an important feature for overlaying with thin veneers.

Particleboard is well worked with machines, but the quality of edges depends on board type and manufacturing characteristics (board quality). Generally, tools dull faster when cutting particleboard than solid wood, due to the presence of resin; the dulling effect is greater in boards glued with urea than phenolic resin. Hard (tungsten carbide) teeth are needed for sawing.

Depending on use, particleboard may be required to resist fire (thickness and density are important factors), or fungal or insect attack. These properties are improved by additives during production.

NEWER DEVELOPMENTS

by various factors, such as type of catalyst (salts increase the equilibrium), and addition of wax (small effect) and spent pulp liquors (equilibrium increases at high relative humidities) (48).

Particleboard is made for interior use (furniture, etc.), or for structural purposes (i.e., to bear loads). Two products, *waferboard* and *oriented strand board* (OSB), together with

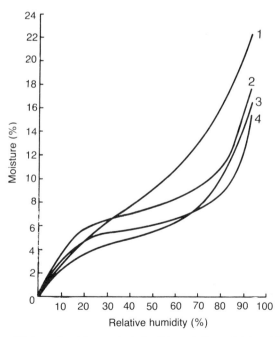

Figure 24-29. Changes of equilibrium of moisture content of solid wood (1), particleboard (2, interior; 3, exterior) and medium-density fiberboard (4) in relation to relative humidity at a constant temperature of 20°C (68°F). Average adsorption–desorption values. (Adapted from Ref. 54.)

COM-PLY (a composite of particleboard and plywood, see Chapter 22), are relatively new developments of importance, especially in countries, such as the United States and Canada, where great quantities of wood are used in housing and other buildings. Both waferboard and OSB are used as nonveneered panels (12, 17, 18, 41).

Waferboard is composed of wafers—that is, large, nearly square flakes (Figure 24-30A and C), made preferably from low-density hardwoods, such as aspen, but spruce has also been used.[6] Waferboard is structurally inferior in comparison to plywood (it has low bending strength and stiffness), and this drawback has spurred the evolution of technology that led to the development of OSB (Table 24-1). OSB is a three-layer panel, made of strands (Figure 24-30B and D), which in the surface layers are placed parallel to the direction of panel production and in the core crosswise (Figure 24-31); in essence, OSB is similar in construction to plywood, and its bending strength, stiffness (modulus of elasticity), and dimensional stability are similar. Waferboard is made without alignment of particles. Both waferboard and OSB are glued with phenolic resin, which is powder in waferboard, and powder or liquid in OSB. Due to different particle geometry, the surface appearance of these two types of particleboard is different (Figure 24-30C and D).

MOLDED AND MINERAL-BONDED PARTICLEBOARD

In addition to flat boards, boards of various shapes are produced by mixing particles and resin (usually in powder form), and applying

Figure 24-30. (A) Wafers, (B) strands, (C) waferboard, (D) oriented strand board (OSB). (Courtesy of A, and B, Pallmann; C, and D, Würtex)

Table 24-1. Comparative Properties of Panel Products (Plywood, Particleboard, COM-PLY)[a,b]

Property	Unit	Plywood	Particleboard			COM-PLY
			F/B	W/B	OSB	
Density	g/cm³	—	0.40–0.80	—	—	—
MOR	N/mm²	45	10–50	20	50	45
	psi	6500	1500–7000	3000	7000	6500
MOE	N/mm²	8500	1000–5000	3500	8500	8500
	1000 psi	1250	150–700	500	1250	1250
Tension ∥	N/mm²	—	5–25			
	psi	—	700–3500	—	—	—
Tension ⊥ (IB)	N/mm²	0.8	0.2–1.2	0.5	0.8	0.6
	psi	120	30–170	70	120	90
Water absorption, 24 h immersion						
weight	%	—	20–75	—	—	—
volume	"	—	10–50	—	—	—
Linear expansion, 50–90(97%) RH	"	0.2	0.2–0.6	0.25	0.2	0.25
Thickness swelling, 24 h immersion	"	10	5–15	30	25	15
Thermal conductivity[c]						
SI		—	0.05–0.12	—	—	—
English		—	0.4–0.95	—	—	—

[a]Plywood (Douglas-fir), F/B = flakeboard, W/B = waferboard (aspen), OSB (aspen), COM-PLY (Southern pine).
[b]For fiberboard, see Table 25-2.
[c]SI units: kcal/h/m²/°C; English: Btu/ft²/h/°F.
Source: Refs. 15 and 41.

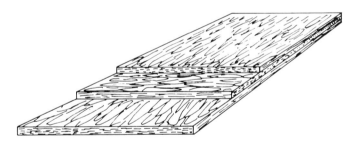

Figure 24-31. Three-layer particleboard with crossing orientation of particles. (After Ref. 11, reproduced by permission from Springer-Verlag.)

high temperature and pressure in molds; the proportion of resin is high (21).

Particleboard is also made by use of wood and mineral binders, mainly cement (5, 59). Wood is usually in the form of excelsior, although particles may also be used. Particle geometry is an important factor affecting properties (5). The product has good sound- and thermal-insulating properties, and is resistant to fungi and fire, lighter in weight than cement but heavier than particleboard glued with synthetic resins. It is used in building constructions (interior and exterior paneling, partitions, ceiling linings, etc.) and for decorative purposes. Not all wood species are suitable, however; problems of retardation of cement-setting (hydration) exist with certain species (e.g., oak), and such problems are caused by the presence of soluble sugars and other extractives (7, 24). Steam or hot-water treatment or chemical additives, such as calcium chloride, may be needed to overcome this incompatibility. In general, heartwood, due to its higher extractive content, is more incompatible than sapwood, as is decayed wood, but blue stain (in pine) was not found to have an inhibiting effect. Species

Table 24-2. Density, Bending Strength (MOR), and Dimensional Changes of Particleboard Bonded with Organic and Inorganic Agents

Product	Density (g/cm³)	Bending strength N/mm² (psi)	Dimensional changes[a] (%)
Particleboard organic bonding	0.61–0.73	8–20 (1200–2900)	3–5
Cement particleboard	1.10–1.30	9–15 (1300–2200)	3–5
Magnesite particleboard	0.85–0.90	7–14 (1000–2000)	—
Gypsum particleboard	1.10–1.20	4–10 (600–1500)	0.3–0.9
Gypsum fiberboard	1.10–1.15	6–8 (900–1200)	0.3–0.6
Excelsior board, light, inorganic bonding	0.36–0.57	0.4–1.7 (60–250)	1.0–2.4

[a]Dimensional changes of length or width corresponding to changing conditions of exposure from 20°C (68°F) and 30% RH to 20°C and 85% RH. Note that comparative properties of cement particleboard, gypsum particleboard and fiberboard, and organic (synthetic resin)-bonded particleboard are also given by Lemper, K., Hilbert, T., and H. Günzerodt, *For. Prod. J.* 40(6):37-40, 1990.
Source: Refs. 19 and 31.

used are poplar, spruce, pine, and others. Other mineral binders are magnesite and gypsum; wood is used in the form of particles or fibers. Some properties of boards made with such binders are shown in Table 24-2. Gypsum boards are increasingly produced in Europe, and some are made by use of industrial gypsum, which is a by-product of chemical factories. Gypsum boards work well with tools and machines, they are suitable for many uses, and like other inorganic-bonded boards, perform better when exposed to fire, in comparison to conventional (organic-bonded) particleboard.

THE PROBLEM OF FORMALDEHYDE

Particleboards glued with urea resins present problems due to release of formaldehyde vapors in factories, homes, and other places, where boards are used as partitions, paneling, etc. (26, 36, 37). Such vapors are irritating to eyes, nose, and the respiratory system, and may cause allergy and perhaps more serious health hazards in high doses. The amount of vapors is affected by various factors, such as room temperature, board moisture content, amount of resin, and mainly resin composition as ex-

Figure 24-32. Large decorative flakes on the surface of fiberboard: (A) cedar (*Thuja*), (B) mahogany.

pressed by the ratio of urea and formaldehyde (Figure 24-25). All these variables, if high, increase the release of vapors. Release is also higher from the edges of boards than from large surfaces, and is affected by board construction (particle geometry, porosity, density).

Formaldehyde release may be reduced in various ways, such as proper ventilation in factories, complete coating of boards with veneer or special paints which reduce the rate of release, use of resins with less formaldehyde (38), and special treatment of the boards after they come out of the press—namely, spray with ammonia vapors, removal of the excess of ammonia by ventilation, and exposure to vapors of formic acid to remove the last traces of ammonia (47, 61). Release of vapors is reduced with the passage of time (see also Chapter 21).

Formaldehyde vapors are also released by other thermosetting resins. The problem is greater with urea-formaldehyde, because it is widely used due to low cost. Formaldehyde problems also exist with melamine-formaldehyde. Phenol-formaldehyde releases few such vapors.

REFERENCES

1. Akers, L. E. 1966. *Particle Board and Hardboard*. Oxford, New York: Pergamon Press.
2. Allan, D. 1979. Manufacturing considerations of particleboard requirements for furniture application. In *Particle Board—Today and Tomorrow*, pp. 304–323, ed. R. Weinbrenner, Stuttgart, DRW Verlag.
3. Autorenkollektiv. 1975. *Werkstoffe aus Holz*. Leipzig. VEB Fachbuchverlag.
4. Ayla, C. and G. Weissmann. 1982. Verleimungsversuche mit Tanninformaldehydharzen aus Rindenextrakten von *Pinus brutia* Ten. *Holz Roh- Werkst.* 40:13–18.
5. Badejo, S. O. O. 1988. Effect of flake geometry on properties of cement-bonded particleboard from mixed tropical hardwoods. *Holz Roh-Werkst.* 22: 357–370.
6. Biblis, E. J. 1979. Effect of moisture on softwood plywood, particleboard and composite wood panels. In *Symposium on Wood Moisture Content—Temperature and Humidity Relationships*, pp. 84–91. U.S. For. Prod. Lab.
7. Biblis, E. J., and Cheng-Fan-Lo. 1968. Sugars and other extractives: Effects on the setting of southern pine-cement mixtures. *For. Prod. J.* 18(8):28–34.
8. Biblis, E. J., and G. E. Coleman. 1974. Sawdust/bark particleboard makes structural panels with veneer faces. *Forest Industries* July. 70–71.
9. Bodig, J., and B. A. Jayne. 1982. *Mechanics of Wood and Wood Composites*. New York. Van Nostrand Reinhold.
10. Bryan, E. L., and A. P. Schniewind. 1965. Strength and rheological properties of particleboard. *For. Prod. J.* 15(4):143–148.
11. Deppe, H.-J. 1981. Zum Stand der Bauspanplattenherstellung. *Holz Roh-Werkst.* 39(10):425–432.
12. Deppe, H.-J., and K. Ernst. 1977. *Taschenbuch der Spanplattentechnik*. Stuttgart. DRW-Verlag.
13. Dinwoodie, J. M. 1986. Waferboard and OSB. BRE Information Paper IP 5/86.
14. F. A. O./U. N. 1958. *Fiberboard and Particle Board*. Rome.
15. F. A. O./U. N. *Plywood and Other Wood-Based Panels*. Rome.
16. F. A. O./U. N. 1976. *Proceedings of the World Consultation on Wood-Based Panels, Brussels*. San Francisco. Miller Freeman.
17. Forintek Canada Corporation. 1980/1982. Proceedings Canadian Waferboard Symposia. Special Bull. SP 505E and SP 508E.
18. Greten, R. 1979. Neuentwicklungen und Verfahrenstechniken mit spezieller Betrachtung von Spanplatten mit orientierten Eigenschaften (OSB) in europäischer Raum. Intern. Particleboard Symposium FESYP 78, pp. 85–101. Stuttgart. DRW-Verlag.
19. Gregoriou, A. 1989. Technology of Glued Products. (Class Notes, mimeo, Greek). Thessaloniki.
20. Heller, W. 1980. Die Herstellung von Spanplatten aus unkonventioneller Rohstoffen. *Holz Roh-Werkst.* 38:393–396.
21. Ingenieurschule f. Holztechnik, Dresden. 1965. *Taschenbuch der Holztechnologie*. Leipzig. VEB Fachbuchverlag.
22. International Organization for Standardization. I.S.O. Standards 820, 821, 822, 823.
23. Johnson, E. S., ed. 1956. *Wood Particle Board Handbook*. Raleigh, N. C.: North Carolina State College, School of Engineering.
24. Kavvouras, P. K. 1987. Suitability of *Quercus conferta* wood for the manufacture of cement-bonded flakeboards. *Holzforschung* 41:159–163.
25. Kelly, M. W. 1977. Critical Literature Review of Relationships between Processing Parameters and Physical Properties of Particleboards. U.S. For. Prod. Lab. Gen. Tech. Report FPL-10.
26. Kelly, M. W. 1977. Review of particleboard manufacture and processing. In *Wood Technology: Chemical Aspects*, ed. I. S. Goldstein, pp. 220–234. Washington, D.C., Am. Chem. Soc.
27. Klauditz, W. 1955. Entwicklung, Stand und holzwirtschaftliche Bedeutung der Holzspanplattenherstellung. *Holz Roh-Werkst.* 13:405–421.
28. Klauditz, W., and G. Stegmann. 1957. Über die Eignung von Pappelholz zur Herstellung von Holzspanplatten. *Holzforschung* 11:174–179.

29. Kollmann, F., ed. 1966. *Holzspanwerkstoffe*. Berlin: Springer Verlag.
30. Kollmann, F. F. P., E. W. Kuenzi, and A. J. Stamm. 1975. *Principles of Wood Science and Technology, II. Wood Based Materials*. Berlin. Springer Verlag.
31. Kossatz, G., K. Lempfer, and H. Sattler. 1982. Anorganish gebundene Holzwerkstoffplatten. *FESYP Geschäftbericht* 83:98-108.
32. Lewis, W. C. 1967. Thermal Conductivity of Wood-Base Fiber and Particle Panel Materials. U. S. For. Prod. Lab. Res. Paper FPL 77.
33. Maloney, T. M. 1977. *Modern Particleboard and Dry-Process Fiberboard Manufacturing*. San Francisco. Miller Freeman.
34. Mari, E. L. 1983. Effect of formaldehyde to urea mole ratio on the properties of UF resins and particleboard. *FPRDI J.* 12(3/4):20-34 (Philippines).
35. Marian, J. E. 1967. Wood, reconstituted wood and glued laminated constructions. In *Adhesion and Adhesives*, Vol. 2, eds. R. Houwink, and G. Salomon, pp. 166-280. Amsterdam. Elsevier.
36. Marutzky, R., Mehlhorn, L., and H. A. May. 1980. Formaldehydemissionen beim Herstellungprozess von Holzspanplatten. *Holz Roh-Werkst.* 38(9):329-335.
37. Marutzky, R., L. Mehlhorn, and W. Menzel. 1981. Verminderung der Formaldehyd-Emission von Möbeln. *Holz Roh- Werkst.* 39(1):7-10.
38. Mayer, J. 1979. Chemische Aspekte bei der Entwicklung formaldehydarmer Klebstoffe für die Holzwerkstoffindustrie. Intern. Particleboard Symposium FESYP 78, pp. 102-111. Stuttgart. DRW-Verlag.
39. Mitlin, L., ed. 1969. *Particleboard Manufacture and Application*. U.K. Pressmedia Ltd.
40. Modlin, B. D., and I. A. Otlev. 1975. *Herstellung von Spanplatten*. Leipzig: VEB Fachbuchverlag.
41. Montrey, H. M. 1983. Emerging structural panels for light-frame construction. In *Wall and Floor Systems: Design and Performance of Light-Frame Structures* (pp. 25-40). Proceeding 317, Forest Products Research Society, Madison, Wisc.
42. Moslemi, A. A. 1974. *Particleboard*, Vols. 1 and 2. Illinois: Illinois University Press.
43. Passialis, C., A. Grigoriou, and E. Voulgaridis. 1988. Verleimung von Spanplatten mit Rindenextrakten der *Pinus halepensis* Mill. *Holzforsch. Holzverwert.* 40(3):50-52.
44. Paxton, B. H. 1980. Impact testing and requirements of chipboard for flooring. *J. Inst. Wood Sci.* 8(5):208-213.
45. Plath, E., and E. Schnitzler. 1974. Das Rohdichteprofil as Beurteilungsmerkmal von Spanplatten. *Holz Roh-Werkst.* 32:443-449.
46. Pozgaj, A. 1980. Einfluss der Feuchte auf die Zug- und Druckfestigkeit von Spanplatten. *Holztechnologie* 21(4):200-206.
47. Roffael, E. 1982. *Die Formaldehyd-Abgabe von Spanplatten und anderen Werkstoffen*. Stuttgart: DRW Verlag Weinbrenner.
48. Roffael, E., and A. Schneider. 1981. Zum Sorptionsverfahren von Holzspanplatten (4). *Holz Roh-Werkst.* 39:17-23.
49. Sandermann, W. 1963. *Chemische Holzverwertung*. München: BLV Verlag.
50. Soiné, H. 1983. Kontinuierliche Pressverfahren in der Spanplatten-industrie—die Anfänge bis zum Mende-Verfahren. *Holz Roh-Werkst.* 41:485-487.
51. Starecki, A. 1979. Spanplatten aus Holz mit Rindenanteil. *Holztechnologie* 20(2):108-115.
52. Sudhakara Reddy, B., and J. Durst. 1981. Panel boards from agricultural wastes for developing countries. A review. *Holzforsch. Holzverwert.* 33(6):110-114.
53. Stegmann, G., and J. Durst. 1966. *Grunlagenforschung über die technische Nutzbarmachung von geringwertigem Wald- und Abfallholz—Nutzbarmachung von Eichenholz zur Herstellung von Holzspanwerkstoffen*. Köln. Westdeutscher Verlag.
54. Suchsland, O. 1972. Linear hygroscopic expansion of selected commercial particleboards. *For. Prod. J.* 22(11):28-32.
55. Tang, R. C., C. Y. Hse, and Z. J. Zhou. 1984. Effect of flake-cutting patterns and resin contents on dimensional changes of flakeboard under cyclic hygroscopic treatment. In *Durability of Structural Panels*, ed. E. D. Price, pp. 43-51. Gen. Tech. Report SO-53, USDA, So. For. Exp. Sta., New Orleans, Louisiana.
56. Tsoumis, G., C. Passialis, and Ph. Siamidis. 1977. Experimental particleboard from oak fuelwood. *Technica Chronica* 3:12-18 (Greek, English summary).
57. U.S.D.A., Forest Service—Forest Products Research Society. 1978. Structural Flakeboard from Forest Residues. Symposium Proc., Kansas City, Missouri.
58. Vital, B. R. and J. B. Wilson. 1980. Water adsorption of particleboard and flakeboard. *Wood Fiber* 12(4):264-271.
59. Wienhaus, O. 1979. Werkstoffe aus Holz und Zement. *Holztechnologie* 20(4):207-215.
60. Youngquist, J. A., R. M. Rowell, and A. Krzysik. 1986. Mechanical properties and dimensional stability of acetylated aspen flakeboard. *Holz Roh-Werkst.* 44:453-457.
61. Ziegler, R. D. 1981. Review: 1980 particleboard literature in non-U.S. journals. *For Prod. J.* 31(8):55-58.
62. Dexin, Y. and B. A.-L. Oestman. 1983. Tensile strength properties of particle boards at different temperatures and moisture contents. *Holz Roh-Werkst.* 41:281-286.

FOOTNOTES

[1]Tannin-formaldehyde, spent pulp liquors, isocyanates, and bark extractives (4, 43) have also been used (see Chapter 21), but on a rather experimental basis.

[2]Research has shown the importance of the diameter of the sprayed resin droplets. Optimal dimension is that of 8-35

μm, but actual diameters in practice are up to 100 μm. Large droplets result in insufficient coating of the particles, and this affects adversely the mechanical, as well as other, properties of particleboard. The size and distribution of droplets are affected by the air pressure of spraying, and by resin formulation, viscosity, solids, and temperature. Small particles (fines), due to their greater surface, hold more adhesive. For this reason, they are coated separately, or are transported faster through the mixing equipment (42).

[3]Particleboards are commonly produced in the following dimensions (width × length): 4×6 ft (122 × 183 cm), 4×10 ft (122 × 305 cm), 5×9 ft (152 × 274 cm), 6 × 8 ft (183 × 244 cm), 6 × 12 ft (183 × 366 cm), and some are 7 × 17 ft (213 × 518 cm) and longer; thicknesses range from 8 to 25 mm (0.32–1 in.). "Mende" boards are thin, down to 2 mm (less than 0.1 in.). Compact extruded boards range in thickness from 8 to 22 mm (0.32–0.88 in.), and perforated from 23 to 120 mm (about 1–5 in.).

[4]A measure of the dimensional stability of particleboard (and other products or solid wood) is the "coefficient of dimensional stability" (CDS), which is determined from the relationship:

$$\text{CDS} = \frac{\%\text{ change of linear dimension}}{\%\text{ change of moisture content}} \times 100 \quad (24\text{-}1)$$

Some coefficients are as follows: solid wood 10–30 (radial–tangential), fiberboard 4–5, plywood (lumber core) 2–3, plywood (all-veneer) 1.5–2.5, particleboard 1.5–5. A smaller coefficient means higher dimensional stability (2).

[5]According to some experiments, the following results were obtained for screw holding power in comparison to solid wood (100%): plywood (all-veneer and core) 75%, particleboard 50%, hardboard 45% (oil-tempered 70%). These figures are only indicative, as this property depends on constructional characteristics of each product (30).

[6]Large, decorative wafer-like flakes are sometimes placed on the surface of common particleboard or fiberboard (Figure 24-32).

25

Fiberboard

Fiberboard is different from particleboard because wood, or other ligno-cellulosic material, is used in the form of fibers instead of particles, and an adhesive is not always needed for bonding. The fibers are held together by the development of hydrogen bonds, flow of lignin, interweaving, or addition of a synthetic resin; the relative importance of these factors depends on the type of fiberboard and manufacturing procedure (1, 3, 21, 23, 34).

Manufacturing of fiberboard started at about the beginning of this century, and developed successively in England (1898), the United States (1908), France (1928), Sweden (1929), and other countries (11, 19). Initially, there was a similarity to paper manufacture—mechanical or chemical pulping and forming by suspension of fibers in abundant water. Later, pulping by "explosion" was invented (Masonite process, 1924), and dry mat forming (i.e., forming without water) was introduced. Fiberboard manufacture is now an important sector of the wood-using industry, but lags in dynamism of development in comparison to particleboard.

TYPES OF FIBERBOARD

There are two types of fiberboard—*insulation* and *compressed* (Table 25-1); this distinction is based on density and method of production (i.e., whether hot pressure is applied or not).

Insulation board is semirigid and rigid. *Semirigid* boards are very low in density (0.02–0.15 g/cm^3) and are primarily used as insulation and cushioning, placed around curves or between framing members. The production of such fiberboard is quite limited. *Rigid* boards (0.15–0.40 g/cm^3) have structural uses, such as in wall assemblies, sheathing, roof insulation, and ceiling tiles (10, 35). The latter are plain or perforated, and have a paint finish or decorative imprints.

The main representative of compressed fiberboard is *hardboard* (0.80–1.20 g/cm^3). Large quantities of hardboard are produced, and uses include house siding, floor underlayment, concrete forms, prefinished paneling, furniture and kitchen cabinetwork, and many others. Some hardboards are made with striated, relief, leather-like, perforated, and other textures, finished with lacquers or plastic films, or grooved to simulate lumber.

For many years, hardboard dominated fiberboard production. A new product, called *medium-density fiberboard* or *MDF* (0.65–0.85 g/cm^3), entered the market in the United States in the 1960s (4, 6, 16, 18, 34), and is now made in large and increasing quantities in many countries. MDF has tight edges and a practically homogeneous texture, can be machined like solid wood, and is suitable even for carving; it has a smooth surface that can be directly coated, painted, or grain-printed, but is also manufactured in textured surfaces simulating rough sawn wood or raised latewood grain, and is available finished with plastic films. MDF is produced in thicknesses ranging from 6 to 40 mm (0.4–1.6 in.), and has many uses including furniture, paneling, door and window frames, doors, and siding.

High-density fiberboard (1.20–1.45 g/cm^3)

Table 25-1. Types of Fiberboard[a]

Type	Density (g/cm^3)
Insulating (noncompressed)	
1. Semirigid	0.02–0.15
2. Rigid	0.15–0.40
Compressed	
1. Medium density	0.40–0.80
2. Hardboard	0.80–1.20
3. High density	1.20–1.45

[a]Another classification, proposed for international standardization (ISO-818), recognizes three types of fiberboard: soft (density ≤0.35 g/cm^3), medium hard (>0.35 ≤ 0.80 g/cm^3), and hard (>0.80 g/cm^3) (15).
Source: Ref. 12.

is an expensive product (requires a high proportion of resin and special, high-capacity presses); its production is justified only for special purposes (11).

According to method of production, fiberboard is classified into *wet-process* and *dry-process*; the former resembles the production of paper and the latter of particleboard.

RAW MATERIALS

The woods used for fiberboard are both softwoods and hardwoods. Species with thin-walled fibers are preferred, because such fibers bend and collapse easily, present a larger area of contact, and contribute to development of more and stronger hydrogen bonds; long length of fibers is also important especially for insulation boards (12).

However, availability and cost, due to competition from the pulp and paper industry, are leading to utilization of a great variety of species and more hardwoods, including high-density woods, such as oak, beech, birch, eucalypts, tropical woods, and mixed species (1, 8, 27). There is also an increasing use of low-grade woods and residues from sawmills (including sawdust) (9, 12), veneer and plywood plants, logging residues, chips, and waste paper and paperboard. Sawdust (up to 20%) is considered not to have a substantial effect on hardboard quality (12).[1]

Green and debarked wood is preferred, but bark is not a limiting factor. Bark in proportions of 15–20% (i.e. wood with its own bark) may be utilized without adverse effects on board properties (12). As previously mentioned, extractives of bark and wood have adhesive properties, because, under hot pressure, they polymerize to substances that are similar to synthetic resins (24) (Chapter 21); low melting-point waxes contained in some barks, also have an adhesive effect (2). Other ligno-cellulosic materials, such as bagasse (sugar cane residues), flax, straw, and bamboo, are sometimes used to make fiberboard, but the main source is wood.

Additives include materials to increase strength, resistance to moisture, microorganisms or fire, or improve some other property. Among the materials used are resin, paraffin (wax), asphalt, alum, insecticides, fungicides, fire retardants, starch, synthetic resins, and drying oils.

The production of wet-process fiberboard requires the consumption of large quantities of water, 10–15 tons of water per ton of fiberboard (12, 19).

PREPARATION OF WOOD

The wood is seldom debarked, but if desired, the bark is removed by friction or knife-cutting mechanisms. Next, the wood (or other lignocellulosic material) is reduced to chips by disk- or drum-type chippers (Chapter 24). The chips are similar to those used for particleboard or pulp, and are stored out-of-doors (for a limited period of time) or in silos; they are screened and transported by conveyors equipped with magnetic devices to detect and separate metallic objects (12).

PULPING

Pulping is mechanical. The main method is the *thermomechanical* process, where the chips are subjected to steaming under pressure, and reduced to fibers and fiber bundles by the attrition action of defibrators or attrition mills. These are machines equipped with two disks with coarse surfaces; one is stationary and the other revolves, or both revolve in opposite directions and at a very small distance from one another. Steaming is more effective on hardwoods with respect to the energy needed to achieve separation of fibers. In pulping dense hardwoods, defibering is facilitated by chemical treatment (cold soda or hot sodium sulfate) (11, 19).

Another pulping method used in some factories is the *Masonite* or *explosion* process (accidentally invented by Mason in the U.S.). This is basically thermomechanical and is applied as follows: A quantity of chips is placed in a vertical cylinder, called a "gun," and steamed under pressure. The pressure is raised and then quickly released by opening a valve. As a result, the softened, steam-impregnated chips explode and are transformed into fibers and fiber bundles (17).

A few factories apply the *groundwood* process. Short, debarked logs are pressed against a "stone," which revolves rapidly and is continuously treated with abundant water to keep the temperature low and to remove the detached fibers and fiber bundles. This method is applied with softwoods and light hardwoods.

These methods, except explosion, are also used in making pulp for paper; in some cases, the chips are chemically treated to facilitate pulping.

After the primary breakdown of chips to fibers by pulping, the pulp is subjected to further mechanical treatment by disk refiners (see Chapter 26).

FIBER DRYING

Fibers intended for dry-process fiberboard are dried. Dryers can be classified into two principal groups—drum-type and tube-type—which are generally similar to equipment used in drying particles for particleboard.

BLENDING

Considerable quantities of wet-process boards are made without the addition of resin, but in some wet- and dry-process hardboards, small amounts of resin are added (1–3% of dry fiber weight). Dry-process boards may take up to 8% of resin depending on board density and desired properties. Phenol-formaldehyde, urea-formaldehyde, and sometimes (in MDF) melamine-urea-formaldehyde are used. In boards produced by the wet-process, special formulations of liquid phenolic resin are added in the attrition mills, and defibering and resin blending take place simultaneously (12). Other additives are also blended with the fibers prior to mat forming.

FORMING

Mat forming, also called "felting," is done by two methods—wet and dry. In both cases, the goal is to achieve a uniform distribution of material (i.e., boards of uniform density and desired thickness).

Wet Forming

In wet forming, the fibers are transported in a low-consistency water suspension, about 1–2%, on a wire screen. The method applied in modern factories is similar to papermaking (use of a modified Fourdrinier paper machine; see Chapter 26). The mat is formed on an oscillating endless wire screen, moving more slowly than in papermaking, and the water excess is removed by vacuum and pressure between rolls or platens.

Older methods include:

- *Deckle box.* Forming is done by use of a bottomless, four-sided frame, which may be raised or lowered on top of a wire screen. A certain quantity of pulp, by volume or weight, sufficient for one board, is pumped inside the frame. By application of vacuum underneath and pressure from above, the water excess is removed, and the thickness of the fiber mat is reduced; the frame is then raised again, and so on. The procedure is slow, and for this reason it is preferred to apply the following method where forming is continuous.
- *Cylinder machine.* The equipment consists of a large drum (single cylinder), about 2–4 m (6–12 ft) in diameter, the surface of which is a wire screen. The drum revolves inside a tank containing a suspension of pulp at a level that is a little higher than the axis of the

drum. A mat is formed on the screen, and the water excess is removed by application of vacuum and pressure. A double-cylinder machine also exists, made of two drums revolving in opposite directions. Each cylinder forms half of the total mat thickness, and the two halves meet and are "laminated" in the nip between the cylinders; there is no vacuum. The procedure is more productive and gives a mat of symmetrical fiber structure—which is not the case with all other forming machines (34).

Dry Forming

In dry forming, the fibers are transported by air. This is a relatively newer method, which is advantageous because no water is needed—an important consideration with regard to environmental pollution.

A new development in mat formation is orientation of fibers; the fibers are aligned in the machine direction (i.e., the general direction of production flow), instead of being randomly oriented. Fiber orientation results in boards of physical and mechanical properties approaching those of solid wood.

MAT DRYING

Wet-formed mats are dried at temperatures ranging from 120 to 190°C (250–370°F); there are three types of dryers: tunnel, drum, and platen. Respectively, the mat passes through a heated tunnel, or heat is provided by drums, placed in one or more levels, or by heated platens.

PRESSING

Insulation boards are not hot-pressed. After wet forming, the mat is brought to the desired thickness by cold roll pressure and then dried.

Compressed boards are subjected to either *wet* or *dry* pressing. In both cases, the pressure is hot and applied by presses that usually have many openings, and platens heated by steam, water or oil, and rarely (single-opening presses) by high-frequency electricity.[2] Machines also exist that produce fiberboard by continuous forming and pressing (Figure 25-1); the panels are sectioned after pressing.

Pressing is "wet" or "dry" depending on the moisture content of the mat of fibers. The pressure is considered wet when a mat is brought in the press at about 20–30% consistency of fibers in water; in dry-formed mats,

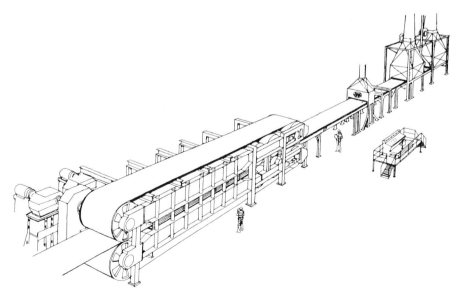

Figure 25-1. Continuous belt-type hot press for production of thin particleboard or fiberboard. (Courtesy of Washington Iron Works.)

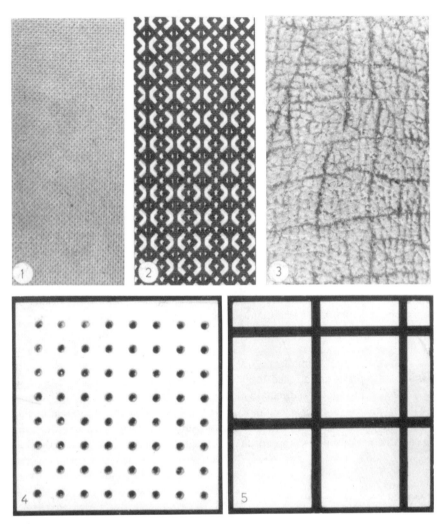

Figure 25-2. Fiberboard: 1, back of hardboard produced by wet pressing (see the imprints of the screen); 2 and 3, decorative hardboard (2, perforated; 3, imitation of leather); 4 and 5, decorative insulation fiberboard (tiles).

the moisture content of the fibers is 6–12% (34). The products of the one or the other method can be recognized, because in wet pressing the screen is printed on the underside of the boards (Figure 25-2).

In wet pressing hardboard, maximum pressure is on the order of 5 N/mm^2 (750 psi) and temperature 180–210°C (360–410°F), and respective figures for dry pressing are 6–8 N/mm^2 (900–1200 psi) and 170–220°C (340–430°F). Pressure/time diagrams are shown in Figure 25-3.

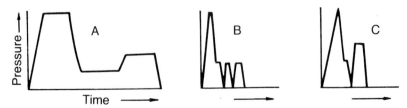

Figure 25-3. Pressure-time diagrams for production of compressed fiberboard by (A) wet forming-wet pressure, (B) wet forming-dry pressure, and (C) dry forming-dry pressure. (After Ref. 30.)

TEMPERING

Heat or oil treatment of hardboard coming out of a hot press is often applied to improve certain properties, such as strength, hygroscopicity, dimensional stability, and surface hardness. Such treatment is called "tempering," and the product tempered hardboard.

Thermal treatment involves exposure of boards to a high temperature, as high as 180°C (350°F), and is applied to wet-process boards (26). In dry-process boards, there is little improvement (12). Boards are treated in kiln-like chambers with air circulation to avoid ignition. The effects are attributed to physicochemical changes, such as differentiation and flow of lignin, and increased cross-linking of cellulose and other polymers contained in wood (14), but prolonged exposure to high temperatures has opposite results due to decomposition of cellulose.

Oil treatment is applied by immersion of boards, or spraying or brushing their surface with drying oils (linseed oil, soya oil, tall oil, etc.); sometimes the mat is sprayed before pressing. The amount of oil varies from 2–12% of dry wood, depending on oil viscosity and method of application. The boards are subsequently exposed to high temperatures, 160–170°C (320–340°F), for a few hours (14). This treatment is applied instead of heat treatment in special cases [e.g., boards used for flooring or exterior paneling (siding) (20)]. The effects are likewise attributed to complex physicochemical changes of the constituents of wood (14, 19, 20). Oil-tempering results in air pollution, and is presently avoided; property improvement can be simply achieved by adding more resin (17).

CONDITIONING, TRIMMING, AND OTHER PROCESSING

Immediately after their exit from a hot press, or tempering treatment, compressed fiberboards are hot, have a low moisture content (1–2%) and are fragile. In order to avoid problems in subsequent use, they are subjected to conditioning; this involves their transportation, in slow pace and on edge, through a chamber of controlled environmental conditions, or rotating in a "cooler," as in the case of plywood. After conditioning, the boards are trimmed by sawing (trim residues are burned or repulped). Panel sizes vary from a standard 122 × 244 cm (4 × 8 ft) to as much as 244 × 3000 cm (8 × 100 ft).[3] Thicknesses also vary from 3 to 50 mm. (0.1–2 in.) (12). Dry-process MDF boards are sanded.

Production may end here, but the tendency is for fiberboards (rigid insulation, hardboard, MDF) to be finished by various surface modifications, as previously explained. The steps of fiberboard production are shown in Figure 25-4. In summary, and according to previous discussion, the following general remarks apply to production of insulation board, hardboard, and MDF.

Insulation board is produced by wet forming. The mat is roll-cold pressed and subsequently dried. Hot pressing is not applied. Bonding is based on the formation of hydrogen bonds. No resin is added, but other additives, such as starch, are often used to enhance bonding (17).

Hardboard is made either by the wet or the dry process. The wet process involves wet forming. Excess water is drained by prepressing. The prepressed mat is transported in a hot press and pressed on a screen, which is imprinted on the underside of the boards (type S-1-S). In the dry process, the fibers are dried after pulping, dry (air) forming of the mat and dry pressing follow; both faces of the boards are smooth (type S-2-S). Fiber bonding is achieved by lignin flow, some hydrogen bonding in the wet process, and by the addition of some resin, 1–2% of dry board weight in the wet process and 2–3% in the dry process.

Medium-density fiberboard (MDF) is usually made by the dry process, and resin is added in a proportion of 8–12%.

PROPERTIES OF FIBERBOARD

The properties of fiberboard depend mainly on its density. Resin content and manufacturing modifications are also important. Important properties are shrinkage and swelling, mechanical properties, thermal conductivity, and oth-

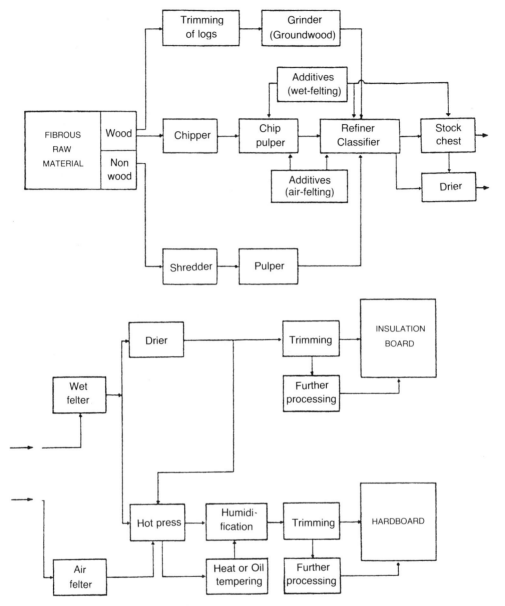

Figure 25-4. Steps in the production of fiberboard (lower half continues after the upper half). (Courtesy of FAO.)

ers. Characteristic differences exist between types of fiberboard, as shown in Table 25-1.

Shrinkage and Swelling

In all types of fiberboard, and because of a preferred orientation of fiber length in the direction of production, shrinkage and swelling along the length of panels is smaller than thickness shrinkage and swelling. With intentional orientation of the fibers, longitudinal shrinkage and swelling should be smaller, approaching that of solid wood, but even without such orientation, linear dimensional changes are very small. This is shown in Table 25-2, and this situation is supported by other data. In commercial medium-density fiberboards (0.60–0.75 g/cm^3) of variable origin, an increase of relative humidity from 30 to 90% resulted in a 0.19–0.28% increase of length and a 4.3–

Table 25.2. Properties of Fiberboard

Property	Unit	Insulation	Medium density	Hardboard (standard)	MDF
Density	g/cm^3	0.25–0.40	0.42–0.80	0.90–1.05	0.70–0.87
MOR	N/mm^2	1.5–5.5	10–28	30–55	25–42
	psi	210–780	1500–4000	4200–8000	3600–6000
MOE	N/mm^2	170–180	1400–4900	2800–5600	2600–4800
	1000 psi	25–125	200–700	400–800	380–700
Tension ∥	N/mm^2	1.5–3.5	8.5–21	21–40	—
	psi	200–500	1200–3000	3000–5700	—
Tension ⊥ (IB)	N/mm^2	0.07–0.17	0.2–0.6	—	0.5–1.1
	psi	10–24	28–85	—	72–160
Water absorption, 24-h immersion					
weight	%	15–60	6–40	10–30	—
volume	"	5–15	—	10–30	—
Linear expansion, 50–90(97)% RH	"	0.5	0.2–0.4^a	0.6	0.25
Thickness swelling, 24-h immersion	"	—	4	3–20	2–6^b
Thermal conductivity^c					
SI		0.035–0.056	0.06–0.10	0.13	—
English		0.27–0.45	0.50–0.80	1.10	—

^{a}30–90% RH.
^{b}Two-hour immersion.
^{c}SI units: kcal/h/m^2/°C; English: Btu/ft^2/h/°F.
Source: Data from Ref. 12, except MDF from Refs. 6, 16, and 18.

15.0% thickness swelling (4). Similar results were observed in other studies. A "springback" (i.e., permanent increase of thickness after reduction of moisture to the original level) was also observed, which in the above case was 1.4–4.5% (4) and in another study up to 10% (31). An increase of the density of fiberboard and of the yield of pulp (i.e., yield from a certain quantity of wood) was found to be the cause of greater dimensional change and greater springback (i.e., the boards had a lower dimensional stability) (19). In contrast, refining the pulp, increasing the proportion of resin, and increasing temperature and pressure, up to a certain limit, had a reducing effect on shrinkage and swelling. Some factors may interact—for example, wet-process hardboard of high density, produced from high-yield pulp and the addition of 1% phenolic resin, was found to have a low-dimensional stability; however, this was improved by increasing the proportion of resin to 2%. Differences in dimensional changes exist among boards made from different species (31, 32). Tempering has a reducing effect on thickness swelling (20).

Mechanical Properties

The main mechanical properties of fiberboard are strength in static bending (modulus of rupture) and modulus of elasticity. Research has shown that the properties are affected by such factors as board density, kind of furnish (fine fibers, fiber bundles), pulp yield and refining, the addition and kind of resin, manufacturing variables, etc. Some findings are as follows: there is a strong relationship between bending strength and density (Figure 25-5), but modulus of elasticity and internal bond are not always related as well (5); boards made of only fine fibers or fiber bundles are generally brittle; flexibility can be improved by mixing fine fibers with coarse fibers or fiber bundles (23); phenol-formaldehyde was found to improve the strength of wet-formed hardboard in comparison to thermoplastic resin (24); reduction of pulp yield of oak from 88% to 80% improved the mechanical properties, and a similar effect was achieved by increasing the proportion of resin (31); temperature at pressure up to 190°C (370°F) had a favorable effect but a higher

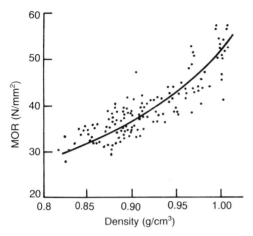

Figure 25-5. Relationships of bending strength and density in hardboard. (After Ref. 3.)

temperature (220°C or 430°F) was adverse[4]; prepressing improved the bending properties, but faster application of pressure had an adverse effect (more at higher temperatures); boards made of mature wood are generally tougher than those made of juvenile wood (23); tempering improves all mechanical properties, but the effect depends on the quality of treated hardboard: lower-quality boards are improved more; oil-tempered boards have a higher bending strength but break more easily (20); wet-process boards have a more uniform distribution of density, because water is a more efficient forming medium, and such boards have been reported to be somewhat superior in properties (17, 25); addition of bark, above a certain limit was found to reduce bending strength (29) (Figure 25-6), although other studies have shown satisfactory results with high proportions (2, 28), and experimental boards made from bark only were found to be suitable for normal uses (33); kind of bark (tree species) is important, and adverse effects of bark inclusion may be counteracted by the addition of more resin (9, 36). In comparison to solid wood, fiberboard exhibits more creep and a greater reduction of strength with increasing moisture content (7).

Other Properties

Other properties of fiberboard include hygroscopicity, thermal conductivity, quality of surface, nail- and screw-holding capacity, surface hardness, weatherability, etc. Figure 24-30 shows that, under the same environmental conditions, the equilibrium moisture content of fiberboard (and particleboard) is lower in comparison to solid wood. The following values were observed with hardboard from various sources: at relative humidities of 30%, 65%, and 90%, equilibrium moisture contents were 4.3–5.2%, 6.7–8.1%, and 10.2–14.5%, respectively (4). Water absorption is reduced by tempering: oil-tempering reduced absorption up to half. Thermal conductivity is considerably lower in insulation boards, and generally increases with density (22) (Figure 24-29). Quality of surface refers to smoothness (important in accepting overlays without sanding), hardness, and paintability. Nail and screw withdrawal resistance, weatherability, machinability, and quality of edges are also important properties.[5]

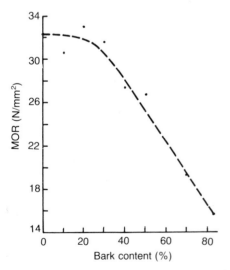

Figure 25-6. Influence of bark content on the bending strength (modulus of rupture) of hardboard made of poplar wood. (After Ref. 29.)

REFERENCES

1. Akers, L. E. 1966. *Particleboard and Hardboard*. Oxford, New York: Pergamon Press.

2. Anderson, A. B., and H. Helge. 1959. Utilization in Norway: bark in hardboard. *For. Prod. J.* 4:31A.
3. Autorenkollektiv. 1975. *Werkstoffe aus Holz*. Leipzig: VEB-Fachbuchverlag.
4. Bennet, G. A. 1969. An Investigation of Some Properties of Medium Hardboard. For. Prod. Res. Lab., England, Timberlab Paper No. 4.
5. Boehme, C. 1980. Untersuchungen über den Zusammenhang zwischen Rohdichte und Querzugfestigkeit, Biegefestigkeit sowie Biege-E-Modul bei veschiedenen Holzfaserplattentypen. *Holzforsch. Holzverwert.* 32(5):109-113.
6. Borchgrevink, G. 1980. Fabricating and using MDF. *Proceedings 14th Intern. Particleboard Symposium*, W.S.U. pp. 291-311.
7. Chan, L. W. W. 1979. Strength properties and structural use of tempered hardboard. *J. Inst. Wood Sci.* 8(4):147-160.
8. Chow, P. 1976. Properties of medium-density, dry-formed fiberboard from seven hardwood residues and bark. *For. Prod. J.* 26(5):48-55.
9. Chow, P. 1979. Phenol adhesive bonded medium-density fiberboard from *Quercus rubra* L. bark and sawdust. *Wood Fiber* 11(2):92-98.
10. Dinwoodie, J. M. 1986. Use of fibre building board. BRE Information Paper IP 23/86.
11. F.A.O./U.N. 1958. *Fibreboard and Particle Board*. Rome.
12. F.A.O./U.N. 1966. *Plywood and Other Wood Based Panels*. Rome.
13. F.A.O./U.N. 1976. *Proceedings of the World Consultation on Wood-Based Panels*, Brussels. San Francisco: Miller Freeman.
14. Ingenieurschule f. Holztechnik, Dresden. 1965. *Taschenbuch der Holztechnologie*. Leipzig: VEB Fachbuchverlag.
15. International Organization for Standardization. I.S.O. Standards 766, 768, 769, 818, 819.
16. Johansson, L. N. A. 1983. MDF-Mitteldichte Faserplatten. *Holz Roh- Werkst*. 41:255-260.
17. Haygreen, J. G., and J. L. Bowyer. 1982. *Forest Products and Wood Science*. Ames: Iowa State University Press.
18. Kehr, E. 1977. Verfahren zur Herstellung von Faserplatten mittleres Dichte-MdF. *Holztechnologie* 18(2):67-76.
19. Kollmann, F. F. P., E. W. Kuenzi, and A. J. Stamm. 1975. *Principles of Wood Science and Technology. II. Wood Based Materials*. Berlin, New York: Springer Verlag.
20. Kumar, V. B. 1961. Neuere Untersuchungen an ölgehärteten Faserplatten. *Holz Roh- Werkst*. 19:15-21.
21. Langendorf, G., E. Schuster, and R. Wagenfuehr. 1972. *Rohholz*. Leipzig: VEB Fachbuchverlag.
22. Lewis, W. C. 1967. Thermal Conductivity of Wood-Base Fiber and Particle Panel Materials. U.S. For. Prod. Lab. Res. Paper FPL 77.
23. Maloney, T. M. 1977. *Modern Particleboard and Dry Process Fiberboard Manufacturing*. San Francisco: Miller Freeman.
24. Marian, J. E. 1967. Wood, reconstituted wood and glued laminated structures. In *Adhesion and Adhesives*, eds. R. Houwink and G. Salomon, pp. 166-280. Amsterdam, London, New York: Elsevier.
25. Mataki, Y. 1972. Internal structure of fiberboard and its relation to mechanical properties. In *Theory and Design of Wood and Fiber Composite Materials*, ed. B. A. Jayne, pp. 219-252. Syracuse, New York: Syracuse University Press.
26. McMillin, W. C. 1969. Fiberboards from loblolly pine refiner groundwood: Aspects of fiber morphology. *For. Prod. J.* 7:56-61.
27. Myers, G. C. 1979. Hardboards from mixed tropical hardwoods. *For. Prod. J.* 29(5):44-48.
28. Semana, A. J., and A. B. Anderson. 1968. Hardboard from benquet pine bark-wood compositions. *For. Prod. J.* 18(7):28-32.
29. Sinclair, G. D., and D. K. Dymond. 1968. Hardboard from poplar wood. *Tappi* 51(9):108-111A.
30. Spalt, H. A. 1977. Chemical changes in wood associated with wood fiberboard manufacture. In *Wood Technology: Chemical Aspects*, ed. I. S. Goldstein, pp. 193-219. Washington, D.C.: Am. Chem. Soc.
31. Steinmetz, P. E. 1973. Producing Hardboards from Red Oak. U.S. For. Prod. Lab. Res. Paper FPL 219.
32. Steinmetz, P. E., and O. J. Fahey. 1971. Effect of Manufacturing Variables on Stability and Strength of Wet-Formed Hardboards. U.S. For. Prod. Lab. Paper FPL 142.
33. Stewart, L. D., and D. L. Bulter. 1968. Hardboard from cedar bark. *For. Prod. J.* 18(12):19-23.
34. Suchsland, O. and G. Woodson. 1986. *Fiberboard Manufacturing Practices in the United States*. USDA, Forest Service, Ag. Handbook No. 640.
35. U.S. Forest Products Laboratory. 1987. *Wood Handbook*. USDA Agr. Handbook 72. (revised)
36. Woodson, G. 1976. Effects of bark, density profile, and resin content on medium-density fiberboards from southern hardwoods. *For. Prod. J.* 26(2):39-42.
37. Yao, J. 1978. Hardboard from municipal solid waste using phenolic resin or black liquor binder. *For. Prod. J.* 28(10):77-82.

FOOTNOTES

[1] A Swedish plant reportedly utilized 80% sawdust (13). Municipal solid waste has also been used experimentally to produce (''wet-process'') hardboard of standard quality; in addition to paper, the material contained small pieces of aluminum foil, and both firm and foam plastics (added: 1% wax sizing, 3% phenolic resin) (37).

[2] Some presses are very large. A wet-process hardboard plant is reported to use a 30-opening hot press with 2.4 × 7.3 m (8 × 24 ft) platens; a single-opening, dielectrically heated MDF press has a platen size of 2.4 × 19.5 m (8 × 64 ft) (12).

[3] Proposed international (ISO) standard sizes, regarding all panels (plywood, particleboard, fiberboard) are given in a footnote in Chapter 22 (Plywood).

[4] The fibers are plasticized by high temperatures and, therefore, their interconnection becomes better. At temperatures of 130–190°C (270–370°F), lignin flows and acts as an adhesive (30). The reduction of strength at 220°C (430°F) is attributed to reduction of strength of fibers than of the bond.

[5] Property values are specified by standards. For example, according to U.S. standards, hardboard siding is required to have: maximum linear expansion (between 30% and 90% relative humidity) 0.36–0.40%, maximum thickness swelling 10%, minimum modulus of rupture 1800–3000 psi (12.4–20.7 N/mm^2), minimum impact 9 in. (3.6 cm), minimum hardness 450 lb (205 kg), minimum nail resistance 150 lb (68 kg), maximum water absorption 15%, weatherability of substrate (maximum residual swell—i.e., springback) 25%, moisture content 2–9%, and not more than 3% variance between any two boards in any one shipment or order. Insulation boards (rigid) are required to have a thermal conductivity value (k, English) of 0.38–0.48, deflection of dry board at a specified minimum load 0.16–0.25 in. (4–31 mm), according to type of board, a certain flame spread resistance, etc.

26

Paper

Paper is produced from wood fibers[1]; use of other plant or synthetic fibers is limited. Plant fibers are advantageous because they do not need an adhesive for bonding. Various additives, such as starch, gums, and synthetic polymers, are used to improve adhesion and paper strength.

This product was originally made in China[2] (105 A.D.) from bark fibers (mulberry inner bark), hemp residues, and rags. The art was transferred from China to Japan (610 A.D.), Samarkand (Central Asia, 751), Baghdad (793), Spain (1100), and elsewhere. "Factories" making paper were later established in Italy (1260), Germany (1389), England (1494), and North America (1690) (28).

The invention of typography (Gutenberg Bible, 1445), and of specialized paper machines, together with the use of wood as a raw material, were decisive factors for the development of the production of paper. The use of wood started relatively recently (about 1850); until then, rags (cotton and linen) were the main material. The abundant availability of wood made possible the present huge development of the paper industry, which was previously facing difficult problems of raw material supply. Up to the nineteenth century, paper was made by hand (using a frame screen submerged in a fiber suspension), whereas today machines produce up to 1500 running meters of paper in one minute (39).

RAW MATERIALS

In addition to wood (or other materials), the production of paper requires chemicals for pulping, various additives, and water.

WOOD

Only fir and spruce wood were used initially, but now almost all wood species—softwoods and hardwoods[3]—can be utilized. Availability in needed quantities and cost are the decisive factors for selection of species (12).

The wood is transported to the factory in round form, or in the form of residues of other wood-using industries (sawmill slabs, veneer residues), or as chips produced from industrial or logging residues. Utilization of residues shows an increasing trend. Wood in round form comes in varying lengths, from about 1–2 m (3–6 ft) to whole tree trunks. Minimum diameter varies; the possibility of utilizing small diameters is limited by the need for debarking. Sawmill slabs and veneer residues from the exterior part of logs is a favored material; it is sapwood which is more easily pulped (includes few extractives), and contains long fibers in comparison to "juvenile" wood[4] (near the pith; see Chapter 6). Certain production methods permit the use of sawdust.

The wood is transported by trucks, railroad, or ships. Water transportation is also used—logs transported by the flow of rivers, or bundled in rafters and towed.[5]

In the factory, the wood is stored on the ground or in water. Out-of-door storage of chips presents both advantages (better utilization of the storage ground, lower cost of handling) and disadvantages (risk of fungal attack or fire, loss due to wind).

The quantity of wood needed to establish a pulp factory depends mainly on the pulping process. The lowest quantity varies from about

60,000 tons per year for mechanical pulping to 500,000 tons for chemical pulping (32).

OTHER PLANT FIBERS

Fibers suitable for papermaking may be obtained from other (usually annual) plants, such as wheat, straw, sugar cane residues (bagasse), esparto, bamboo, reeds, flax, and cotton. These sources, except bamboo, have the disadvantage that their supply is seasonal, and the raw material is bulky, which makes their transportation to long distances uneconomical. Also, their fibers differ from wood fibers with regard to chemical composition and morphology. With the exception of cotton (which contains cellulose in a proportion of about 95%), fibers from other plants contain less cellulose. Some values are as follows: holocellulose 47–58%, cellulose 30–38%, lignin on average 17%, and hemicelluloses 18–30%. Fiber morphology is also variable; certain plants (flax, hemp) have long fibers and a high ratio of length to diameter, whereas the fibers of others are short (see Table 26-1).

Cotton, flax, and hemp fibers (mainly used as rags[6]) produce paper of high strength and low weight—a combination suitable for special uses, such as currency, carbon paper, cigarettes, and large-volume editions (e.g., telephone directories). Esparto fibers produce soft paper with very good printing properties. Straw is used to make paperboard, writing and book paper, and other papers usually by mixing long-fibered, softwood pulp, because otherwise the paper is rigid (difficult to bend and fold), heavy, low in strength, and of undesired opacity. Bagasse and bamboo produce paper of satisfactory quality for various uses (7, 19).

SYNTHETIC FIBERS

In general, plant fibers have advantages, because their cellulosic skeleton provides self-adhesive ability and strength; disadvantages also exist, however, including heterogeneity of size and shape, change of dimensions with varying moisture content (due to shrinkage and swelling), limited durability (they are attacked by microorganisms, chemicals, and high temperatures), and presence of other substances (lignin, extractives, hemicelluloses), which must be removed, partly or entirely. For these reasons, it would be desirable to replace them with synthetic fibers, but they also have disadvantages: their cost is much higher (10–20 times in comparison to wood fibers), and they possess no ability of self-adhesion—therefore, adhesives should be added. Thus, the use of such fibers (glass, nylon, dacron, etc.) is limited to specialized products.

CHEMICALS

Several chemicals are used for the production of paper. They are added at various stages of

Table 26-1. Dimensions of Plant Fibers

Kind	Length (mm)	Width (μm)
Cereal straws		
(wheat, barley, etc.)[a]	1.5 (0.7–3.1)	13 (7–24)
Rice[a]	1.4 (0.6–3.5)	8 (5–14)
Sugar cane (bagasse)[a]	1.7 (0.8–2.8)	20 (10–34)
Esparto[a]	1.1 (0.5–1.6)	9 (7–14)
Bamboo[a]	2.7 (1.4–4.3)	14 (7–27)
Hemp	25 (5–55)	25 (10–51)
Flax	33 (9–70)	19 (5–38)
Cotton	18 (10–40)	20 (12–38)

Source: Data derived from Ref. 31.
[a]These plants contain smaller-sized elements (i.e., parenchyma cells, vessel members, and epidermal cells or trichomes). The dimensions of the main cells of wood are 3–5 mm in length for softwood tracheids and 1–2 mm for hardwood fibers; wood also contains parenchyma cells and vessel members in hardwoods (see Chapter 3). Nonwood fibers derive from grasses (cereal straws, rice, bagasse, esparto, bamboo, corn, etc.), bark ("bast fibers"—i.e., hemp, flax, kenaf, etc.), and seed hairs (cotton, etc.). Each category includes plants of lesser importance for papermaking. Papermaking fibers are also mineral (asbestos, glass, etc.), organic man-made fibers (rayon, nylon, etc.), and animal fibers (wool, silk).

production with the purpose of pulping, bleaching the pulp, and improving fiber adhesion and paper properties. Such additives provide protection from diffusion of solutions, especially ink, in the mass of paper, improvement of printing properties, control of color, reduction of hygroscopicity, etc. The particular chemicals are named in the discussion of production processes.

WATER

Paper production requires the consumption of large quantities of water (for pulping, treatment of the pulp, and forming the paper sheet). The water is obtained from rivers, lakes, streams, or wells, and its availability is a decisive factor for the site selection of such an industry. Depending on the method of production and type of product, the quantities of needed water vary from about 40,000 to 400,000 l/t of paper. The water should be not only abundant but also of good quality, as salts (calcium, magnesium, etc.), gases (oxygen, carbon dioxide), and foreign matter or dirt have an adverse effect on the machines (clogging, erosion) and generally on production. Discarding the refuse (cooking liquors) creates problems of environmental pollution if proper measures are not taken, but if such refuse is cleaned, most of the water (up to 95%) may be reused.

PREPARATION OF WOOD

The preparation of wood includes debarking, chipping, screening, and storage of chips, and is generally similar to fiberboard production (see Chapter 25). *Debarking* is an almost general requirement for paper production, because bark differs from wood with regard to cellular structure (see Appendix). Thus, the presence of bark causes a high consumption of chemicals, affects digester capacity, and has adverse effects on the strength and cleanliness of the pulp. The bark is removed by various methods as applied in sawmills or veneer mills, but some special methods are also used (i.e., debarking in a revolving drum or by hydraulic power) (Figure 26-1).

Drum debarking is a common method. The

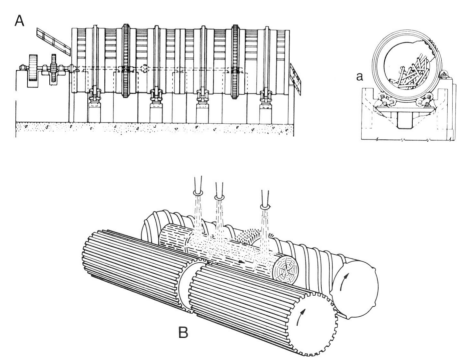

Figure 26-1. Debarking by (A) revolving drum (a, transverse section of the drum), and (B) water. (A, Courtesy of Canadian Ingersoll Rand and B, Streambreaker; after Ref. 30).

logs, about 1–2.5 m. (3–8 ft) long, are transported by conveyor inside a large metallic drum, 3 m (10 ft) or more in diameter and 15–20 m (50–65 ft) long, which has open ends and perforated sides. The drum revolves slowly, and the logs rub one another and against the sides of the drum, as they move from entrance to exit. Thus, the bond between wood and bark is loosened, parts of the bark are detached, fall through the openings of the drum on a conveyor belt, and are removed. Part of the drum, or all of it, may be submerged in water to facilitate debarking, and occasionally the water is heated if the wood is frozen.

Hydraulic debarking is applied in certain regions of the United States and Canada on large-dimension logs. The bark is removed by a "jet" of water, which is directed on the log under high pressure (about 10 N/mm^2 or 1500 psi) (5, 19). This force breaks and removes the bark, but large quantities of water and high consumption of energy are required. The high moisture of the removed bark is also a drawback, because it is more difficult to burn moist bark for production of steam.

Chipping of wood to particles is mainly applied when pulping with chemicals, in order to facilitate their entrance into the mass of wood. The chips are produced by a revolving disk with knives attached in a radial arrangement (Figure 26-2). The size of the disk varies (up to 3.8 m or 153 in. in diameter) (5), and the number and dimensions of the knives also vary. The protrusion of the knives from the plane of the disk may be regulated; thus, it is possible to control the length of the chips, which is about 1.5–2 cm (about 0.5–1 in.). The logs are fed at an angle of 45°, and the chips are cut by the knives of the disk, in combination with knives placed at the base and sides of the enclosure. Instead of logs, it is possible to chip residues from other wood industries, as well as whole tree trunks by horizontal feeding. Efforts to achieve a more complete wood utilization include research on chipping logs with bark and removing it from the chips.

The length of chips is based on considerations regarding fiber length in the pulp and paper, facilitation of the entrance of chemicals in the mass of wood, and energy consumption for production of chips. Entrance of chemicals is also facilitated by an inclined cut (an angle of 45°) of the ends (transverse surfaces) of the chips. With regard to energy, reduction of length to half doubles the energy. Chip length varies from 15–30 mm ($\frac{5}{8}$–$1\frac{1}{4}$ in.) and thickness from 3–9 mm ($\frac{1}{8}$–$\frac{3}{8}$ in.); width is less critical (Figure 26-2B).

After chipping, the chips are screened to remove fines and unsuitable dimensions. There are various systems for screening—usually perforated and vibrating frames placed one above the other. The openings of the upper screen are such that it retains oversize chips, whereas a lower screen retains suitable chips and allows passage of fines (Figure 26-3). Large chips are

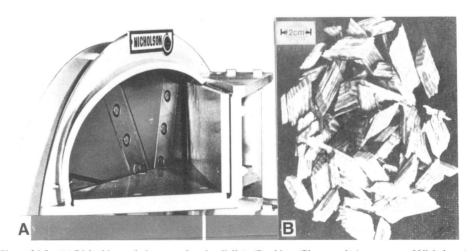

Figure 26-2. (A) Disk chipper (knives are placed radially), (B) chips. (Photograph A, courtesy of Nicholson.)

Figure 26-3. A system for screening chips. (Courtesy of Allis-Chalmers.)

brought back to the chipping machines to be rechipped, and fines are transported to be burned for steam production. Use of uniform chips is necessary in order to ensure a uniform action of chemicals during pulping.

The screened chips are transported to storage, either out-of-doors or in closed silos, and then to digesters for pulping. Transportation is performed by conveyor belts or by air, and seldom by specialized arrangements if the chips are moved in a vertical direction.

PULPING

There are three pulping methods—*mechanical*, *chemical*, and *chemical-mechanical*.

Mechanical Pulping

The traditional method of mechanical pulping is the *groundwood process*, where roundwood is ground against a "stone." Logs of spruce and fir are the usual species, but other softwoods (including pines) are also used. The adverse effects of resin are neutralized by additives, such as alum and alkali, in combination with control of the pH of the pulp. Suitable hardwoods are poplar and species of similar structure, including some eucalypts. In general, there is a preference for woods of young age and fast growth, without heartwood. The logs should be straight and about equal in length, 40 cm–1.60 m (1.5–5 ft). Pulp is produced by grinding against a stone, which is cylindrical in shape and large in diameter and length (e.g., 1.55 × 1.65 cm or 62 × 66 in., and some larger); this revolves at a high peripheral speed (50 m/s or 10,000 ft/min), with its axle parallel to log length. The moisture content of wood is a very important factor, and should be sufficiently high, higher than 30% (45–50%), to facilitate defibering. However, wood is not actually reduced to fibers, because the greater part of the pulp is made of fiber bundles, and damaged and broken fibers. Grinding is completed by the simultaneous action of abundant water to avoid excessive increase of temperature. Heat is, however, a pulping factor, because it softens lignin, and acts in combination with pressure and water. Various mechanical systems are used for grinding, and they differ depending on manner of pressing the logs against the grinding stone (hydraulically, by chains), type of feeding (from one or more inlets, separately or contin-

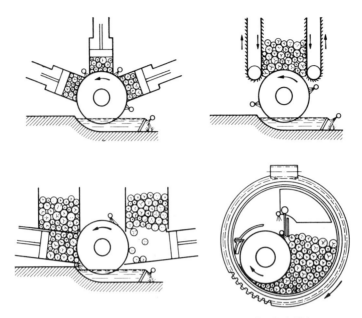

Figure 26-4. Various types of wood grinders. (After Ref. 27.)

uously and automatically), and arrangement of parts (revolution of the disk inside water or not, inside a drum, etc.; see Figure 26-4). The yield of pulp is higher in comparison to all other methods; theoretically, it is about 100%, but actually about 95% due to loss of some extractives and fines (5, 6).

The process is old (since 1850), but still used to a considerable extent to produce paper for newsprint and less for other uses (books, napkins, etc.). The paper has good printing and other properties (e.g., opacity), but low strength, and becomes yellow with time. Strength may be improved by adding long-fibered chemical pulp, sulfite, or Kraft (13) (see below).

Mechanical pulp is also produced from wood chips, which are ground in a "disk refiner" (see under "Beating and Refining"). *Thermomechanical* pulping (TMP) is the main process in this category. The chips are presteamed (saturated steam at a temperature of 130°C, i.e., 250°F, or higher), and ground under pressure. This is a relatively new method (after 1960), but it is fast-growing in worldwide application. The method has several advantages, such as utilization of a variety of species and residues (including limited amounts of sawdust), relatively small units, high yield, (90–95%), relatively low water requirements (facilitating the selection of a plant site), and less pollution in comparison to chemical pulping (10). A relatively high consumption of energy (about double in comparison to chemical pulping, and 1.5 times higher in comparison to grinding logs) (32) is a disadvantage of the process. The pulp is suitable for various kinds of paper (newsprint, books, periodicals, tissue) and paperboard.[7]

Unsteamed chips are used in *refiner mechanical* pulping; they are carried in water and disk-refined at atmospheric pressure. The properties of such pulp are intermediate between stone groundwood and thermomechanical pulp, and the energy consumed is also intermediate (38).

Chemical Pulping

In chemical pulping, the fibers of wood are separated by dissolving the lignin of the middle lamella. However, delignification also takes place in the secondary wall, which allows the stiff cylindrical fibers to collapse into ribbons that provide greater surface contact for hydrogen bonding. Delignification results from the

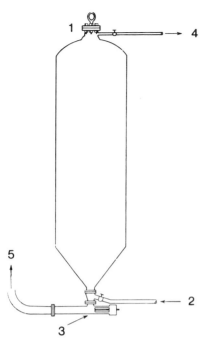

Figure 26-5. A common type of digester (vertical, permanent). Chips and pulping chemicals are introduced from above (*1*) and steam underneath (*2*). Blow valve (*3*), exit of gases (*4*), and exit of pulp (*5*). (After Ref. 19.)

action of acidic or alkaline chemicals in digesters (Figure 26-5) under high temperature and pressure. Reduction to pulp is accomplished by blowing out the contents of the digester, and usually no other mechanical processing is applied except mild stirring. Depending on the specific pulping method, almost all lignin and hemicelluloses or part of them are removed; for this reason, pulp yields differ and range from about 40 to 60%. Yields higher than 50% may require mechanical action to help defibering (5).

There are two chemical pulping processes: *sulfite* (acidic) and *alkaline*.

Sulfite Process. For many years (from about 1870 to 1940), this was the most important method of pulping wood (7). Later, the construction of new plants stopped and the process remains static or declining in importance. Chemical pulping is now dominated by the alkaline Kraft process, for reasons explained further in this chapter.

Initially, pulping was performed by sulfurous acid (H_2SO_3) and calcium bisulfite $Ca(HSO_3)_2$. This is also done today, but aside from calcium other bases are used—magnesium $Mg(HSO_3)_2$, sodium ($NaHSO_3$), or ammonium (NH_4HSO_3).[8]

Calcium bisulfite successfully pulps softwoods without resin or with low resin content (fir, spruce, hemlock), but it is not suitable for resinous softwoods, such as pines. Also, pulping with this base produces refuse (spent liquors) from which usable pulping chemicals cannot be reclaimed; therefore, it contributes to water pollution (rivers, lakes, sea), where such liquors are discharged (5, 13, 19, 37).

These drawbacks may be limited by change of base and other modifications (pulping in two stages at different pH) (13). Thus, it is possible to pulp a greater variety of wood species (including pines, Douglas-fir, and hardwoods), reclaim and reuse part of the chemicals from spent liquors (in combination with steam production by their burning), reduce environmental pollution, and produce pulp of higher purity (more α-cellulose) and higher yield (5, 6, 13).

The cooking liquor is prepared by dissolving sulfur dioxide, obtained by burning sulfur, and allowing this solution to react with limestone. This is introduced in a digester containing the chips. The digester is a large "pressure-cooking," cylindrical vessel with a conical base and vaulted top (Figure 26-5), usually 5 m (15 ft) in diameter and 15 m (50 ft) high, and having a capacity of 12–15 tons of paper (digesters exist that have a capacity up to 35 tons). Digesters are made of stainless steel or have an interior lining resistant to the action of acids. Pulp mills usually have many digesters.

The cooking liquor is introduced hot until the digester is filled. If the wood is dry and the air contained in cell cavities hinders the entrance of the liquor, the chips are first steamed or, after the digester is filled, the pressure is successively raised and lowered (19). Pulping conditions vary, but an indicative set is maximum pressure 0.6–0.7 N/mm² (90–100 psi), maximum temperature 125–160°C (260–320°F), and total digestion time 6–12 h or more (5).[9] The content of the digester is then forcibly emptied into a vessel (blow pit), which is about double in capacity in comparison to the digester, and is made of wood, concrete lined with wood, or stainless steel. Thus, the softened chips are reduced to fibers. Finally, the pulp is repeatedly washed with water.

The yield of pulp may be influenced, within limits, by regulation of temperature, acidity, and time. Under usual production conditions, using calcium bisulfite and spruce wood, the total yield is about 50% (at 90% delignification, 3% residual lignin). The unbleached pulp has a light color and, in considerable proportion, it is widely used in mixture with mechanical pulp. If treatment in the digester is briefly prolonged, and the remaining lignin is removed by bleaching, the pulp becomes white and is suitable for book, writing, and other papers. A variation in pulping conditions and supplementary chemical purification, resulting in reduction of yield, gives "dissolving" pulp, which is used for the production of rayon, plastics, tire cord, cellulose esters, and other products.

Sulfite pulp has lower strength (tensile, bursting, tearing) in comparison to alkaline Kraft pulp, but its bleaching is easier (19).

Alkaline Process. Alkaline pulping started in the 1850s by use of a soda (sodium hydroxide, NaOH) solution. Soda pulping had a wide industrial application in the past, but it is now almost entirely replaced[10] by the "sulfate" or Kraft process, which in addition to soda, utilizes sodium sulfide (Na_2S) as an active pulping chemical.

Pulping by the *soda* process is limited to hardwoods only, such as poplar, birch, maple, and other wood species, as well as agricultural residues. The pulp has low strength, and for production of printing paper, it is mixed with long-fibered softwood pulp. Its color is dark, but may be bleached. The yield is 45% or less (24).

This method consumes large quantities of soda, although methods have been devised to reclaim a high proportion (up to 85%) of the chemicals from spent liquors. The cooking liquor is supplemented by adding sodium carbonate (Na_2CO_3) during the reclaiming process. Observations that if sodium sulfate (Na_2SO_4) is added instead of sodium carbonate, the pulp is higher in strength and yield, and may be made from a wider selection of wood species at a faster rate and lower cost, has led to the domination of this process (from about 1880) over soda pulping. The process is known as *sulfate* or *Kraft*,[11] although the acting chemical is not sodium sulfate but sodium hydroxide and sodium sulfide. For a long time, the dark color of this pulp was a disadvantage, but this has also been corrected (since 1930), and the process was unobstructed in its development. Today soda pulping is virtually replaced by Kraft, and for this reason the following discussion is limited to the latter process.

Any wood species may be pulped by the Kraft process, but this is especially suitable for resinous woods, such as pines. The resin is dissolved, and separated from the pulp as *tall oil*, which constitutes an important by-product.[12] Aside from pines, other softwoods (fir, spruce, Douglas-fir, redwood, etc.) and hardwoods (oak, maple, birch, etc.) may be used. In comparison with softwoods, hardwoods produce short-fibered pulp, low in strength, require less time for pulping and give higher yields. In general, selection of species depends on availability and cost, as well as on the intended use of the pulp.

As previously mentioned, the cooking liquor contains sodium hydroxide and sodium sulfide. The proportion of sodium sulfide is 15-35%. Three to five kilograms (6-10 lb) of liquor are used for every kilogram of dry wood, containing 40-80 g/l of alkali. Pulping is done in steel digesters, usually without a special inner lining. However, erosion develops, and for this reason the walls of the digesters are made thick, and sometimes a stainless-steel lining is added. Digesters are usually about 10 ft (3 m) in diameter and 40 ft (12 m) high, and pulping conditions are as follows: maximum pressure 0.7-1.0 N/mm^2 (100-150 psi), maximum temperature about 180°C (360°F), and total time varies from 2.5 h for pulp for fiberboard or paperboard and up to 5 h for bleached pulp. Cooking time is 0.5-2 h, and varies depending on temperature, proportion of alkali, and proportion of sodium sulfide in the liquor.[13] An increase of one or more of these variables results in reduction of the time required to produce pulp of a certain yield and quality (30). Denser woods need more time (19).

Alkaline pulping is carried out in "batch,"

Figure 26-6. Two continuous (Kamyr) digesters. (A) 100 ft (30 m) high, 14 ft (4.2 m) diameter, semichemical pulping hardwoods. (B) 280 ft (85 m) high, 18 ft (5.5 m) diameter, Kraft pulping pine. (Courtesy of McMillan-Bloedel.)

or in "continuous" digesters (Figure 26-6). The latter (developed in the 1950s) are high (200 ft or 60 m and higher with a ratio of height to diameter 15:1), and pulping is successive—in zones. Daily production varies from 200 to 1000 tons of air-dry pulp,[14] although small units also exist. Continuous digesters are advantageous due to lower cost of production, in combination with higher yields and better pulp quality (4, 19, 37).

In pulping by batch digesters, the chips are introduced first and then a suitable amount of cooking liquor. Heating is done initially by the introduction of steam until temperature reaches the desired high point. Continued heating is by forced liquor circulation and the use of external heat exchangers. Alkalis cause a swelling of the wood, and for this reason alkaline cooking liquors enter its mass readily. Because the time required for alkaline pulping is considerably less than acid sulfite pulping, the method is adaptable to continuous pulping. Delignification is more rapid in Kraft pulping compared to acid sulfite, and the middle lamella (which is essentially only lignin) is readily removed and fiberization occurs. Kraft fibers exhibit a higher strength compared to acid sulfite fibers, because there is little hydrolysis of the cellulose under alkaline conditions. The retention of hemicelluloses is also greater in Kraft pulping; hemicelluloses act as a hydrogen-bonding agent, which, together with greater individual fiber strength, results in stronger paper. The extent of lignin removal, and consequently yield, depends on the intended use of the pulp, which is employed for various purposes—from production of dark-colored paperboard and bags to highly bleached writing paper and books of high quality. For rayon and other cellulosic products, the Kraft process requires a prehydrolysis stage to reduce the retention of hemicelluloses. In the preparation of this type of pulp, the presence of lignin and hemicelluloses in the material must be kept to a minimum; therefore, the yield is low, often less than 40%.

After cooking, the contents of the digester are emptied and, under the action of out-going steam, the chips are reduced to fibers. If high yields are sought, defibering is mechanically assisted. Chips that are not defibered (knots, etc.) are separated on a screen. The spent cooking liquor (black liquor), which contains the dissolved constituents of wood, is condensed by evaporation until its content of solids is 50–60%, and the residue is burned in furnaces. Burning produces heat, which is utilized in the production of steam. Sodium salts (Na_2CO_3 and Na_2S), remaining in molten form, are dissolved in water. The solution is called "green liquor." Under the action of calcium hydroxide, sodium carbonate is changed to sodium hydroxide (caustic soda) as follows:

$$Ca(OH)_2 + Na_2CO_3 \longrightarrow CaCO_3 + 2NaOH$$

Sodium sulfide remains unchanged. Calcium carbonate settles out (as sediment), and the remaining clear liquor (sodium hydroxide and sodium sulfide), called "white liquor," is reused in the digesters (5, 30). Losses of cooking chemicals are replaced by adding sodium sulfate to the black liquor before burning, which gives sodium sulfide. Thus,

$$Na_2SO_4 + 2C \longrightarrow Na_2S + 2CO_2$$

Reclamation of chemicals is better in a large-scale process, and partly for this reason there is a tendency to establish large units of alkaline pulping, some of which are in excess of 2000 tons of pulp per day.[15]

The importance of reclaiming chemicals is obvious with respect to economic considerations and environmental pollution. This is an advantage of this process, in addition to high strength of pulp, possibility of utilizing a variety of species, and faster rate of production (5, 18). Disadvantages include air pollution, dark color of unbleached pulp, and high bleaching cost. Air pollution is caused by sulfur compounds, which have an unpleasant odor and characterize such pulp plants from a great distance. Pollution is also caused by bleaching wastes (18), although this happens in all cases where bleached pulp is produced.

Chemical-Mechanical Pulping

These processes apply both chemical treatment and mechanical energy for pulping, and include: *semichemical*, *chemimechanical* (mechanochemical), and *chemiground* pulping. Semichemical is the main process; the others are of minor importance.

In *semichemical* pulping, defibering is done partly by chemical and partly by mechanical energy—in two successive stages. This process has been in use since about 1930, but it has limited industrial application. Yields are high (65–85%) (6). The species are mainly hardwoods (from poplar to oak, eucalypts, and tropical species) with little use of softwoods (fir, Douglas-fir) (5).

The wood is debarked and chipped, but the chips are cut smaller in comparison to previous processes (length is 1.2 cm or 0.5 inch or less) (5), in order to facilitate entrance of the chemicals and mechanical defibering that follows. In the digesters, the treatment is less drastic in comparison to chemical pulping. Temperatures are lower and lesser quantities of chemicals are used. An increase of the quantity of chemicals results in reduction of yield (6).

Pulping may be carried out with the chemicals used in sulfite or alkaline pulping. A usual method is the so-called "neutral sulfite," where the chemical is sodium sulfite (Na_2SO_3) with added sodium carbonate (Na_2CO_3) or sodium bicarbonate ($NaHCO_3$), or "green liquor" from alkaline pulping to avoid an excessive fall of pH below the neutral level (13).

Pulping may be done in batch digesters (permanent or revolving); however, the tendency is to use continuous digesters of various systems (horizontal-flow, upflow, downflow, downflow-upflow in inclined digesters). Conditions vary depending on wood species, desired yield, pulp quality, and type of digester; maximum temperature is 160–185°C (320–365°F), and pressure 0.6–1.1 N/mm^2 (90–160 psi). Yields vary but they are higher than those obtained by chemical pulping, because only part of lignin and hemicelluloses are dissolved in the digesters. At the exit, the chips continue to keep their solid form. Defibering is performed mechanically by disk attrition mills.

The chemical and mechanical properties of this pulp are intermediate between mechanical softwood pulp and chemical pulp, but resembles more closely chemical pulp, and its macroscopic appearance is more like that of chemical pulp. Major use of semichemical pulp is the production of corrugated paperboard for packaging. It is also used for newsprint, packaging paper, and absorbent paper, but also for books and periodicals after bleaching.

Chemimechanical (or mechanochemical) pulping was developed in the 1950s.[16] It is characterized by high yields (85–95%), which are close to the yields of mechanical pulping. Pulp is mainly produced by action of cold soda (sodium hydroxide) on chips under the following conditions: soda 2–6%, temperature 30–50°C (85–120°F), time 30–240 min. The chips are small, like match sticks (5). The equipment (digesters) is similar to that used in semichemical pulping, although special systems also exist. Softening of chips in the digesters is followed by mechanical defibering, as with the previous process. Chemical action does not cause a substantial change of lignin, and defibering takes place mainly by mechanical energy.

Finally, the *chemiground* process is a variation of mechanical pulping. The wood is used in round form, subjected to the action of neutral sodium sulfite, and mechanically defibered. The method enables production of pulp from hardwoods, which do not produce pulp of good quality by grinding (13).

The wastes of semichemical pulping pollute the environment, but methodologies exist for reclamation of chemicals (19).

TREATMENT OF THE PULP

The pulp produced by the above methods is subjected to a series of treatments, which include *screening* and *cleaning*, *thickening*, *bleaching*, *beating* and *refining*, *coloring*, and addition of various chemicals (*additives*) to improve self-adhesion of the fibers.

Screening and Cleaning

The purpose of these treatments is to remove wood which has not been pulped (knots, fiber bundles) and foreign matter (sand, soil dirt, metals, conglomerates of chemicals, etc.). They are removed by screens and cleaners; in the latter, separation is based on density differences between pulp and foreign matter (19).

Screening is "coarse" or "fine," depending on the size of screen openings. Coarse screening is usually followed by one or more stages of fine screening, which is performed by flat or cylindrical screens. Screening may also be applied after bleaching, Cleaning is usually done by centrifuging.

The problems of screening and cleaning are reduced if proper attention is given to cleanliness of the raw material. Proper debarking and removal of metallic objects and packaging materials are very important steps in that respect.[17]

Thickening

During screening and cleaning, the pulp is in the form of a low-consistency suspension of fibers in water (0.1–0.6% and up to 1.5% when pressure is applied). Before any further treatment, the consistency is raised to 3–6% (19, 30). Such "thickening of the stock" is accomplished by equipment usually consisting of a cylindrical frame covered by fine screen and revolving inside a tank containing a fiber suspension of low consistency. Pulp is retained on the screen and water is removed by application of vacuum. The water is reused for dispersion of pulp.

Bleaching

The principal aim of pulp bleaching is to increase brightness. Pulp is bleached in two ways—by destruction of chromophoric groups (lignin-retaining bleaching) or by removal of lignin residues (lignin-removing bleaching). The first method is used with pulps of high lignin content, produced by mechanical or semichemical pulping, where removal of lignin is not desirable, because it means reduction of yield. The second method is applied to chemical pulps, and is the only one that gives permanent bleaching (5, 14, 19).

However, it should be noted that:

- Pulp produced by the sulfite process has a light color, whereas alkaline pulp is fairly dark, and semichemical and mechanical pulps are of intermediate color.
- Color is attributed to lignin, some extractives (tannins, etc.), metallic ions (iron, copper) reacting with phenolic substances contained in the pulp, and bark particles.

Lignin-removing (delignifying) bleaching is usually done with chlorine (Cl_2) and chlorine-based chemicals, and also with oxygen (O_2); whereas the most common lignin-retaining bleaching chemicals are sodium dithionite and sodium peroxide; oxygen bleaching is also gaining in importance (29).

The bleaching process is completed in one or more stages, depending on the type of pulp. The simplest technique (one-stage) involves use of sodium or calcium hypochlorite on chemical pulp. Pulp produced by alkaline pulping (Kraft or soda) may be bleached in one stage (up to a brightness of 78–80)[18] without a substantial loss of fiber strength. When bleaching is done in many stages, the coloring materials are removed gradually, chemical action on the fibers is the lowest possible in each stage, and the bleaching chemicals are better utilized (5). After each stage, and at the end, the pulp is washed with water; therefore, large quantities of water are needed. Lignin-removing bleaching is usually multistage, and oxidative stages are normally combined with at least one alkaline extraction (NaOH) step.

If the pulp is not immediately used for paper production, it is prepared for storage or transportation to other factories. Preparation entails making rolls, or separate sheets (about 40 × 60 cm or 16–24 in.), stacked in packages, and this is done by pressing and drying in machines that are similar to papermaking machines.

Beating and Refining

This is a very important stage of pulp preparation for paper production. Pulp intended for other products is not treated in this manner.

A certain mechanical preparation of the fibers is prerequisite to the production of strong, dense, uniform, and hard paper. Paper produced from fibers retaining their natural morphology has low strength and nonuniform texture, it is fuzzy and porous, and unsuitable for most uses (19).

Mechanical treatment (beating, refining) involves crushing, cutting, and splitting the fibers. For this purpose, the fibers pass, in the form of a water suspension, between two rough surfaces of special construction, which are at a small but regulated distance from one another. One of these surfaces is revolving and the other is stationary, or rarely both revolve. The results are: partial or total collapse of cell cavities, swelling of cell walls, higher flexibility of the fibers, reduction of their length, and partial de-

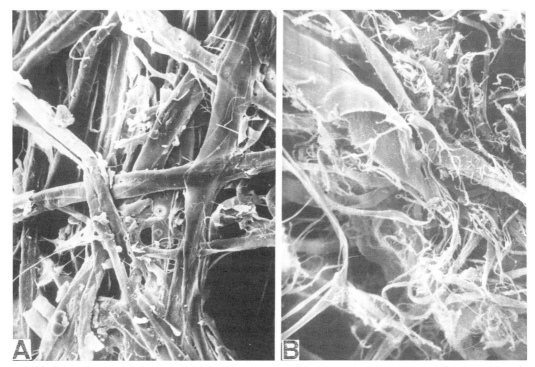

Figure 26-7. Wood fibers (white spruce and Douglas-fir tracheids) of thermomechanical (TMP) pulp; (A) before and (B) after beating, first stage (100×). (Courtesy of I. Sachs.)

tachment of microfibril bundles (Figure 26-7); the latter contributes to an increase of the exterior surface of the fibers (6, 7, 19, 25, 39). These changes are controlled by the morphology of the treating surfaces (blunt, sharp) and the extent (duration) and rate of treatment, and they affect to a considerable degree the properties of pulp and paper. For this reason, such treatment should be handled carefully, and guided by experience and laboratory test results of pulp samples.

The equipment belongs to four general types—*Hollander*, *Jordan*, *conical*, and *disk* (Figure 26-8); the first is a beater, and the others are refiners.

The *Hollander beater* (invented in Holland) has been servicing the pulp and paper industry for more than 200 years. It is made of a large tub, oval in shape, and a heavy (several tons) steel roll with a surface made of inserted bars (blades), which may differ in number, shape, and material. The roll revolves above one or more bedplates of similar construction. The pulp circulates peripherally and passes repeatedly between the two surfaces. In contrast to the other three types of equipment, the Hollander beater treats separate batches of pulp, and may accept directly dry pulp as well as paper—which are first dispersed and then beaten. Today, this device is considered insufficient for large factories. Disadvantages also exist with regard to space, maintenance, and labor cost, but the Hollander beater continues to be valued for quality of treatment and for pulp made of rags or pulp for special paper. However, the trend is to replace this equipment.

Hollander beaters have the capability of accepting dry pulp in contrast to the refiners, where dry pulp should first be dispersed in water. This is done in "pulpers" (i.e., large cylindrical tanks equipped with one or more revolving, propeller-like elements, placed at the bottom or side-wise). This equipment is also used to disperse used paper.[19] Another type of machine, called a "deflaker," is used to defiber chips that have not been completely defibered.

The *Jordan* refiner (after the name of its inventor) treats the pulp continuously. It differs from the Hollander beater and resembles the two other types (conical, disk refiner). The device consists of a conical (truncated cone), solid

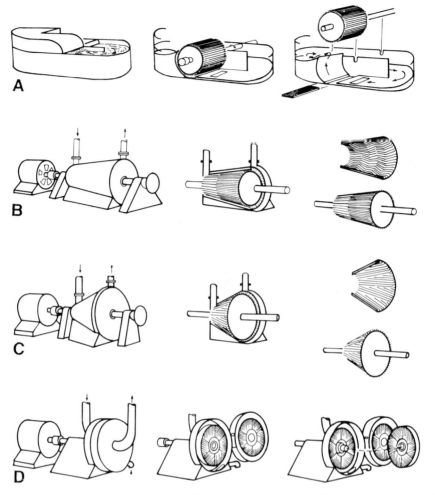

Figure 26-8. Beaters-refiners: (A) Hollander beater, (B) Jordan, (C) conical, and (D) disk refiners. (After Ref. 11.)

steel element, with longitudinal (parallel to length) bars mounted on its surface. This part (rotor, plug) revolves inside a stationary conical shell (stator, shell), from the interior surface of which bars also protrude. Pulp enters from the small end and comes out from the other side, passing between rotor and stator through a clearance that can be regulated.

The *conical* refiner is similar to the Jordan. The difference is that the interior cone has a wider angle (e.g., 60°) in comparison to the Jordan (10–20°), and the entire rotor/stator assembly is replaceable, whereas in Jordans the bars are mounted individually in slots provided in the rotor. Thus, conical refiners are capable of mounting different types (morphology) of treating elements. Also, conical refiners work more by brushing the fibers, and keep cutting to a minimum (as is also the case with the Hollander beater), whereas in Jordans there is more cutting and less brushing (6).

The disk refiner is made of a disk equipped with bars, and revolving parallel to another stationary disk, or there are two revolving disks, or two revolving disks are placed on both sides of a stationary disk. In contrast to the other machines, the flow of fibers is vertical. This means less load per fiber and less cutting. A variation of this refiner allows treatment of pulp in high consistency (30–40%) (6), whereas the usual consistency in all machines is 3–6%. There are various types of disks, which differ with respect to design and bar placement. Paper mills usually have a series of the above equipment, and there may be combinations (Hollander–Jordan, disk–Jordan, etc.) (19).

Defibrilization (detachment of microfibrils from cell walls), resulting from beating or refining, favors the development of molecular attractive forces (hydrogen bonds, van der Waals forces) by increasing the contacting surfaces. These forces hold wood (and other plant) fibers together without the need for an adhesive; interweaving of fibers is of little importance.

ADDITIVES

Additives are used to improve certain properties of paper, such as resistance to entrance of liquids, brightness, opacity and surface smoothness, printing properties, color, and strength (in dry condition or after wetting).

Resistance to entrance of liquids (ink, water, etc.) is sought by sizing (i.e., addition of rosin, wax, asphalt, or synthetic materials) (6, 19). The main additive is rosin (colophony), which is a constituent of pine resin. Rosin is preferred due to low cost and satisfactory results; it is added in a partially or completely saponified (neutralized) form, and for this reason alum (alum sulfate) is also added. Additives contribute to limited waterproofing of the surface of paper; they retard but are not able to prevent the entrance of liquids.

Other chemicals, called "fillers," are added to improve the quality of paper (or paperboard) in the following sense: a network made only of fibers usually has a rough surface with elevations, cavities, and empty spaces. Fillers can fill such irregularities and, therefore, reduce surface heterogeneity. At the same time, brightness, opacity, printing, and other properties are improved—and in the case of cigarette paper, porosity and rate of burning (19). These additives include: kaolin (aluminum pyrate, $Al_2O_3 \cdot 3SiO_2$), titanium dioxide (TiO_2), calcium carbonate ($CaCO_3$) and other calcium compounds, barium sulfate ($BaSO_4$), zinc sulfide (ZnS), etc. (7, 19). They are added in a proportion of 1–10% of fiber weight, in combination with alum or other chemicals that help these additives adhere to the fibers. Aside from the addition to pulp, fillers may be added to the surface of finished paper or during its production. Synthetic polymers are also used as fillers.

Prior to adding alum, and preferably prior to the addition of water-proofing materials, coloring additives are mixed in the pulp to ensure color uniformity. Color is the most important characteristic that attracts attention to a sheet of paper, and it is increasingly used to promote paper consumption. Most paper is colored, or some light shade of color is imparted, and this is usually done by adding water-soluble coloring agents to the pulp. Rarely, paper is colored by use of insoluble coloring "pigments" in the form of fine particles. White paper also incorporates color. Almost all bleached pulps have a yellowish shade, and a small quantity of reddish-black color is added to counterbalance it. White paper is in fact ash-white in color.

Additives are also used to improve bonding of the fibers. Starch (modified), natural glues, or synthetic resins are usually added to increase the strength of paper in a dry state. As mentioned earlier, bonding of fibers is attributed to molecular attractive forces, known as hydrogen bonds and van der Waal forces. However, such forces dissolve or are neutralized in water, and for this reason the strength of wet paper is practically zero. This property is an advantage for the recycling of paper, although in certain cases the preservation of strength in a wet state is desirable.[20] This is sought by the addition of synthetic resins (urea-formaldehyde, melamine-formaldehyde, polyamide-type resins, etc.); they are added in the form of aqueous solutions in a proportion of 0.5–1% and rarely up to 5%, after the pulp leaves the last refining machine.

FORMING—PRODUCTION OF PAPER

The next stage is forming the fibers to a continuous sheet of desired thickness and weight, and then making paper by the removal of excess water by pressure and heat. These processes are carried out by complex machines, which may produce a continuous sheet of paper, 1.5–10 m (5–30 ft) wide, 10–500 g/m² in weight, at a speed of up to 1500 m (5000 ft)/min, and in excess of 1000 tons/day.

There are two traditional types of papermaking machines—Fourdrinier and cylinder. They differ in the manner of forming the fiber mat.

In the Fourdrinier machine, forming is accomplished on an endless metallic web (wire mesh belt) equipped with many openings. This is made like a table top, and moves at a high speed in a horizontal-longitudinal direction, while at the same time it oscillates side-wise. Pulp, in a very low consistency of fibers in water (0.1–1.0%) (6, 19) is piped into a mechanical system (flowspreader), where it is transformed to a stream, equal in width to the width of the machine. Subsequently, it is led to a system of flow control (headbox) and then to the above-mentioned wire mesh belt, where the fibers are laid, and most of the water (95% or more) is removed by gravity and the application of vacuum by pump systems (suction boxes) installed under the belt. The "press section" of the machine is next. There, the mat, transferred on a cloth (blanket-like) belt, is pressed by press drums (rolls) in order to reduce its moisture. A supplementary, drastic reduction of moisture (down to 7–10%) is made in the "dryer section," which follows, and is made of a series (40–70) of revolving drums (rolls) heated by steam. Next, the paper passes through a series of revolving drums (calender rolls), which smooth surface irregularities by an ironing action, and regulate its density and thickness. Finally, the paper is wound into rolls (Figures 26-9 and 26-10).

The *cylinder machine* is different, mainly because the fibers are laid on the surface of one or more porous cylindrical structures, the periphery of which is covered by wire mesh. The mat is formed on the wire by revolution of the cylinder, which is partially immersed in a tank (vat) containing fibers in low consistency, while a reduced pressure is created inside the cylinder. The mat is received on a cloth belt, and is pressed and dried in a manner generally similar to the Fourdrinier machine (5).

In addition to these two basic types of machines (there are variations), other machines have been designed in recent years. They differ mainly in forming the fiber mat—for example, by horizontal movement of a wire mesh and cylinder, or two parallel wires, where the mat is formed like a "sandwich" (7, 39).

FINAL TREATMENTS

The paper becomes available for consumption usually after a series of other treatments, such as supplementary smoothing between rolls (supercalendering), splitting and rewinding to produce rolls of the desired size, and cutting to sheets. Other treatments include improvement of properties by coating or impregnation of chemicals, lamination (e.g., to produce a three-layer corrugated paperboard), and finally conversion to various products. All this is done by specialized machines.

One of the most important treatments is coating the surface of paper to improve its printing and writing properties and appearance. This is a supplementary treatment because, as mentioned earlier, materials to achieve such goals are added to the pulp. Coating is applied by mechanical devices, which also smooth the coating material between pressure rolls. Coat-

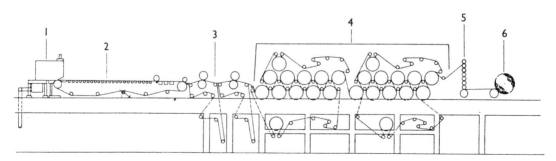

Figure 26-9. Fourdrinier machine: 1, flow spreader (and headbox); 2, wire mesh belt (Fourdrinier table); 3, press section; 4, dryer section; 5, calender rolls (stack); 6, paper roll (reel). (Courtesy of Beloit.)

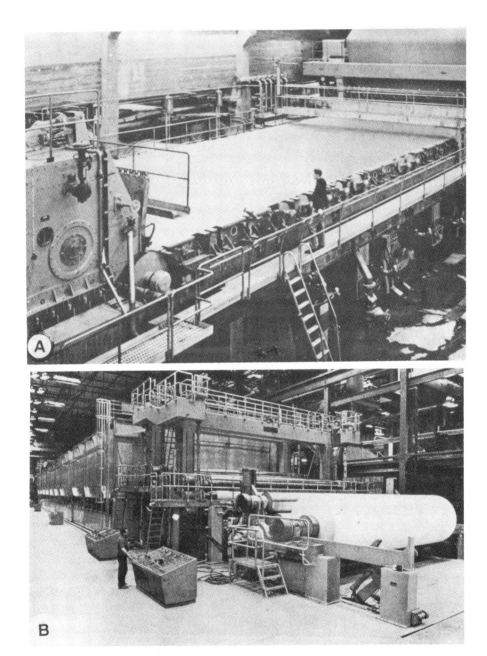

Figure 26-10. Fourdrinier machine: (A) front (flow spreader and mat formation) and (B) back (paper roll) of machine. (Courtesy of A, Nash and B, Valmet.)

ing materials include kaolin, titanium dioxide, calcium carbonate, and others. They are in the form of aqueous suspension in combination with a natural or synthetic substance possessing adhesive properties (starch, latex) (7). Sometimes, such treatment is applied during the production of paper, near the completion of the dryer section.

Coating with plastics (e.g., polyethylene resin) waterproofs paper (and paperboard), and makes it suitable for containers (milk, etc.) and other products.

PROPERTIES OF PULP AND PAPER

The properties of pulp and paper vary due to differences in raw materials and production methodologies. For wood (and other plant fibers), cell morphology is a fundamental factor of influence; thus, fiber length, fiber diameter, lumen diameter (diameter of cell cavity), and cell-wall thickness are very important (1, 2, 3, 16, 22, 36). For example, research has shown that the ratio of lumen diameter to fiber diameter[21] (an expression of the tendency of a fiber to collapse) is strongly related to pulp and paper sheet density. Sheet density, combined with individual fiber strength, relates highly to sheet tensile and bursting strength. In a more complex manner, tear factor is related to sheet density, fiber length, and fiber strength: tear factor at first increases with increasing sheet density, and subsequently decreases at higher sheet densities; fiber length and fiber strength both have a positive influence on tear factor—up to a critical level of sheet density corresponding to the level of bonding between fibers, beyond which fiber rupture prevails as the mechanism of failure (35). Another study showed that of the morphological characteristics of hardwood fibers, those contributing the most were fiber length, length/wall thickness (L/T) ratio, and fibril angle. Parenchyma cells were detrimental to bursting and tensile strength, but vessel elements had no effect on tensile strength (36).

Pulp is evaluated by chemical and physical tests.[22] Chemical tests include determination of cellulose and noncellulosic (mainly lignin) content, and physical tests measure fiber characteristics and resistance to water flow through the pulp ("freeness"). Pulp is also evaluated from pulp sheets made in the laboratory. Such sheets are used to determine the following properties: weight per unit of surface, specific volume (cm^3/g, ft^3/lb), coefficient of light diffusion, tensile strength, stretch, bursting, tearing, folding, zero-span tensile strength, and moisture content.

Paper is produced in virtually thousands of grades and products, which present minor or very important differences.[23] The evaluation is also based on specifications that apply to the following properties[24]: dimensional stability, weight (g/m^2), thickness, density, tensile strength, bursting, tearing, folding, bending, abrasion resistance, porosity, surface texture, optical properties (brightness, color, opacity, gloss), and chemical properties (aging, flammability, acidity). Identification of additives and wood species in pulp or paper is made by use of chemical reagents and microscopes (31).

REFERENCES

1. Amidon, T. E. 1981. Effect of the wood properties of hardwoods on Kraft paper properties. *Tappi* 64(3):123-126.
2. Anonymous. 1981. [Comments on H. H. Holton's Marcus Wallenberg Award]. *For. Prod. J.* 8:24.
3. Barefoot, A. C., R. G. Hitchings, E. L. Ellwood, and E. H. Wilson. 1970. The Relationship Between Loblolly Pine Fiber Morphology and Kraft Paper Properties. N.C. Agr. Expt. Station, Tech. Bull. No. 202.
4. Bialkowsky, H. W., and P. S. Billington. 1961. Pulp and paper processes and their characteristic differences. *Tappi* 44(3):195-201.
5. Britt, K. W., ed. 1964. *Handbook of Pulp and Paper Technology*. New York: Reinhold.
6. Britt, K. W. ed. 1970. *Handbook of Pulp and Paper Technology*, 2nd ed. New York: Van Nostrand Reinhold.
7. Britt, K. W. 1974. Paper and paper production. *Encyclopaedia Britannica* 13:966-977.
8. Casey, J. P., ed. 1980. *Pulp and Paper: Chemistry and Technology*, Vol. 1, 3rd ed. New York: Wiley.
9. Casey, J. P. 1984. *Casey's Reports on Paper and the Paper Industry*. New York, Basel: Marcel Dekker.
10. Chiang, T. I. 1978. Economics of thermomechanical market pulpmills in the southeastern part of the United States. *For. Prod. J.* 28(7):18-23.
11. Danforth, D. W. 1970. Beating and refining—Equipment. In *Handbook of Pulp and Paper Technology*, ed. K. W. Britt, pp. 333-344. New York: Van Nostrand Reinhold.
12. F.A.O./U.N. 1973. *Guide for Planning Pulp and Paper Enterprises*. Rome.
13. Farmer, R. H. 1967. *Chemistry in the Utilization of Wood*. Oxford, New York: Pergamon Press.
14. Fengel, D., and G. Wegener. 1984. *Wood: Chemistry, Ultrastructure, Reactions*. Berlin, New York: Walter de Gruyter.
15. Hoffmann, P., R. Patt, and M. Stamm-Lueders. 1978. Some investigations into the suitability of young *Pinus caribaea* (Morelet) from plantations in Venezuelan Guayana for pulp production. *Holzforschung* 32(4):138-141.
16. Horn, R. A. 1978. Morphology of Pulp Fiber from Hardwoods and Influence on Paper Strength. U.S. For. Prod. Lab. FS Res. Paper 312.
17. Katzen, R., R. Frederickson, and B. F. Brush. 1980.

The alcohol pulping and recovery process. CEP (Chemical Engineering Progress), February, pp. 62–67.
18. Kleppe, P. J. 1970. Kraft pulping. *Tappi* 53(1):35–47.
19. Libby, C. E., ed. 1962. *Pulp and Paper Technology*, Vols. I and II. New York: McGraw-Hill.
20. McGovern, J. N. 1981. Hardwood pulpwood utilization in the United States. *Tappi* 64(3):129–132.
21. Myerly, R. C., M. D. Nicholson, R. Katzen, and J. M. Taylor. 1981. The forest refinery. *Chemtech* 11:186–192.
22. Palisoc, J. G. 1979. Fiber Cell Wall Thickness: How It Affects Pulp and Paper Quality. FORPRIDECOM Tech. Note 201, Philippines.
23. Palmer, E. R., and P. Greenhalgh. 1983. The Production of Pulp and Paper on a Small Scale. Tropical Development and Research Institute, London.
24. Panshin, A. J., E. S. Harrar, J. S. Bethel, and W. J. Baker. 1962. *Forest Products*. New York: McGraw-Hill.
25. Parham, R. A., and H. M. Kaustinen. 1974. *Papermaking Materials: An Atlas of Electron Micrographs*. Appleton, Wisconsin: Inst. Paper Chemistry.
26. Perry, J. H. 1970. Groundwood–Stone. In *Handbook of Pulp and Paper Technology*, ed. K. W. Britt, pp. 179–188. New York: Van Nostrand Reinhold.
27. Sandermann, W. 1963. *Chemische Holzverwertung*. München: BLV Verlags.
28. Sandermann, W. 1988. *Die Kulturgeschichte des Papiers*. Berlin, New York: Springer Verlag.
29. Sjostrom, E. 1981. *Chemistry: Fundamentals and Applications*. New York, London: Academic Press.
30. Stephenson, J. N., ed. 1950. *Pulp and Paper Manufacture*, Vols. 1 and 2. New York: McGraw-Hill.
31. Strellis, I., and R. W. Kennedy. 1967. *Identification of North American Commercial Pulp Woods and Pulp Fibers*. Toronto: University of Toronto Press.
32. Styan, G. E. 1978. TMP—Attractive outlet for sawmill residues. *For. Prod. J.* 28(7):13–17.
33. Technical Association of the Pulp and Paper Industry. TAPPI Standards.
34. Thiesmeyer, L. R. 1967. More and more for less and less. *For. Prod. J.* 17(9):13–17.
35. Wangaard, F. F. 1973. Fiber characteristics in relation to paper properties. *Proceedings IUFRO Meeting* (So. Africa), 2:1128–1148.
36. Wangaard, F. F., R. M. Kellogg, and A. W. Brinkley. 1966. Variation in wood and fiber characteristics and pulp-sheet properties of slash pine. *Tappi* 49(6):263–277.
37. Wenzl, H. F. J. 1970. *The Chemical Technology of Wood*. New York, London: Academic Press.
38. West, W. B. 1979. Mechanical pulping as of today. *TAPPI* 62(6):19–22.
39. Whitney, R. P. 1980. *The Story of Paper*. Atlanta, Georgia: TAPPI.
40. Zarges, R. V., R. D. Neuman, and J. B. Christ. 1980. Kraft pulp and paper properties of *Populus* clones grown under short-rotation intensive culture. *Tappi* 63(7):91–94.

FOOTNOTES

[1] In the pulp and paper (and fiberboard) industry, the term "fiber" is not used in the botanical sense (see Chapter 3), but as a cumulative term for all cells of wood.

[2] Before the invention of paper, man used to write on stones, shells, ceramic tablets, animal skins, and papyrus. Papyrus was made in ancient Egypt about 5000 years ago. Raw material for this product were stems of *Cyperus papyrus*, a plant growing on the shores of the Nile and reaching a height of 1–3 m (3–10 ft). The production of papyrus included detachment of fibrous layers from the stems, their cross-wise arrangement one on top of another, and pressure. Adhesion was assisted by the sap exuded after crushing the stems. After drying, papyrus was smoothed by rubbing with a stone or ivory (28). A very interesting account of papermaking in ancient times is given by Pliny in his "Natural History" (Book XIII).

[3] The use of hardwoods is extensive. For example, in the United States, hardwoods are used in proportions that range from 25 to 70%. In some regions, oak is the main species (20).

[4] Experiments have shown that wood from plantations [young softwoods (15) and hardwoods (40)] produce a satisfactory pulp, if cooking is milder in comparison to "adult" wood. Whole-tree-pulping presents problems due to presence of bark, foliage, and foreign matter.

[5] In Canada and the U.S.S.R., chips have been experimentally transported, from forest to factory, inside piping—with water (34).

[6] Wastes from manufacturing textiles, clothing, rope, etc.

[7] Thermomechanical pulping has also been discussed under fiberboard; the conditions of pulping for fiberboard are different (temperature, plate design, plate clearance, etc.).

[8] Sulfurous acid is produced by burning sulfur or pyrites (e.g., iron sulfide, FeS_2) and dissolving the resulting sulfur dioxide (SO_2) in water. Calcium bisulfite is produced by reacting sulfur acid with calcium carbonate ($CaCO_3$) or milk of lime—$Ca(OH)_2$—and magnesium bisulfite by similar reactions. Sodium bisulfite is made from sodium carbonate (Na_2CO_3), and ammonium hydrogen sulfite from ammonia (5, 19).

[9] The rate of diffusion of chemicals in the chips is faster with sodium and ammonium bases, slower with calcium, and intermediate with magnesium (5).

[10] This applies to pulping wood. The method is used for pulping other plant materials, such as straw, bamboo, and esparto (13).

[11] Kraft means strength in German and Swedish. The term "Kraft" or "Kraft pulp" is used internationally in reference to its high strength.

[12] Tall oil contains resin acids (*rosin*), fatty acids, and neutral substances. The volatile terpenes of wood are condensed as sulfate *turpentine*. Yields of tall oil usually vary from about 30 to 50 kg (60–110 lb)/ton of pulp; much higher yields of more than 100 kg (220 lb)/ton have been reported for slash pine (*Pinus elliottii*) in the United States.

The yield of turpentine is 3–6 kg (6.5–13 lb)/ton of pulp, but this is greatly reduced after extended storage of chips; after 30 weeks of storage, about 80% of the turpentine is lost. Kraft pulping is the dominant source of rosin and turpentine; production by tapping living trees or wood extraction is stagnant or declining.

[13]Improvement of alkaline pulping is possible by adding a small quantity of anthraquinone, which increases the rate of delignification and the yield.

[14]Chemical pulping factories are generally large. In a report investigating the establishment of paper mills in developing countries (23), the smallest economically viable units are assigned the following capacities (tons/day): mechanical 10 t/d, semichemical 20 t/d, chemical from wood more than 100 t/d (chemical from straw with chemical recovery 30–40 t/d and without recovery 1 t/d), recycled waste paper 1 t/d. (Relationships of tons of paper/tons of wood are given in a Table of Equivalents at the end of this book).

[15]A nonconventional chemical process uses alcohol as a pulping agent (alcohol pulping, organosolv pulping). The method may be used to pulp biomass, is suitable for small units (e.g., production of 5000 tons/year, equivalent to about 15 t/d), is applicable to both softwoods and hardwoods, gives pulp of high strength (similar to Kraft pulp) and high purity (suitable for paper, synthetic fibers, plastics, etc.); pulping is fast, and the alcohol may be reused. Solvent recovery is very critical due to high cost of the solvent (17, 21).

Examples of other processes suitable for small-scale operations are: ammonia-based pulping; oxygen (air) and alkali processes (i.e., the "Hope" process and the "Doddel" process), universal process (acid-based), and other nonconventional methods (23).

[16]Some authors include this method under semichemical pulping, and others in a separate category between mechanical and semichemical pulping (8, 9).

[17]The need for good debarking is greater in acid (sulfite) than Kraft pulping, where bark can be pulped to a great extent. Wood pulped by the Kraft process for paperboard (corrugated, etc.) may not be debarked at all. Packaging materials (wires, metallic bands, strings) are often present when used paper (newspapers, magazines, etc.) is utilized.

[18]Brightness is measured with a reflection meter, which is equipped with photoelectric cells and measures the quantity of (blue) light (in percent) reflected from a pulp sheet in comparison to reflectance of a "standard" plate of magnesium oxide (100% brightness). Unbleached pulp gives readings that range from 25–65%. After bleaching, brightness may reach values higher than 90%. Maximum lignin-retaining bleaching brightness can be as high as 80–85%, but reversion is a problem in any lignin-retaining pulp (7).

[19]Used paper constitutes an important source of pulp, and it is usually mixed with "virgin" pulp. Reuse (recycling) includes the following steps: classification by kind, de-inking (usually by caustic soda or sodium carbonate), pulping, screening, and bleaching the pulp (19). In 1989, used (waste) paper covered 35% of the needs for pulp in western Europe (EEC countries 32%), 45% in Japan, 45% in Sweden, and 34% in the U.S. (A. Anders, *Scandinavian Pulp and Paper Magazine* No. 5/6:26–31, 1990, and *Annual Report, Swedish Pulp and Paper Ass'n*, 1990).

[20]A paper is considered to possess "wet strength" if it keeps 15% or more (in exceptional cases up to 30–40%) of its dry strength (19).

[21]The ratio of lumen diameter to fiber diameter is called "coefficient of flexibility." A related concept is "wall fraction," which is the ratio of double-cell wall thickness to fiber diameter. Both indicate fiber plasticity and are expressed as percentages (lumen diameter $\times$ 100/fiber diameter). Wall fraction varies from about 10% (thin-walled fibers with a wide lumen) to 90% (thick-walled fibers with a narrow lumen). The average wall fraction should not exceed 50% in pulpwood of good quality, and preferably should be below 40%.

[22]There are no internationally accepted test specifications. TAPPI (Technical Association of the Pulp and Paper Industry, USA) standards are mainly used for testing the properties of pulp and paper (33).

[23]The following general types are included among paper products: newsprint, printing, writing, package, absorbent, decorative, etc. Paper thicker than 0.3 mm and stiff is classified as paperboard, and used for packaging, building constructions, etc.

[24]Both pulp and paper are hygroscopic and for this reason their testing is conducted under constant hydrometric conditions. Relative humidity is usually 65% (in the U.S., 50 $\pm$ 2%) and temperature 23 $\pm$ 2°C (71–77°F) (5).

APPENDIX

This appendix contains three chapters I. *Other Wood and Forest Products*, II. *Temperate and Tropical Woods*, and III. *Bark as a Material*. The first chapter is a synoptic supplement on utilization of wood and other forest products; the second chapter refers to wood identification with some additional information on geographical source, properties, and uses; and the third chapter deals with the structure, properties, and utilization of bark.

Other Wood and Forest Products

Previous chapters of this book (Chapters 16-26) have been devoted to the utilization of wood for products of primary manufacture. In an effort to present a wider picture of the value of wood as a raw material and of the forest as a source of materials other than wood, this chapter deals, in a concise manner, with other wood products—products of the foliage of forest trees, pine resin and other exudates and extractives, and other forest products.

OTHER WOOD PRODUCTS

Other products of wood include products of secondary manufacture (by mechanical and chemical processing); wood is also examined as a source of energy.

Products of Secondary Mechanical Processing

Such products include: furniture, building components and structures (doors, windows, wooden houses, flooring), containers, railroad ties, means of transportation, musical instruments, athletic equipment, wood carvings, excelsior, wood flour, and others (9, 16, 20, 24, 27, 33, 38, 39).

All these products are produced by previously described basic processes (sawing, cutting, drying, gluing, and sometimes preservation), and they are made by further processing of lumber, veneer, plywood, laminated wood, particleboard, or fiberboard. Additional processing includes planing, turning, carving, shaping and routing, drilling, connecting (gluing, nailing, screwing), bending, sanding, coloring (dyeing, painting), finishing, etc.

Planing is performed by machines consisting of a steel cylinder to which two or more knives are securely and axially fastened; their protrusion from the surface of the cylinder may be regulated according to the desired depth of cut. In some cases there are two cylinders, placed parallel—one above the other; the wood is directed to pass between them, and is simultaneously planed on both sides. Planing is also done by straight hand planers.

Planing produces smooth surfaces, wood figure is accentuated, defects appear clearly, and desired final dimensions of thickness are given to lumber or members of wooden structures.

Turning is done by turning machines or lathes, to which knives of various shapes are attached (21). The wood is supported in a horizontal position, between two centering arms, and revolves on its axis, while the cutting attachment (knife) moves along its length and produces the desired shape of turned surface. Special machines may reproduce several (up to 12) similar turned members from a prototype.

Molding is performed by machines (molders, shapers, routers), equipped with knives of desired shape and attached to a vertical revolving axis. Edges of furniture and other items are shaped in this manner. The wood is pressed to the cutting mechanism. Shaping of edges is also done for the purpose of joining wooden elements (e.g., by tongue and groove, mortise and tenon, and dovetailing).

Drilling or *boring* is done by various types of hand tools or machines. One or more suitable accessories ("bits") are attached to horizontal or vertical "heads," and by a revolving action they cut holes in the wood to facilitate the assembly of parts.

Joining of parts is done by gluing, nailing, or screwing, or by special metallic connectors (staples, etc.), or wooden dowels (i.e., cylindrical wooden rods inserted between the joints). The relative ease of such joining is an advantage for wood. There are various types of nails and screws, depending on their material (metal), size, shape, etc. The nail- or screw-holding capacity depends on their dimensions (length, diameter) and surface morphology, and on the density and moisture content of wood.

Bending is performed after steaming or other treatment of wood (e.g., exposure to ammonia or urea) (34). Curved members are also produced by sawing, but this method has the disadvantage that the wood is cut diagonally or transversely to its grain, and waste is high.

Steaming is applied to air-dry wood (moisture content about 15%), and the wood is bent on forms of desired curvature. Species such as ash, elm, oak, maple, beech, sycamore, and others have good bending properties, but the wood should be of good quality, without defects at least at the points where high stresses develop during bending.

Ammonia, in liquid or gaseous condition, applied under pressure and temperature, plasticizes the wood, and in that condition it may be bent easily to various shapes. The method is applicable to all wood species, but hardwoods bend more easily than softwoods, and heavy woods better than light woods. Ammonia is also applied to air-dry wood (3).

Thin lumber, veneer and plywood may be bent without preparation, to a limiting radius of curvature, which depends on thickness, density, grain direction, and moisture content of the wood. In some cases, such products are simultaneously bent and glued by cold or hot pressure (e.g., for making laminated wood). Curved members made from particleboard or fiberboard are pressed in forms (see Chapters 22, 23, and 24).

Sanding of wood is done by suitable tools and machines, and the purpose is to reduce surface coarseness and prepare the wood for finishing. Sanding materials are hard minerals in fine-particle form (silica, aluminum oxide). Particles of such materials are laid and glued on paper or cloth, then attached onto drums or endless belts.

Finishing is applied to decorate or protect wood products for interior (furniture, flooring, etc.) or semi-exterior use (doors, windows, etc.). (31) The materials used are dyes, transparent varnishes, and organic solutions containing coloring, water repellent (resin, wax, etc.), or toxic substances (insecticides, fungicides).

The main products of secondary manufacturing follow.

Furniture. Although some furniture is made of solid wood, the wood is usually in the form of veneer, plywood, particleboard, or fiberboard (9, 24, 37). Softwoods (pine, spruce, fir, larch, Douglas-fir, Mediterranean cypress, etc.) are used less than hardwoods (oak, walnut, maple, beech, and tropical woods, such as Palissander, Padauk, Limba, Sapele, Tiama, Teak, etc.).

Use of veneer instead of solid wood does not make inferior furniture, assuming that gluing is done by suitable adhesives and techniques. Adhesives used for making furniture are usually polyvinyl emulsions (they have replaced natural glues used in older times), but the constituent parts (products of primary manufacture, such as plywood or particleboard) are usually glued with urea-formaldehyde.

As explained elsewhere (Chapter 10), solid wood of large dimensions, especially flat-sawn, may develop defects (warping, etc.) due to differential radial and tangential shrinkage. In addition, it is a fact that there is a limit to large dimensions of solid wood, or its price may be prohibitively high. Use of veneer gives the opportunity for most people to own furniture made of valuable woods (oak, mahogany, etc.), which previously was an exclusive privilege.

In addition to wood, other materials are used in furniture (metals, plastics, glass), sometimes in combination with wood, but in most cases wood is the preferred material. This preference is related to attractive features of wood, such as variety of colors and figures, and a feeling of "warmth" in appearance and touch.

Furniture is made in two main styles—traditional and modern. Traditional furniture is associated with kings, nobles, designers or makers, and is characterized by complex carvings and curvatures; whereas modern furniture has simple lines.

Buildings. Use of wood in buildings includes doors and windows, and other parts of structures, such as homes, stores, warehouses, gymnasiums, schools and churches, sheds, roofs, ceilings, exterior staircases, balcony railings, porches, etc. There is a certain replacement of wood by other materials (aluminum, plastics), but its use continues to be extensive. Replacement of wood is mainly due to insufficient measures of protection, when used in structures exposed to the weather (exterior doors, windows, etc.) and, therefore, subjected to deterioration (decay, erosion by weathering). Wood intended for such use should be treated with preservatives or water repellents prior to placement in service. After treatment, and air-drying, the wood may be painted. Fire retardants are also employed in some cases.

The species used are usually softwoods (pine, fir, spruce, Douglas-fir, larch, and others), because they are more easily worked with tools and machines, and less expensive and available in a variety of dimensions. Hardwoods include oak, elm, chestnut, other species, and tropical woods. The wood is often used in the form of reconstructed products, such as plywood and particleboard, and sometimes as laminated wood (beams, arches, etc.).

Use of wood in buildings is traditional and extensive in suburban and rural areas of North America and Northern Europe; however, in the Mediterranean area and other countries, where the local supply is limited, stones, bricks, and cement are the main materials of construction.

Flooring. This is also a building component, but is treated here separately as a product of specific requirements (42).

Wooden floors are made of an upper part (main floor) and a supporting part (subfloor). There are two types of flooring: (a) floors made of lumber boards of long length, which are placed side by side or interconnected by tongue-and-groove, and (b) floors made of short, 25-80 cm (about 0.75–2.5 ft) long elements called "parquet." Parquet flooring is of two kinds—thick (classical type) and thin (mosaic type). In the classical type, the elements are connected by tongue-and-groove, whereas mosaic elements have straight edges.

Species used for flooring are oak, beech, chestnut, maple, birch, ash, walnut, spruce, fir, Douglas-fir, larch, and various tropical woods, such as Azobe, Teak, Afrormosia, Iroko, Bete, Makoré, Wenge, Bubinga, Opepe, Sapele, and others. Hardness plays an important role in selecting woods for flooring.

Parquet flooring is classified according to dimensions and wood quality, which is judged on the basis of color (uniform or nonuniform, presence of heartwood and sapwood in the same element), figure (e.g., "silver grain" in oak, see Chapter 1), and defects (small knots, etc.). In Italy and France there are 6 grades of parquet, in Greece 3, in Yugoslavia 5, and in the United States 7 (3 with radial and 4 with tangential figure).

Parquet is placed on a subfloor and in various arrangements (parallel, orthogonal, at an angle, "herring-bone," etc.; Figure I-1). Mosaic parquet comes in squares, with the elements glued on paper or interconnected with adhesive. Parquet is nailed to the subfloor (usually made of softwood lumber), or glued on concrete.

Wooden floors are sometimes treated with preservatives for protection against insects or fungal attack. After installation, they are sanded and coated with varnish.

Containers. Boxes, crates, barrels, and other containers of liquids or solid materials are made of wood in various sizes and shapes, depending on intended use. Boxes or crates are used for solid materials (food, appliances, ammunition, etc.), while liquids are mainly stored in barrels. The wood should have sufficient strength, and should not contaminate the contents (e.g., by disagreeable odor). Woods

Figure I-1. Patterns of parquet flooring. (Courtesy of Shelman.)

used for boxes and crates include pine, spruce, poplar, beech, and other species, while barrels are made of Douglas-fir, white oak, ash, and other species. There is a certain competition by metals (steel, aluminum) and plastics, but wood is relatively inexpensive and for certain uses (fermentation of wine, aging of whiskey, etc.) it is considered irreplaceable (27).

Railroad Ties. Most ties (up to 90% worldwide) are made of wood, but steel and concrete ties are also used (8). Wood has important advantages for use as a tie: it is easily worked with tools and machines, does not rust, it may be treated with preservatives to prolong its durability, has light weight and high elasticity, and makes traveling more comfortable. Ties are made of softwoods (pine, Douglas-fir, larch) and hardwoods (oak, beech, black locust, elm, hornbeam). Wood with some defects (blue stain and knots) may be used. "Red heart" of beech (see Chapter 7) cannot be treated with preservatives, and it is recommended that affected wood should not cover more than one third of the cross-sectional area of a tie, and should not appear at all on upper (exposed) surfaces. Fungal attack, insect holes, knot holes, frost cracks, and ring shakes are not allowed. The service life of ties is greatly increased by preservative treatment (see Chapter 19); thus, it is possible to utilize species that have low natural durability. The wood should be air-dried (for 6 months or more depending on climatic conditions), and properly prepared prior to treatment by cutting rail-seat positions, drilling holes, and incising. There are two dimensions of ties—standard and narrow gauge.

Mine Timbers. In mines, timbers are used to support the roofs and sides of galleries. The timbers are either round and usually debarked, or sawn, and they are placed vertically and horizontally, forming supports in a double T shape. In vertical members, the ratio of length to least dimension in width should not exceed 11:1 (see Chapter 11).

Wood is preferred due to low initial cost, favorable ratio of strength to weight, ease of working with tools and machines, and a "warning" capacity (characteristic noise before failure). Species without satisfactory natural durability may be treated with preservatives or fire-retardants in order to increase their resistance to fungi, insects, and fire. Suitable woods should have good strength (especially in axial compression), no important defects (knots, spiral grain), and they should be placed in service after air-drying. Chestnut, oak (preferably white), other hardwoods, hard pines, larch, Douglas-fir, and other species, depending on local availability and preservative treatment, are used as mine timbers (see Chapter 16).

Transportation. Wood is currently used as supplementary material for construction of ships, boats, railroad cars, and other means of transportation, in contrast to previous periods when such use was extensive or exclusive.

In large ships, wood finds uses in interior paneling, flooring, recreation rooms or furniture—in the form of lumber, veneer, plywood, laminated wood, particleboard, or fiberboard—together with other materials, the use of which may be more common than wood. Wood is also used in some pleasure boats, and some fishing boats are still constructed of wood, although that art is gradually vanishing; the main species is pine, but several others are also used, including oak, elm, ash (rarely beech and sycamore), Mediterranean cypress, Douglas-fir, larch, cedar, and olive tree wood. Traditional species in the construction of large ships are oak, elm, teak, and others.

The use of wood in passenger railroad cars is limited except in special cases (sleeping and restaurant cars). In freight cars for transporting materials, wood has a wider application (flooring, sides, roofs, doors). Floors are made with treated beech, softwoods, or oak. Wood is also used in trucks and wagons; species include elm, black locust, hornbeam, olive, oak, eucalypts, and others. In older times, wood was also used in private automobiles, both in their interior and exterior.

The use of wood in airplanes was also extensive at one time. Up to the second World War wooden airplanes existed (e.g., the so-called "Mosquitos" in Britain). Today, wood may be present in various forms (veneer, plywood, sandwich constructions), coated with aluminum or other materials.

Matches. Matches are usually made from poplar wood, but willow, birch, spruce and other species are also used. Diffuse-porous woods do not break easily, and they are more uniformly treated. Matches are produced by various methods, usually by peeling logs. The technique of production may be summarized as follows: after peeling, to a thickness of about 2 mm, the veneers are cut to strips that have a width equal to the length of the match sticks, and by another system they are shaped to final dimensions. The sticks are then washed with water, and they are impregnated with monoammonium phosphate, so that flame and glowing charcoal are quickly extinguished after the match is used, in order to reduce the danger of fire. The sticks are subsequently transported to a dry kiln, and are dried for 10–15 min at a temperature of 80–100 °C (175–212°F). Sanding follows inside a revolving drum, where the sticks are rubbed together and against the sides of it. The next steps include the parallel arrangement of the sticks, removal of defective sticks, vertical placement on a metallic base, preheating, formation of heads by immersion in suitable chemicals, and slow drying. (Safety matches usually have antimony sulfide, oxidizing agents such as potassium chlorate, and sulfur or charcoal in the heads, and red phosphorus on the striking sides of match boxes; nonsafety matches usually have phosphorus sesquisulfide in their heads.) The final step is to place a number of matches in a small wooden box made of thin peeled veneer. Wood is often replaced by paperboard in match-making.

Musical Instruments. Many musical instruments (string, wind, percussion) are made entirely or partly of wood. Both softwoods (mainly spruce) and hardwoods (beech, maple, sycamore, hornbeam, etc.), including tropical woods (ebony, rosewood, and others), are used. The wood should have normal structure (without defects, such as knots and grain deviations), and medium width of growth rings (see Chapter 13 and Ref. 30).

Athletic Equipment. Such products include baseball bats, tennis rackets, skis, bows and arrows, and others. There is a tendency for replacement by other materials (metals, plastics), but wood is still finding uses, due to advantageous properties, such as elasticity, toughness, and favorable strength-to-weight ratio. Toughness is the most important property for selection of species, which include oak, ash, birch, hickory, and others; they are often used in the form of laminated wood (7). Gun stocks may also be included in this category; they are made of walnut and other hardwoods.

Carvings. Wood has been carved since ancient times (32). Wooden statues of ancient Egypt are exhibited in many museums. In ancient Greece, statues were originally wooden. In the case of large products, carving may be performed in strips of wood which are subsequently glued together. Suitable species are: pines, juniper, Mediterranean cypress, yew, spruce, Douglas-fir, ash, elm, oak, lime (basswood), birch, maple, boxwood, tropical woods (mahogany, ebony, teak, etc.).

Excelsior. Excelsior is made of ribbons of wood, 0.03–0.5 mm thick, 0.5–4 mm wide, and 8–25 cm (3–10 in.) long. They are produced by peeling as in making veneer. The difference is that, in addition to the main knife, there is a number of small knives which cut the wood parallel to the direction of peeling. These knives make incisions prior to the main cut, and their distance controls the width of the ribbons. The wood is cut at a moisture content of about 20%. Species used are: spruce, pine, and light hardwoods, such as poplar; resin in pines may limit the uses of excelsior. Excelsior is mainly used in packaging and building boards (e.g., cement-excelsior boards).

Wood Flour. Wood flour is used as an additive to reduce cost, modify properties, or replace some product. Specifically, wood flour is added as a filler (to plastics, linoleum, insulating bricks, and moldings), is used as an oil absorbent, in the production of dynamite (together with nitroglycerine and other chemicals), in making soap, and as a mild abrasive (for cleaning silverware, etc.). Wood flour is produced in three sizes of particles (the smallest are 0.25 mm in diameter).

Other Products. In addition to those previously mentioned, some other important mechanically derived products are: artificial limbs, baseball bats, baskets, blackboards, blinds (Venetian), bobbins, bowling pins, brush blocks, caskets, clothes pins, concrete forms, crutches, dowels, farm tools, fishing rods, garden tools, golf clubs, handles, ice-cream spoons, insulating pins, iron boards, ladders, novelties, oars, pallets, patterns and models, paving blocks, pencils (26), picture frames, playground structures (36, 46), rulers, scaffolding, scientific instruments, shingles, shoe lasts, sleighs, smoking pipes, spools, tanks, tongue depressors, toothpicks, toys (46), vats, and many others.

Products of Chemical Utilization

Products of chemical processing (2, 17, 18, 23) are those produced by chemical modification of wood. Such products are pulp and paper, products of cellulose and other chemical constituents of wood, products of pyrolysis, hydrolysis, gasification, and other products.

Pulp is a chemical product when produced by chemical or semichemical pulping, as previously mentioned (see Chapter 26). During pulping by chemical processes, the middle lamella, the greater part of which is lignin, is dissolved, and the fibers (cells) of the wood are separated. Semichemical pulping is performed by a combination of chemical and mechanical action, whereas in mechanical pulping the fibers are separated by mechanical action (grinding, "explosion")—that is, mechanical pulp is not a chemical product.

Pulp produced by chemical processing, so that all noncellulosic constituents are removed, is the raw material of many products, such as synthetic fibers, photographic films, varnishes, explosives, etc. Such pulp is cellulose (α-cellulose), produced either by acid sulfite pulping or prehydrolysis Kraft, and containing the cellulosic part of wood in the form of fibers, from the walls of which the other chemical constituents have been removed. During its utilization for making other products, cellulose is chemically modified (i.e., its molecules are subjected to chemical changes and dissolved), with the result that the fibers are morphologically destroyed.

Synthetic fibers, such as rayon, are produced from solutions of cellulose derivatives (cellulose nitrate, cellulose acetate, cellulose xanthate), which are solidified into continuous or noncontinuous filaments used for weaving fabrics. *Films* are made by a similar chemical processing, but the solution is passed through wide slits (to produce ribbons or sheets), instead of tiny holes as in the case of filaments. *Varnishes* are solutions of cellulose nitrate, acetate, or butylate. *Plastics* are also made from such solutions (by use of a suitable plasticizer and coloring additives), and they find many uses (e.g., eyeglass frames, toys, tools, appliances, etc.). Explosives are produced from cellulose nitrate.

Lignin is available in very large quantities in mills that produce pulp by chemical processes. In most (alkaline) pulp mills, lignin is used as a fuel (wastes are burned to recover pulping chemicals); however, lignin is also utilized in making various products, such as synthetic vanillin, pharmaceuticals, plastics, solvents, ceramics, binding material in road construction, in drilling oil wells, production of tannin substitutes and adhesives, synthetic rubber, etc. Yet only a small proportion of lignin is utilized; huge quantities are wasted, primarily because of its chemical structure.

Other products are made by *pyrolysis*, *gasification*, and *hydrolysis* of wood. *Pyrolysis* (under exclusion of air) includes *carbonization*, *destructive distillation*, and *liquification*.

Carbonization gives charcoal, which is used as fuel, as well as in the production of activated (highly absorptive) charcoal, dynamite, fireworks, pharmaceutical and chemical preparations, as a filtration material, and for other uses.

Destructive distillation is different from carbonization because the liquids and gases produced are not lost but are condensed and separated into useful products. In addition to charcoal, other products include tar, acetic acid, methanol, acetone, phenols, etc. The method is applied by heating wood (at 400–900°C or 750–1650°F) in large retorts made of stainless steel or ceramic material. The yield depends on the conditions of distillation (temperature, speed and duration of heating, application of vacuum), and kind of wood. Hardwoods are preferred because they give an increased yield of acetic acid (wood vinegar) and methanol (wood alcohol).

Liquification provides liquids and gases used as fuels (see below).

Gasification is the production of gases from wood; CO, CO_2, HN_4 (methane), and H are the main products. This process takes place at high temperatures, 1000°C (1800°F) or higher, in the presence of oxidizing agents (primarily air or pure oxygen) in quantities smaller than those required for the complete burning of wood. The gases produced are utilized for energy purposes, or making various organic compounds—primarily methanol, methane, ammonia, and related chemicals, such as ethylene, acetylene, propylene, benzol, and toluene.

Hydrolysis (*saccharification*) gives sugars and other products. Cellulose is hydrolyzed, and hemicelluloses more easily so, but lignin remains undissolved. Cellulose yields glucose by action of dilute sulfuric acid at a temperature of 170–190°C (340–375°F), or concentrated acid at room temperature. The product is used as animal feed or, by fermentation, gives ethanol (ethyl alcohol), which may be used in making various products, including plastics, such as polyethylene and polysterene, and ethanol. Other products of hydrolysis are yeast and glycerine. Hydrolysis of hemicelluloses gives ethanol, furfural, levulinic acid (the latter two are used in the plastics industry), and others. In contrast to hydrolysis of cellulose and hemicelluloses, hydrolysis of wood produces lignin residues.

Other products of chemical wood utilization are rosin and terpentine produced by steam distillation of pine stumps or as by-products of alkaline or Kraft pulping (12).

Wood as a Source of Energy

Wood is both an old and modern source of energy (14, 15, 28, 29, 35, 40). In addition to its use for heating and cooking, wood was used in ancient times for smelting metal ores, and in this manner many forests were destroyed in Europe and elsewhere. Today about half of the world production of wood is used as fuel.

The use of wood as a source of thermal energy is based on the production of heat when it burns. On average, wood produces 4500 kcal/kg in oven-dry condition, and about 15% less when air-dry (see Chapter 12).

Aside from fuelwood in round form, as it comes from the forest, residues from wood-using industries and workshops (sawmills, veneer plants, etc.) are also used as fuel. They are burned for heating, drying, production of steam, electrical energy, and other industrial needs. The burning of chemical residues (e.g., spent liquors of Kraft pulping) produces energy (steam), while at the same time pulping chemicals are reclaimed.

Residues are also burned after compression (densification) and shaping into round form, briquettes, or pellets (30). Briquettes are also made from carbonized residues, and they are utilized in barbecues and related uses. In such products, additives are included, such as wax to enhance burning, and other substances to give a pleasant odor and uniform color.

In times of energy crises, wood attains a greater importance as a source of energy, because it is a renewable material. Aside from fuelwood obtained from forests or industrial residues, wood is also produced in "energy plantations" from fast-growing species, which are harvested like agricultural crops and utilized as total biomass (wood, bark, foliage).

However, the needs of man for energy are great, and wood from forests (or plantations) cannot satisfy but a very small part[1]. Also, burning wood may be a waste of material, if such wood can be utilized in making more valuable products (particleboard, fiberboard, paper, etc.). Prerequisites for such utilization are sufficient quantities, as well as economical collection and transportation to factories. This is not always the case, and burning for steam production is often the best solution. For example, in sawmills or other wood-using factories, residues may contribute, in part or total, to the energy needed to run the machines[2].

Aside from direct *burning*, utilization of wood as a source of energy is possible by *pyrolysis* (carbonization, destructive distillation, liquification), *gasification*, and *hydrolysis*, as previously mentioned.

Carbonization gives charcoal. Charcoal is produced in traditional pits (the wood is piled and covered with earth), or in portable or permanent kilns made with bricks or metal. The heating value of charcoal is higher than that of wood, about 7000 kcal/kg (5500–8000 kcal/kg), but the yield of charcoal is low, about 20% of the weight of air-dry wood. In modern installations, the yield may reach 40%, and the gases produced are not lost in the atmosphere; they are used to heat the kiln, and this contributes to better utilization of thermal energy.

Destructive distillation produces charcoal and combustible (noncondensable) gases, mainly CO, H_2, CO_2, and hydrocarbons.

Liquification gives pyrolytic oil (liquid fuel), charcoal, and wood gas (producer gas). A yield of pyrolytic oil of 28–58% may be obtained, with a heating value of 5800–8900 kcal/kg.

Gasification gives wood gas, mainly CO and H_2, with a heating value of 1300–3000 kcal/kg. The gas is suitable for heating or as fuel for internal combustion engines (45). Such usage was common during the second World War. In Europe alone, more than one million vehicles and other transportation means were adapted to use producer gas from wood, charcoal, and crop residues. Current efforts are directed to apply gasification to residues of wood-using industries or harvesting residues in forests, instead of burning them for production of steam. Wood gas may also be utilized to produce methanol, which is used alone or in mixture with gasoline to run car and other engines.

Hydrolysis and subsequent fermentation yield ethanol (ethyl alcohol) which, like methanol, may be used alone or in mixture with gasoline (as "gasohol") to run cars; ethannol and methanol are cleaner burning fuels (cause less air pollution) than gasoline.

Wood as a source of energy is advantageous, because it is a renewable material. It is also important to note that processing wood in industries consumes less energy in comparison to other materials; 50 times more energy is required to produce an equal quantity of aluminum, 30 times more to produce plastics, and 10 times more to produce steel. Wood is also a heat-insulating material.

UTILIZATION OF FOLIAGE

Foliage (needles, leaves)[3] of forest trees constitutes a raw material with possibilities of utilization in nutrients (for animals and humans),[4] pharmaceuticals, aromas, and other products. Uses of foliage range from their direct consumption by grazing to preparations made by chemical processing. The latter include feeds containing vitamins, etherial oils, products of chlorophyll, and other products (4-6, 10, 11, 13, 22, 25)

Utilization of foliage of coniferous and broad-leaved species (spruce, birch, etc.) to produce vitamin-containing feeds has been developed mainly in the Soviet Union. The product is called "Muka"; trials with various animals (cattle, poultry, sheep, goats, pigs, rabbits) have shown that it contributes to better growth, higher productivity, better health, and better reproduction.[5] Such favorable effects are attributed to vitamins, carotine, trace elements, and other biologically active ingredients contained in small quantities. This product may replace an equal weight of alfalfa (up to 20–30%). In 1976, about 300 plants making "Muka" existed in the Soviet Union.

Etherial oils are produced by extraction and used in pharmaceuticals and aromas. An example is eucalyptol, a monoterpene from eucalypts, used in various cold medicines, such as cough drops and inhalators.

Other products are: a chlorophyll-carotine preparation used in tooth pastes, soaps, ointments, medicinals for treatment of burns and wounds, and for improvement of health and productivity of animals, etc. A product of chlorophyll (containing potassium) is used in pharmaceuticals and aromas; and a condensate of provitamin as animal (cattle, poultry, etc.) feed has been reported to improve their vitality and growth.

Foliage (needles mixed with wood and bark in the form of logging residues) has been tried for production of particleboard without satisfactory results, but considerable improvement is possible by mixing regular wood particles. Experimentally, foliage has also been used in making fiberboard, a packaging material (similar to excelsior) and briquettes, and as additive to thermosetting resins (after treatment). Coniferous foliage has been shown to have potential as an additive in adhesives (filler, extender) (10, 13).

Aside from technological problems, utilization of foliage depends on the economics of collection and processing of required quantitites; the adverse effect on site productivity for future tree crops should also be taken into consideration.

UTILIZATION OF RESIN AND OTHER EXUDATES AND EXTRACTIVES

Resin is produced by tapping living trees, or from the distillation of wood, or as a by-product of alkaline (Kraft) pulping. Sources of resin are mainly various species of pine: *Pinus halepensis*, *P. brutia*, *P. maritima*, *P. pinea* (Europe), *P. elliottii*, *P. palustris* (U.S.A.), *P. massoniana*, *P. yunnanensis* (China), *P. roxburghii* (India, Pakistan), and *P. merkusii* (Indonesia, Philippines).

Distillation of resin obtained from living trees (oleoresin) gives rosin (colophony) and turpentine. These are also produced by distillation of stumps and as by-products of pulping.[6] Rosin has wide industrial applications; it is used raw or after modification to improve its properties. It is available as solid in various grades, and its uses include: sizing in papermaking (to control water absorptivity of paper), production of ink, linoleum, electrical appliances and insulators, as additive in lubricants, in soaps, matches, building materials, adhesives, production of synthetic rubber, shoe polishes, chewing gums, cleaning metals, etc. Turpentine is a liquid, insoluble in water but soluble in organic solvents (alcohol, ether, benzol, etc.). It dissolves easily inorganic (sulfur, phosphorus, etc.) and organic materials (rubber, resins, varnishes, paints, etc.). It is used as a solvent in the preparation of paints, varnishes, and synthetic camphor, in making aromas and pharmaceuticals, in insecticides, for adulteration of etherial oils, and preparation of lubricants, among others.

Aside from pines, resin is produced in small quantitites, from spruce and Douglas-fir, and also from the bark of fir. Such resins find application in pharmaceuticals and aromas.

Other *exudates* (excretions) include maple syrup, rubber, styrax (storax), and mastick. Respective sources are the following species: *Acer saccharum* and *A. nigrum*, *Hevea braziliensis*, *Liquidambar orientalis*, and *Pistacia lenstiscus* var. *chia* or *latifolia*.

Extractives are organic substances (tannins, coloring materials, etc.), which are contained in trees, shrubs, or lower plants (in wood, bark, foliage, and fruits), and may be removed (extracted) by use of various solvents (water, alcohol, acetone, ether, etc.). Extractives influence the utilization of wood (they affect color, resistance to fungi and insects, flammability, heating values, and technical properties), but some extractives are themselves important forest products.

Main extractives of practical importance are tannins and coloring materials. (Resin is also an extractive, when produced by extraction from wood and not by tapping living trees.) Extractives are also certain etherial oils produced from fruits and foliage, and sometimes from bark and wood.

Tannins are phenolic substances that are important mainly in leather-making. They react with "collagen" (a protein contained in the skin of animals), and produce an insoluble substance which gives to skin desirable leather properties. Tannins are also used in the preparation of ink, and form adhesives, when reacted with formaldehyde. They are contained in considerable quantities in the bark of certain species (oak, pine, spruce, larch, eucalypt, acacia,[7] etc.), in wood (chestnut, *Schinopsis lorentzii*, etc.), foliage (*Rhus coriaria*), and fruits (*Quercus aegilops*, *Q. infectoria*, *Q. cerris*). Also, tannins (and coloring materials) are contained in outgrowths (galls), appearing on leaves or fruits of certain species (e.g., *Q. coccifera*) and caused by insects. Coloring materials are also produced from bark, wood, foliage, roots, and flowers. Such substances, known since ancient times, have been widely used mainly for dyeing textiles; now, such use is rare, due to replacement by synthetic dyes. Coloring materials are contained in considerable quantitites in species of *Quercus*, *Rhamnus*, *Cotinus*, *Haematoxylon*, and others.

OTHER FOREST PRODUCTS

Other forest products are Christmas trees (27), basket-weaving (willow) sticks, fruits, briarwood (an outgrowth of *Erica arborea*) (1, 41), ornamental, pharmaceutical, and aromatic plants (19), organic soil (humus), etc. Lignites (and perhaps petroleum) may also be considered forest products (deriving from forests of old geological periods). In addition, forests contribute to animal production by providing food to wildlife and grazing livestock, production of honey, etc.

REFERENCES

1. Alexandrian, D. 1981. Une étude sur la bruyère arborescente pour la fabrication des pipes. *Rev. For. Fr.* 111 (2):128–132.
2. Anonymous. 1980. Valorisation chimique du bois. *Rev. Bois Appl.* 12:18–21.
3. Bariska, M., and C. Schuerch. 1977. Wood softening and forming with ammonia. In *"Wood Technology: Chemical Aspects*, ed. I.S. Goldstein, pp. 326–347. Washington, D.C.: Am. Chem. Soc.
4. Barton, G.M. 1976. Foliage. Part II. Foliage chemicals, their properties and uses, in *Applied Polymer Symposium*, No. 28 pp. 465–484. New York: John Wiley & Sons.
5. Barton, G.M., and B.F. MacDonald. 1977. A new look at foliage chemicals. W. For. Prod. Lab. Canada (reprint), pp. 47–52.
6. Barton, G.M., J.A. McIntosh, and S. Chow. 1978. The present status of foliage utilization. Amer. Inst. Chem. Eng. (AOChE) Symposium Series No. 177, 74:124–131.
7. Capron, J.H. 1963. *Timber Laminating*. Bloomington, Illinois: McKnight Publ. Co.
8. Centre Technique du Bois. 1973. *Le Supports en Bois pour Voies Ferrées*. Paris.
9. Champion, F.J. 1961. Products of American Forests. U.S.D.A. Misc. Publ. 861.
10. Chauhan, B.R.S. 1981. Pine needles as filler and extender for phenol-formaldehyde resin in plywood making. *Holzforsch. Holzverwert.* 33(6):114–118.
11. Chawia, J.S. 1981. Industrial products from pine needles. *Holzforsch. Holzverwert.* 33(6):114–118.
12. Chiang, T.I., W.H. Burrows, W.C. Howard, and G.D. Woodard. 1971. *A Study of the Problems and Potentials of the Gum Naval Stores Industry*. Atlanta: Georgia Institute of Technology.
13. Chow, S., P.R. Steiner, and L. Rozon. 1979. Efficiency of coniferous foliage as extenders for powdered phenolic resin. Intern. Symp. Particleboard Proc., pp. 329–342. Washington.
14. Forests Products Research Society. 1976. Wood Residue as an Energy Source. FPRS Energy Workshop Proc. No. P-75-13.
15. Forest Products Research Society. 1976. Energy and the Wood Products Industry. Proceedings No. P-76-14.
16. Giordano, G. 1976. *Tecnologia del Legno*. Vol. 3. Torino: Unione Tipografico-Editrice.
17. Goldstein, I.S. 1981. Chemicals from biomass: Present status. *For. Prod. J.* 31(10):63–68.
18. Hajny, G.J. 1981. Biological Utilization of Wood for Production of Chemicals and Foodstuffs. U.S. For. Prod. Lab. Res. Paper FPL 385.
19. Hill, A. 1937. *Economic Botany*. New York: McGraw-Hill.
20. Hoadley, R.B. 1980. Understanding Wood: A Craftsman's Guide to Wood Technology. Newton, Connecticut: The Tauton Press.
21. James, G.T. 1958. *Woodturning*. London: J. Murray.
22. Keays, J.L. 1976. Foliage. Part I. Practical utilization of foliage. In *Applied Polymer Symposium*, No. 28, pp. 445–464. New York: Wiley.
23. McIntosh, J.A., and D.M. Wright. 1977. Are the forests new protein source? B.C. Lumberman (reprint, p. 3).
24. Murphey, W.K., and R.N. Jorgensen. 1974. Wood as an Industrial Arts Material. New York, Oxford: Pergamon Press.
25. Nielson, R.W. 1977. Wood and foliage for animal fodder. Forest Utilization Symposium Proc., pp. 43–49. Forintek Canada Corp.
26. Olson, J.R., and F.C. Fackler. 1986. Utilization of hardwoods for pencil manufacture. *For. Prod. J.* 36(3): 44–50.
27. Panshin, A.J., E.S. Harrar, J.S. Bethel, and W.J. Baker. 1962. *Forest Products*. New York: McGraw-Hill.
28. Philippou, J.L. 1981. Conversion of biomass into energy and chemicals by thermolytic methods. In *Components of Productivity of Mediterranean-Climate Regions—Basic and Applied*, ed. N.S. Margaris and H.A. Mooney, pp. 243–255. The Hague: Junk Publ.
29. Resch, H. 1979. Energy recovery from wood residues. *Holzforsch. Holzverwert.* 31(4):79–82; and 1982. Densified wood and bark fuels. *Holzforsch. Holzverwert.* 34(4): 69–74.
30. Richter, H. G. 1988. *Holz als Rohstoff für den Musikinstrumentenbau*. Celle: Moeck Verlag.
31. Scharf, R. 1958. *Complete Book of Wood Finishing*. London: Faber &Faber.

32. Skinner, F. 1961. *Wood Carving*. New York: Bonanza Books.
33. Spencer, A.G., and J.A. Luy. 1975. *Wood and Wood Products*. Columbus, Ohio: Merrill Publ.
34. Stevens, W.C., and N. Turner. 1970. *Wood Bending Handbook*. London: H.M.S.O.
35. Tillman, D.A. 1978. *Wood as an Energy Resource*. New York, London: Academic Press.
36. Tsoumis, G. 1974. Wooden structures in playgrounds. *Dassica Chronica* 181/183:8–10 (Greek).
37. Tsoumis, G. 1979. Wood as a material for furniture and other uses. *To Dasos* 85:8–11 (Greek).
38. Tsoumis. G. 1983. Forest biomass utilization in Greece. In *Forest Biomass*, ed. W.A. Côté, pp. 109–115. New York, London: Plenum Press.
39. Tsoumis, G. 1987. Investigation of the utilization potentials of the wood of Mediterranean shrubs and coppice forests. Proceedings EEC Seminar on "Wood Technology," pp. 16–23, Munich.
40. Tsoumis, G., and J. Philippou. 1982. Wood as a renewable source of energy. Proceedings Conf. "Mild Forms of Energy," Vol. B, pp. B10-31-40, Thessaloniki (Greek).
41. Tsoumis, G., N. Kezos, J. Fanariotou, E. Voulgaridis, and C. Passialis. 1988. Characteristics of briarwood. *Holzforschung* 42(2):71–77.
42. U.S. Forest Products Laboratory. 1961. *Wood Floors for Dwellings*. Ag. Handbook No. 204.
43. Weichun, Z., et al. 1981. Estimation and utilization of nutritive ingredients of the needles of some conifer species. *Chemistry and Industry of Forest Products* 1(4):31–39 (People's Republic of China).
44. Weichun, Z., and Y. Quinrong. 1989. Preparation and application of pine chlorophyll-carotene paste microcapsules as complex feed additives for minks. *Chemistry and Industry of Forest Products* 9(2):56–64, (People's Republic of China).
45. Wendorff, G.B. 1980. Moteurs à gaz de bois. *Rev. Bois Appl.* 12:25–28.
46. Wodarz, S. (undated). *Wie man Spielgeräte aus Holz baut*. München: BLV Verlag.

FOOTNOTES

[1] In the United States, the contribution of wood to total energy requirements was estimated at about 3%; this is largely in the form of mill residues (sawmill, etc.), including pulping residues. The wood-using industry meets at least 50% of its energy needs by burning residues.

[2] Widely practiced burning (in fireplaces, stoves, etc.) may pollute the environment due to smoke. However, wood produces very low (acceptable) emission of SO_2 (sulfur dioxide) and nitrogen oxides compared to other fuels (coal, fuel oil, etc.). Smoke may be eliminated by use of catalytic afterburners.

[3] The term "technical foliage" includes needle- or leaf-bearing twigs and shoots up to 0.25 in. (6 mm) in diameter.

[4] During the second World War, a vitamin-containing drink was prepared for the inhabitants of Leningrad from pine needles. At that time, 10% of wood sawdust was added to bread.

[5] The addition of chemical derivatives from pine needles to animal feeds, in a proportion of essential substance of 0.2–0.5% by weight, was reported to increase the weight of poultry (8–17%) and egg production (9–19%), and to decrease diseases and consumption of feeds (8–28%). The addition of 2.5–4.5% needle flour to feeds of pigs increased their weight by 15–30% (43). A study with minks, also showed an increase of body weight, 7.7–16.1% (more in females), better quality fur, and more resistance to diseases (44).

[6] Resin obtained from living trees is called *oleoresin* or *tree-gum resin*, and from distillation of resinous wood *wood resin*. Components of resin (rosin, turpentine) are obtained as by-products of alkaline (Kraft) pulping. Resin is also termed *naval stores*, because it was used to preserve wood in shipbuilding; resin produced by tapping living trees is also called *gum naval stores* and from wood (distillation or pulping) *wood naval stores* (12, 27).

[7] Wattle tannin is the most important commercial tannin and comes from the bark of certain *Acacia* species.

11

Temperate and Tropical Woods

This chapter deals with the identification of European, North American, and tropical woods, and presents some general information on provenance (geographical source), properties, and uses.

IDENTIFICATION

Identification of a sample of wood—with regard to the genus or species of the tree that produced it, presents not only a plant taxonomic interest. It is also a helpful tool in commerce of timber and wood products, and sometimes in art, archaeology, and criminology (63, 69, 74).

Identification of wood is comparatively more difficult than identification of the tree that produced it. Wood is the most conservative component of a tree. It should be remembered that dendrological classification of trees into separate taxonomic groups is based on differences in leaves, flowers, fruits, and other morphological characteristics (76). Wood does not provide analogous differences in its structure. In several cases, only genera or groups of species can be distinguished macroscopically or microscopically. Even the application of electron microscopy to wood anatomy has not provided significant features that could be used for recognition of certain species.

There are two common procedures for wood identification—by *keys* and by *cards*.[1] A key is used when dealing with a relatively small number of species—for example, the species of a certain geographical region or country. All keys are dichotomous (a Greek word meaning dividing into two). The basic procedure in constructing such a key is to make an opening statement (e.g., softwoods with resin canals) which, further down in the key, is contrasted by an opposite statement (i.e., softwoods without resin canals). Thus, in this example, all softwoods are divided into two groups. Within each group, other macroscopic and/or microscopic features are dealt with in the same manner, until progressively one arrives at features that are characteristic of a single genus or species.

When a large number of species is involved, anatomical features are recorded on cards, which include information on botanical families and geographical origin. Each species is provided with a separate, perforated card, in which the perforations opposite the features that are characteristic of this species are clipped off. Thus, on inserting a needle into the pack of cards through any perforation, cards with clipped-off perforations will fall out. This process is repeated until (theoretically) a last card remains; this should identify the sample in question. This idea has been expanded to mechanically punched cards which, in addition to wood structure, include information on wood properties and uses, dendrological characteristics of the parent tree, and so on. Such cards are sorted electronically (82).

In some specialized laboratories, data of wood characteristics are stored in a computer and utilized for identification. The procedure is useful especially in cases of many and difficult hardwood samples, but there is a problem of gathering and preparing the data in a standard format for computer handling (46, 84). So far, key identification remains the common method.

This chapter includes keys for North American, European, and tropical species. These keys may be called descriptive dichotomous because, in addition to their usefulness for identification, they provide some continuity of description of the genera and species involved. In other types of dichotomous keys, contrasting statements are often far apart, and it is not easy to reconstruct the macroscopic and microscopic features that are characteristic of a given genus or species.

Although existing literature was helpful, especially for North American woods, these keys are largely based on examination of material from the wood collections in the School of Forestry at Yale University (S. J. Record Memorial Collection, now at the U.S. Forest Products Laboratory in Madison, Wisconsin), the School of Forestry at the Pennsylvania State University, the Smithsonian Institution in Washington, D.C., the Laboratory of Forest Utilization at Aristotelian University in Thessaloniki, and the Institute of Forest Research in Athens, Greece.

Common keys could be justified for many North American and European species, especially hardwoods. However, all species are treated separately, according to geographical provenance, for the purpose of facilitating instruction.

Macroscopic keys are provided for all species examined. Separate microscopic keys are given for softwoods, but for hardwoods only a few microscopic characteristics are included in the macroscopic keys. Most hardwoods may be identified without using the microscope, because they possess a much greater variation of macroscopic characteristics.

There are no perfect tools for wood identification, and this applies especially to identification of softwoods. For best possible results, a "seeing eye" able to recognize often slight differences is a necessary precondition in all cases, including these keys. Most characteristics are described in a comparative manner, and certain descriptions are subjective. For example, there are not enough words to describe the color of a piece of wood. The task is complicated by possible changes of color due to exposure or treatment (see Chapter 2), and also due to natural variation of color within a species. Usually, sapwood is of little help; it has about the same light color in several species. Therefore, the color of heartwood is preferred for identification purposes. Odor is another feature that cannot be easily described. This is more pronounced on freshly exposed surfaces, because it is due to volatile substances which tend to evaporate with time.

The keys apply to typical samples. Wood obtained from young trees, seedlings, and near the pith (juvenile wood), from the outer growth rings of very old trees, and from branches and roots will show considerable differences from the characteristics described. Also, variability within a species may present difficulties in identification, especially in softwoods. Different samples from the same species, and sometimes adjacent growth rings may differ in structure, such as in distribution of resin canals, relative width and transition from earlywood to latewood, dentation of ray tracheids, tyloses, and so on. Differences are often induced by rate of growth (ring width). In oaks and other ring-porous hardwoods, the structure of growth rings may vary markedly with rate of growth; very narrow rings have little latewood and appear diffuse-porous. Presence of reaction wood, especially mild compression wood, may be misleading.

To facilitate comparison, data on specific gravity are included in the keys. The values given are average, and they are based on oven-dry weight and air-dry volume (at 12% moisture content—see Chapter 8). Because the purpose is only indicative, all values were rounded. It should be remembered, that different samples of the same species and of the same tree may differ in specific gravity; for this reason, the values given by different literature sources differ, sometimes quite considerably. In most cases, this is also the basic reason why identical genera or species may be shown to have a different specific gravity in the two sets of keys.

This chapter is illustrated with photographs of cross-sections at low magnifications (approximately $5\times$ to $10\times$). Such photographs are not provided for all species. The purpose is to demonstrate characteristic features rather than individual species. This approach is considered sufficient to support both sets of keys without undue increase of the number of illustrations. It is also realistic, in that individual photographs are seldom helpful in demonstrating actual differences within a genus, and sometimes between genera. For example, all softwoods without resin canals appear in photographs about the same. It should also be noted, as mentioned elsewhere, that many species belonging to the same genus are similar in their basic structure, irrespective of their growing in different geographical regions.

Several of the species included in the keys (e.g., yew, osage-orange, honeylocust, catalpa, persimmon, hazel, carobtree, and others) have little or no commercial value as sources of wood. They are included for instructional purposes, being often subjects of comparative exercise by students of wood structure.

For accurate observation of macroscopic characteristics, with the naked eye or a hand lens, it is indispensable to have a clean-cut surface (as a rule, a cross-section), prepared with a sharp knife. Moistening the wood may facilitate observation in some species, while in others splitting may provide distinctive characteristics. For microscopic work, it is necessary to have properly prepared samples and to make proper use of the microscope. Identification may be based on very small samples—including sawdust (73, 77).

Macroscopic Key to North American Softwoods with Resin Canals

A Vertical resin canals always present.
    a Vertical resin canals comparatively large (Figure II-1), mostly numerous, solitary, or in small tangential groups of 2 to 3, visible to naked eye as small dots in cross-section and as fine lines on longitudinal surfaces. Heartwood colored with more or less pronounced resinous odor.
    Pine—*Pinus*
    b Vertical resin canals medium-sized to small (Figure II-2), invisible without lens; relatively few, solitary or in small tangential groups of 2 to 5. Heartwood colored or indistinct.
        a_1 Wood moderately hard and heavy. Transition from earlywood to latewood generally abrupt. Heartwood colored.
            a_2 Heartwood light reddish-brown when freshly cut, becoming upon exposure reddish to reddish-brown. Sapwood whitish to yellowish or reddish-white. Freshly cut surfaces have char-

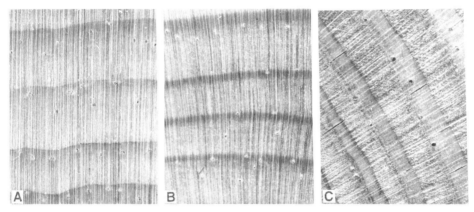

Figure II-1. Softwoods with resin canals. Canals relatively large, mostly numerous, solitary or in small tangential groups of 2 to 3. (A) Transition from earlywood to latewood gradual (White pine—*Pinus strobus*). (B) Transition from earlywood to latewood fairly abrupt to abrupt (Scots pine—*Pinus silvestris*). (C) Transition from earlywood to latewood very abrupt, contrast striking (American southern pines—*Pinus palustris*, etc.).

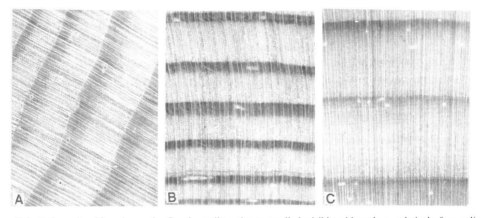

Figure II-2. Softwoods with resin canals. Canals medium size to small, invisible without lens, relatively few, solitary or in small tangential groups of 2 to 5. (A) Transition from earlywood to latewood abrupt (Douglas-fir—*Pseudotsuga menziesii*). (B) Transition from earlywood to latewood abrupt (Larch—*Larix*). (C) Transition from earlywood to latewood relatively gradual (Spruce—*Picea*). (Photograph B Courtesy of H. Sachsse.)

acteristic odor. Resin canals rather numerous, mostly in small tangential groups. (Sp.gr. 0.45).

Douglas-fir—*Pseudotsuga menziesii* (Mirb.) Franco (Figure II-2A)

b_2 Heartwood tending to a brown rather than red tone of color. Sapwood whitish. Wood surfaces have oily feel. Resin canals relatively few, mostly solitary or in small tangential groups. (Sp.gr. 0.55.)

Larch—*Larix* (Figure II-2B)

Tamarack (Eastern larch)—*Larix laricina* (Du Roi) K. Koch.

Western larch—*Larix occidentalis* Nutt.

b_1 Wood comparatively soft and light (sp. gr. 0.40). Transition from earlywood to latewood generally gradual. Heartwood not colored or little different from sapwood. Resin canals mostly solitary or in small tangential groups (Figure II-2C).

Spruce—*Picea*

a_2 Heartwood not distinct; wood nearly white or pale yellowish brown, lustrous, relatively fine textured. Resin canals frequently difficult to find with naked eye.

White spruce—*Picea glauca* (Moench) Voss

Red spruce—*Picea rubens* Sarg.

Black spruce—*Picea mariana* (Mill.) B.S.P.

Engelmann spruce—*Picea engelmannii* Parry

b_2 Heartwood light pinkish-yellow to pale brown. Sapwood whitish to yellowish. Wood less lustrous, medium textured. Resin canals more distinct.

Sitka spruce—*Picea sitchensis* (Bong.) Carr.

Macroscopic Key to North American Pines

A Wood comparatively soft and light (sp. gr. 0.35). Transition from earlywood to latewood gradual (Figure II-1A). Texture uniform, fine to coarse.

Soft pines

a Heartwood pale brown to reddish-brown, darkening upon exposure. Sapwood white to pale yellowish-white. Texture relatively fine. Resin canals large, distinct to naked eye. Resinous odor mild.

White pines

Eastern white pine—*Pinus strobus* L. (Figure II-1A).

Western white pine—*Pinus monticola* Dougl.

b Heartwood yellowish to pale brown, darkening upon exposure to light or reddish-brown. Sapwood white to yellow-white. Texture relatively coarse. Resin canals large, conspicuous to naked eye. Resinous odor pronounced. Fresh lumber exudes a sugary substance.

Sugar pine—*Pinus lambertiana* Dougl.

B Wood moderately soft and light to hard and heavy. Transition from earlywood to latewood more or less abrupt to quite abrupt. Texture medium to coarse.

Hard pines

a Wood comparatively hard and heavy (sp. gr. 0.55). Transition from earlywood to latewood abrupt to quite abrupt. Latewood dense, thorny. Heartwood yellowish-brown to golden- or reddish-brown, with strong resinous odor when freshly cut. Sapwood whitish to yellowish. Resin canals large, distinct to naked eye.

Southern pines (Figure II-1C).

Longleaf pine—*Pinus palustris* Mill.

Shortleaf pine—*Pinus echinata* Mill.

Loblolly pine—*Pinus taeda* L.

Slash pine—*Pinus elliottii* Engelm., etc.

Pitch pine—*Pinus rigida* Mill.

b Wood moderately soft and light (sp. gr. 0.40). Transition from earlywood to latewood more or less abrupt to abrupt. Latewood not thorny. Heartwood varies in color. Sapwood whitish to yellowish.

a_1 Resin canals small, mostly solitary, relatively inconspicuous to naked eye.

a_2 Heartwood light brown to yellowish-brown. Sapwood narrow. Latewood not pronounced.

Transition from earlywood to latewood more or less abrupt. Split tangential surfaces frequently dimpled (with numerous slight dentations).
Lodgepole pine—*Pinus contorta* Dougl.
- b_2 Heartwood light brown to light golden-brown. Sapwood wide. Latewood pronounced (dense). Transition from earlywood to latewood abrupt. Split tangential surfaces not dimpled.
Jack pine—*Pinus banksiana* Lamb.
- b_1 Resin canals relatively large, solitary, or in groups of 2 or 3, distinct to naked eye.
  - a_2 Heartwood light reddish, yellowish-brown, or golden-brown. Sapwood wide. Transition from earlywood to latewood abrupt. Split tangential surfaces occasionally dimpled.
  Ponderosa pine—*Pinus ponderosa* Laws.
  - b_2 Heartwood reddish-brown, or golden-brown. Sapwood narrow. Transition from earlywood to latewood fairly abrupt. Split tangential surfaces not dimpled.
  Red pine—*Pinus resinosa* Ait.

Macroscopic Key to American Softwoods Without Resin Canals

A Wood without aromatic odor.
- a Without colored heartwood. Axial parenchyma not visible.
  - a_1 Color whitish to light brown. Earlywood whitish, latewood brownish. Transition from earlywood to latewood gradual, sometimes tending to abrupt (sp. gr. 0.40).
  Fir—*Abies* (Figure II-3A).
    Eastern balsam firs
      Balsam fir—*Abies balsamea* (L.) Mill.
      Fraser fir—*Abies fraseri* (Pursh) Poir.
    Western firs
      Pacific silver fir—*Abies amabilis* (Dougl.) Forbes
      White fir—*Abies concolor* (Gord. & Glend.) Lindl.
      Grand fir—*Abies grandis* (Dougl.) Lindl.
      Subalpine fir—*Abies lasiocarpa* (Hook.) Nutt.
      California red fir—*Abies magnifica* A. Murr.
      Noble fir—*Abies procera* Rehd.
  - b_1 Color light brown with a slight reddish tinge in earlywood and latewood. Transition from earlywood to latewood abrupt, sometimes tending to gradual (sp. gr. 0.40).

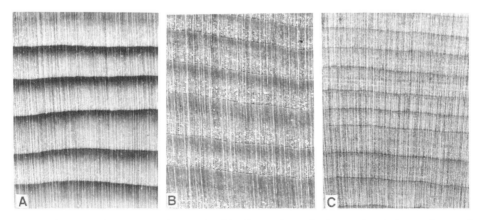

Figure II-3. Softwoods without resin canals. (A) Transition from earlywood to latewood gradual, sometimes tending to abrupt (Fir—*Abies*). (B) Transition from earlywood to latewood abrupt, sometimes tending to gradual (Hemlock—*Tsuga*). (C) Transition from earlywood to latewood gradual, dark-colored spots (axial parenchyma) may be seen without lens, false rings may be present (Juniper, Eastern redcedar—*Juniperus*).

        Hemlock—*Tsuga* (Figure II-3B).
           Eastern (Canada) hemlock—*Tsuga canadensis* (L.) Carr.
           Western (Pacific) hemlock—*Tsuga heterophylla* (Raf.) Sarg.
           Mountain hemlock—*Tsuga mertensiana* (Bong.) Carr.
- b With colored heartwood. Sapwood pale white. Dark-colored axial parenchyma visible with lens and often to naked eye, especially in sapwood.
  - a_1 Heartwood light cherry-red to dark brownish-red. Wood moderately soft and light (sp. gr. 0.40).
    Redwood—*Sequoia sempervirens* (D. Don) Engl.
  - b_1 Heartwood varies in color, light to dark brown. Wood moderately hard and heavy (sp. gr. 0.45). Longitudinal surfaces have waxy feel, especially in dark specimens. Odor somewhat unpleasant.
    Baldcypress—*Taxodium distichum* (L.) Rich.
- B Wood with aromatic odor.
  - a Heartwood dark-colored, sapwood whitish.
    - a_1 Wood moderately hard and heavy (sp. gr. 0.45). Heartwood bright red or dull red. Odor and taste mild, not spicy.
      Eastern red cedar—*Juniperus virginiana* L. (Figure II-3C).
    - b_1 Wood soft to moderately soft, light.
      - a_2 Heartwood pale reddish-brown, sometimes with a red tinge. Odor characteristic of most pencils, taste spicy. Wood soft and light (sp. gr. 0.35).
        Incense cedar (Pencil cedar)—*Libocedrus decurrens* Torr.
      - b_2 Heartwood reddish-brown. Odor fragrant, taste faintly bitter. Wood moderately soft, light (sp. gr. 0.35).
        Western red cedar (Giant arborvitae)—*Thuja plicata* Donn.
  - b Heartwood relatively light-colored.
    - a_1 Heartwood yellow or yellowish-brown. Sapwood whitish to yellowish-white. Taste somewhat bitter. Wood relatively hard and heavy (sp. gr. 0.45).
      - a_2 Heartwood yellow. Odor mild but unpleasant.
        Alaska cedar—*Chamaecyparis nootkatensis* (D. Don) Spach
      - b_2 Heartwood light yellow to yellowish-brown. Pleasant spicy odor.
        Port-orford cedar—*Chamaecyparis lawsoniana* (A. Murr.) Parl.
    - b_1 Heartwood light brown or pinkish-brown. Sapwood whitish. Wood relatively soft and light (sp. gr. 0.30).
      - a_2 Heartwood light brown. Odor mild.
        Northern white cedar (Eastern Arborvitae)—*Thuja occidentalis* L.
      - b_2 Heartwood pinkish-brown. Wood somewhat oily. Odor more pronounced.
        Atlantic white cedar—*Chamaecyparis thyoides* (L.) B.S.P.

Microscopic Key to North American Softwoods

- A Vertical resin canals always present. Horizontal canals small, present in fusiform rays.
  - a Cross-field pits window-like or pinoid. Resin canals comparatively large (70–200 µm in diameter), mostly numerous, solitary or in groups of 2 or 3. Epithelial cells thin-walled. Ray tracheids comparatively large and numerous, occurring in one to several marginal rows, frequently interspersed in high rays and often composing entire low rays.
    Pine—*Pinus*
  - b Cross-field pits piceoid. Resin canals medium-sized to small (50–90 µm in diameter), comparatively few, solitary or in small tangential groups of 2 to 5. Epithelial cells thick-walled. Ray tracheids small and few, usually in single marginal rows.
    - a_1 Tracheids with spiral thickenings, seen in longitudinal sections. Resin canals mostly in small tangential groups.
      Douglas-fir—*Pseudotsuga menziesii* (Mirb.) Franco
    - b_1 Tracheids without spiral thickenings, or only exceptionally and sporadically. Resin canals mostly solitary.
      - a_2 Transition from earlywood to latewood abrupt. Bordered pits on radial walls of vertical

tracheids often in pairs (biseriate). Ray tracheids with smooth inner walls, only sporadically with small dentations.
Larch—*Larix*
  Tamarack (Eastern larch)—*Larix laricina* (Du Roi) K. Koch
  Western larch—*Larix occidentalis* Nutt.

b_2 Transition from earlywood to latewood gradual. Bordered pits in a single row, very seldom in pairs. Ray tracheids slightly dentate.
Spruce—*Picea*
  White spruce—*Picea glauca* (Moench) Voss
  Red spruce—*Picea rubens* Sarg.
  Black spruce—*Picea mariana* (Mill.) B.S.P.
  Engelmann spruce—*Picea englemannii* Parry
  Sitka spruce—*Picea sitchensis* (Bong.) Carr.

B Vertical and horizontal resin canals absent; traumatic vertical canals may be present.
  a Vertical tracheids with spiral thickenings. Vertical parenchyma absent. Cross-field pits cupressoid.
    Pacific yew—*Taxus brevifolia* Nutt.
  b Vertical tracheids without spiral thickenings.
    a_1 Vertical parenchyma absent or very sparse terminal.
      a_2 Ray tracheids absent from most rays or very sparse. Cross-field pits taxodioid. Rays high, sometimes over 30 cells high.
        Fir—*Abies*
          Eastern balsam firs
            Balsam fir—*Abies balsamea* (L.) Mill.
            Fraser fir—*Abies fraseri* (Pursh) Poir.
          Western firs
            Pacific silver fir—*Abies amabilis* (Dougl.) Forbes
            White fir—*Abies concolor* (Gord. & Glend.) Lindl.
            Grand fir—*Abies grandis* (Dougl.) Lindl.
            Subalpine fir—*Abies lasiocarpa* (Hook.) Nutt.
            California red fir—*Abies magnifica* A. Murr.
            Noble fir—*Abies procera* Rehd.
      b_2 Ray tracheids present, usually marginal. Cross-field pits cupressoid. Rays low, as a rule not over 15 cells high.
        Hemlock—*Tsuga*
          Eastern (Canada) hemlock—*Tsuga canadensis* (L.) Carr.
          Western (Pacific) hemlock—*Tsuga heterophylla* (Raf.) Sarg.
          Mountain hemlock—*Tsuga mertensiana* (Bong.) Carr.
    b_1 Vertical parenchyma present to abundant.
      a_2 End walls of ray parenchyma cells nodular. Cross-field pits cupressoid.
        a_3 Indentures (at the corners of ray parenchyma cells) present.
          a_4 Rays seldom over 6 cells high. Vertical parenchyma abundant, in bands.
            Eastern red cedar—*Juniperus virginiana* L.
          b_4 Rays up to 20 cells high. Vertical parenchyma sparse or absent in some rings, abundant in others.
            Alaska cedar—*Chamaecyparis nootkatensis* (D. Don) Spach
        b_3 Indentures absent. Rays up to 15 cells high. Vertical parenchyma diffuse.
          Incense cedar—*Libocedrus decurrens* Torr.
      b_2 End walls of ray parenchyma cells smooth.
        a_3 Cross-field pits taxodioid (except in *Taxodium* where cupressoid pits are also present).
          a_4 Indentures present. Pits taxodioid. Vertical parenchyma shows great variability; it may be abundant in some rings, but altogether absent in others.
            Western red cedar (Giant arborvitae)—*Thuja plicata* Donn
            Northern white cedar (Eastern arborvitae)—*Thuja occidentalis* L.
          b_4 Indentures absent. Vertical parenchyma diffuse. Vertical tracheids with 2 to 4 vertical rows of opposite bordered pits.

        a_5 Transverse walls of vertical parenchyma nodular. Pits taxodioid or cupressoid. Rays more than 15 (up to 60) cells high.
            Baldcypress—*Taxodium distichum* (L.) Rich.
        b_5 Transverse walls of vertical parenchyma smooth. Pits taxodioid. Rays 10 to 15 (up to 40) cells high.
            Redwood—*Sequoia sempervirens* (D. Don) Endl.
    b_3 Cross-field pits cupressoid.
        a_4 Transverse walls of vertical parenchyma nodular. Vertical parenchyma diffuse, sometimes tending to banded. Rays 1 to 6 cells high.
            Port-orford cedar—*Chamaecyparis lawsoniana* (A. Murr.) Parl.
        b_4 Transverse walls of vertical parenchyma smooth. Vertical parenchyma diffuse. Rays 1 to 12 cells high.
            Atlantic white cedar—*Chamaecyparis thyoides* (L.) B.S.P.

Microscopic Key to North American Pines

A Cross-field pits large, window-like. Ray tracheids with inner walls smooth or dentate.
    a Ray tracheids smooth. Numerous bordered pits on the tangential walls of latewood tracheids. Transition from earlywood to latewood gradual.
        a_1 Cross-field pits 1 to 2, large, tending to rectangular. Resin canals 90–150 μm in diameter.
            White pines
                Eastern white pine—*Pinus strobus* L.
                Western white pine—*Pinus monticola* Dougl.
        b_1 Cross-field pits 2 to 4 oval. Resin canals 150–200 μm in diameter.
            Sugar pine—*Pinus lambertiana* Dougl.
    b Ray tracheids dentate. Cross-field pits 1 to 2, large, tending to rectangular. Occasional bordered pits on the tangential walls of latewood tracheids. Transition from earlywood to latewood fairly abrupt. Resin canals 80–120 μm in diameter.
        Red pine—*Pinus resinosa* Ait.
B Cross-field pits 1 to 6, small, pinoid. Ray tracheids dentate to reticulate (net-like; teeth of opposite walls frequently meet).
    a Transition from earlywood to latewood abrupt to quite abrupt. Ray tracheids reticulate. Resin canals 90–150 μm in diameter.
        Southern pines
            Longleaf pine—*Pinus palustris* Mill.
            Shortleaf pine—*Pinus echinata* Mill.
            Loblolly pine—*Pinus taeda* L.
            Slash pine—*Pinus elliottii* Engelm.
            Pitch pine—*Pinus rigida* Mill., etc.
    b Transition from earlywood to latewood more or less abrupt to abrupt.
        a_1 Ray tracheids heavily dentate, often reticulate
            a_2 Resin canals mostly solitary, 80–90 μm in diameter.
                Lodgepole pine—*Pinus contorta* Dougl.
            b_2 Resin canals solitary or in groups of 2 to 3, 150–180 μm in diameter.
                Ponderosa pine—*Pinus ponderosa* Laws.
        b_1 Ray tracheids shallowly dentate, seldom reticulate. Resin canals mostly solitary, 70–90 μm in diameter.
            Jack pine—*Pinus banksiana* Lamb.

Key to North American Ring-Porous Hardwoods[2]

A Latewood with a radial, flame-like design (Figure II-4).
    a Rays broad, very conspicuous; up to 1 in. or more in height on tangential surfaces, forming lustrous flecks on radial surfaces. Fine (uniseriate) rays also present, visible with lens. Fine, tangential lines of parenchyma distinct. Wood hard, heavy (sp. gr. 0.70), and lustrous.

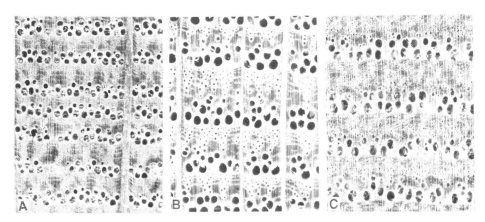

Figure II-4. Ring-porous hardwoods. Latewood with a radial, flame-like design. (A) Latewood pores small and numerous, difficult to count, abundant tyloses in earlywood pores of heartwood, broad rays present (White oak—*Quercus* spp.). (B) Latewood pores relatively large, readily counted, little or no tyloses, broad rays present (Red oak—*Quercus* spp.). (C) Rays fine, pores in earlywood often plugged with tyloses (Chestnut—*Castanea*).

    Oak[3]—*Quercus*
        a_1 Latewood pores (vessels) very small and numerous, barely visible with lens, exceedingly difficult or impossible to count (thin-walled, angular in outline, mostly in large groups). Abundant tyloses in heartwood. Broad rays commonly exceeding 1 in. in height. Heartwood golden brown. Sapwood light brown.
            White oaks (Figure II-4A).
                White oak—*Quercus alba* L.
                Bur oak—*Quercus macrocarpa* Michx.
                Overcup oak—*Quercus lyrata* Walt.
                Post oak—*Quercus stellata* Wangenh.
                Swamp chestnut oak—*Quercus michauxii* Nutt.
                Chestnut oak—*Quercus prinus* L.
                Swamp white oak—*Quercus bicolor* Willd., etc.
        b_1 Latewood pores few, plainly visible with lens, readily counted (thick-walled, rounded in outline, solitary). Little or no tyloses. Broad rays rarely exceeding 1 in. Heartwood reddish-brown, sometimes light brown. Sapwood pale reddish to grayish.
            Red oaks (Figure II-4B).
                Northern red oak—*Quercus borealis* Michx. f.
                Black oak—*Quercus velutina* Lam.
                Shumard oak—*Quercus shumardii* Buckl.
                Southern red oak—*Quercus falcata* Michx.
                Scarlet oak—*Quercus coccinea* Muenchh.
                Pin oak—*Quercus palustris* Muenchh.
                Willow oak—*Quercus phellos* L., etc.
      b Rays all fine (uniseriate), visible only with lens. Pores in earlywood often plugged with tyloses. Wood relatively soft and light (sp. gr. 0.45), not lustrous. Heartwood golden brown. Sapwood light brown.
            American chestnut—*Castanea dentata* (Marsh.) Borkh. (Figure II-4C).
  B Latewood without a radial, flame-like design (Figure II-5).
      a Pores in latewood numerous.
        a_1 Pores in latewood forming mostly continuous bands
            a_2 Pores in earlywood usually in a single row. Heartwood light brown to brown, often with a reddish tinge. Sapwood light brown.
                a_3 Pores in earlywood regularly spaced, with few tyloses. Transition to latewood abrupt.

Figure II-5. Ring-porous hardwoods. Latewood with a tangential design. (A) Pores in latewood forming continuous bands (Elm—*Ulmus*). (B) Pores in latewood arranged in clusters, abundant tyloses obscuring the outline of earlywood pores (Black locust—*Robinia pseudoacacia*). (C) Pores in latewood few, solitary or in small multiples, tyloses present but not abundant, parenchyma forming bands near the end of growth rings (Ash—*Fraxinus*).

        Rays not distinct to naked eye. Wood moderately hard and heavy (sp. gr. 0.50).

        American elm—*Ulmus americana* L.

    b_3 Pores in earlywood variable in size, irregularly spaced, and mostly plugged with tyloses. Transition to latewood more or less gradual. Rays distinct to naked eye. Wood hard and heavy (sp. gr. 0.60).

        Hard elm

            Rock elm—*Ulmus thomasii* Sarg.

            Winged elm—*Ulmus alata* Michx.

            Cedar elm—*Ulmus crassifolia* Nutt.

  b_2 Pores in earlywood in more than one, up to 4, rows. Transition to latewood tending to gradual. Wood moderately hard and heavy (sp. gr. 0.50).

    a_3 Heartwood dark reddish-brown or chocolate-brown. Sapwood grayish-white to light brown. Rays not distinct to naked eye.

        Slippery elm (Red elm)—*Ulmus rubra* Michx.

    b_3 Heartwood gray to yellowish-gray. Sapwood greenish to grayish. Rays distinct to naked eye.

        Hackberry—*Celtis occidentalis* L.

        Sugarberry—*Celtis laevigata* Willd.

b_1 Pores in latewood in groups (clusters)

  a_2 Wood hard and heavy (sp. gr. 0.55–0.80). Transition from earlywood to latewood abrupt.

    a_3 Tyloses conspicuous.

      a_4 Tyloses not abundant; outlines of earlywood pores distinct with lens. Rays distinct to naked eye on cross-sections, conspicuous on radial surfaces. Axial parenchyma not visible. Heartwood orange–yellow to yellow–brown, becoming reddish upon exposure. Sapwood yellowish. (Sp.gr. 0.55.)

          Red mulberry—*Morus rubra* L.

      b_4 Tyloses abundant; outlines of earlywood pores not distinct with lens. Rays distinct to naked eye on cross-sections, sometimes inconspicuous on radial surfaces. Axial parenchyma surrounds groups of latewood pores and tends to unite laterally (paratracheal confluent).

        a_5 Latewood pores small, mostly indistinct with lens. Heartwood golden–brown when fresh, becoming dark orange–brown upon exposure, frequently with reddish streaks. Coloring matter readily soluble in water. Sapwood yellowish. (Sp.gr. 0.80.)

Osage-orange—*Maclura pomifera* (Raf.) Schneid.

  b_5 Latewood pores large, mostly distinct with lens. Heartwood golden-brown with a greenish tinge; without reddish streaks. Coloring matter not readily soluble in water. Sapwood yellowish. (Sp.gr. 0.70.)
  Black locust—*Robinia pseudoacacia* L. (Figure II-5B).

 b_3 Tyloses absent or sparse. Axial parenchyma surrounds groups of latewood pores and tends to unite laterally. Rays distinct to naked eye.

  a_4 Tyloses absent. Heartwood pale reddish-brown. Sapwood yellowish. (Sp.gr. 0.70.)
  Honeylocust—*Gleditsia triacanthos* L.

  b_4 Tyloses sparse. Heartwood yellowish-gray, sometimes greenish. Sapwood yellowish. Growth rings as a rule wide. (Sp.gr. 0.65.)
  Tree-of-heaven—*Ailanthus altissima* Swingle (*A. glandulosa* Desf.)

 b_2 Wood soft and light (sp. gr. 0.40–0.45). With odor. Tyloses present but not abundant. Transition from earlywood to latewood gradual.

  a_3 With characteristic odor and spicy taste. Rays visible on cross-sections without lens. Heartwood dull gray–brown, or orange–brown. Sapwood yellowish. Axial parenchyma surrounding the outer latewood pores (vasicentric), and tending to unite laterally (confluent); visible with lens.
  Sassafras—*Sassafras albidum* (Nutt.) Nees

  b_3 Odor mild, said to suggest kerosene. Without taste. Rays not visible without lens. Heartwood pale gray–brown. Sapwood grayish. Axial parenchyma not associated with latewood pores, but forming bands, visible with lens, near the growth ring boundaries.
  Catalpa
   Northern catalpa—*Catalpa speciosa* Warder
   Southern catalpa—*Catalpa bignoniodes* Walt.

b Pores in latewood few, solitary or in small multiples (made of 2 to 3 vessels, thick-walled). Rays not visible without lens.

 a_1 Earlywood pores closely spaced, usually in several rows. Tyloses present but not abundant. Transition from earlywood to latewood abrupt. Axial parenchyma visible.
 Ash—*Fraxinus* (Figure II-5C).

  a_2 Heartwood grayish–brown, sometimes with a reddish tinge. Sapwood whitish. Wood hard and heavy (sp. gr. 0.60). Axial parenchyma surrounding latewood pores, sometimes united into bands near the end of wide rings.
  White ash—*Fraxinus americana* L.
  Green ash (Red ash)—*Fraxinus pennsylvanica* Marsh.
  Oregon ash—*Fraxinus latifolia* Benth.

  b_2 Heartwood grayish–brown, darker. Sapwood whitish to light brown. Wood moderately hard and heavy (sp. gr. 0.50). Axial parenchyma seldom united into bands.
  Black ash (Brown ash)—*Fraxinus nigra* Marsh.

 b_1 Earlywood pores not closely but irregularly spaced usually in a single row. Transition from earlywood to latewood gradual, often creating the impression of a semi-ring-porous wood. Wood generally hard and heavy (sp. gr. 0.70).

  a_2 Heartwood pale brown to brown, or yellowish–brown. Sapwood whitish to pale brown. Axial parenchyma distinct with lens, arranged in fine, continuous tangential lines in the outer half of the growth ring, between the latewood pores. Tyloses present, but not abundant.
  Hickory—*Carya*
  True hickories—*Eucarya*
   Shagbark hickory—*Carya ovata* (Mill.) K. Koch
   Shellbark hickory—*Carya laciniosa* (Michx. f.) Loud.
   Mockernut hickory—*Carya tomentosa* Nutt.
   Red hickory—*Carya ovalis* (Wangenh.) Sarg.
   Pignut hickory—*Carya glabra* (Mill.) Sweet
  Pecan hickories—*Apocarya*
   Pecan—*Carya illinoensis* (Wangenh.) K. Koch

Bitternut hickory—*Carya cordiformis* (Wangenh.) K. Koch
Nutmeg hickory—*Carya myristicaeformis* (Michx. f.) Nutt.
Water hickory—*Carya aquatica* (Michx.f.) Nutt.

    b_2 Heartwood small, blackish-brown. Sapwood whitish to grayish-brown. Ripple marks present. Axial parenchyma lines short, difficult to distinguish with lens. Tyloses absent.
Persimmon—*Diospyros virginiana* L.

Key to North American Diffuse-Porous Hardwoods[4]

A Pores not uniformly distributed throughout a growth ring.
    a Pores show a tendency to diagonal arrangement; solitary or in multiples (made of 2 to 4 vessels); gradually diminishing in size, with the result that a semi-ring-porous effect is produced. Rays fine, barely visible with lens (uni- to 5-seriate). Tyloses fairly abundant.
        a_1 Heartwood light chocolate-brown, sometimes with darker streaks. Sapwood whitish to light yellow-brown. Wood moderately hard and heavy (sp. gr. 0.55).
Black walnut—*Juglans nigra* L. (Figure II-6A).
        b_1 Heartwood pale chestnut-brown. Sapwood whitish to light gray-brown. Wood soft and light (sp. gr. 0.40).
Butternut—*Juglans cinerea* L.
    b Pores not scattered but forming radial, flame-like bands which cross from ring to ring. Growth rings not clearly distinct. Rays numerous, broad (mostly aggregate), and fine (mostly uniseriate). Heartwood dull brown to gray-brown. Sapwood whitish to light gray-brown. Tyloses usually absent. Wood hard and heavy (sp. gr. 0.90).
Live oak—*Quercus virginiana* Mill. (Figure II-6B).
B Pores more or less uniformly distributed throughout a growth ring.
    a Rays of variable width, some appearing as wide as in oaks.
        a_1 Broad rays conspicuous, well-demarcated, lustrous, and somewhat enlarged at the growth-ring boundaries. (Rays vary from uni- to multiseriate.)
            a_2 Broad rays few, occupying a relatively small surface area—about one tenth. Fine rays numerous. Growth rings end with a distinctly darker zone of latewood. Heartwood whitish with a reddish tinge to reddish-brown. Sapwood whitish. Wood hard and heavy (sp. gr. 0.65).
American beech—*Fagus grandifolia* Ehrh. (Figure II-6C).
            b_2 Broad rays numerous; nearly all rays are broad, occupying a much larger area—about one third. Fine rays few. Growth rings end with a narrow, dark, or whitish band of latewood.

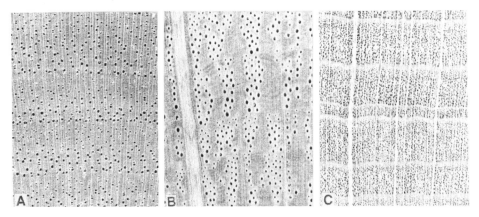

Figure II-6. Diffuse-porous hardwoods. (A) Pores gradually diminishing in size, rays fine (a semi-ring-porous or semi-diffuse-porous wood; Walnut—*Juglans*). (B) Pores not scattered, forming radial bands, growth rings not clearly distinct, broad rays present (Live oak, Holm oak—*Quercus ilex*). (C) Pores uniformly distributed within a growth ring, typical diffuse-porous, broad rays present, few (Beech—*Fagus*).

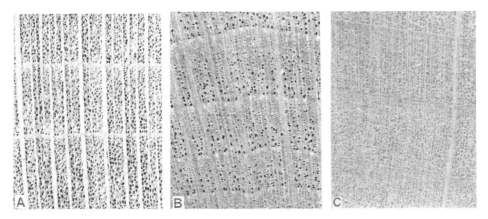

Figure II-7. Diffuse-porous hardwoods. Broad rays present. (A) Broad rays numerous (compare to beech), nearly all rays broad (Sycamore, Plane—*Platanus*). (B) Broad rays not well demarcated, dull, growth rings wavy (Hornbeam—*Carpinus*). (C) Broad rays at wide intervals, dull (Alder—*Alnus*).

        Heartwood light to dark brown or reddish-brown, not always distinct. Sapwood whitish to yellowish or reddish-brown. Wood moderately hard and heavy (sp. gr. 0.50).
        American sycamore—*Platanus occidentalis* L. (Figure II-7A).
- b_1 Broad rays dull, not well-demarcated (aggregate).
  - a_2 Growth rings wavy. Axial parenchyma in fine tangential lines visible with lens. Aggregate rays closely spaced. Pores not too numerous. Heartwood yellowish-white, usually not distinct from sapwood. Wood hard and heavy (sp. gr. 0.65).
    American hornbeam (Blue beech)—*Carpinus caroliniana* Walt. (Figure II-7B).
  - b_2 Growth rings not wavy. Lines of axial parenchyma not visible. Aggregate rays often at wide intervals, relatively inconspicuous to naked eye. Pores very numerous. Heartwood not colored. Original color of wood—in freshly felled trees—whitish, quickly changing to reddish upon exposure. Wood relatively soft and light (sp. gr. 0.40).
    Red alder—*Alnus rubra* Bong. (Figure II-7C).
- b Broad rays absent; all rays narrow.
  - a_1 Rays fairly distinct to naked eye.
    - a_2 Pores very numerous, usually crowded together.
      - a_3 Heartwood rich reddish-brown. Sapwood whitish to light reddish-brown. Traumatic gum canals sometimes present. Wood moderately hard and heavy (sp. gr. 0.50).
        Black cherry—*Prunus serotina* Ehrh.
      - b_3 Heartwood variable in color, from bright greenish or olive to purplish, frequently streaked. Sapwood whitish, often variegated. Growth rings terminated by light-colored lines or bands of axial parenchyma. Wood relatively soft and light (sp. gr. 0.45).
        Yellow poplar—*Liriodendron tulipifera* L.
    - b_2 Pores relatively few to fairly numerous, but not crowded; solitary or in small multiples.
      - a_3 Rays not very close together, of two widths (i) as wide as the largest pores (3- to 8-seriate), and (ii) fine, scarcely visible with lens (uni- to 3-seriate). Heartwood uniform light reddish-brown. Sapwood white with a reddish tinge. Growth rings well distinct and terminated by a narrow, darker band of latewood. Wood hard and heavy (sp. gr. 0.60). Pith flecks rare.
        Hard maple
          Sugar maple—*Acer saccharum* Marsh.
          Black maple—*Acer nigrum* Michx. f.
      - b_3 Rays very close together, graduating in width from as wide as the largest pores to fine. In general, less distinct (uni- to 5-seriate). Growth rings not well distinct. Wood moderately hard and heavy (sp. gr. 0.50).

        a_4 Heartwood pale brown, often with a greenish or grayish tinge. Growth rings terminated with a narrow, darker band of latewood. Pith flecks common.
            Soft maple
                Red maple—*Acer rubrum* L.
                Silver maple—*Acer saccharinum* L.
        b_4 Heartwood pinkish-brown. Sapwood whitish—with a reddish or grayish tinge. Growth rings terminated with a narrow, lighter band of latewood. Pith flecks rare.
            Bigleaf maple (Oregon maple, Pacific maple)—*Acer macrophyllum* Pursh.
  b_1 Rays mostly not visible to naked eye but distinct with lens.
      a_2 Pores relatively few, not crowded.
          a_3 Pores scattered, wider than the rays. (Rays uni- to 5-seriate.) Heartwood light to dark brown or reddish-brown. Sapwood whitish, yellowish, or light yellowish-brown. Wood hard and heavy (sp. gr. 0.60).
              Birch—*Betula* (Figure II-8B).
                  Yellow birch—*Betula alleghaniensis* Britton
                  Sweet birch (Black birch)—*Betula lenta* L.
                  Paper birch (White birch)—*Betula papyrifera* Marsh.
                  Gray birch—*Betula populifolia* Marsh.
                  River birch (Red birch)—*Betula nigra* L.
          b_3 Pores show a tendency to form short radial or oblique lines, creating the impression of a design. (Rays uni- to 3-seriate.) Heartwood whitish to light brown with a red tinge. Sapwood whitish. Wood hard and heavy (sp. gr. 0.70).
              Hophornbeam—*Ostrya virginiana* (Mill.) K. Koch
      b_2 Pores numerous, crowded.
          a_3 Wood soft, readily dented with thumbnail, and light (sp. gr. 0.40). Heartwood pale white or creamy-brown, not always distinct from sapwood. (Rays uni- to 6-seriate.)
              Basswood—*Tilia* (Figure II-8C).
                  American basswood—*Tilia americana* L.
                  White basswood—*Tilia heterophylla* Vent.
          b_3 Wood moderately hard and heavy (sp. gr. 0.50). (Rays uni- to 4-seriate.)
              a_4 Heartwood variable in color, from light to dark brown, often with darker streaks. Sapwood pale brown, sharply demarcated. Traumatic gum canals sometimes present.
                  Sweetgum—*Liquidambar styraciflua* L.
              b_4 Heartwood greenish or brownish-gray. Sapwood not sharply demarcated. Traumatic gum canals absent.
                  Tupelo—*Nyssa*

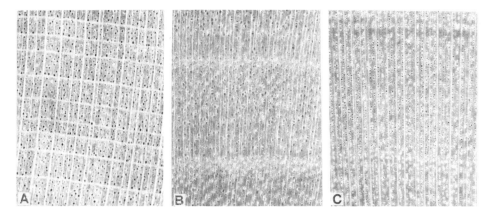

Figure II-8. Diffuse-porous hardwoods. Rays narrow but distinct. (A) Pores relatively few, the largest equal in width to rays (Maple—*Acer*). (B) Pores wider than the rays (Birch—*Betula*). (C) Pores numerous, crowded (Basswood—*Tilia*).

Water tupelo—*Nyssa aquatica* L.
Black tupelo (Black gum)—*Nyssa sylvatica* Marsh.
c Rays indistinct or invisible with lens (uniseriate). Wood soft and light (sp. gr. 0.40).
   a_2 Pores tiny, texture fine. Growth rings discernible with difficulty under lens. Wood yellowish-white often with reddish or brownish streaks. (Vessels thick-walled with spiral thickenings.)
   Buckeye—*Aesculus*
     Yellow buckeye—*Aesculus octandra* Marsh.
     Ohio buckeye—*Aesculus glabra* Willd.
   b_2 Pores somewhat larger, texture coarser. Growth rings mostly distinct, due to a darker zone of latewood. (Vessels thin-walled without spiral thickenings.)
     a_3 (Rays homocellular, as a rule; marginal cells brick-shaped or squarish, very rarely upright). Heartwood pale creamy-white. Sapwood not clearly defined. Wood lustrous.
     Poplar—*Populus* (Figure II-9A).
       Aspen
         Quaking aspen—*Populus tremuloides* Michx.
         Bigtooth aspen—*Populus grandidentata* Michx.
       Cottonwood
         Eastern cottonwood (Eastern poplar)—*Populus deltoides* Bartr.
         Balsam poplar (Tacamahaca poplar)—*Populus balsamifera* L.
         Black cottonwood—*Populus trichocarpa* Torr. and Gray
         Swamp cottonwood (Swamp poplar)—*Populus heterophylla* L.
     b_3 (Rays heterocellular; marginal cells upright or squarish.) Heartwood dull reddish-brown. Sapwood whitish.
     Black willow—*Salix nigra* Marsh. (Figure II-9B).

Macroscopic Key to European Softwoods

A Vertical resin canals always present.
   a Vertical resin canals comparatively large (Figure II-1), mostly numerous, solitary or in small tangential groups of 2 to 3, visible to naked eye as small dots in cross-section and as fine lines on longitudinal surfaces. Heartwood colored, with more or less pronounced resinous odor.
   Pine—*Pinus*
   b Vertical resin canals medium-sized to small (Figure II-2), invisible without lens; relatively few, solitary or in small tangential groups of 2 to 5.
     a_1 Wood moderately hard and heavy (sp. gr. 0.60). Transition from earlywood to latewood gen-

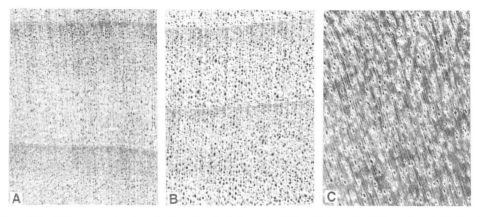

Figure II-9. Diffuse-porous hardwoods. Rays fine, indistinct to invisible with lens. (A, B) Pores fine (A, Poplar, Aspen, Cottonwood—*Populus*; and B, Willow—*Salix*). (C) Pores relatively large, not crowded (Eucalypt—*Eucalyptus camaldulensis*).

erally abrupt. Heartwood colored with a brown tone. Sapwood whitish. Wood surfaces have oily feel.
European larch—*Larix europaea* Lam. et DC. (*L. decidua* Mill.) (Figure II-2B).

b_1 Wood moderately soft and light (sp. gr. 0.45). Transition from earlywood to latewood generally gradual. Heartwood not colored; wood pale yellowish-brown, lustrous.
European spruce (Norway spruce)—*Picea abies* Karst. (*P. excelsa* Link) (Figure II-2C).

B Vertical resin canals normally absent (Figure II-3).
- a Wood moderately soft and light (sp. gr. 0.45). Without colored heartwood; wood light yellow-brown. Transition from earlywood to latewood gradual, sometimes tending to abrupt.
  Fir—*Abies* (Figure II-3A).
  Silver fir (White fir)—*Abies alba* Mill. (*A. pectinata* Lam. et DC.)
  Grecian fir—*Abies cephalonica* Loud., etc.
- b Wood hard and heavy. With colored heartwood.
  - a_1 Without aromatic odor. Wood hard and heavy (sp. gr. 0.65). Heartwood brown, sapwood light brown. Transition from earlywood to latewood gradual.
    Yew—*Taxus baccata* L.
  - b_1 With aromatic odor. Wood somewhat less hard and heavy (sp. gr. 0.60). Growth rings often irregular—false, discontinuous. Dark-colored axial parenchyma visible with lens and often to naked eye, especially in sapwood.
    - a_2 Heartwood yellow-brown, sapwood light yellow-brown. False rings may be present.
      Mediterranean cypress—*Cupressus sempervirens* L.
    - b_2 Heartwood brown to brownish-red, sapwood light brown. False and discontinuous rings may be present. Characteristic cedar-like odor.
      Juniper (Cedar)—*Juniperus* (Figure II-3C).
      Common juniper—*Juniperus communis* L.
      Syrian juniper—*Juniperus drupacea* Labill.
      Grecian juniper—*Juniperus excelsa* Bieb.
      Phoenician juniper—*Juniperus phoenicea* L.[5]

Microscopic Key to European Softwoods

A Vertical resin canals always present. Horizontal canals small, present in fusiform rays.
- a Cross-field pits window-like (fenestriform, fenestrate) or pinoid. Resin canals comparatively large (70–200 µm in diameter), mostly numerous, solitary or in groups of 2 or 3. Epithelial cells thin-walled. Ray tracheids comparatively large and numerous, occurring in one to several marginal rows, frequently interspersed in high rays and sometimes composing entire low rays.
  Pine—*Pinus*
- b Cross-field pits piceoid. Resin canals medium-sized to small (50–90 µm in diameter), comparatively few, solitary or in small tangential groups of 2 to 5. Epithelial cells thick-walled. Ray tracheids small and few, usually in single marginal rows.
  - a_1 Transition from earlywood to latewood abrupt. Bordered pits on radial walls of vertical tracheids often in pairs (biseriate). Ray tracheids with smooth inner walls, only sporadically with small dentations.
    European larch—*Larix europaea* Lam. et DC. (*L. decidua* Mill.)
  - b_1 Transition from earlywood to latewood gradual. Bordered pits in a single row, very seldom in pairs. Ray tracheids slightly dentate.
    European spruce (Norway spruce)—*Picea abies* Karst. (*P. excelsa* Link)

B Vertical and horizontal resin canals absent; traumatic vertical canals may be present.
- a Vertical tracheids with spiral thickenings. Vertical parenchyma absent. Cross-field pits cupressoid.
  Yew—*Taxus baccata* L.
- b Vertical, tracheids without spiral thickenings.
  - a_1 Vertical parenchyma absent or very sparse terminal. Cross-field pits taxodioid. Ray parenchyma cells sometimes containing crystals of calcium oxalate. Rays high, sometimes over 30 cells high, as a rule uniseriate.
    Fir—*Abies*

Silver fir (White fir)—*Abies alba* Mill. (*A. pectinata* Lam. et DC)
Grecian fir—*Abies cephalonica* Loud., etc.

b_1 Vertical parenchyma abundant—in most rings.[6] False or discontinuous rings may be present. Cross-field pits cupressoid.

    a_2 Rays up to 40 cells high, uniseriate, sometimes in parts biseriate.
        Mediterranean cypress—*Cupressus sempervirens* L.

    b_2 Rays seldom exceeding 10–12 cells in height,[7] uniseriate.
        Juniper (Cedar)—*Juniperus*
            Common juniper—*Juniperus communis* L.
            Syrian juniper—*Juniperus drupacea* Labill., etc.

Macroscopic Key to European Pines

A Wood comparatively soft and light (sp. gr. 0.40). Heartwood golden-brown, sapwood light golden-brown. Latewood not pronounced. Transition from earlywood to latewood gradual (Figure II-1A). Texture uniform, fine. Resin canals large and numerous, distinct to naked eye, mostly solitary.
Macedonian pine—*Pinus peuce* Griseb.

B Wood comparatively harder and heavier (sp. gr. 0.50–0.60). Transition from earlywood to latewood fairly abrupt to abrupt (Figure II-1B).

    a Heartwood strongly resinous. Resin canals large and numerous, conspicuous to naked eye. Sapwood whitish to yellowish.

        a_1 Heartwood reddish-brown, with latewood dark reddish-brown. Resin canals solitary. Tendency to form false rings.
            Maritime pine—*Pinus pinaster* Sol. (*P. maritima* Poir.)

        b_1 Heartwood brown with latewood yellowish-brown. Sapwood relatively narrow. Transition from earlywood to latewood sometimes variable. Resin canals solitary or in groups of 2 or 3.
            Aleppo pine—*Pinus halepensis* Mill.
            Calabrian pine—*Pinus brutia* Ten.

    b Heartwood not strongly resinous. Resin canals distinct to inconspicuous with naked eye.

        a_1 Latewood pronounced. Transition from earlywood to latewood abrupt. Resin canals relatively large and numerous, distinct to naked eye.

            a_2 Heartwood pale reddish-brown—after exposure. Sapwood wide.
                Scots pine—*Pinus sylvestris* L. (Figure II-1B).

            b_2 Heartwood pale brown. Sapwood relatively narrow. Texture coarser.
                Austrian pine (Black pine)—*Pinus nigra* Arn. (*P. laricio* Poir.)

        b_1 Latewood not so pronounced. Transition from earlywood to latewood fairly abrupt. Resin canals relatively small and few, not well distinct to naked eye.

            a_2 Heartwood golden-brown. Sapwood yellowish to brownish. Growth rings narrow; texture uniform, fine—wood used for carvings.
                Whitebark pine—*Pinus leucodermis* Ant. (*P. heldreichii* Christ.)

            b_2 Heartwood reddish-brown. Sapwood reddish-white. Growth rings often wide; tendency to form false rings.
                Umbrella pine (Parasol pine, Stone pine)—*Pinus pinea* L.

Microscopic Key to European Pines

A Cross-field pits 1 to 2, large, window-like (fenestriform). Resin canals numerous, 90–150 μm in diameter.

    a Ray tracheids with inner walls smooth. Horizontal walls of ray parenchyma cells thick. Numerous bordered pits on the tangential walls of latewood tracheids. Transition from earlywood to latewood gradual.
        Macedonian pine—*Pinus peuce* Griseb.

    b Ray tracheids dentate to reticulate (net-like; teeth of opposite walls frequently meet). Horizontal walls of ray parenchyma cells thin or mainly thin. Transition from earlywood to latewood more or less abrupt.

Scots pine—*Pinus sylvestris* L.
Austrian pine (Black pine)—*Pinus nigra* Arn. (*P. laricio* Poir.)

B Cross-field pits 1 to 6, small, pinoid. Resin canals numerous or relatively few.
- a Ray tracheids moderately dentate to dentate, teeth mostly sharp. Horizontal walls of ray parenchyma cells mainly thin. Resin canals mostly solitary, numerous, 100–180 μm in diameter.
  Maritime pine—*Pinus pinaster* Sol. (*P. maritima* Poir.)
- b Ray tracheids moderately to slightly dentate, teeth mostly blunt and short.
  - a_1 Resin canals numerous, solitary or in groups of 2 or 3, 120–200 μm in diameter, located in latewood and near the zone of transition. Horizontal walls of ray parenchyma cells mainly thick.
    Aleppo pine—*Pinus halepensis* Mill.
    Calabrian pine—*Pinus brutia* Ten.
  - b_1 Resin canals relatively few, 70–120 μm in diameter, located mostly in latewood.
    - a_2 Horizontal walls of ray parenchyma cells mainly thick. False rings may be present.
      Umbrella pine (Parasol pine, Stone pine)—*Pinus pinea* L.
    - b_2 Horizontal walls of ray parenchyma cells mainly thin.
      Whitebark pine—*Pinus leucodermis* Ant. (*P. heldreichii* Christ.)

Key to European Ring-Porous Hardwoods[8]

A Latewood with a radial, flame-like design.
- a Rays broad, very conspicuous; up to 1 in. or more in height on tangential surfaces, forming lustrous flecks on radial surfaces. Fine (uniseriate) rays also present, visible with lens. Fine, tangential lines of parenchyma distinct. Wood hard, heavy (sp. gr. 0.70), and lustrous.
  Oak[9]—*Quercus*
  - a_1 Latewood pores (vessels) very small and numerous, barely visible with lens, exceedingly difficult or impossible to count (thin-walled, angular in outline, mostly in large groups). Abundant tyloses in heartwood. Broad rays commonly exceeding 1 in. in height. Heartwood golden-brown to brown. Sapwood light brown. Wood generally similar to white oak (Figure II-4A).
    Sessile oak (Durmast oak)—*Quercus petraea* Liebl. (*Q. sessiliflora* Salisb.)
    English oak—*Quercus robur* L. (*Q. pedunculata* Ehrh.)
    Pubescent oak—*Quercus pubescens* Willd. (*Q. lanuginosa* Thuill.)
    Broadleaved oak—*Quercus conferta* Kit. (*Q. farnetto* Ten.)
    Valonian oak (Grecian oak)—*Quercus aegilops* L., etc.
  - b_1 Latewood pores few, plainly visible with lens, readily counted (thick-walled, rounded in outline, solitary). Little or no tyloses. Broad rays rarely exceeding 1 in. Heartwood reddish-brown to brown. Sapwood pale reddish to grayish. Wood generally similar to red oak (Figure II-4B).
    Turkey oak (Mossy oak)—*Quercus cerris* L.
    Cork oak—*Quercus suber* L., etc.
- b Rays all fine (uniseriate), visible only with lens. Pores in earlywood often plugged with tyloses. Wood moderately soft and light (sp. gr. 0.50), not lustrous. Heartwood golden-brown. Sapwood light brown.
  European chestnut—*Castanea sativa* Mill. (*C. vesca* Gaertn., *C. vulgaris* Lam.) (Figure II-4C).

B Latewood without a radial, flame-like design (Figure II-5). Wood hard and heavy.
- a Pores in latewood numerous.
  - a_1 Pores in latewood forming mostly continuous bands.
    - a_2 Pores in earlywood in 1 to 2 rows. Rays indistinct or sometimes only slightly distinct to naked eye (sp. gr. 0.65).
      Elm—*Ulmus* (Figure II-5A).
      - a_3 Heartwood chocolate-brown. Sapwood light brown. Latewood pores in fairly straight, tangential bands. Rays in tangential section short, never exceeding 1 mm in height (commonly 5- to 6-seriate, and sometimes wider).
        English elm (Field elm)—*Ulmus procera* Salisb. (*U. campestris* L.)
      - b_3 Heartwood yellowish-brown, often streaked with green. Latewood pores tending to form wavy bands. Rays often exceeding 1 mm in height (rarely more than 5-seriate).[10]
        Wych elm (Mountain elm)—*Ulmus glabra* Huds. (*U. montana* With.)

b_2 Pores in earlywood in 1 to 4 rows. Rays distinct to naked eye (uni- to 8-seriate). Heartwood yellowish-gray. Sapwood greenish to grayish. (Sp. gr. 0.70.)

Hackberry—*Celtis australis* L.

b_1 Pores in latewood in groups (clusters) (Figure II-5B).

a_2 Wood dark-colored. Tyloses present.

a_3 Tyloses not abundant; outlines of earlywood pores distinct with lens. Rays distinct to naked eye on cross-sections, conspicuous on radial surfaces (uni- to 6-seriate). Axial parenchyma not visible. Heartwood orange–yellow to yellow–brown, becoming reddish upon exposure. Sapwood yellowish (sp. gr. 0.60).

White mulberry—*Morus alba* L.

Black mulberry—*Morus nigra* L.

b_3 Tyloses abundant; outlines of earlywood pores not distinct with lens. Rays distinct to naked eye on cross-sections, sometimes inconspicuous on radial surfaces (uni- to 5-seriate). Axial parenchyma surrounds groups of latewood pores and tends to unite laterally (paratracheal confluent). Heartwood golden-brown with a greenish tinge. Sapwood yellowish (sp. gr. 0.70).

Black locust—*Robinia pseudoacacia* L. (Figure II-5B).

b_2 Wood light-colored. Tyloses sparse. Axial parenchyma surrounds groups of latewood pores and tends to unite laterally (paratracheal confluent). Rays distinct to naked eye (uni- to 12-seriate). Heartwood yellowish-gray, sometimes greenish. Sapwood yellowish. Growth rings as a rule wide. (Sp. gr. 0.60.)

Tree-of-heaven—*Ailanthus altissima* Swingle (*A. glandulosa* Desf.)

b Pores in latewood few, solitary or in small multiples (made of 2 to 3 vessels, thick-walled). Earlywood pores closely spaced, usually in several rows. Tyloses present but not abundant. Rays not visible without lens (uni- to biseriate, seldom 3-seriate). Axial parenchyma surrounding latewood pores, sometimes united into bands near the end of wide rings. Heartwood yellowish-white, little different in color from sapwood. (Sp. gr. 0.70.)

Ash—*Fraxinus* (Figure II-5C).

European ash—*Fraxinus excelsior* L.

Flowering ash (Manna ash)—*Fraxinus ornus* L.

Key to European Diffuse-Porous Hardwoods[11]

A Pores not uniformly distributed throughout a growth ring.

a Pores show a tendency to diagonal arrangement; solitary or in multiples (made of 2 to 4 vessels); gradually diminishing in size, with the result that a semi-ring-porous effect is produced (Figure II-6). Rays fine, barely visible with lens (uni- to 4-seriate). Tyloses fairly abundant. Heartwood light chocolate-brown, sometimes with darker streaks. Sapwood whitish to light yellow–brown. Wood moderately hard and heavy (sp. gr. 0.60).

European walnut—*Juglans regia* L. (Figure II-6A).

b Pores not scattered but forming radial, flame-like bands which cross from ring to ring. Growth rings not clearly distinct. Rays numerous, broad (mostly aggregate), and fine (mostly uniseriate). Heartwood golden-brown to brown. Sapwood light brown. Wood hard and heavy (sp. gr. 0.90).

Evergreen oaks (Figure II-6B).

Holm oak—*Quercus ilex* L.

Kermes oak—*Quercus coccifera* L.

B Pores more or less uniformly distributed throughout a growth ring.

a Rays of variable width, some appearing as wide as in oaks.

a_1 Broad rays conspicuous, well demarcated, lustrous, and somewhat enlarged at the growth ring boundaries. (Rays vary from uni- to multiseriate).

a_2 Broad rays few, occupying a relatively small surface area—about one tenth. Fine rays numerous. Growth rings end with a distinctly darker zone of latewood. Wood reddish-white, becoming darker upon exposure; "red-heartwood" frequently present. Wood hard and heavy (sp. gr. 0.70).

European beech—*Fagus sylvatica* L. (Figure II-6C).

b_2 Broad rays numerous; nearly all rays are broad, occupying a much larger area—about one third. Fine rays few. Growth rings end with a narrow dark or whitish band of latewood. Heartwood light to dark brown or reddish-brown, not always distinct. Sapwood whitish to yellowish- or reddish-brown. Wood moderately hard and heavy (sp. gr. 0.55).
Oriental plane—*Platanus orientalis* L. (Figure II-7A).

b_1 Broad rays dull, not well demarcated (aggregate). Without colored heartwood.

a_2 Growth rings wavy. Axial parenchyma in fine tangential lines visible with lens. Wood yellowish-white to grayish, hard and heavy (sp. gr. 0.80). (Perforations simple.)
Hornbeam—*Carpinus* (Figure II-7B).
   European hornbeam—*Carpinus betulus* L.
   Eastern hornbeam—*Carpinus orientalis* Mill. (*C. duinensis* Scop.)

b_2 Growth rings not wavy. Lines of axial parenchyma not visible. (Perforations scalariform.)

a_3 Pores numerous. Original color of wood—in freshly felled trees—whitish, quickly changing to reddish upon exposure. Wood relatively soft and light (sp. gr. 0.50). (Perforations with 18–22 bars.)
European alder (Black alder)—*Alnus glutinosa* Gaertn. (Figure II-7C).

b_3 Pores not so numerous, with a more or less pronounced tendency toward a radial arrangement. Components of broad (aggregate) rays sometimes discernible with lens. Wood harder and heavier (sp. gr. 0.65). (Perforations with 2–7 bars.)
European hazel—*Corylus avellana* L.
Mediterranean hazel—*Corylus colurna* L.

b Broad rays absent; all rays narrow.

a_1 Rays fairly distinct to just visible to naked eye (uni- to 7-seriate, seldom wider). Wood hard and heavy (sp. gr. 0.60), yellow-reddish to gray-reddish, without colored heartwood. (Perforations simple.)
Maple—*Acer* (Figure II-8A).

a_2 Rays fairly distinct, variable in width, mostly wider than the pores (uni- to 7-seriate, seldom 10-seriate).
Sycamore plane (Great Maple)—*Acer pseudoplatanus* L.
Monspessulanian maple—*Acer monspessulanum* L.

b_2 Rays just visible to naked eye with proper lighting, appearing practically uniform in width and equal to or narrower than the pores (uni- to 6-seriate).
Norway maple (Bosnian maple)—*Acer platanoides* L.
Field maple—*Acer campestre* L.
Balkans maple—*Acer heldreichii* Orph., etc.

b_1 Rays fine (uni- to 4-seriate), visible with lens, as wide or narrower than the pores. With or without colored heartwood.

a_2 Pores scattered, quite numerous to sparse.

a_3 Pores more or less numerous. (Rays as a rule uni- and biseriate, seldom 3- and 4-seriate.)

a_4 Pores quite numerous and tiny, often as wide as the rays. Heartwood reddish-brown. Sapwood pale brown. Wood hard and heavy (sp. gr. 0.75). (Perforations multiple—not scalariform.)
Wild service tree—*Sorbus torminalis* Crantz.
Service tree (European Mountain Ash)—*Sorbus domestica* L.
Rowantree—*Sorbus aucuparia* L.
Whitebeam tree—*Sorbus aria* Crantz.

b_1 Pores not too numerous, somewhat larger and wider than the rays. Wood yellowish-brown, without colored heartwood. Pith flecks often present. Wood hard and heavy (sp. gr. 0.60). (Perforations scalariform.)
Swedish birch (English birch)—*Betula verrucosa* Ehrh. (*B. pendula* Roth.) (Figure II-8B).

b_3 Pores sparse and large, uniformly scattered. Tyloses abundant. Heartwood yellowish-brown, variegated. Sapwood yellowish. Wood hard and heavy (sp. gr. 0.80). (Rays mostly biseriate to 4-seriate, seldom 5-seriate. Vestured pits present.)
Carobtree—*Ceratonia siliqua* L.

b_2 Pores not scattered, numerous to less numerous.

a_3 Pores numerous with a more or less pronounced radial arrangement, which appears to cross from ring to ring. Rays as wide as the pores (uni- to 6-seriate). Wood readily dented with thumbnail, relatively light (sp. gr. 0.50), reddish-white, without colored heartwood.
Lime (Linden)—*Tilia* (Figure II-8C).
Silver limetree—*Tilia tomentosa* Moench. (*T. argentea* Desf.)
Large-leaved limetree—*Tilia grandfolia* Moench. (*T. platyphyllos* Scop.)
Small-leaved limetree—*Tilia parvifolia* Ehrh. (*T. cordata* Mill.)
Common limetree—*Tilia vulgaris* Hayne.
b_3 Pores not in continuous radial lines or bands.
a_4 Pores relatively few, small, somewhat larger in earlywood, with a tendency to form short radial or oblique lines, creating the impression of a design. Axial parenchyma in fine, tangential lines. Rays barely visible with lens (uni- to 3-seriate, seldom 4-seriate). Wood hard and heavy (sp. gr. 0.80), reddish-brown, without colored heartwood.
European hophornbeam—*Ostrya carpinifolia* Scop.
b_4 Pores more numerous, relatively large, solitary or in short radial lines. Growth rings not well demarcated; successive rings may show differences in arrangement of pores. Axial parenchyma abundant (paratracheal). Rays quite conspicuous with lens (uni- to biseriate, seldom 3-seriate). Wood hard and heavy (sp. gr. 0.80), with an oily feel; usually spirally grained, especially in old trees. Heartwood greenish-brown, variegated. Sapwood greenish.
European olive-tree—*Olea europaea* L.
c Rays indistinct or invisible with lens (Figure II-9) (uniseriate). Wood generally soft and light (sp. gr. 0.40).
a_2 Pores tiny, texture fine. Growth rings discernible with difficulty under lens. Wood yellowish-white, often with reddish or brownish streaks. (Vessels thick-walled with spiral thickenings.)
Common horse-chestnut—*Aesculus hippocastanum* L.
b_2 Pores somewhat larger. Texture coarser. Growth rings mostly distinct, due to a darker zone of latewood. (Vessels thin-walled without spiral thickenings.)
a_3 Rays homocellular, as a rule; marginal cells brick-shaped or squarish, very rarely upright.)
Poplar—*Populus* (Figure II-9A).
a_4 Without colored heartwood. Wood yellowish-white with a reddish tinge. Pith flecks may be present.
European aspen—*Populus tremula* L.
b_4 With colored heartwood.
a_5 Heartwood yellowish-brown, sometimes with reddish streaks. Sapwood whitish to yellowish.
White poplar—*Populus alba* L.
b_5 Heartwood light greenish-brown, not too different from sapwood.
Black poplar—*Populus nigra* L.
b_3 (Rays heterocellular; marginal cells upright or squarish.) Heartwood light-brown to red-brownish. Sapwood whitish. Pith flecks may be present.
White willow—*Salix alba* L.
Crack willow—*Salix fragilis* L., etc. (Figure II-9B).

Key to Tropical Woods

A Ring-porous. Growth rings distinct, pores (vessels) large at the beginning, gradually become smaller and very small toward the end of the growth ring, single and in multiples. Abundant tyloses. Rays visible to naked eye. Parenchyma initial, paratracheal vasicentric, in large pores visible to naked eye, in others not seen even with a hand lens. Color of wood golden-brown with continuous, practically parallel, axial, brown stripes. Wood medium in weight to heavy (sp. gr. 0.55–0.80), medium in hardness to hard.
Teak—*Tectona grandis* L.f.

B Diffuse-porous.
  a Parenchyma paratracheal, aliform or confluent (boundary, banded, or vasicentric parenchyma may coexist).
    a_1 Growth rings easy to see with naked eye. Wood heavy and hard with continuous, practically parallel, axial, colored stripes, pronounced or not pronounced.
      a_2 Color of wood pale brown with pronounced brown stripes. Parenchyma paratracheal aliform, sometimes confluent or apotracheal diffuse-in-aggregates (in discontinuous and wavy tangential lines), visible to naked eye. Pores large, mostly single, sometimes in multiples, easy to see with naked eye. Tyloses rare. Rays visible with lens. (Sp. gr. 0.70–0.85).
      Zebrano, Zingana—*Microberlinia brazzavillensis* A. Chev.
      b_2 Color of wood golden-brown. Axial stripes pronounced or less pronounced.
        a_3 Stripes brown, not pronounced. Parenchyma paratracheal vasicentric to slightly aliform, abundant, also boundary, visible to naked eye. Growth rings darker toward the end, with a narrow zone without pores. Pores rather medium in size, single, seldom in multiples, visible to naked eye, without tyloses, few with yellowish inclusions. Rays visible to naked eye. (Sp. gr. 0.75–0.90).
        Amazakoué, Ovangkol—*Guibourtia ehie* J. Léonard.
        b_3 Stripes brown-black, pronounced or less pronounced. Parenchyma paratracheal aliform, not pronounced, connected to wavy lines, also vasicentric and unilateral, easy to see with lens. Growth rings wavy. Pores medium to large, single and in multiples, easy to see with naked eye, many with tyloses. Rays visible with lens. (Sp. gr. 0.85–1.00).
        Palissander, Indian Rosewood—*Dalbergia latifolia* Roxb.
      c_2 Color of wood red with darker stripes, not pronounced. Parenchyma paratracheal aliform, sometimes confluent, visible to naked eye. Pores not numerous, large, single and in multiples, easy to see with naked eye, without tyloses. Rays visible with lens. (Sp. gr. 0.65–0.85).
      Padauk—*Pterocarpus soyauxii* Taub.
    b_1 Growth rings not clearly delineated. Wood without axial stripes.
      a_2 Color of wood brown-red or red. Parenchyma paratracheal aliform and fine boundary. Wood heavy and hard.
        a_3 Color of wood brown-red, not uniform, with discontinuous and irregular darker linear markings. Parenchyma visible to naked eye. Pores medium, not numerous, mostly single, also in multiples, visible to naked eye. Without tyloses. Pores visible to naked eye. (Sp. gr. 0.80–0.95).
        Bubinga—*Guibourtia tessmannii* J. Léonard.
        b_3 Color of wood red, relatively uniform. Parenchyma not so pronounced, easy to see with lens. Pores medium to large, few single, mostly in multiples, easy to see with naked eye. Many pores with yellowish inclusions. Rays visible with lens. (Sp. gr. 0.75–0.90).
        Afzelia, Lingue—*Afzelia africana* Smith
      b_2 Color of wood golden-brown, dark. Parenchyma paratracheal aliform, pronounced, easy to see with naked eye, also boundary not pronounced. Pores large, not numerous, mostly single, also in multiples, almost all with tyloses, easy to see with naked eye. Rays difficult to see with naked eye. Wood heavy (sp. gr. 0.60–0.75) and hard.
      Iroko, Kambala, Odoum—*Chlorophora excelsa* Benth. et Hook f.
      c_2 Color of wood yellowish. Wood medium in weight and hardness.
        a_3 Parenchyma paratracheal aliform, often confluent, visible to naked eye. Pores not numerous, medium in size, single and in multiples, easy to see with naked eye. Tyloses sparse. Rays difficult to see with lens. (Sp. gr. 0.45–0.65).
        Afara, Limba, Fraké—*Terminalia superba* Engl. et Diels
        b_3 Parenchyma paratracheal aliform, not pronounced, sometimes confluent, visible with lens. Pores not numerous, small to medium, single and in multiples, visible to naked eye, without tyloses. Rays visible with lens. (Sp. gr. 0.60–0.70).
        Ramin—*Gonystylus bancanus* Baill.
  b Parenchyma paratracheal vasicentric or unilateral, not pronounced, visible with lens. Rays difficult to see with naked eye.
    a_1 Color of wood brown-golden (bronze) with axial bands of contrasting luster (on radial surfaces).

Growth rings distinct. Pores large, rather numerous, single and in multiples, easy to see with naked eye, many with tyloses. (Boundary parenchyma seldom present, visible to naked eye.) Wood medium in weight (sp. gr. 0.45–0.60) and hardness.

Idigbo, Framiré—*Terminalia ivorensis* A. Chev.

b_1 Color of wood yellowish. Growth rings not clearly delineated. Pores medium in size, not numerous, single and in multiples, easy to see with naked eye, without tyloses. Wood medium in weight (sp. gr. 0.45–0.64) and hardness.

Antiaris, Ako—*Antiaris welwitschii* Engl.

c Parenchyma boundary or banded (seldom paratracheal, not pronounced).

a_1 Parenchyma banded.

a_2 Parenchyma pronounced, in wide and wavy bands, easy to see with naked eye. Growth rings not clearly delineated. Pores small to medium, visible to naked eye, not numerous, with abundant tyloses. Rays difficult to see with naked eye. Wood heavy (sp. gr. 0.80–0.90), hard, brown in color.

Difou—*Morus mesozygia* Stapf.

b_2 Parenchyma not pronounced, in narrow bands or lines. Color of wood brown with reddish tinge, and axial bands of contrasting luster.

a_3 Parenchyma in narrow bands, visible to naked eye. Pores large, mostly single, seldom in multiples, not numerous, easy to see with naked eye, some with tyloses. Rays visible to naked eye. Axial bands of contrasting luster pronounced. Wood medium in weight to heavy (sp. gr. 0.60–0.75), hard.

Kosipo, Omu—*Entandrophragma candollei* Harms

b_3 Parenchyma in fine lines, not visible with naked eye. Pores usually small, not numerous, usually in multiples, some with tyloses, difficult to see with naked eye. Rays difficult to see with lens. Axial bands of contrasting luster not pronounced. Wood medium in weight to heavy (sp. gr. 0.60–0.80), hard.

Makoré—*Dumoria heckelii* A. Chev.

b_1 Parenchyma boundary and in-between banded, not pronounced, visible to naked eye. Rays difficult to see with naked eye. Color of wood brown with reddish tinge and axial bands of contrasting luster. Growth rings not clearly delineated. Without tyloses. Wood medium in weight to heavy (sp. gr. 0.60–0.80), hard.

a_2 Pores medium in size to small, single and in multiples, with radial arrangement, rather numerous, visible to naked eye.

Sapele—*Entandrophragma cylindricum* Sprague

b_2 Pores rather medium in size, single and in multiples, not numerous, visible to naked eye.

Sipo, Utile—*Entandrophragma utile* Sprague

c_1 Parenchyma boundary (indistinct apotracheal may also be present).

a_2 Color of wood brown-red with axial bands of contrasting luster. Parenchyma boundary pronounced, visible to naked eye.

a_3 Rays visible to naked eye. Pores medium in size, not numerous, mostly single, seldom in multiples, visible to naked eye, without tyloses. Wood medium in weight to heavy (sp. gr. 0.55–0.90), hard.

Tiama, Gedu Nohor—*Entandrophragma angolense* C.DC.

b_3 Rays difficult to see with naked eye. Pores medium in size, not numerous, mostly in multiples, some single, visible to naked eye. Wood medium in weight to heavy (sp. gr. 0.40–0.85), medium in hardness.

Mahogany, American—*Swietenia macrophylla* King

b_2 Color of wood ash-brown or yellowish. Parenchyma boundary not pronounced.

a_3 Color of wood ash-brown, rather uniform. Parenchyma apotracheal, visible with lens. Pores and rays difficult to see with naked eye. Pores not numerous, small, single and in multiples, few with tyloses. Wood medium in weight (sp. gr. 0.60–0.70) and hardness.

Bété, Mansonia—*Mansonia altissima* A. Chev.

b_3 Color of wood yellowish. Wood rather light (sp. gr. 0.35–0.50), not too soft. Pores not numerous, large or medium in size, single, seldom in multiples, easy to see with naked eye, many with tyloses.

Obeche—*Triplochiton scleroxylon* K. Schum.

454 APPENDIX II

d   Parenchyma not visible with lens.
- a_1 Color of wood yellowish with reddish tinge.
  - a_2 Wood very light, spongy (sp. gr. 0.10–0.20). Pores single and in multiples, large and small, easy to see with naked eye.
    Balsa—*Ochroma lagopus* SW.
  - b_2 Wood medium in weight and hardness.
    - a_3 Color of wood with pronounced reddish tinge. Wood medium in weight (sp. gr. 0.40–0.50) and hardness. Rays visible with lens. Pores numerous, small to medium in size, single and in multiples, visible to naked eye.
      Okoumé, Gaboon—*Aucoumea klaineana* Pierre
    - b_3 Color of wood relatively light, ash-yellow with light reddish tinge. Wood medium in weight (sp. gr. 0.50–0.60) and hardness. Rays visible to naked eye. Pores numerous, small to medium in size, most with tyloses or whitish inclusions, visible to naked eye.
      Aiélé, Canarium—*Canarium schweinfurthii* Engl.
- b_1 Color of wood golden-brown or reddish-brown, with axial bands of contrasting luster.
  - a_2 Rays very easy to see with naked eye. Pores relatively few, rather nonuniform in distribution, single and in multiples, large, with few tyloses, easy to see with naked eye. Wood medium in weight to heavy (sp. gr. 0.60–0.80), hard, golden-brown in color.
    Niangon—*Tarrietia densiflora* Aubrev. et Norm.
  - b_2 Rays not very easy to see with naked eye. Pores uniformly distributed, numerous.
    - a_3 Color of wood red-brown. Pores medium in size, single and in multiples, easy to see with naked eye, many with tyloses. Wood medium in weight (sp. gr. 0.45–0.55) and hardness.
      Acajou, Khaya—*Khaya ivorensis* A. Chev.
    - b_3 Color of wood golden-brown. Pores medium in size or large, single and in multiples, many with tyloses, easy to see with naked eye.
      - a_4 Wood medium in weight (sp. gr. 0.50) and hardness, and relatively light in color. Pores medium to large. Rays difficult to see with naked eye.
        Seraya (Lauan) white—*Parashorea plicata* Brandis
      - b_4 Wood medium in weight to heavy (sp. gr. 0.60–0.80), hard, dark in color. Pores large and medium. Rays not easy to see with naked eye.
        Meranti (Lauan) dark-red—*Shorea pauciflora* King
- c_1 Color of wood brown, lighter or darker, without axial bands of contrasting luster.
  - a_2 Color dark, brown. Pores medium in size, numerous, single and in multiples, few with black inclusions (tyloses). Rays difficult to see with naked eye. Wood medium in weight (sp. gr. 0.45–0.60) and hardness.
    Dibetou—*Lovoa trichilioides* Harms
  - b_2 Color brown, lighter. Pores variable in size (small to large), some with tyloses. Rays visible with lens. Wood heavy (sp. gr. 0.80–0.90), hard.
    Opepe, Bilinga—*Nauclea trillesii* Merr.

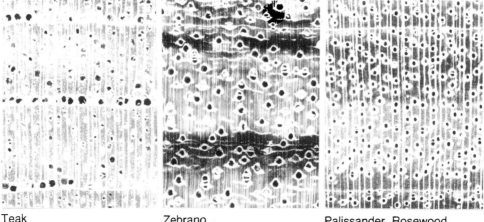

Teak
Tectona grandis

Zebrano
Microberlinia brazzavilensis

Palissander, Rosewood
Dalbergia latifolia

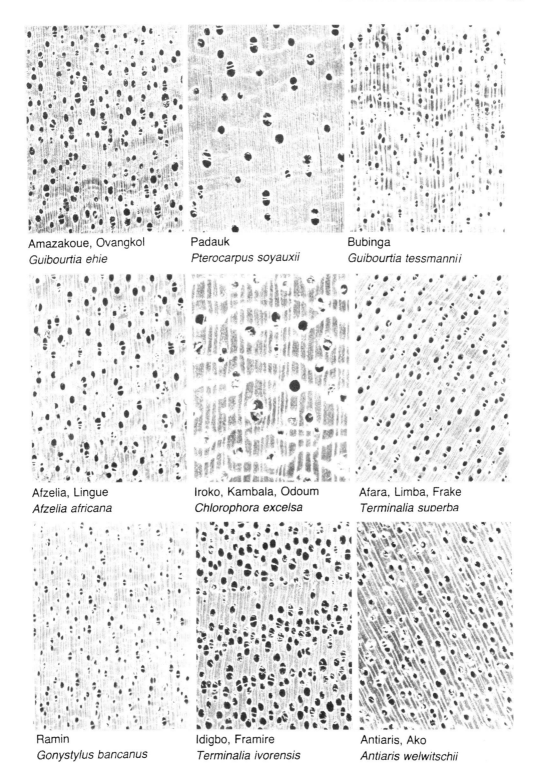

Amazakoue, Ovangkol
Guibourtia ehie

Padauk
Pterocarpus soyauxii

Bubinga
Guibourtia tessmannii

Afzelia, Lingue
Afzelia africana

Iroko, Kambala, Odoum
Chlorophora excelsa

Afara, Limba, Frake
Terminalia superba

Ramin
Gonystylus bancanus

Idigbo, Framire
Terminalia ivorensis

Antiaris, Ako
Antiaris welwitschii

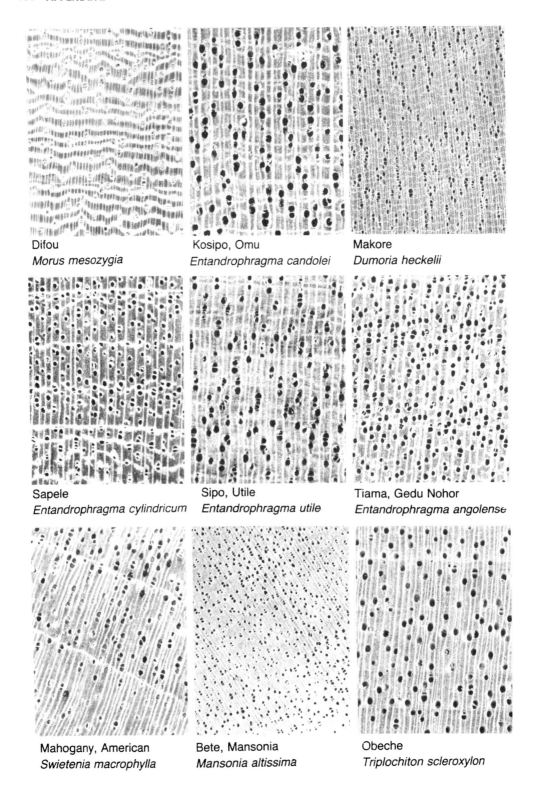

TEMPERATE AND TROPICAL WOODS 457

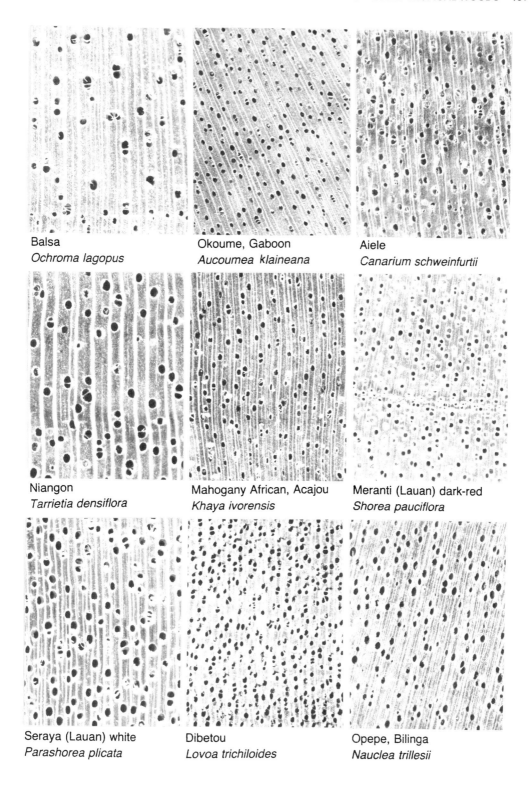

Balsa
Ochroma lagopus

Okoume, Gaboon
Aucoumea klaineana

Aiele
Canarium schweinfurtii

Niangon
Tarrietia densiflora

Mahogany African, Acajou
Khaya ivorensis

Meranti (Lauan) dark-red
Shorea pauciflora

Seraya (Lauan) white
Parashorea plicata

Dibetou
Lovoa trichiloides

Opepe, Bilinga
Nauclea trillesii

GEOGRAPHICAL SOURCE, PROPERTIES, USES

Information on geographical source, properties, and uses of the European and tropical woods included in the preceding keys is presented here in a general manner in order to give a better overall picture of these woods. Geographical source refers to the location of respective forests. The information on properties is supplementary to that given in other chapters (Chapters 8–15) or scattered throughout this book. With regard to uses, note may be made that, in addition to traditional, new uses may develop due to technological advancements while others may cease due to replacement of wood by other materials. Tree dimensions, quality, and available quantity, as well as existing industries and economics, are decisive factors regarding usage possibilities in a certain region.

European[12]

Pine (*Pinus* spp.). *Source.* Several pine species grow naturally in the forests of Europe—from the Mediterranean region to the Scandinavian countries and the USSR. The main species are: *Pinus halepensis*, *P. brutia*, *P. pinea*, *P. maritima* (*P. pinaster*), *P. nigra*, and *P. silvestris*; and minor species, *P. peuce*, *P. leucodermis* (*P. heldreichii*), *P. cembra*, *P. mugo*, etc. The main exotic (planted) species is *P. radiata*.

Properties. Pines are divided into "soft" (wood relatively soft with little resin) and "hard" (wood relatively hard, often resinous). Practically all European pines are hard pines (exception is made by the Balkan pine—*P. peuce*). In general, pine wood is easily worked with tools and machines, dries easily, and finishes well. Sapwood readily impregnated with preservatives, heartwood difficult to treat but naturally more durable especially when resinous. Sapwood very susceptible to stain fungi (blue stain).

Uses. Building constructions, flooring, crates, masts, poles, railroad ties (treated), barrels, furniture, boats, mine timbers, plywood, particleboard, fiberboard, pulp.

Spruce (*Picea abies* = *P. excelsa*). *Source.* From Greece to Northern Europe.

Properties. Wood softer than pine wood, less durable, easily worked with tools and machines, dries easily, finishes well, preservative treatment difficult.

Uses. Building constructions, masts, poles, flooring, musical instruments, crates, boats, particleboard, fiberboard, pulp.

Fir (*Abies* spp.). *Source.* Main species is *Abies alba*, extending from Greece to France, Italy, Austria, Germany. Indigenous to Greece are *A. cephalonica* and a natural hybrid (*A. alba* × *A. cephalonica*).

Properties and Uses. Wood similar to spruce, less easy to work with tools and machines (has harder and larger knots) but easier to treat with preservatives.

Larch (*Larix europea* = *L. decidua*). *Source.* Austria, Germany, Switzerland, Czechoslovakia, Poland.

Properties. Wood heavy and hard (relative to other European softwoods, especially when rings are narrow, 1–2 mm), and, respectively, with good strength, elasticity, and toughness. Moderate shrinkage, good dimensional stability. Durable heartwood even under water (similar to oak), preservative treatment difficult. Careful (slow) drying necessary (tends to split and warp more than spruce and pine). Working with tools and machines may be hindered by exit of resin, especially in sawing.

Uses. Building constructions (both interior and exterior), flooring (parquet), veneer, furniture, cabinets, barrels, silos, masts, boats and ships, garden utensils.

Douglas-Fir (*Pseudotsuga menziesii* = *P. Douglasii*). *Source.* Canada (British Columbia), Montana, California, Colorado, Texas, and Mexico. Wood is produced in Europe (Great Britain, etc.) from planted forests.

Properties. Good strength in relation to density, especially in static bending. Low shrinkage. Medium resistance to attack by insects, fungi, etc. Resistant to preservative treatment. Easy to work with tools and machines (more difficult than pine). Dries well and fast without serious degrade. Knotty wood tends to split. Good nail-holding, satisfactory gluing.

Uses. Building constructions, poles, posts, veneer and plywood, naval constructions, railroad ties, barrels, crates, furniture, pulp, and paper.

Mediterranean Cypress (*Cupressus sempervirens*). *Source.* Syria, Asia Minor, Cyprus, Greece.

Properties. High durability and dimensional stability. Moderate strength, dries easily, worked easily with tools and machines.

Uses. Furniture, masts, building components, poles, boxes, railroad ties, pulp.

Yew (*Taxus baccata*). *Source.* Limited availability (trees scattered in the forests of Europe).
Properties. High durability and dimensional stability. High elasticity and toughness. Best softwood for steam bending. Good machining properties, dries relatively fast with little degrade.
Uses. Furniture, archery, measuring devices, novelties.

Juniper (*Juniperus* spp.). *Source.* Mediterranean countries; some species reach Northern Europe and Central Asia.
Properties. High durability and dimensional stability. Moderate strength. Machines well (some dulling effect on cutting tools), and tends to split in nailing. Dries slowly and tends to check.
Uses. Turnery, carvings, novelties, musical instruments, toys and (when available in large dimensions) furniture, building components, flooring, poles, veneer.

Oak (*Quercus* spp.). *Source.* Throughout Europe, some species limited to the southern part.
Properties. High strength, moderate dimensional stability. Good for steam bending. Resistant to attacks except sapwood, which is usually treated for exterior use. Heartwood resistant to preservative treatment. "Red oaks" (*Q. cerris*) have a lower durability (resistance to fungi and insects), but are treated more easily. Machined well and glued well, good nail- and screw-holding. Drying slow and difficult. Evergreen oaks (*Q. ilex, Q. coccifera*, etc.) are moderately durable, and machined with difficulty (dulling tools).
Uses. Oak is a well-liked wood, used from ancient times for construction of buildings and ships. Other uses: furniture, parquet flooring, barrels (except red oaks), railroad ties. Evergreen oaks: agricultural tools, turnery, charcoal. Cork oak (*Q. suber*) is the source of commercial cork (and cork products).

Chestnut (*Castanea vesca = C. sativa*). *Source.* Southern Europe to Caucasus, Asia, and North Africa. (In most of the area of occurrence, chestnut forests have been killed by a fungus, *Endothia parasitica*).
Properties. High durability, relatively soft, strength lower than oak but better dimensional stability. Machined well, dries slowly.
Uses. Parquet flooring, furniture, posts, poles, beams, mine timbers, house utensils, barrels, particleboard, fiberboard.

Elm (*Ulmus* spp.). *Source.* Southern Europe to Scandinavia and Northern Soviet Union, East Asia, and North Africa (depending on species).
Properties. High elasticity and toughness, suitable for steam bending. Good dimensional stability. Heartwood durable in water or soil. Preservative treatment easier than (white) oak. Dulling cutting tools. Good nailing, screwing, and finishing. Attractive figure.
Uses. Furniture, veneer, boats and ships, musical instruments, agricultural tools, sporting goods, wagons.

Ash (*Fraxinus* spp.). *Source.* Southern to northern Europe, parts of Asia.
Properties. Good strength, especially toughness, very suitable for steam bending, less durable than oak and not suitable for exterior use, moderately resistant to treatment, machined well, dries well and fairly fast. The properties of brown heartwood are not inferior.
Uses. Sporting goods, handles of tools and rifles, wagons, boats and ships, furniture, parquet flooring, agricultural tools, artificial limbs, ladders, beams.

Walnut (*Juglans* regia). *Source.* Indigenous to the Balkans, Southwest and perhaps Central Asia, now extending to a larger area toward north, east, and west.
Properties. Good strength, similar to oak, bends well, dries well but rather slowly, finishes well, medium dimensional stability and medium durability (relatively durable under water), color improves by steaming, sapwood easily treated.
Uses. Furniture (solid or veneered), interior paneling, turnery, carvings, sporting goods, rifle handles, musical instruments.

Black Locust (*Robinia pseudoacacia*). *Source.* Not indigenous to Europe (introduced from North America).
Properties. High durability. Strength similar to oak (toughness similar to ash). Excellent for bending. Resistant to abrasion. Good dimensional stability. Satisfactory machining but nailing difficult. Dries slowly, care is needed. High resistance to preservative treatment.
Uses. Furniture, posts, farm tools, ladders, railroad ties, sporting goods, mining timbers, carriages, wagons.

Mulberry (*Morus alba, M. nigra*). *Source.* Indigenous to China and Northern Iran.
Properties. High durability. Suited for steam

bending. Machines well. Dries with relative difficulty.

Uses. Posts, furniture, farm tools, barrels, turnery.

Tree-of-Heaven (*Ailanthus altissima*).
Source. Indigenous to China.

Properties. Similar to ash in many mechanical properties. Satisfactory machining, but often contains tension wood. Fast growth.

Uses. Simple furniture, pulp, etc.

Beech (*Fagus silvatica*).
Source. Central to Southern Europe from Spain to Crimaea, Scotland and Scandinavia to Sicily and Greece. Two other secondary species are recognized extending *F. orientalis* to Northern Iran and *F. moesiaca* in the Balkans; the wood has no practical differences.

Properties. High shrinkage, low dimensional stability under changing humidity conditions. Susceptible to fungal attack, but easily treated with preservatives (except "red heart"). Wood suited to steam bending. Steaming improves the color of beech and diminishes the adverse visual effect of "red heart," raising the value of the wood. Good machining properties (including peeling), glued and finished well. Fast drying results in splitting and warping. Steaming considered to exercise an improving effect on properties and workability.

Uses. Furniture (chairs, etc.), railroad ties, parquet flooring, veneer, plywood, particleboard, fiberboard, pulp, tool handles, piano parts, boats, toys, textile weaving shuttles, fuelwood.

Plane, Sycamore (*Platanus orientalis*).
Source. Southeastern Europe to Asia Minor.

Properties. Similar to beech but lower in strength. Not durable in contact with the soil. Machined with relative difficulty. Tends to warp in drying.

Uses. Crates, small items, turnery, furniture, tools, fuelwood.

Hornbeam (*Carpinus* spp.).
Source. European hornbeam (*C. betulus*) in the forests of Europe up to Northern Sweden and Southwest Soviet Union, in Caucasus and Asia Minor. Eastern hornbeam (*C. orientalis*) in Southeast Europe, from Italy to Caucasus, in Asia Minor and Iran.

Properties. Good strength mainly in toughness and splitting. Good for steam bending. Resistant to abrasion. Susceptible to fungal attack. Preservative treatment relatively easy. Machined well in green condition. Nailed, painted, and finished well. Dries well and fairly fast.

Uses. Tool handles, wooden parts of vehicles, flooring, piano parts, textile shuttles, fuelwood.

Alder, European (*Alnus glutinosa*).
Source. In forests of Europe up to Caucasus and Siberia, North and West Asia, North Iran and Northwest Africa.

Properties. Moderate shrinkage, low strength. Not suited for steam bending. Susceptible to insect attack. Easy preservative treatment. Dries well and fairly fast. Machined well (when dry).

Uses. Plywood (interior plies), furniture (dyed to resemble walnut, etc.), crates, pencils, toys, turnery, small items, pulp.

Maple (*Acer* spp.).
Source. Main species are Great maple (*A. pseudoplatanus*) in forests of South and Central Europe, Crimaea, Asia Minor and West Caucasus; Norway maple (*A. platanoides*) in Europe and Caucasus, and Field maple (*A. campestre*) in Europe, Caucasus, North Iran and Algeria.

Properties. Moderate shrinkage. Strength similar to beech. Moderate elasticity. High resistance to abrasion. Wood suited to steam bending. Susceptible to fungal attack. Sapwood attacked by insects. Unsuitable for exterior uses without preservative treatment. Easy to treat. Good machining properties. Glues and finishes without difficulty. Care needed in drying to avoid discoloration. Mineral streaks may be present.

Uses. Furniture, veneer, flooring, sporting goods and small items, carvings, musical instruments, tool handles, textile shuttles, toys.

Service Tree, Mountain Ash, etc. (*Sorbus* spp.).
Source. In forests of Europe, some species (*S. aucuparia*) up to Siberia, others (*S. domestica*) in South and Central Europe, North Africa, and Asia Minor.

Properties. High hardness and elasticity, similar strength to oak but better in toughness and splitting. Moderate durability (should be treated for exterior use). Machined with difficulty (turns well). Good dimensional stability after careful drying.

Uses. Carvings, tools, turnery, carriages.

Birch (*Betula verrucosa* = *B. pendula*).
Source. Europe (including Greece) and Asia Minor.

Properties. Strength better than oak. High elasticity. Wood suited to steam bending. Susceptible to fungi, attacked by insects. Easy preservative treatment. Machined well after drying. Good gluing, dyeing, and finishing properties. Dries fairly well and fast with little degrade.

Uses. Furniture, veneer, plywood, turnery, sporting goods, musical instruments, toys, pulp, boats, crates.

Lime, Basswood (*Tilia* spp.).
Source. In most forests of Europe.

Properties. Good strength in relation to light weight. Moderate shrinkage. Susceptible to insects. Easy to treat with preservatives. Machined easily. Dries well and fast with little degrade.

Uses. Carvings, turnery, toys, pencils, sporting goods (tennis rackets), beehives, musical instruments, piano parts, plywood, matches, furniture (blind parts), charcoal (drawing, etc.).

Hophornbeam (*Ostrya carpinifolia*).
Source. Southern Europe and Asia Minor.

Properties and Uses. Similar to hornbeam.

Olive Tree (*Olea europaea*).
Source. Mediterranean countries to Southern Crimaea, Canary Islands.

Properties. High durability. Machines well and dries well.

Uses. Furniture, carriages, turnery, tool handles, small items, fuelwood.

Poplar (*Populus* spp.).
Source. European aspen (*P. tremula*) in Europe, Siberia, Asia, and Northwest Africa, white poplar (*P. alba*) in South, Central, and East Europe, North Africa, and West Asia, black poplar (*P. nigra*) also in South, Central, and East Europe, North Africa, and Asia. Hybrid poplars (mainly Euramerican) widely cultivated due to fast growth.

Properties. Moderate shrinkage. Good strength in relation to light weight. Not suited to steam bending. Machined well but difficulties often arise due to presence of tension wood. Dries well and fairly fast (sometimes wet spots remain in the interior). Dyes and glues well. Susceptible to fungal and insect attack.

Uses. Boxes and crates, furniture (blind parts), matches, artificial limbs, toys, cheap structures, small items, battery (separation) sheets, veneer, plywood, excelsior, particleboard, fiberboard, pulp, and paper.

Willow (*Salix* spp.).
Source. In all or most of Europe (wet sites).

Properties. Moderate strength (good toughness), susceptible to fungal and insect attack. Sapwood is treated well, but not heartwood. Machines well (presence of tension wood creates difficulties). Glues well. Dries well and fairly fast without special difficulties (wet spots sometimes remain).

Uses. Boxes and crates, furniture (blind parts), artificial limbs, cheap structures, veneer, plywood, particleboard, fiberboard, pulp, baskets, farm tools, posts, sporting goods.

Horse Chestnut (*Aesculus hippocastanum*).
Source. Greece, Bulgaria, Albania.

Properties. Low shrinkage. Suited to steam bending. Attacked by fungi and insects. Treated easily with preservatives. Machines well. Dries well with little degrade. Glues, dyes, and finishes well.

Uses. Crates, toys, cigar boxes, artificial limbs, furniture (blind parts), drawing tables, tennis rackets, turnery.

Tropical Woods

Teak (*Tectona grandis*).
Source. Native to Burma and Thailand, India, Indochina, Java. Planted in tropical areas of Africa and Latin America.

Properties. Dries slowly with little or no degrade. High dimensional stability. Heartwood very durable to fungi, termites, and marine borers, very resistant to preservative treatment. Resistant to chemicals. May cause dermatitis. Nails and paints well. Machines well to smooth finish but is abrasive. Glues satisfactorily.

Uses. Shipbuilding, building constructions, flooring, furniture, veneer, plywood, musical instruments, poles, mine timbers, joinery, toys, turnery, paneling, carvings.

Zebrano, Zingana, African Zebrawood (*Microberlinia brazzavillensis*).
Source. West Africa, mainly Cameroon and Gabon.

Properties. Difficult to dry, tends to warp (should be quarter-sawn and dried carefully). Good shock resistance. Heartwood durable, resistant to termite attack, very resistant to preservative treatment. Machines well. Good gluing and painting properties. Finishing sometimes difficult due to interlocked grain. Veneer sometimes fragile.

Uses. Decorative veneers and plywood, paneling, building constructions, flooring, shipbuilding, mine timbers, furniture, tool handles, ladders, skis, toys, turnery, joinery.

Amazakoue, Ovangkol, Ehie, Mozambique (*Guibourtia ehie*).
Source. West Africa (Ivory Coast, Ghana, Gabon)

Properties. Difficult to dry. After felling, it may

split and be attacked by insects. May discolor in contact with metals. Resistant to preservative treatment. Saws with relative difficulty, planes to a good finish, but wears cutting tools due to silica content. Glues and paints well.

Uses. Decorative veneers, plywood, paneling, furniture (a walnut-like wood), building constructions, flooring, mine timbers, musical instruments, joinery, turnery, toys, carvings, poles.

Palissander, Indian Rosewood, Shisham (*Dalbergia latifolia*). *Source.* India, Brazil, Madagascar, Honduras, and other Central American countries.

Properties. Dries well with little degrade. Working with tools and machines not easy, needs care; turns well, and slices well for veneer. Glues well and takes a good finish. Resistant to attack by fungi and termites, durable in water.

Uses. Furniture, turnery, carvings, paneling, building constructions, decorative veneers, toys, musical instruments.

Padauk (*Pterocarpus soyauxii*). *Source.* West Africa (Cameroon, Nigeria, Andaman Islands)

Properties. Dries slowly without degrade. High dimensional stability and durability. Saws slowly but well—sawdust may cause respiratory problems. Resistant to preservative treatment. Holds nails and screws well, seldom splits.

Uses. Furniture, flooring, boat-building, decorative veneer, tool handles, sporting goods, turnery, toys, joinery, ladders, building constructions, carvings, toys.

Bubinga (*Guibourtia tessmannii*). *Source.* Tropical Africa (Cameroon, Gabon)

Properties. Dries slowly but well. High dimensional stability and durability. Saws well, but wears cutting tools due to silica content, and fine sawdust often irritates skin, eyes, and throat. Resistant to preservative treatment. Moderately resistant to marine borers and termites. Glues well, gives a good finish, difficult to stain.

Uses. Durable building constructions, flooring, furniture, decorative veneers, joinery, plywood, handles, ladders, paneling, railroad ties, particleboard, fiberboard, poles, toys, turnery.

Afzelia, Lingue (*Afzelia africana*). *Source.* West, Central, and East Africa (Sierra Leone, Ivory Coast, Sudan, Uganda, Ghana, Angola, Mozambique, Zambia, Zimbabwe).

Properties. Dries well with little or no degrade. Very good dimensional stability. Difficult to machine, wears tools. Gives a good finish, but sometimes difficult to stain. Very durable. Sawdust may be irritating. Resistant to acids and fire. Moderate in bending properties.

Uses. Exterior building uses including harbor work, mine timbers, furniture, veneer and plywood, railroad ties, pulp, particleboard, fiberboard, poles, turnery.

Iroko, Kambala, Odoum (*Chlorophora excelsa*). *Source.* Tropical Africa (Ivory Coast to Angola, Sudan to Mozambique)

Properties. Good dimensional stability and durability. Does not bend easily after steaming. Resistant to marine borers and fire. Wet sawdust may cause dermatitis. Dries easily. Very resistant to preservative treatment. Machines well but wears cutting tools. Holds nails and screws. Glues satisfactorily. Stains and finishes well. Discolors in contact with iron.

Uses. Building constructions, flooring, boat-building, mine timbers, furniture, veneer and plywood, pulp, paneling, railroad ties, joinery, particleboard, fiberboard, poles, toys, turnery.

Afara, Limba, Fraké, Korina (*Terminalia superba*). *Source.* West Africa.

Properties. Dries easily with little or no degrade. Dimensionally stable. Machines well, takes a good finish. Heartwood nondurable, readily attacked by insects and fungi. Glues well and nails satisfactorily but tends to split. Very resistant to preservative treatment. Moderate in strength, sometimes brittle.

Uses. Veneer, plywood, furniture, interior joinery, flooring, boats, musical instruments, paneling, turnery, toys, poles, pulp, particleboard, fiberboard.

Ramin (*Gonystylus bancanus*). *Source.* Philippines, Malaysia, Indonesia, Borneo, Sumatra, Java.

Properties. Dries easily but tends to split, end-check, and discolor. Low dimensional stability and low durability. Susceptible to attacks by fungi (including blue stain) and insects (including termites), not suitable for exterior use and in contact with water. Easily treated with preservatives. Machines easily, and glues, stains, and finishes well. Strength similar to beech and used as alternative.

Uses. Furniture, flooring, paneling, doors, toys, plywood, particleboard, fiberboard, picture frames, joinery, turnery.

Idigbo, Framiré (*Terminalia ivorensis*). *Source.* West Africa (Nigeria, Ghana, Ivory Coast).
Properties. Dries easily with little degrade and has good dimensional stability. Easily worked with tools and machines, but sawdust irritates the skin. Turns, glues, stains, and finishes well. Holds nails and screws. Moderately resistant to termites. Does not steam-bend easily. Sometimes brittle (brittle heart common). Discolors in contact with iron, and may stain fabrics under moist conditions. Resistant to preservative treatment.
Uses. Light constructions, flooring, furniture, sporting goods, veneer, plywood, paneling (oak-like appearance), matches, boxes and crates, joinery, turnery, poles, particleboard, fiberboard.

Antiaris, Ako (*Antiaris welwitschii*). *Source.* West, Central, and East Africa (Senegal, Nigeria, Cameroon, Kenya, Tanzania).
Properties. Dries rapidly but tends to warp and end-split. Machines easily, glues well, nails with some difficulty. Low durability, attacked by insects and blue stain. Easily treated with preservatives.
Uses. Light constructions, furniture parts, veneer, plywood, boxes and crates, matches, joinery, carvings, paneling.

Difou (*Morus mesozygia*). *Source.* Central Africa (Senegal to Cameroon and Gabon).
Properties. Wood heavy and hard. Dries and saws well but wears cutting tools. Holds nails and screws. Glues and finishes well. Attacked by fungi, insects, marine borers. Resistant to fire. Very resistant to preservative treatment.
Uses. Building constructions, flooring, furniture, veneer, paneling, joinery, turnery, toys.

Kosipo, Omu (*Entandrophragma candolei*). *Source.* Central Africa, Ivory Coast.
Properties. Dries slowly and tends to warp. Machines well, holds nails and screws, stains and glues well, finishes well. Low strength in relation to weight. Does not steam-bend easily. Attacked by insects and marine borers. Resistant to preservative treatment.
Uses. Light constructions, flooring, furniture, joinery, boats, veneer, plywood, boxes and crates, paneling, toys, turnery.

Makoré (*Dumoria heckelii* = *Tieghmella heckelii*). *Source.* West Africa (Ghana, Ivory Coast).
Properties. Dries slowly with little degrade. Good dimensional stability. Good mechanical properties. Resistant to attack by insects and fungi. Blunts cutting edges, especially when working with dry wood, due to high silica content, and produces dust which irritates the nose, throat, and skin. Glues well, takes good finish, sometimes splits in nailing.
Uses. Attractive wood mainly used in furniture, solid or in veneer form; high-quality joinery, boats, flooring, marine plywood, cabinetwork.

Sapele (*Entandrophragma cylindricum*). *Source.* West, Central, and Southern Africa (Cameroon, Uganda, Congo, Zaire, Ivory Coast).
Properties. Drying and machining influenced by presence of interlocked grain. Flat-sawn lumber tends to warp and check. Careful drying required. Moderate dimensional stability and durability. Discolors in contact with iron. Good gluing, nailing, and finishing properties. Resistant to preservative treatment. Mahogany-like wood, heavier, harder, and stronger than African mahogany (*Khaya ivorensis*), more difficult to work and less dimensionally stable.
Uses. Veneer, furniture, paneling, boats, plywood, flooring, joinery, toys, musical instruments.

Sipo, Utile (*Entandrophragma utile*). *Source.* West and Central Africa (Ivory Coast, Cameroon, Zaire, Ghana, Uganda, Angola, Sierra Leone).
Properties. Generally similar to Sapele but relatively easier to dry (unless it has pronounced interlocked grain); more durable and less decorative.
Uses. Generally similar to Sapele.

Tiama, Gedu Nohor (*Entandrophragma angolense*). *Source.* West, Central, and East Africa (Nigeria, Ivory Coast, Cameroon, etc.).
Properties. Dries rapidly but may warp and split due to interlocked grain. Machines well but tends to wear cutting tools. Sawdust irritating to the nose and throat. Good gluing, nailing, and finishing properties. Good dimensional stability, moderate durability. Resistant to preservative treatment. Veneer may be fragile. A walnut-like wood.
Uses. Furniture, joinery, cabinetwork, building constructions, flooring, paneling, sporting goods, veneer, plywood, musical instruments, boxes and crates, toys, boats, carvings.

American Mahogany, (*Swietenia macrophylla*). *Source.* Mexico, Honduras, Brazil, Peru, Venezuela, Bolivia, Colombia.

Properties. Dries easily without appreciable degrade and has good dimensional stability. Machines easily, steam-bends well, and gives excellent finish. Moderately durable to fungi and insects, little resistance to marine borers, and resistant to preservative treatment.

Uses. Valued as the best of the commercial mahoganies[13] for furniture, fine joinery, veneer, paneling, musical instruments, boats, turnery, carvings.

Bete, Mansonia (*Mansonia altissima*).

Source. West Tropical Africa (Ivory Coast, Cameroon, Ivory Coast, Ghana, Nigeria, Namibia).

Properties. Dries easily with little degrade, but tends to split near knots. Moderate dimensional stability. Machines easily with little dulling effect, has good nailing, gluing, steam-bending, and finishing properties. Durable to fungi and termites, very resistant to preservative treatment. Sawdust causes irritation to the nose, throat, and skin. Attractive—walnut-like wood.

Uses. High-quality furniture and cabinetwork, veneer, plywood, paneling, flooring, musical instruments, turnery, building contructions. The bark contains a poisonous substance.

Obeche, Samba, Abachi (*Triplochiton scleroxylon*).

Source. West Africa (Guinea, Liberia, Ivory Coast, Nigeria, Cameroon).

Properties. A very light hardwood. Dries rapidly with little or no degrade, and has good dimensional stability. Machines well but sharp cutting edges are necessary, and end-grain tends to crumble. Not durable. Readily attacked by insects and fungi, including blue stain. Good gluing, nailing, and finishing properties. Resistant to preservative treatment. Grain typically interlocked, gives a striped figure.

Uses. Light constructions, furniture parts, cabinetwork, veneer, plywood, boxes and crates, blockboard (core plywood), musical instruments, artificial limbs, particleboard, fiberboard, pulp, toys, turnery.

Balsa (*Ochroma lagopus*).

Source. Tropical America (Bolivia, Ecuador, Colombia, Southern Mexico, Guatemala, Honduras, Panama, Brazil, Costa Rica, Nicaragua, Venezuela).

Properties. Balsa is the lightest wood in use. Dries easily (kiln-drying preferred) and has good dimensional stability. Easy to work with sharp cutting tools. Does not hold nails and screws, but glues, paints, and finishes satisfactorily. Low thermal conductivity, high sound absorption. Low durability, susceptible to attack by insects and fungi, including blue stain. Resistant to preservative treatment.

Uses. For thermal insulation, sound absorption, antivibration, and buoyancy purposes (rafts, lifebelts, floats), core in metal-faced sandwich constructions, in aircraft floors and partitions, model airplanes, etc.

Okoumé, Gaboon (*Aucoumea klaineana*).

Source. West tropical Africa (Gabon, Equatorial Guinea, Congo).

Properties. Dries easily with little degrade. Abrasive to saws due to silica content, but easily peeled to veneer. Holds nails and screws, and glues well. Susceptible to fungi (including blue stain) and insects. Resistant to preservative treatment.

Uses. Veneer, plywood, furniture, flooring, light constructions, musical instruments, boxes and crates, joinery, paneling, turnery, toys, carvings.

Aiélé, Canarium (*Canarium schweinfurthii*).

Source. East, Central, and West Africa (Gabon, Congo, Zaire, Uganda, Nigeria).

Properties. Dries slowly and may check, warp, or collapse. Moderate dimensional stability. Saws with difficulty due to silica, but working with other (sharp) tools is satisfactory. Holds nails and screws, glues well, and finishes well. Not resistant to fungi (including blue stain) and insects, very resistant to preservative treatment.

Uses. Veneer, plywood, flooring, furniture, paneling, joinery, light constructions, boxes and crates, poles, toys, turnery, particleboard, fiberboard, pulp.

Niangon (*Tarrieta densiflora*).

Source. West Africa (Ivory Coast, Gabon, Cameroon, Sierra Leone, Ghana, Liberia).

Properties. Dries without difficulty, but sometimes tends to warp. Moderate dimensional stability. Moderate blunting of cutting tools. Difficult to work when "gummy"; excess of gum may cause stain and gluing problems (improvement is possible by treatment of surfaces with chemicals—e.g., caustic soda or ammonia). Durable, very resistant to preservative treatment. Wood similar to African mahogany.

Uses. Light building constructions, flooring, boatbuilding, furniture, veneer, joinery, turnery, toys, poles, particleboard, fiberboard, pulp.

African Mahogany, Acajou (*Khaya ivorensis*).

Source. West Africa (Guinea, Gabon, Cameroon, Ghana, Ivory Coast, Liberia, Congo, Sierra Leone).

Properties. Dries well with little degrade. Machines well, but presence of tension wood creates problems. Good dimensional stability, holds nails and screws, glues, stains, and finishes well. Moderately durable to fungi and insects. Very resistant to preservative treatment. Sawdust may cause dermatitis.

Uses. Furniture, joinery, doors, boatbuilding, veneer, plywood, paneling, flooring, cabinetwork.

White Seraya, White Lauan (*Parashorea plicata*). *Source.* Borneo, Philippines.

Properties. Dries slowly with some tendency to warp and split. Good dimensional stability. Machining satisfactory, holds nails and screws, and glues well. Staining and finishing satisfactory. Low resistance to insect and fungal attack. Very resistant to preservative treatment.

Uses. Furniture, boatbuilding, flooring, veneer, plywood, musical instruments, light constructions.

Dark Red Meranti, Red Lauan, Philippine Mahogany (*Shorea pauciflora*). *Source.* Malaysia, Indonesia, Philippines.

Properties. Dries slowly with some tendency to degrade (warp, split). Good dimensional stability. Machines without difficulty but not recommended for turnery. Holds nails and screws, glues and finishes well. Moderately durable, resistant to preservative treatment.

Uses. Veneer, plywood, paneling, furniture, boatbuilding, building constructions, flooring, cabinetwork, joinery.

Dibetou, African Walnut (*Lovoa trichilioides*). *Source.* West Africa (Sierra Leone, Gabon, Nigeria, Zaire).

Properties. Dries rapidly and well, and has good dimensional stability. Machines well but interlocked grain requires sharp tools. Holds nails and screws, glues and finishes well. Moderate steam-bending properties. Moderately durable, very resistant to preservative treatment.

Uses. Furniture, joinery, veneer, plywood, cabinetwork, turnery, paneling, gunstocks, building constructions, boatbuilding, boxes and crates, toys, poles.

Opepe, Bilingua (*Nauclea trillesii = N. diderichii*). *Source.* West Africa (Ivory Coast, Gabon, Ghana, Nigeria, Angola, Uganda, Congo, Cameroon, Guinea, Zaire).

Properties. Dries slowly with little degrade. Machines well but care is needed due to interlocked grain. Good dimensional stability. Some tendency to split in nailing. Glues and finishes well. Poor steam bending. Resistant to fungi and marine borers, moderately resistant to termites. Resistant to preservative treatment.

Uses. Marine constructions, structural work, poles, flooring, furniture, railroad ties, wagons, mine timbers, veneer.

REFERENCES

1. Beekman, W. B., ed. 1964. *Elsevier's Wood Dictionary, Vol. I.* Amsterdam, New York: Elsevier.
2. Begemann, H. F. 1977. *Was ist Holz?* Gernsbach: DBV Verlag.
3. Boas, I. H. 1947. *The Commercial Timbers of Australia: Their Properties and Uses.* Melbourne, Australia: C.S.I.R.O.
4. Bolza, E., and W. G. Keating. 1972. *African Timbers—The Properties, Uses and Characteristics of 700 Species.* Melbourne, Australia: C.S.I.R.O.
5. Bosshard, H. H. 1983. *Holzkunde*, Vol. 1. Basel, Stuttgart: Birkhäuser.
6. Boutelje, J. B. 1980. *Encyclopedia of World Timbers.* The Swedish Forest Products Research Laboratory.
7. Brazier, J. B., and G. L. Franklin. 1961. Identification of Hardwoods: A Microscopic Key. For. Prod. Res. Lab., England, Bull. No. 46.
8. Brown, H. P., A. J. Panshin, M. Seeger, and R. Trendelenburg. 1932. *Das Holz der forstlich wichtigsten Bäume Mitteleuropas.* Hanover: Verlag Schaper.
9. Burgess, P. F. 1966. Timbers of Sabah. For. Dept., Sabah, Malaysia.
10. Cenerini, M., and M. L. Edlmann Abbate. 1984. Usi e Proprieta Tecnologiche di Legni di Latifoglie Americane. Inst. Ric. Legno XXXII, Firenze.
11. Centre Technique du Bois. 1958. Principaux Bois Indigènes et Exotiques en France: Carateristiques, Sommaires et Emplois. Paris.
12. Centre Technique Forestier Tropical. 1958. Bois Tropicaux. Nogent s/Marne, France.
13. Centre Technique Forestier Tropical. 1966. Bois Tropicaux. Nogent s/Marne, France.
14. Chudnoff, M. 1984. *Tropical Timbers of the World.* U.S.D.A., Forest Service. Ag. Handbook 607.
15. Clifford, N. 1953. *Commercial Hardwoods: Their Characteristics, Identification and Utilization.* London: Pitman and Sons.
16. Core, H. A., W. A. Côté, and A. C. Day. 1978. *Wood Structure and Identification.* Syracuse, New York: Syracuse University Press.
17. Dahms, K. G. 1968. *Afrikanische Exporthölzer.* Stuttgart: DRW-Verlags.
18. Dreisbach, J. F. 1952. *Balsa Wood and Its Properties.* Columbia, Connecticut: Columbia Graphs.
19. Edlin, H. L. 1969. *What Is That? A Manual of Wood Identification.* London: T. and H. Ltd.
20. Farmer, R. H. 1972. *Handbook of Hardwoods* (2nd ed., revised). London: Dept. Environment, For. Prod. Res. Lab., England, HMSO.
21. Forest Products Research Laboratory, England. 1953.

An Atlas of End-Grain Photomicrographs for the Identification of Hardwoods. Bull. No. 26, HMSO, London.
22. Forest Product Research Laboratory, England. 1960. Identification of Hardwoods—A Lens Key. Bull. No. 25, HMSO, London.
23. Gayer, S. 1954. *Die Holzarten und ihre Verwendung in der Technik*. Leipzig: Fachbuchverlag.
24. Gerry, E. 1957. Key for Identification of Woods Used for Box and Crate Construction. U.S. For. Prod. Lab. Report No. 258.
25. Giordano, G. 1976. *Tecnologia del Legno*, Vol. 3. Torino: Unione Tipografico-Editrice.
26. Giordano, G., L. Demi, G. Rapi, and L. Demi. 1957. Del riconoscimento della provenienza dei legni di alcuni pini (*Pinus sylvestris*, *P. laricio*, *P. montana*, etc.) in base a caratteri macroscopici, microscopici, chimici e chimico-fisici. *Cons. Naz. d. Ricerche, Centr. Nation. Legno* 1:9–65, Firenze.
27. Giordano, G., E. Orlandi, and G. Rapi. 1965. Sur l'emploi de la gas-chromatographie pour reconnaitre les pins des differentes espèces. *Holz Roh-Werkst.* 23:31.
28. Gottwald, H. 1958. *Handelshölzer*. Hamburg: Holzmann Verlag.
29. Gregory, M. 1980. Wood identification: An annotated bibliography. *IAWA Bull.* n.s. 1(1–2):3–41.
30. Greguss, P. 1955. *Identification of living gymnosperms on the basis of xylotomy*. Budapest: Akademiai Kiadó.
31. Greguss, P. 1959. *Holzanatomie der Europäischen Laubhölzer und Sträucher*. Budapest: Akademiai Kiadó.
32. Harrar, E. S., ed. 1957. *Hough's Encyclopedia of American Woods*, Vols. I–XI. New York: Speller and Sons.
33. Hartig, R. 1890. *Die anatomischen Unterscheidungsmerkmale der wichtigeren in Deutschland wachsenden Hölzer*. München: M. Reiger.
34. Huber, B., and Ch. Rouschal. 1954. *Mikrophotographischer Atlas Mediterraner Hölzer*. Berlin: Haler Verlag.
35. Hudson, R. H. 1960. The anatomy of the genus *Pinus* in relation to its classification. *J. Inst. Wood Sci.* 6:26–46.
36. Jane, F. W. 1970. *The Structure of Wood*. London: A&C Black.
37. Jaquiot, C. 1955. Atlas d'anatomie des bois conifères. Centre Technique du Bois, Paris.
38. Jay, B. A. 1968. *Timbers of West Africa*. High Wycomb, Bucks, England: TRADA.
39. Knigge, W., and H. Schulz. 1966. Grundriss der Forstbenutzung. Berlin: P. Parey.
40. Knuchel, H. 1954. *Das Holz*. Frankfurt: Verlag Sauerländer & Co.
41. Koehler, A. 1926. The Identification of Furniture Woods. U.S.D.A. Misc. Circ. No. 66.
42. Kriebs, D. A. 1950. Commercial Foreign Woods on the American Market. Dept. Botany, Pennsylvania State University.
43. Kukachka, B. F. 1960. Identification of coniferous woods. *Tappi* 43:887–96.
44. Kutsha, N. P., and L. L. Emery. 1970. Foreign Woods in Maine—1969. Tech. Bull. 45, University Maine.
45. Miles, A. 1978. *Photomicrographs of World Woods*. London: BRE, HMSO.
46. Miller, R. B. 1980. Wood identification via computer. *IAWA Bull.* n.s. 1(4):154–160.
47. Nairn, P. (ed.). (Undated). *Wood Specimens—100 Reproductions in Color*. London: The Nema Press; *A Second Collection of Wood Specimens—100 Reproductions in Color*. London: Tothill Press (1957).
48. Mullins, E. J., and T. S. McKnight, eds. 1981. *Canadian Woods* (3rd ed.). University of Toronto Press.
49. Normand, D. 1950, 1955, 1960. *Atlas des Bois de la Côte d'Ivoire*, Vols. 1–3. France: Centre Technique Forestier Tropical.
50. Normand, D., 1972, 1976. *Manuel d'Identification des Bois Commerciaux*, Vols. 1–3. France: Centre Technique Forestier Tropical.
51. Panshin, A., and C. De Zeew. 1980. *Textbook of Wood Technology* (4th ed.). New York: McGraw-Hill.
52. Petric, B. 1964. Microscopic identification of important native and in Yugoslavia cultivated conifer woods. *Drvna Ind.* 15:178–191.
53. Phillips, E. W. J. 1959. Identification of Softwoods by Their Microscopic Structure. Dept. Sci. Ind. Res., For. Prod. Res. Bull. No. 22, London.
54. Record, S. J. 1934. *Identification of the Timbers of Temperate North America*. New York: John Wiley & Sons.
55. Rendle, B. J. 1969. *World Timbers*, Vol. 1. Europe and Africa. London: E. Benn Ltd.
56. Reno, J., and B. F. Kukachka. 1950. Wood Identification Chart, No. 1: Commercial Hardwoods. Chicago: Vance Publ. Corp.
57. Roth, F. 1895. Timber (An elementary discussion of the characteristics and properties of wood). U.S. Govern. Printing Office, Washington, D.C.
58. Sachsse, H. 1984. *Einheimische Nutzhölzer*. Berlin: P. Parey.
59. Scheiber, C. 1965. *Tropenhölzer*. Leipzig: VEB Fachbuchvertag.
60. Schimich, P. 1951. The anatomical structure of wood of some *Juniperus* species. *Ann. Fac. Agr. Silv.* 1949–50, 3:181. University Skopje, Yugoslavia.
61. Schoonover, S. E. 1951. *American Woods*. Santa Monica, California: Walting & Co.
62. Schwankl, A. 1959. *Welches Holz ist das?* Stuttgart: Frankhsche Verlagshandlung.
63. Scott, H. M. 1952. Burglars and bordered pits. *J. So. African For. Ass.* 22:2–6.
64. Swan, E. P. 1966. Chemical methods of differentiating the wood of several western conifers. *For. Prod. J.* 16(1):51–54.
65. Titmuss, F. H. 1971. *Commercial Timbers of the World*. London: Technical Press.
66. Trendelenburg, R., and H. Mayer-Wegelin. 1955. *Das Holz als Rohstoff*. München: Hanser Verlag.
67. Tsoumis, G. 1964. Microscopic identification of the

wood of the pine species of Greece. *To Dasos* 35:34–38 (Greek, English summary).
68. Tsoumis, G. 1968. *Wood as Raw Material*. Oxford, New York: Pergamon Press.
69. Tsoumis, G. 1974. Practical application of wood identification to detection of crimes. *Dassica Chronica* 16(3):77–82, 84. (Greek).
70. Tsoumis, G. 1979. Greek and Tropical Woods. Thessaloniki (Greek).
71. Tsoumis, G. 1980. Adverse health effects associated with mechanical processing of wood, *Iatriki* 38:75–77. (Greek).
72. Tsoumis, G. 1983. SEM observations on irradiated old wood. *IAWA Bull.* 4(1):41–45.
73. Tsoumis, G. 1985. Identification of European conifers from sawdust. In *Xylorama*, ed. L. J. Kucera, pp. 198–203. Basel: Birkhäuser.
74. Tsoumis, G. 1990. Wood identification in the detection of crimes. *J. Inst. Wood Sci.* 12(2):93–95.
75. Tsoumis, G., E. Voulgaridis, and P. Noulopoulos. 1978. A key for identification of 30 tropical woods imported to Greece. *Annals School Agr. For.*, Aristotelian University K:241–260.
76. Tsoumis, G., and N. Athanasiadis. 1981. *Systematic Forest Botany: Trees and Shrubs of the Forests of Greece* (2nd ed.). Thessaloniki (Greek).
77. Tsoumis, G., and N. Kezos. 1981. Identification of conifers from sawdust. *Annals Dept. Agr.*, Aristotelian University 24:49–61 (Greek, English summary).
78. U.S.D.A., Forest Products Laboratory. 1956. *Wood—Colors and Kinds*, Agr. Handbook No. 101.
79. U.S.D.A., Forest Products Laboratory. Technical Notes Nos. 116/1952, 103/1953, 215/1957, 198/1958, D-11/1958, and 125/1961.
80. Vasiljevic, S. 1950. About the width of xylem rays on certain species of the genus *Acer. Bull. Coll. For.* I: 117–151. University of Belgrade.
81. Vorreiter, L. 1949. *Holztechnologisches Handbuch*, Vol. I, Vienna.
82. Wagenführ, R. 1964. Die Bestimmung von Holzarten mit Hilfe von Machinenlochkarten. *Holztechnologie* 5:14–16.
83. Wagenführ, R., and C. Scheiber. 1974. *Holzatias*. Leipzig: VEB Fachbuchverlag.
84. Wheeler, E. A., C. A. LaPasha, T. Zack, and W. Hatley. 1986. Computer-aided wood identification: Reference manual. N. Carolina Ag. Res. Serv. Bull. No. 474.
85. Zucconi, L., G. Rapi, and E. Orlandi. 1957. Del riconoscimento del legno dei pini Mediterranei—Indagine macroscopica. Indagini chimicofisiche. *Cons. Naz. d. Ricerche, Centr. Nation. Legno* 2:1–23, Firenze.

FOOTNOTES

[1] Methods of chemical analysis have also been applied in a few cases, but these need specialized equipment.

[2] This is a macroscopic key. Microscopic characteristics are enclosed in parentheses.

[3] Evergreen (live) oaks are included in the following key.

[4] This is a macroscopic key. Microscopic characteristics are enclosed in parentheses.

[5] Junipers seldom attain tree size. Additional species include *Juniperus foetidissima* Willd., *J. macrocarpa* Sip., *J. nana* Willd., and *J. oxycedrus* L.

[6] Amount and arrangement of parenchyma may vary in adjacent rings. Parenchyma may be zonate or scattered.

[7] Of eight *Juniperus* species examined (see first key), only *J. communis* L. was found to have rays exceeding that height, reaching up to 25 cells, in exceptional cases.

[8] This is a macroscopic key. Microscopic characteristics are enclosed in parentheses.

[9] Evergreen oaks are included in the following key.

[10] Elm samples often exhibit deviations from the above characteristics—with regard to number of rows of earlywood pores and appearance of rays on tangential surfaces. Differences often exist among adjacent growth rings.

[11] This is a macroscopic key. Microscopic characteristics are enclosed in parentheses.

[12] Information pertaining to geographical source, properties and uses of important commercial woods grown in the United States, and of tropical woods imported in that country, is found in *Wood Handbook* (USDA, Forest Service, Ag. Handbook No. 72/1987). See Ref. 48 for Canadian woods. The tropical woods discussed in this book are the main species imported in Europe.

[13] In addition to this and to African mahogany (*Khaya ivorensis*), other woods marketed as mahoganies include: Sapele, Sipo (Utile), Tiama, Makoré, Meranti, Seraya (Lauan), Niangon, etc.

III

Bark as a Material

Bark participates in the total volume of a tree by an average proportion of about 10–15%, but there are considerable differences depending on species, age (dimensions) of tree, and other factors.[1] As a material, bark is usually considered undesirable and is discarded; however, lately, and in the context of efforts made to achieve a better utilization of the biomass produced by forest trees, bark is attracting intense research interest, and optimistic perspectives exist for its wider use as a source of mechanical products or chemical processing, the production of energy, as well as other uses.

Bark is different from wood in *structure*, *chemical composition*, and *properties*.

STRUCTURE

Bark is composed of tissues lying outside the cambium. In young stems, these tissues include, originally, *primary phloem*, *cortex*, and *epidermis*, and subsequently (after addition of cambium derivatives) *secondary phloem*, *primary phloem*, and *epidermis*. In most species, epidermis is subjected to inner pressure and ruptures during the first season of growth, but only after an underlying protective layer, called *periderm*, is formed (see Chapter 5, Figure 5-2).

Each periderm is composed of three layers: *phellem* or *cork layer*[2] (outside), *phellogen* or *cork cambium*, and *phelloderm*. Phellogen is a "meristematic" (capable of division) tissue, which produces the other two by dividing "periclinally"; phellogen may also divide "anticlinally" and, in this manner, each periderm may accommodate, within limits, the growth of tree diameter.

The first periderm protects the underlying tissues from environmental exposure and cell death. Such protection is afforded by the phellem, the cells of which are impregnated with waterproof substances (i.e., suberin and waxes). Under the pressure exerted by growth added inside it, this first periderm ruptures, but only after a new periderm is formed underneath.[3] New periderms are successively formed by living parenchyma cells, originally situated in the primary and later in the secondary phloem.

Thus, in older age, the bark is composed of *secondary phloem* and *periderms* (Figure III-1), because, in the meantime, the rest of the tissues (primary phloem, cortex, epidermis) have been sloughed-off (fallen on the ground). The last (innermost) periderm separates the bark into *inner* (living) and *outer* bark; the outer layers of inner bark are transformed into outer bark, and outer parts of the latter are gradually lost (36, 44, 74, 79).

The cells of bark have certain similarities but they have also important differences in comparison to the cells of wood (14, 22, 36, 37, 41, 76, 100).

In *softwoods*, bark is composed of *sieve*, *parenchyma*, and *albuminous cells*; in some species, *fibers* are also included. *Sieve cells* correspond to axial tracheids of wood; they resemble each other morphologically but sieve cells are shorter. The walls of sieve cells are thin (as a rule, there is only primary wall) and cellulosic (nonlignified). When serving as conductive elements, usually only for one season of growth, these cells contain a thin protoplasm. Sieve cells are equipped with *sieve areas*, which may be considered as corresponding to pits (Figure III-2). However, such correspondence is only relative because, as previously mentioned, sieve cells have, as a rule, only primary walls. Sieve areas are circular and depressed cell-wall areas, perforated by a sieve-like cluster of minute openings, through which the protoplasms of adjoining cells are connected. *Albuminous cells* are elongated, similar to parenchyma cells, and contain dense protoplasm and nucleus; they are always in contact with sieve cells, connected to them through sieve areas. *Parenchyma cells* are both axial and radial. Axial parenchyma are

BARK AS A MATERIAL 469

Figure III-1. Pine bark (Aleppo pine). The outer bark is shown to contain many periderms, which give the impression of growth rings.

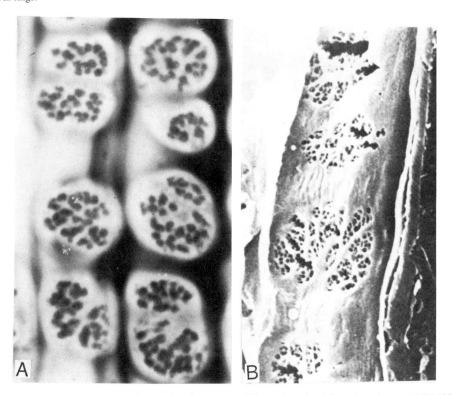

Figure III-2. Sieve areas. (A) *Pinus radiata* (light microscope, 1000×), (B) cedar (*Libocedrus plumosa*, SEM 1850×). (Photograph A Courtesy of L. M. Srivastava; B after Ref. 13, reproduced by permission from the International Association of Wood Anatomists.)

also present in species that have no such parenchyma in their wood (e.g., pine, spruce, and yew). Radial parenchyma constitute bark rays, which usually are uniseriate. On transverse sections of bark and within a short distance from the cambium, the rays are displaced from the radial direction; they are locally dilated (swollen), and finally become entirely indistinct. *Fibers* are present in the bark of many softwood species, although these species (e.g., pine) have no fibers in wood. Fibers form secondary walls, normally three-layered, which are not always lignified. In addition, the bark of some softwoods contains *resin canals*. Axial canals are present in the primary phloem, but they disappear when such phloem is sloughed-off with the passage of time. Secondary phloem may have both axial and radial canals, but not in all species (e.g., axial canals are not present in pines). Radial canals are similar to respective canals of wood. Axial canals are not always typical, and may be present in softwoods which have no such canals in wood [e.g., in Mediterranean cypress and cedar (*Cedrus*)]. The bark of fir has resin cysts. Redwood, baldcypress, hemlock, and yew have no resin canals in their bark (100).

In *hardwoods*, bark is composed of *sieve tubes* (*sieve-tube members*), *companion cells*, *fibers*, and *parenchyma* cells. Sieve tubes are the main conducting elements; they are composed of sieve-tube members, which are counterparts of the vessel members of wood; their walls are equipped with *sieve plates*. Specialized sieve plates exist in end walls, and sometimes in side walls of sieve tube members corresponding to the perforation plates of vessel members. Sieve-tube members, like the sieve cells of softwoods, contain a thin protoplasm without a nucleus; they are thin-walled, nonlignified, and in close association with *companion cells*, which contain dense protoplasm and nucleus. Other cellular components of hardwood bark are *fibers*, which are abundant, thick-walled, and with a varying degree of lignification in different species. In addition, there are axial and ray phloem *parenchyma cells*. Axial parenchyma is more abundant in hardwoods than in softwoods, and ray parenchyma constitutes bark rays which, as in the case of softwoods, at a short distance from the cambium, lose their typical form and gradually become indistinct.

Differentiation of the structure of bark, at a short distance from the cambium, is not limited to the rays. Both sieve cells and sieve tubes serve the transportation of food, after photosynthesis, only for one season of growth (22, 36). Subsequently, they die, and being, as a rule, thin-walled and nonlignified, they are crushed due to pressure exerted from cambial division and dilatation of rays. Pressure is also produced from transformation of phloem parenchyma cells to *sclereids* ("stone" cells) by enlargement, thickening of their walls, and lignification; sclereids are isodiametric or slightly to considerably elongate (Figure III-3), and have a polylamellate wall (81). Formation of sclereids, which gradually enlarge, contributes considerably to the disorganization of bark structure.

As a result of the above changes, it becomes difficult or impossible to distinguish growth rings in the

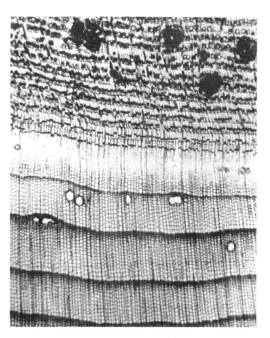

Figure III-3. Sclereids ("stone" cells) in spruce bark (black spots, upper part). Growth rings are also distinct in the bark, defined by tangential rows of parenchyma cells; the cambial region is in a process of growth (division), and the boundaries of three xylem rings (and included resin canals) are also shown (20×). (Reproduced by permission from Pergamon Press.)

bark. Distinction of growth increments is not easy, however, even without changes, because the width of seasonal growth is very small in bark (0.2–1.0 mm), and in most cases there are no clear diagnostic criteria (57). Clear distinction is possible only in few species (e.g., alder and birch) (37), where a typical difference of structure exists between phloem produced at the beginning and at the end of a growing season (early bark and late bark); however, such distinction is possible only under a microscope. The macroscopic impression that there are growth rings in outer bark is due to the presence of dead periderms (Figure III-1), which have no relation to growth rings.[4] For all these reasons, bark structure should be studied microscopically and in the region located near the cambium.

The length of bark cells has not been studied extensively. In pine (*P. halepensis*) and oak (*Q. conferta*) (27), and in some tropical woods (82), a tendency of reduction of length, from the cambium outward, has been observed (maximum length may be found in–between, i.e., between cambium and younger periderm) (27, 82), whereas the length of wood cells is generally reduced from cambium to pith (see Chapter 6).[5] The magnitude of values varies in different species. In pine, sieve cells and fibers were found, on the average, 2.5 (2.1–3.1) mm long, and the length of oak fibers was 1.04 (1.01–1.4) mm (27). In American softwoods (Douglas-fir, larch, cedar), sieve cells were 1.2–6.2 mm long and fibers 0.7–3.0 mm (14), while in tropical woods, fibers ranged from 1.1 to 1.9 mm (82). In pine and oak, bark cells were generally similar to respective wood cells with regard to length (27), but in tropical woods they were about 40–70% longer (82), and in sweetgum (*Liquidambar styraciflua*) 25% shorter (23).

CHEMICAL COMPOSITION

Cellulose and hemicelluloses are found in lower proportions in the bark, whereas extractive content is greater in comparison to wood. The celluloses of bark and wood may differ; in bark, their degree of polymerization and the crystallinity are lower. Hemicelluloses are similar to those in wood. Pectin substances are found in higher proportions in bark than in wood. In general, polysaccharide content varies considerably in the bark, even within a tree (18). Differences in the chemical composition of extractives were found between inner and outer bark, varying in different seasons of a year (8). In different pine species, the acidity (pH) of bark was about 3.5 (67); differences exist between inner and outer bark. Certain extractives are hydrophobic (9, 104). The lignins of bark and wood present important differences with regard to properties (45).

Quantitatively, the chemical constituents of bark are widely variable. Extractive content was found to vary considerably according to tree species and kind of solvent. The following percentages have been reported: 1–14% (extraction with ether), 1–23% (alcohol), 5–30% (water), 0.5–28.7% (benzol) (18, 25), and total extractives up to 46.2% (18). In pine (*P. halepensis*), bark was found to contain 49.9–58.6% total extractives (more in inner bark) and sapwood only 6.1%; respective values for oak (*Q. conferta*) were 16.8 and 9.4%. Ash of bark, especially inner bark, contained more inorganic elements than sapwood (27). Cellulose was found to vary from 16.5% (Scots pine) to 37.6% (*Gingko biloba*), and degree of polymerization from about 500 to 1000 (Scots pine—beech) (18). In different species, holocellulose was 65–68% cellulose, hemicelluloses varied from less than 1% to 26%, and pectin content was 2% (fir, Scots pine), 3–4% (birch) and 7% (spruce) (18). Douglas-fir bark fibers contained 44.80% lignin as compared to 30.15% of wood fibers (45). Bark contains proteins (birch 3.8%, inner bark of birch 5.0%, and of black locust 21.6%) (45);

there is a considerable variation within a yearly period. Some barks contain alkaloids and vitamins in very small quantities. Ash content varies widely, from 0.6 to 10.7% according to some measurements (42). Silica also varies within and between species: pines 0.03–0.16%, larch 0.26%, spruce 0.08–0.14%, Douglas-fir 0.06%, poplar 0.08% (25).

PHYSICAL AND MECHANICAL PROPERTIES

Bark properties that have been subjects of study are density, hygroscopicity, shrinkage and swelling, mechanical properties, and thermal properties.

Density varies between species, between trees of the same species, and within a tree. Aside from structural differences, density values are influenced by the proportion of inner and outer bark, because they differ in density; as a rule, inner bark has a lower density than outer bark. In certain species, the density of bark is lower and in others higher in comparison to wood; the situation is considerably affected by the amount of extractives (51). Mean values of basic density (outer bark) were found to vary in softwoods from 0.29 to 0.70 g/cm^3, and in hardwoods from 0.28 to 0.81 g/cm^3; some values (North American species) are as follows: fir 0.38–0.43, spruce 0.34–0.41, pine 0.25–0.71, beech 0.53, oak 0.49–0.80, and poplar 0.37–0.60 g/cm^3. There is a large variation both in inner and outer bark (25, 62, 66, 70).

Inner and outer bark contain very different quantities of *moisture* in living trees. Outer bark contains little moisture, 25–35%, with small seasonal variations. In contrast, the moisture content of inner bark and its seasonal variation are high. Inner bark was measured to contain 205% moisture in poplar (91) and 250% in pine (*Pinus caribaea*) (64). There are seasonal differences between new and old inner bark; in poplar, new bark had a higher moisture content in June and old bark in March (91). In inner bark of yellow poplar, moisture content was higher in June and considerably lower in October, whereas wood did not show large differences (87). After a protracted immersion in water, the moisture content of bark may reach very high values depending on density and extractives (see Chapter 9). The fiber saturation point of bark of various softwoods and hardwoods was found to vary from about 20 to 30% (63). A study of barks from tropical and temperate species showed a similarity in behavior; equilibrium moisture content values exhibited a much wider variation, however, in comparison to respective woods (Figure III-4); the differences were found to be

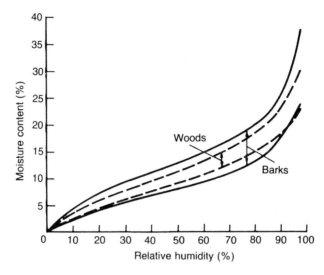

Figure III-4. Variation of equilibrium moisture content of 16 European tree barks and woods at 20°C (68°F). (After Ref. 93.)

greater in desorption (78). Inner and outer barks of western hemlock showed an equivalent emc, but higher than that of sapwood (70). Differences among barks were attributed to very different proportions of water-soluble extractives (93).

Bark is anisotropic with regard to *shrinkage* and *swelling* but not as much as wood. Axial shrinkage is higher in comparison to wood (66, 70). Tangential and radial shrinkage do not show a clear difference. Volumetric changes are similar to those of wood. Measurements in bark of softwood and hardwood species gave the following mean values of shrinkage (variation is shown in parentheses) (66):

   axial: 3.3% (0.4–6.2)
   radial: 7.6% (1.4–11.7)
   tangential: 6.3% (3.1–10.8)
   volumetric: 13.6% (9.5–18.5)

In contrast to wood, there is no apparent relationship between density and shrinkage or swelling of bark, presumably due to its structural disorganization and generally high extractive content.

The *strength* of bark is lower or similar to that of wood. A study of Douglas-fir bark and wood showed about the same strength in transverse compression, but the strength in axial compression was only about one third to one sixth that of wood. Modulus of elasticity was lower in bark (58). Bark, like wood, is anisotropic. Some studies have shown that the strength of bark is affected by the same factors that affect the strength of wood (e.g., density, moisture, temperature), but such information is limited. In Douglas-fir, it was observed that the strength of bark is drastically reduced by high temperatures (150°C, 300°F).

The *thermal properties* of bark and wood present similarities. Thermal conductivity was found to be lower up to a moisture content of 20%, but it tends to approach the thermal conductivity of wood at higher moisture contents due to the high thermal conductivity of water (25). Axial thermal conductivity was found higher and radial lower in comparison to wood (65). The specific thermal conductivity of bark is about equal to that of wood, and it is also influenced by moisture and temperature (see Figure 12-4). The heating values of bark and wood are generally similar; softwood barks were reported to range from about 4600 to 5400 kcal/kg and hardwood barks from 4000 to 5700 kcal/kg [both in oven-dry condition (53); see Chapter 12 about the heating value of wood]. Differences exist depending on species and extractive content (e.g., resin in pines) (31, 39, 53).[6]

UTILIZATION OF BARK

The utilization of bark presents problems not only because its structure, chemical composition, and properties are different from wood, but also because its separation from wood is difficult—when bark should be separated after chipping unbarked wood (harvesting residues, slabs, etc.). Various methods have been proposed for such separation, and such methods are based on density differences (46), differential water absorption (46), floating in air, compression (6), and gyratory screening (32)—or separation is sought by electrical means (electrical field) (95), vacuum (43), steaming and pressure, and other methods. Barks of different species may differ in their response to the same treatment, but differences also exist within a species, depending on harvesting time (101) (e.g., the compression method gives better results if the trees are felled during the growing season) (6) and the time interval between felling and debarking (72). A combination of methods may give better results. Methods have also been proposed to separate bark, foliage, and wood; in all cases, the problem is not only technical but mainly economical.

Bark may be utilized in products (particleboard, fiberboard), or as source of adhesives, other chemicals and energy (fuel), as soil conditioner, absorbent of pollutants, and for other uses (12, 50).

Particleboards with satisfactory properties have been made by adding up to 25% bark by weight (2, 102); higher proportions (up to 40%) may be used with proper selection of particle geometry, density of boards, and kind and proportion of adhesive. However, the quality of boards diminishes with increasing proportions of bark (15, 20). Aside from mixing bark and wood chips, three-layer particleboard of good quality has been made with a middle layer from bark (and outer layers from wood), and use of bark extractives as adhesives. Several studies have shown that bark extractives (with the addition of formaldehyde) give good adhesion, similar or better in comparison to the use of phenol-, or urea-formaldehyde (3, 4, 16). Particleboard has also been made solely from bark (103) with very good internal bond strength and dimensional stability of thickness, but low strength in bending and high longitudinal swelling (2). Presence of bark in outer layers (use of unbarked wood) adds dark spots and contributes to reduction of bending strength of panels; however, up to 50% of bark may be added in the middle layers (some species reduce internal bond) (49). Three-layer boards, with a middle layer from bark, have shown good thermal insulating properties, better than boards made only from wood (88). However, bark is variable depending on species (61), age, thickness, and position in the tree (28, 34, 55, 83), and such variability may create problems in bark utilization. Problems may also arise from foreign matter (sand, soil dirt, etc.) attached to bark during harvesting trees in a forest (101), or by bark reduced to dust during processing (75). Particleboards produced from logging residues had better properties when particles from residues were placed in the core and conventional furnish is surface layers (86). Inclusion of bark requires more adhesive and, therefore, a higher expense.

Other products of bark utilization include *fiberboard*, with or without mixture of wood (94, 99) (see, however, Figure 25-6), *moldings* (wood carving substitutes) (54, 71), *novelties*, and *reinforced plastics* (73, 97). Bark *pellets* are incorporated in concrete to produce building boards, they are mixed with asphalt, and, by the addition of chemicals and fertilizers, they find applications as soil conditioners (102).

Bark may contribute to the production of *pulp* for paper and other uses (48), but yields are low. The proportion of useful fibers varies in different species, but generally bark may add 2–6% fibrous cells (21). However, the acceptability of bark is related to the pulping method. In mechanical and sulfite pulping, bark is almost entirely undesired (a proportion lower than 1% in mechanical pulping and less than 0.5% in sulfite pulping is acceptable), although semi-chemical pulps have a greater tolerance—up to 12% of bark was found to have a very small influence on the strength of paperboard. The influence of bark in Kraft pulping depends on conditions of production (equipment for cleaning the pulp, etc.); under favorable conditions, up to 12% bark had no significant effects on the strength of pulp (48).[7] In another study (90), substitution of 15% of wood by bark was found to have a deleterious effect on pulp yield, ash, and lignin content.

Chemicals produced from bark include tannins, dyes, anti-oxidants, substances with pharmaceutical properties, waxes, and others (30, 52, 96). Important possibilities are recognized in this area, but better knowledge of the chemistry of bark is needed, and difficulties arise due to differences that exist among barks of different tree species.

Bark is also a source of energy. As previously mentioned, the heating value of bark is similar to that of wood. Bark is used for steam production in wood-using industries, but high moisture content drastically reduces its heating value (10, 31, 40, 77, 89).

Other uses of bark include *production and packaging of ornamental plants in nurseries and greenhouses* (26, 47) (tannins may inhibit germination of some seeds), *surfacing recreation trails and highways*, and *landscaping* (38, 68). Bark is also finding applications in *environmental protection* (5, 7, 11, 29, 59, 60) (absorption of spilled oil, metallic ions from industrial wastes, bacteria and other microorganisms from sewage effluents, etc.), and is also used as *poultry litter* (1), and *plastic reinforcement* (by incorporation of bark fibers). Efforts continue to find new uses for bark, because large quantities of this material are available (102).[8]

CORK

Cork (17, 24, 35, 56, 69) is the outer bark of cork oak (*Quercus suber*); it is removed every 8–10 (6–11) years, without adverse effects on the trees, which gradually replace it. Cork oak constitutes natural forests in maritime regions of western Mediterranean countries (Portugal, Spain, France, Italy, Algeria, Morocco, Tunisia).

Cork is collected from May to August (usually June and July) by workers using a two-headed hatchet (long-handled, light axe) or curved (crescent-shaped) hand saw. Two peripheral cuts are made on the trunk of the tree, one near the ground and the other under the crown (usually up to a height of 2 m), and the cork (outer bark) is separated by two vertical cuts, and pried out by use of the hatchet handle. Sometimes, large branches are also debarked. Debarking starts when the trees have a diameter of about 20 cm (20–40 years of age), and is continued at intervals of about 10 years, up to an age of about 200 years. Differences in cork production exist among trees due to genetic differences, although soil conditions and tree age are also important factors. Trees that do not produce satisfactorily are harvested for wood (the wood belongs to the "red oak" group). Lower sites with more fertile soil give thicker and more spongy cork, lower in value; whereas higher and drier sites give typical, heavier cork of higher value, although less thick. Cork collected from young trees (first collection) is rough and low in value. Cork from a second collection is better—but highest quality is produced from a third collection and later. Yield per tree varies from about 20–200 kg (50–500 lb), and cork thickness from about 2 to 6 cm (1–2.5 in., see Figure III-5).

The cork collected is temporarily piled, and subsequently placed in boiling water (in large copper vessels for about 30 min) to remove tannins and foreign matter. This treatment is also helpful to soften the exterior, fissured bark, which is thus more easily removed. After that, the cork is air-dried until its weight is reduced to about two thirds of green weight. Finally, the cork pieces are trimmed and graded according to quality and thickness.

Cork is made of very small cells (about 30 million are contained in 1 cm^3 or about 450 million in 1 in.^{3}), and typically 14-sided (6 four-sided and 8 hexagonal surfaces). Cork tissue presents (in tan-

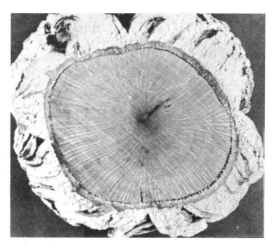

Figure III-5. Cork oak (*Quercus suber*). Transverse section of a tree trunk showing (from outside): outer bark (cork), inner bark, and wood. (Reproduced by permission from Pergamon Press.)

gential view) a characteristic honeycombed structure (Figure III-6), first observed by Hooke (1665). Cell dimensions vary between "early" and "late" cork; early cork cells have a prism height of 30–40 μm and cell-wall thickness 1–1.5 μm; however, late cork cells are much shorter (down to 10 μm) and have thicker walls (nearly double). This differentiation is the cause for distinction of growth rings (often visible in cork products—e.g., bottle stoppers), but the appearance of rings is also due to corrugation of the lateral surfaces of cork cells. Corrugations are most frequent in the first early cork cells (deforming or collapsing against the late cork cells of the previous growing season), but they also occur within an annual growth; heavily corrugated cells do not appear to have cell-wall fractures. This is attributed to the chemical composition of cork, containing high amounts of waxes and suberin. Cell walls are made of a lignin and cellulose-rich middle lamella and a thicker secondary wall of alternating lamellae of suberin and waxes (85).

With regard to properties, cork is light and floats (density 0.20–25 g/cm^3), impenetrable by water and other liquids, insulating to heat and sound, with a high coefficient of friction (does not slip), high compressibility, and low modulus of elasticity. Cork also has high durability (e.g., much higher than oak wood); it is not affected by water, oil, benzol and other organic solvents and various gases (CO, H, N), and presents high resistance to weak acidic solutions.

Cork has been known since ancient times (it was known to ancient Greeks, Romans, and others), and

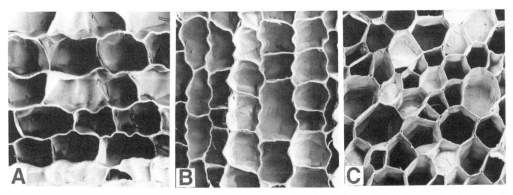

Figure III-6. Cork cells: A, transverse; B, radial; C, tangential section of cork tissue. (SEM photographs: A, 460×; B, C. 330×.) (After Ref. 85; reproduced by permission from the International Association of Wood Anatomists.)

was put to various uses, such as swimming aids (lifebelts), fishing nets, boat accessories, covers of earth jars and other containers, shoes, beehives, thermal insulation of homes, fortifications, flooring, and buildings (mixture of cork particles with mud). The use of cork was greatly expanded after the revival of glass-making in medieval times (Byzantium, Venice).

Today, cork is mainly used for stoppers in glass and other containers, lifebelts, buoys, floats for nets, sporting goods, fishing equipment, shoes and related items, novelties, and other products. A considerable thrust to wider use of cork was given with the development of gluing cork residues after chipping and grinding. In this manner, various composite products (composition cork) are made, such as gaskets (sealing metallic parts), insulating "corkboard" used in refrigerators, covers of pipes and other constructions for thermal and sound insulation, flooring (used mainly in churches, concert halls, libraries, and hospitals), coating materials for floors and walls, bulletin boards, novelties (carvings, etc.), and many others. Such products are either new or replace products of solid cork, and generally contribute to a wider and better utilization of this valuable forest product.

The annual production of cork is about 350 thousand metric tons. About half of this quantity is produced in Portugal. The production of Spain is also high (75–100 thousand tons).

REFERENCES

1. Alison, R. C., and H. C. Jordan. 1973. Criteria and evaluation of hardwood bark as a poultry litter. *For. Prod. J.* 23(8):46–48.
2. Anderson, A. B., A. Wong, and K.-T. Wu. 1974. Utilization of white fir bark in particleboard. *For. Prod. J.* 24(1):51–54.
3. Anderson, A. B., K.-T. Wu, and A. Wong. 1974. Utilization of ponderosa pine bark and its extract in particleboard. *For. Prod. J.* 24(8):48–53.
4. Anderson, A. B., A. Wong, and K.-T. Wu. 1975. Douglas fir and western hemlock bark extracts as bonding agents for particleboard. *For. Prod. J.* 25(3):45–48.
5. Anonymous. 1968. Redwood bark. An aid in pollution control. *For. Prod. J.* 18(8):64.
6. Arola, R. A., and J. R. Erickson. 1973. Compression Debarking of Wood Chips. North Cent. For. Exp. Sta. Res. Paper NC-85.
7. Bermann, R. L., and A. C. Waiss, Jr. 1974. Use of bark to remove heavy metal ions from waste solutions. *For. Prod. J.* 24(9):80–84.
8. Binotto, A. P., and W. K. Murphey. 1975. Season and height variation in extractives and cell wall components of chestnut oak bark. *Wood Sci.* 7(3):185–190.
9. Borgin, K., and K. Corbett. 1971. The hydrophobic properties of bark extractives. *Wood Sci.* 5:190–199.
10. Burnet, D. J. 1974. Steam generation from bark. *Pulp Paper Mag. Canada* 75(5):86–91.
11. Canadian Forestry Service. 1971. The use of bark for controlling oil pollution on water. *Res. News* 14(6):2–3.
12. Carroll, M. N. 1977. Bark utilization. In *Forest Utilization Symposium* (pp. 7–8). Forintek Canada Corp.
13. Chan, L.-L. 1985. The anatomy of the bark of *Libocedrus* in New Zealand. *IAWA Bull* n.s. 6(1):23–34.
14. Chang, Y. P. 1954. Bark Structure of North American Conifers. U.S.D.A. Tech. Bull. No. 1095.
15. Chen, T. V., and M. Paulitsch. 1974. Inhaltstoffe von Nadeln, Rinde und Holz der Fichte und Kiefer und ihr Einfluss suf die Eigenschaften daraus hergestellter Spanplatten. *Holz Roh-Werkst.* 32(10):397–401.

16. Chow, S. 1975. Bark boards without synthetic resins. *For. Prod. J.* 25(11):32–37.
17. Cooke, G. B. 1961. *Cork and the Cork Tree.* New York: Pergamon Press.
18. Dietrichs, H. H. 1975. Polysaccharide der Rinden. *Holz Roh- Werkstoff* 33(1):13–20.
19. Dietz, P. 1975. Dichte und Rindengehalt von Industrieholz. *Holz Roh- Werkst.* 33:135–141.
20. Dost, W. A. 1971. Redwood bark fiber in particleboard. *For. Prod. J.* 21(10):38–43.
21. Einspahr, D. W., and M. L. Harder. 1980. Increasing hardwood fiber supplies through improved bark utilization. *Tappi* 63(9):121–124.
22. Esau, K. 1965. *Plant Anatomy* (2nd ed.). New York: John Wiley & Sons.
23. Ezell, W. A., and J. L. Stewart. 1978. The length of phloem fibers in sweetgum (*Liquidambar styraciflua* L.). *Wood and Fiber* 19(3):186–187.
24. Faubel, A. L. 1938. *Cork and the American Cork Industry.* New York: Cork Institute of America.
25. Fournier, F., and M. Goulet. 1970. Propriétés Physico-Mécaniques de l'Ecorce. Une Etude Bibliographique. Faculté For. Geod., Université Laval, Quebec.
26. Gartner, J. B., and D. J. Williams. 1978. Horticultural uses for bark: A review of current research. *Tappi* 61(7):83–86, 337.
27. Gatsis, C. 1982. A Study of Fiber Length Variation in the Inner Bark, and of Certain Chemical Properties of Bark and Sapwood of *Pinus halepensis* and *Quercus conferta*. Lab. Forest Utilization, Aristotelian University (Greek).
28. Gertjejansen, R., and J. Haygreen. 1973. The effect of aspen bark from butt and upper logs on the physical properties of wafer-type and flake-type particleboards. *For. Prod. J.* 23(9):66–71.
29. Glaser, W. G., and F.-S. Lin. 1974. Removal of emulsified oil by sorption on southern pine bark. *For. Prod. J.* 24(9):87–91.
30. Hall, J. A. 1971. Utilization of Douglas-Fir Bark. Pacific NW For. Range Exp. Sta., Portland, Oregon.
31. Harder, M. L., and D. W. Einspahr. 1976. Bark fuel value of important pulpwood species. *Tappi* 50(12):132.
32. Harkin, J. M., and D. M. Crawford. 1972. Separation of wood and bark by gyratory screening. *For. Prod. J.* 22(5):26–30.
33. Harlow, W. M., and E. S. Harrar. 1968. *Textbook of Dendrology* (5th ed.). New York: McGraw-Hill.
34. Haygreen, J. G., and J. L. Bowyer. 1982. *Forest Products and Wood Science.* Iowa State University Press.
35. Hill, A. 1937. *Economic Botany.* New York: McGraw-Hill.
36. Holdheide, W. 1951. Anatomie mitteleuropäischer Gehölzrinden (mit mikroskopischem Atlas). In *Handbuch der Mikroskopie in der Technik*, ed. H. Freund, Bd.V,T.1, pp. 193–367. Frankfurt: Umschau Verlag.
37. Holdheide, W., and B. Huber. 1952. Aehnlichkeiten und Unterschiede im Feinbau von Holz und Rinde. *Holz Roh- Werkst.* 10:263–268.
38. Holmes, R. L., G. D. Coorts, and P. L. Roth. 1973. Hardwood bark chips as a surfacing material for recreation trails. *For. Prod. J.* 23(5):25–30.
39. Howard, E. T. 1973. Heat of combustion of various southern pine materials. *Wood Sci.* 5(3):194–97.
40. Host, J. R., and D. P. Lowery. 1970. Potentialities for using bark to generate steam power in Western Montana. *For. Prod. J.* 20(2):35–36.
41. International Association of Wood Anatomists (Committee on Nomenclature). 1964. Multilingual Glossary of Terms Used in Wood Anatomy. Verlagsanstalt Buchdruckerei Konkordia Winterthur, Switzerland.
42. Isenberg, I. H. 1980. *Pulpwoods of the United States and Canada.* Vols. I and II. Appleton, Wisconsin: The Institute of Paper Chemistry.
43. Isenbrands, J. G., J. A. Sturos, and J. B. Christ. 1979. Integrated utilization of biomass: A case study of short-rotation intensively cultured *Populus* raw material. *Tappi* 62(7):67–70.
44. Jane, F. W. 1970. *The Structure of Wood.* London: A. & C. Black.
45. Jensen, W., K. E. Kremer, P. Sierilä, and V. Vartiovara. 1963. The chemistry of bark. In *The Chemistry of Wood*, ed. B. L. Browning, pp. 587–666. New York: Interscience Publ.
46. Julien, L. M., J. C. Edgar, and T. M. Conder. 1972. Segregation of aspen, balsam and spruce wood and bark chips based on density differences. *For. Prod. J.* 22(6):56–59.
47. Karagozov, T., and G. Taneva. 1979. Verwertung von Nadelholzrinde in den Betrieben der Holzindustrie. *Holztechnologie* 20(2):116.
48. Keays, J. L., and J. V. Hatton. 1974. The Effect of Bark on Wood Pulp Yield and Quality and on the Economics of Pulp Production. For. Prod. Lab., Vancouver, Brit. Columbia. Inform. Rep. VP-X-126.
49. Krames, U. 1967. Untersuchungen zum Einfluss der Rinde bei Verarbeitung unentrindeten Holzes in der Deckschicht von Spanplatten. *Holztechnologie* 20(1):32–39.
50. Krames, U. 1967. Die Nutzung der Rinde. *Holzforschung u. Holzverwertung* 19(1):1–4.
51. Lai, Y-Z., et al. 1980. Wood and bark specific gravity determination as effected by water-soluble extractives loss. *Wood Sci.* 13(1):47–49.
52. Laver, M. L. 1971. Chemical considerations (and Chemicals from bark). In *Converting Bark into Opportunities*, ed. A. C. van Vliet, pp. 13–14 and 105. Oregon State University, School of Forestry.
53. Laxamana, N. B. 1984. Heating value of some Philippine woods, non-woods and barks. *FPRDI J.* 13(3/4):6–12.
54. Lehmann, W. F. 1968. Molding compounds from Douglas-fir bark. *For. Prod. J.* 18(12):47–53.
55. Lehmann, W. F., and H. E. Wahlgren. 1978. Status and prospects of residue utilization in board product

manufacture in the United States and Canada. *For. Prod. J.* 28(7):24–53.
56. Liese, W. 1978. Research aspects on cork utilization. Paper presented at the CEDULI meeting, Bremen (reprint).
57. Liese, W., and N. Parameswaran. 1971. Über die Rindenanatomie starkborkiger Fichten. *Forstwiss. Clb.* 90(6):370–375.
58. Lin, R. T. 1973. Behavior of Douglas-fir bark components in compression. *Wood Sci.* 6(2):106–111.
59. Liversidge, R. M. 1973. Oil Collection—A Potential Use for Small Particle Wood Residue. CSIRO, For. Prod. Newsletter No. 392.
60. MacDonald, D. C., and T. G. Nguyen. 1974. Activated carbon from bark for effluent treatment. *Pulp Paper Mag. Canada* 75(5):97–101.
61. Maloney, T. M. 1973. Bark boards from four west-coast softwood species. *For. Prod. J.* 23(8):30–38.
62. Manwiller, F. G. 1979. Wood and bark specific gravity of small-diameter-site hardwoods in the South. *Wood Sci.* 11(4):234–240.
63. Martin, R. E. 1968. Interim equilibrium moisture content values of bark. *For. Prod. J.* 18(4):52.
64. Martin, R. E. 1969. Characterization of southern pine barks. *For. Prod. J.* 19(8):23–30.
65. Martin, R. E. 1970. Directional thermal conductivity ratios of bark. *Holzforschung* 24(1):26–30.
66. Martin, R. E., and J. B. Christ. 1968. Selected physical-mechanical properties of eastern tree barks. *For. Prod. J.* 18(11):54–60.
67. Martin, R. E., and G. R. Gray. 1971. pH of southern pine barks. *For. Prod. J.* 21(3):49–52.
68. Mater, J. 1971. Utilization of bark in highway landscaping. *For. Prod. J.* 21(8):17–20.
69. Maydell von, H. J. 1971. *Forstliche Nebennutzungen in den Atlasländern als Grundlage für die Regionalentwicklung.* Hamburg: Kommissionsverlag.
70. Meyer, R. W., R. M. Kellog, and W. G. Warren. 1981. Relative density, equilibrium moisture content, and dimensional stability of western hemlock bark. *Wood and Fiber* 13(2):86–96.
71. Miller, D. J. 1972. Molding characteristics of some mixtures of Douglas-fir bark and phenolic resin. *For. Prod. J.* 22(9):67–70.
72. Miller, D. J. 1975. Treatments to loosen bark. *For. Prod. J.* 25(11):49–56.
73. Miller, D. J., J. D. Wellous, R. L. Krahmer, and P. H. Short. 1974. Reinforcing plastics with Douglas-fir bark fiber. *For. Prod. J.* 24(8):18–23.
74. Morey, P. R. 1973. *How Trees Grow.* London: Arnold.
75. Murphey, W. K., and L. E. Rishel. 1969. Relative strength of boards made from bark of several species. *For. Prod. J.* 19(1):52.
76. Nanko, H., and W. A. Côté. 1980. Bark Structure of Hardwoods Grown on Southern Pine Sites. Syracuse, New York: Syracuse University Press.
77. Newby, W. M. 1973. The changing status of bark. *Pulp Paper Mag. Canada* 74(10):80–82.
78. Okoh, K. I. A., and C. Skaar. 1980. Moisture sorption isotherms of the wood and inner bark of ten southern U.S. hardwoods. *Wood and Fiber* 12(2):98–111.
79. Panshin, A., C. De Zeew, and H. P. Brown. 1964. *Textbook of Wood Technology* (2nd ed.). New York: McGraw-Hill.
80. Panshin, A. J., E. S. Harrar, J. S. Bethel, and W. J. Baker. 1962. *Forest Products.* New York: McGraw-Hill.
81. Parameswaran, N. 1980. Some remarks on the nomenclature of fibres, sclereids and fibre-sclereids in the secondary phloem of trees. *IAWA Bull. n.s.* 1(3):130–133.
82. Parameswaran, N., and W. Liese. 1974. Variation of cell length in bark and wood of tropical trees. *Wood Sci. Tech.* 8(2):81–90.
83. Pardé, J. 1961. *Dendrometrie.* Nancy, France: Impr. Louis-Jean GAP.
84. Passialis, C., and E. Voulgaridis. 1983/84. Bark properties of fir. *Annals Dept. For. Nat. Envir.*, Aristotelian Univ. 26/27:101–125 (Greek, English summary).
85. Pereira, H., M. E. Rosa, and M. A. Fortes. 1987. The cellular structure of cork from *Quercus suber* L. *IAWA Bull.* n.s. 8(3):213–218.
86. Philippou, J. L., and B. M. Collett. 1978. Particleboard from young-growth wood white fir logging residues. *For. Prod. J.* 28(12):35–41.
87. Phillips, D. R., and J. G. Schroeder. 1973. Some physical properties of yellow-poplar wood and bark. Part I. Seasonal moisture content. *Wood Sci.* 5(4):265–269.
88. Place, T. A., and T. M. Maloney. 1975. Thermal properties of dry wood-bark multilayer. *For. Prod. J.* 25(1):33–39.
89. Robertson, J. E. 1968. Bark burning methods. *Tappi* 51(6):90A–98B.
90. Roffael, E., and N. Parameswaran. 1978. Zur Mitverwendung der Rinde beim Sulfataufschluss von Fichtenholz. *Holzforschung* 32(4):113–119.
91. Sachsse, H. 1969. Über die jahreszeitlichen Feuchtigkeitsschwankungen in der Rinde lebender Robusta-Pappeln. *Holz Roh- Werkst.* 27(2):55–66.
92. Schneider, A. 1978. Orientierende Vergleichsuntersuchungen über das Sorptions-verhalten mitteleuropäischer Baumrinden und Hölzer. *Holz Roh- Werkst.* 36(6):235–239.
93. Schneider, A., and N. Parameswaran. 1983. Studies on the sorption behaviour of the barks of tropical trees. *Holzforschung* 37:125–128.
94. Semana, J. A., and A. B. Anderson. 1968. Hardwood from benguet pine bark-wood compositions. *For. Prod. J.* 18(7):28–32.
95. Short, P. H., J. D. Wellous, and R. T. Lin. 1973. Segregating bark fractions electrically. *For. Prod. J.* 23(8):41–45.
96. Soni, P. L., R. Pal, and R. N. Madan. 1980. Utilization of Teak (*Tectona grandis*) bark—Production of pulp for wrapping paper and cellulose derivatives. *Holzfors. Holzverwert.* 32(2):42–48.

97. Soule, E. L., and H. E. Hendrickson. 1966. Bark fiber as a reinforcing agent for plastics. *For. Prod. J.* 16(8):17–22.
98. Srivastava, L. M. 1963. Secondary phloem in the Pinaceae. *Publications in Botany* 36:1–142, University of California Press.
99. Stewart, D. L., and D. L. Butler. 1968. Hardboard from cedar bark. *For. Prod. J.* 18(12):19–23.
100. Tsoumis, G. 1968, 1969. *Wood as Raw Material.* New York: Pergamon Press.
101. Tsoumis, G. 1991. *Harvesting Forest Products.* (In press).
102. Vliet van, A. C., ed. 1971. *Converting Bark into Opportunities.* Oregon State University, School of Forestry.
103. Volz, K. R. 1973. Herstellung und Eigenschaften von Fichten-, Kiefern- und Buchrindenplatten. *Holz Roh–Werkst.* 31(6):221–229.
104. White, M. S., G. Ifju, and J. A. Johnson. 1974. The role of extractives in the hydrophobic behavior of loblolly pine rhytidome. *Wood Fiber* 5(4):353–363.

FOOTNOTES

[1] Sawlogs were reported to contain 5% (beech) to 30% (pine) volume of bark (83). In stacked wood (pulpwood, etc.) of various European species, the proportion of bark was 6.6–29.9% by volume and 6.3–17.2% by weight (19). In general, the proportion of bark varies widely between thin-bark species (beech, birch, poplar, spruce) and thick-bark species (oak, pine, redwood, Douglas-fir); it also increases with tree age, and is higher at the base of trees. In centuries-old trees of giant sequoia (*S. gigantea*) and Douglas-fir, bark thickness may reach 30–60 cm (about 1–2 ft) (33).

[2] In cork oak (*Quercus suber*), abundantly produced phellem is commercial cork.

[3] The length of periderm life varies. In sycamore (plane) a periderm may be active 3–4 years; in oak, 4–6; in maple 5–30; yearly initiation of new periderms, as in poplar and willow is exceptional (36, 37). On the other hand, in beech the original periderm is preserved up to 100 years (74), but in other species (e.g., pine) many periderms are produced and retained dead. In Sequoia, the presence of many dead periderms contributes to the large thickness of bark in very old trees.

[4] Distinct annual rings are formed in the outer bark of cork oak; cork has such rings, which are often visible in cork bottle stoppers.

[5] In both cases (bark, wood) the trend is the same (i.e., increasing length with increasing age of the cambium at the time of formation).

[6] The results of a study of fir (*Abies*) bark may be summarized as follows (84): thickness 5–30 mm (0.20–1.20 in.) and linear relationship between thickness and tree diameter; volume 14% of total trunk volume; moisture content 65–87% (average 75% and increasing with tree height); swelling higher in radial and longitudinal direction in comparison to wood (inner bark: radial 7.5%, tangential 5.7%, longitudinal 3.4%—and outer bark: radial 9.8%, tangential 7.4%, longitudinal 2.8%); permeability to water low (10 times lower in comparison to heartwood and 70 times lower in comparison to sapwood); extractives 3–5 times more than in wood, ash 7–12 times higher, acidity little less (pH of bark 5, sapwood 6, heartwood 6.6), and heating value about 10% lower in comparison to wood.

[7] Thin-bark species (e.g., poplar) may be used without debarking in Kraft and semichemical pulping, but not with the sulfite process. In general, bark may be tolerated, but its presence results in loss of digester capacity, chemicals, and energy.

[8] In North America (U.S., Canada), the quantities of bark available every year, as a by-product of tree harvesting in forests, have been estimated (1979) as 22 million tons (55). Worldwide, the volume of bark brought to wood-processing plants (attached to logs and chips) has been estimated (1972) to be 320 million cubic meters (about 120 million tons) (34).

Unit Equivalents

Length, Area, Volume

1 mm = 0.04 in., 1 cm = 0.4 in., 1 m = 3.28 ft, 1 m^2 = 10.76 ft^2, 1 m^3 = 35.29 ft^3, 1 cm^2 = 0.16 $in.^2$
1 in. = 2.54 cm, 1 ft = 30.48 cm, 1 ft^2 = 0.0929 m^2, 1 ft^3 = 0.0283 m^3
1 bd ft (board foot) = 0.00236 m^3, 1 m^3 = 423.7 bd ft
1 acre = 0.405 hectares (ha), 1 ha = 2.471 acres

Mass, Weight, Force, Stress, Pressure

1 N = 0.102 kp = 0.22 lb = 10^5 dyn
1 N/mm = 1.02 kp/cm = 5.59 lb/in.
1 N/mm^2 = 10.2 kp/cm^2 = 10^6 Pa (1 MPa) = 144.9 psi
1 J = 1 Nm = 0.102 kpm = 10.2 kpcm = 8.82 in. lb
1 J/mm^2 = 10.2 kpm/cm^2 = 5590 in. $lb/in.^2$
1 J/mm^3 = 10,200 $kpcm/cm^3$ = 14.5 × 10^4 in. lb/in^3
1 Pa = 1 N/m^2 = 0.000001 N/mm^2 = 0.0000102 kp/cm^2 = 0.000145 $lb/in.^2$
1 $lb/in.^2$ = 0.0069 N/mm^2 = 0.069 kp/cm^2 = 6900 Pa = 6.9 kPa
1 kg = 2.2 lb, 1 g = 0.0353 oz (ounces), 1 oz = 28.35 g
1 kp = 9.8 N = 2.2 lb, 1 lb = 450 g = 0.45 kg
1 kp/cm^2 = 0.098 N/mm^2 = 98 × 10^3 Pa (98kPa) = 14.2 lb/in^2 (psi)
1 kp/cm = 0.98 N/mm = 5.59 lb/in.
1 kpcm = 0.098 Nm = 0.098 J = 0.864 in. lb
1 kpm/cm^2 = 98.1 kN/m = 9.81 J/cm^2 = 559 in. lb/in^2
1 $kpcm/cm^3$ = 0.000098 J/mm^3 = 14.20 in. $lb/in.^3$
1 mm Hg = 1 Torr = 0.133 kPa = 0.03937 in. Hg

Thermal Units

1 cal = 4.18 J = 3.968 Btu, 1 Btu = 252 cal = 1.055 × 10^3 J = 1.055 kJ
1 cal/cm·s·°C = 1.162 J/m s °K = 3.119 Btu/in.·h·°F
1 kcal/m h °C = 1.162 J/m s °K = 3.119 Btu/in.·h·°F
1 cal/g °C = 1 kcal/kg °C = 4184 J/kg °K = 1.8 Btu·lb·°F, 1 Btu·lb = 0.55 kcal/kg
Temp. °C = (Temp. °F − 32) ÷ 1.8, Temp. °F = (Temp. °C × 1.8) + 32

Equivalents: Products—Roundwood[a]

Product	Unit	Equivalent Roundwood
Lumber, softwood	1 m^3	1.67 m^3
hardwood	1 m^3	1.82
Railroad ties	1 ″	1.82 ″
Veneer	1 ″	1.9 ″
Plywood	1 ″	2.3 ″
Particleboard	1 ton	2.0 ″
Fiberboard	1 ″	2.0 ″
Pulp		
mechanical	1 ″	2.5 ″
chemical	1 ″	4.9 ″
sulfite	1 ″	4.9 ″
Kraft	1 ″	4.8 ″
dissolving	1 ″	5.5 ″
semichemical	1 ″	3.3 ″
Newsprint	1 ″	2.8 ″
Printing and writing paper	1 ″	3.5 ″
Other papers	1 ″	3.25 ″
Paperboard	1 ″	1.6 ″
Charcoal	1 ″	6.0 ″

[a] After FAO/UN.

Indexes

Subject Index

Abnormalities, 84-107
Acoustical properties, 204-207
Adhesion, 327-331
  factors of, 329-331
  mechanical, 327
  mechanism of, 327-329
  specific, 327, 329
Adhesives, 327, 331-336
  animal, 332-333
  assembly time of, 331
  blood albumin, 332-333
  bone or hide, 332
  casein, 332
  elastomers, 335
  inorganic, 336
  isocyanates, 335
  melamine-formaldehyde, 334
  phenol-formaldehyde, 333-334
  plant, 332
  resorcinol-formaldehyde, 335
  soya, 332
  starch, 332
  storage life of, 330, 338
  synthetic, 333-335
  thermoplastic, 335
  thermosetting, 333-335
  urea-formaldehyde, 334
Adsorption, 130-132
Air-drying, 264-272, 283, 284
Aluminum, xii, 172
Amorphous regions, 37, 38
Anisotropy of
  acoustical properties, 205, 206
  electrical properties, 208-209
  fiberboard, 394-395
  mechanical properties, 162-172
  particleboard, 377-378
  plywood, 344-347, 350
  shrinkage—swelling, 145-146, 147, 151
Annulus, 16
Anticlinal division, 61
Apical meristem, 57

Ash of
  bark, 471
  wood, 34
Aspiration, 18, 44, 304
Athletic equipment, 425

Bacteria, 213-215, 233, 262
Bark, 4, 5-7, 468, 478
  chemical composition of, 471
  inner, 3, 6
  outer, 3, 6-7
  pockets, 101
  properties of, 471-472
  proportion of, 468-478
  structure of, 468-471
  uses of, 472-474
Barrier zone, 102-103, 107
Biosynthesis, 50-51
Bleaching, 410
Blue stain, 215-216
Board, lumber, 262
Branch wood, 77
Breaking length, 162
Briarwood, 429
Brightness, 410, 418
Briquettes, 427
Bulking treatment, 156-157
Butt swell, 85

Cambial zone, 58-59, 60
Cambium, 3, 58-61, 82-83
  initials, 59, 82
  stratified, 59, 68
Canals, 30
  gum, 5, 30
  resin, 5, 30
  traumatic, 30, 100-101
Carbonization, 426, 427
Carvings, 425
Casehardening, 154, 155, 282
Cells
  bark, 468-471

483

484 SUBJECT INDEX

Cells (*Continued*)
 companion, 470
 development of, 61–64
 dimensions of, 16, 22–23, 25, 26, 29
 morphology of, 14, 15, 16, 22–30
 mother, 57, 59
 sieve, 468
 wood, 14, 15, 16, 18–30
Cell-wall
 layers, 38–41
 organization, 38–49
 substance (material), 82
 thickness, 41, 55
Cellulose, 34, 55
 in bark, 471
 chains, 34, 35
 in compression–tension wood, 92, 94
 distribution of, 48–49
 proportion in wood, 34, 54
 relation to properties-uses, 49–50
 structure of, 35, 55
Chemical constituents, 34–36
 of bark, 471
 distribution of, 47–49
 effects of, 49–50
Chip-N-Saw, 243, 261
Chipping, 363, 389, 402–403
Collapse, 154–155, 282
Color, 9–10
 abnormal, 98–100
Combustion, 198–199
COM-PLY, 350, 382
Compreg, 156
Compression failures, 97, 181
Conductivity
 electrical, 208–209
 thermal, 196–198, 203
Concrete, xii, 172
Containers, 423–424
Core wood, 70
Cork, 474–475, 478
 cambium, 65, 468
 properties-uses of, 474, 475
 structure of, 474, 475
Cortex, 58
Crassulae, 23–24
Creep, 177–178, 379
Creosote, 295
Cross-field, 18, 21, 32–33
Crown-formed wood, 70
Crystalline regions, 37
Crystallinity, 37, 50, 55
Crystallite, 37

Debarking, 310, 401, 418, 472–473
Decay, 217–222
 brown, 217
 consequences of, 221
 factors of, 217–219
 fungi, 217, 221–222, 233

 resistance to, 218
 soft, 217–218
 white, 217, 218
Defects, 84 (*see also* Abnormalities)
Degradation, 213–233
 biological, 213–229
  bacteria, 213–215
  fungi, 215–222
  insects, 222–226
  marine borers, 226–229
 chemical, 229–230
 climatic, 229
 mechanical, 229
 thermal, 230
Dendrochronology, 8
Dendroclimatology, 8
Density, 111–127
 apparent, 111
 of bark, 471
 basic, 111
 of branch–root wood, 119
 of cellulose, 117
 of cell wall, 127
 of cell-wall material, 113, 127
 of compression–tension wood, 92, 94
 determination of, 121–123, 124
 of earlywood–latewood, 117
 factors affecting, 111–113, 115–118
 of fiberboard, 388–389
 importance of, 123, 125
 of lignin, 117
 of old waterlogged wood, 127
 oven-dry, 111
 of particleboard, 361, 376
 relative, 111
 of temperate woods, 114, 115, 434–451
 of tropical woods, 113, 114–115, 451–454
 variation of, 118–121
 weight (density), 111
Desorption, 130–132
Dielectric properties, 210
Diffuse-porous, 4, 6
Diffusivity, thermal, 198
Dimensional stability, coefficient of, 387
Discoloration, 98–100
Distillation
 destructive, 427
 to determine moisture content, 139
Division, cambial
 anticlinal, 61
 periclinal, 59, 61
Drying
 acoustic emissions in, 292
 fiberboard, 390
 methods
  air-drying, 264–272, 283, 284
  boiling in oils, 287
  chemical, 287
  continuous rising temperature, 286–287
  dehumidification, 284–285

fan, air-drying, 291-292
high frequency electricity, 288
high temperature, 285-286
kiln-drying (conventional), 264-265, 272-283, 284
solar energy, 284, 285
solvent, 287-288
vapor, 287
paper, 414
particles, 367, 368, 369
schedules, 276, 277-279, 280
and weighing, 139
wood, 264-292
Durability, 213, 217, 219, 308

Earlywood, 3, 4
Eccentricity, 88
Edging, 248
Elasticity, 161, 171
Electrical properties, 208-212
Electric moisture meters, 139-142, 144
Electron microscopes, 55
Energy, 62
 wood as source of, 426-427
Environmental pollution, effects on wood, 103
Epidermis, 57-58, 468
Epithelium, 30
Equilibrium moisture content of
 bark, 132-134
 fiberboard, 396
 particleboard, 380-381
 wood, 132-134
European woods
 geographical source of, 458-461
 identification keys of, 445-451
 properties—uses of, 458-461
Excelsior, 366, 383, 384, 425
Extractives, 35, 36, 428, 471

Fiber, 30, 417
 bark, 469-470, 471
 gelatinous, 93-94
 libriform, 30
 other plant, 400
 septate, 30
 synthetic, 400, 426
 tracheid, 30
Fiberboard, 388-398
 panel dimensions of, 393, 397
 production methods of, 389-393
 properties of, 393-396
 tempering of, 393
 types of, 388-389
Fiber saturation point, 134-135
Fibril, elementary, 37
Figure, 11, 12, 87-88
Finger joint, 353, 356, 357
Fire retardants, 295
Flexibility, coefficient of, 418
Floccosoid, 98

Flooring, 423
 parquet, 423
Fluorescence, 13
Folded chain theory, 38, 55
Foliage, 427-428, 430
Formaldehyde emissions, 384-385
Forming
 fiberboard, 390-391
 paper, 413-414
 particleboard, 369-372
Fringe micellar theory, 55
Frost crack, 102, 194, 203
Fuelwood, ix, x, 427
Fumigants, 308
Fungi
 decay, 217-222, 233
 stain, 215-217
Furniture, 422

Gasification, 426, 427
Genetics (and wood), 76, 83
Glues (see Adhesives)
Grading
 appearance, 258, 259-260
 stress, 188, 193
 structural, 189, 193
Grain, 10-11
 cross, 88
 diagonal, 88
 interlocked, 87-88
 loose, 155
 raised, 155
 silver, 7, 11
 spiral, 85-88
Growth
 abnormalities, 84-103
 apical, 57
 in diameter, 57-64
 primary, 57-58
 ring, 3, 4
 secondary, 58-61
 stresses, 75, 83

Hardness, 12, 171-172, 185-186
Hardboard, 388, 389, 393, 394, 395, 398
Hardwood, 3, 7
Heartwood, 4, 69-70
 black, 214, 215
 brown, 100, 214
 formaton of, 69-70
 frost, 100
 red, 99-100
Heat, specific, 198
Heating value, 200-201, 203
Hemicelluloses, 35, 37-38, 48, 49, 51, 55
Holocellulose, 35
Honeycombing, 154, 155, 282
Hydrolysis, 426, 427
Hygroscopicity, 128-144
 reason for, 49, 130
 reduction of, 154-158

Hyphae, fungal, 216, 217, 220
Hysteresis, 133

Identification, 431–467
  keys, European woods, 445–451
    North American woods, 433–445
    tropical woods, 451–457
  procedure for, 431–432
Ignition, spontaneous, 203
Impreg, 156
Incizing, 296, 308, 326
Initials, 59, 61, 68, 82
  fusiform, 59, 61
  ray, 59, 61
Inorganic components, 34, 36, 54
Insects, 222–226
  biological cycle of, 223, 224, 233
  classes of, 224
  control of, 282
Intercellular layer, 15
Intercellular space, 31
Isohygric curves, 134, 137

Kerf, 244–245
Kino vein, 101, 102
Knots, 103–104
  effects of, 104, 107, 178–180
  types of, 103

Laminated wood, 351–360
  advantages of, 351, 354
  production technique for, 356–359
  structures from, 352–353, 359
  uses of, 351, 352–353, 354
Laser cutting, 262
Latewood, 3, 4
Lignification, 63, 65
Lignin, 35–36, 38, 56
  in bark, 471
  in compression–tension wood, 92, 94
  utilization, 426
Liquification, 427
Lumber, 239–263
  dimensions of, 253, 254, 262
  grading of, 258, 259–260, 262–263
  recovery factor, 256
  strip core, 339–340
  yield, 253–254, 256, 257
Luster, 10

Maceration, 32
Macrofibril, 37
Magnetic properties, 211
Margo, 16
Marine borers, 227–229
Matches, 424–425
MDF, 388, 393, 395, 397
Measuring units, 55, 479
Mechanical properties, 160–193
  of bark, 472
  bending strength, 170
  cleavage, 170
  compression strength, 169
  determination of, 181–188
  elasticity, 161, 171
  factors affecting, 172–181
  of fiberboard, 395–396
  hardness, 171–172
  of particleboard, 378–380
  of plywood, 346
  shear strength, 169
  tension strength, 162, 169
  toughness, 170–171
  values of, 163–168
Meristem, 57
Micellae, 36
Microfibrils, 36–38
  biosynthesis of, 50–51
  dimensions of, 36, 38
  orientation of, 38–42, 55
  structure of, 37–38
Microscopes
  electron, 55
  scanning (SEM), 55
  transmission, 55
Middle lamella, 15, 48
Mine timbers, 237, 424
Mineral streak, 98
Modulus of
  elasticity, 161–162, 171, 183–184
  rupture, 170, 184
Moisture content
  of cell wall, 130
  determination of, 139–142
  equilibrium, 133–134
  importance of, 142
  isotherms, 35
  in living trees
    bark, 471
    wood, 128
  maximum, 132
  meters, 139–142, 144
Moisture quotient, 281
Molding fungi, 216–217
Molding (shaping), 421
Moon-ring, 98–99
Multinet growth theory, 50
Musical instruments, 204, 207, 425

Naval stores, 430

Odor, 10
OSB (particleboard), 381, 382

Paper, 399–418 (*see also* Pulp)
  making, 401–415
  products, 418
  properties of, 416
  raw materials for, 399–401
  used, 418

Papyrus, 417
Paracrystalline regions, 37
Parenchyma, 14, 28-30
    axial, 28-29
    in bark, 468, 470
    ray, 29
    types of, 29, 33
Particleboard, 361-387
    adhesives for, 367
    dimensional stability of, 377-378, 387
    extruded, 361, 363, 370, 373, 375, 376, 377
    mineral-bonded, 382-384
    panel dimensions of, 361, 387
    particles for, 362-366
    production data for, 362
    production methods for, 362-375
    properties of, 375-381
    raw materials for, 361-362
    types of, 361, 363, 370, 375, 381-384
Pectins, 35
Pentachlorophenol, 295
Perforation plate, 25, 26, 42
    multiple, 25
    scalariform, 16, 25, 26
    simple, 16, 25, 26
Periderm, 58, 59, 468, 469, 478
Phelloderm, 468
Phellogen, 65, 468
Phloem, 58, 65, 468
Piezoelectric effect, 211
Piling, 237
Pistol butt, 84, 85, 96
Pit, 15-16, 18, 22
    aspiration, 18, 44, 45, 304
    blind, 16
    bordered, 15
    cross-field, 18, 21, 22
    definition of, 15
    field, primary, 62
    formation of, 62-63
    membrane, 43-45
    semi-bordered, 16
    simple, 15
    ultrastructure of, 42-46
    vestured, 18, 45
Pith, 3, 4, 58, 59, 104
    double, 90
Pith fleck, 77, 101
Plasticizing, 350, 398, 422
Plastics, xi, xii, 172, 426
Plywood, 339-350
    adhesives for, 340-342, 349-350
    curved (molded), 347, 348
    lumber core for, 339-340
    panel dimensions of, 350
    production method for, 339-344
    properties of, 344-347
    types of, 347, 349
Poles, 237

Polyethylene glycol, 156
Polymerization, degree, 34, 54
Posts, 237
Preservative treatment, 293-308
    effectiveness of, 302-303
    effects of, 305-306
    factors affecting, 303-305
    methods for, 296-302, 308
    preparation of wood for, 296
    specifications of, 303
Preservatives
    oils, 295
    organic solvent, 295
    water-borne, 294-295
Pressure bar, 313, 314, 315, 320
Procambium, 57, 58
Prosenchyma, 20, 22
Protoderm, 57
Pulp, 403-413, 425-426
    alkaline, 406-408
    chemical, 404-408
    chemiground, 409
    chemimechanical, 409
    Kraft, 406-408, 417, 418
    mechanical, 403-404
    properties of, 416
    refiner mechanical, 404
    semichemical, 408-409
    soda, 406
    sulfite, 405-406
    thermomechanical, 404
Pyrolysis, 200, 203, 426, 427

Railroad ties, 424
Rays, 4, 5, 7, 29
    in bark, 469, 470
    in branch—root wood, 77, 78
    formation of, 59, 61
    types of, 29
Ray crossing, 32 (*see also* Cross-field)
Residues, xi, xii, 262
Resin canals, 5, 30, 31
    axial, 30
    in bark, 470
    cortical, 58
    formation of, 64
    radial, 30
    traumatic, 30, 100-101
Resin
    pine, 428, 430
    pocket, 98
    synthetic, 333-335
Resistance
    electrical, 208-210
    to fungi, 218, 219
    to insects, 218
    to marine borers, 218
Resolving power, 55
Ring-porous, 3-4, 6

Rings
  annual, 3
  in bark, 470
  discontinuous, 89–90
  false, 88–89
  growth, 3, 4
  indented, 90
  indistinct, 3, 6
  in tropical woods, 3, 6, 451–457
Ripple mark, 29
Root wood, 77–79
Rosin, 417–418, 428
Rot (*see also* Decay)
  brown, 217
  soft, 217
  white, 217
Roundwood products, 237–238

Saccharification, 426
Sandwich construction, 348
Sapwood, 4, 5
  included, 98–99
Saws
  band, 239, 240, 241, 245
  blades of, 242–245
  circular, 239, 240, 241, 242, 245
  frame, 239, 240, 241, 245
  gang, 261
  sash gang, 261
Sawing
  machines, 239–242, 243, 250
  methods, 247–250, 252–253
  plan, 248–250
Sawmills, 246, 249, 250, 251, 252
Sclereid, 470
Semi-diffuse porous, 4
Semi-ring porous, 4
Shakes, 97–98
Sieve
  areas, 468, 469
  cells, 468
  tubes, 470
Shrinkage, 145–159
  anisotropy of, 146, 147, 149–151
  of bark, 472
  of compression–tension wood, 92, 94, 150
  control of, 155–158, 159
  determination of, 151–152
  factors, 145–147, 149
  of fiberboard, 394–395
  importance of, 152–155
  of old, waterlogged wood, 147
  of particleboard, 377–378
  of plywood, 344–346
  reasons for, 149–151
Softwood, 3, 7
Spaces
  intercellular, 31, 91, 92
  intermacrofibrillar, 37
  intermicrofibrillar, 37

Specific gravity, 111 (*see also* Density)
Spiral thickenings, 22, 24, 42
Springback of
  fiberboard, 395
  laminated wood, 359
  particleboard, 380
Steaming, 288–290, 292
Steel, xi, xii, 172
Stickers, 267, 291
Stone
  cell, 470
  pocket, 101
Stress
  allowable, 186–187
  basic, 193
  grading, 188
Swelling, 145–159 (*see also* Shrinkage)

Tall oil, 406, 417
Tannins, 36, 428–429
Taper, 10
Tension wood, 90–91, 93–96, 107, 181
Termites, 225–226
Texture, 10
Thermal properties, 194–203
Torus, 16, 42–45, 55
Trabeculae, 23, 24, 33
Tracheids, 14, 22–25
  types of, 22–25, 27
Tropical woods
  geographical source of, 461–465
  identification keys of, 451–454
  properties—uses of, 461–465
Turning, 421
Turpentine, 417–418, 428
Tylosis, 26–27, 33, 47, 55
Tylosoid, 30

Veneer, 309–326
  decorative, 309–310
  drying of, 320–324
  production methods of, 312–320, 326
  utility, 309–310
  wood species for, 310
  yield, 324–325
Vessel, 25
  member, 14, 16, 25–26
Volume-meter, mercury, 121, 122

Waferboard, 381–382
Wall fraction, 418
Warping, 153, 282
Warty layer, 46–47, 63
Weight, 12
  density, 111
Wettability, 329, 330
Wetwood, 213–215, 233
Wood
  adult, 71

SUBJECT INDEX 489

advantages of, ix
branch, 77
compression—tension, 90-96
consumption—production data, x, xi, xiii
core, 70
crown-formed, 70
disadvantages, ix, x
early-, 3, 4
flour, 425
formation mechanism, 57-65
juvenile, 70-72, 175

late-, 3, 4, 82
mature, 71
modified, 159
opposite, 96-97
overmature, 71
protection, 100
reaction, 91
root, 77-79
waterlogged, 127, 233

Xylem, 58, 65

Species Index*

Abachi (*see* Opepe)
Acajou (African Mahogany, Khaya)—*Khaya ivorensis*, 115, 165, 168, 218, 454, 457, 464
Afara—*Terminalia superba*, 114, 165, 168, 218, 452, 455, 462
Afzelia—*Afzelia africana*, 114, 165, 168, 452, 455, 462
Aiélé—*Canarium schweinfurthii*, 115, 165, 168, 454, 457, 464
Ako (*see* Antiaris)
Alder—*Alnus*, 443
Alder, black—*Alnus glutinosa*, 23, 114, 164, 167, 450, 460
Alder, European (*see* Alder, black)
Alder, red—*Alnus rubra*, 23, 114, 163, 166, 443
Amazakoué—*Guibourtia ehie*, 115, 165, 168, 452, 457, 461
Arborvitae, eastern (*see* Cedar, northern white)
Arborvitae, giant (*see* Cedar, western red)
Ash—*Fraxinus*, 440, 459
Ash, black—*Fraxinus nigra*, 441
Ash, brown (*see* Ash, black)
Ash, European—*Fraxinus excelsior*, 449
Ash, European mountain (*see* Service tree)
Ash, flowering—*Fraxinus ornus*, 23, 449
Ash, green—*Fraxinus pennsylvanica*, 441
Ash, manna (*see* Ash, flowering)
Ash, Oregon—*Fraxinus latifolia*, 441
Ash, white—*Fraxinus americana*, 23, 114, 163, 166, 441
Aspen, bigtooth—*Populus grandidentata*, 445
Aspen, European—*Populus tremula*, 164, 167, 451, 461
Aspen, quaking—*Populus tremuloides*, 23, 114, 163, 166, 445

Baldcypress—*Taxodium distichum*, 23, 114, 163, 166, 436, 438
Balsa—*Ochroma lagopus*, 23, 115, 165, 168, 454, 457, 464
Bankia, 227
Basswood—*Tilia*, 444

Basswood, American—*Tilia americana*, 114, 163, 166, 444
Basswood, European—*Tilia* spp., 114, 164, 167, 461 (*see also* Limetree)
Basswood, white—*Tilia heterophylla*, 444
Beech—*Fagus*, 442
Beech, American—*Fagus grandifolia*, 442
Beech, blue (*see* Hornbeam, American)
Beech, European—*Fagus sylvatica*, 23, 114, 164, 167, 449, 460
Beté—*Mansonia altissima*, 23, 115, 165, 168, 453, 456, 464
Birch—*Betula*, 444
Birch, black—*Betula lenta*, 444
Birch, English (*see* Birch, European)
Birch, European—*Betula verrucosa* (*B. pendula*), 23, 118, 164, 167, 450, 460-461
Birch, gray—*Betula populifolia*, 444
Birch, paper—*Betula papyrifera*, 218, 444
Birch, red—*Betula nigra*, 444
Birch, river (*see* Birch, red)
Birch, Swedish (*see* Birch, European)
Birch, sweet (*see* Birch, black)
Birch, yellow—*Betula alleghaniensis*, 23, 114, 163, 166, 444
Black locust—*Robinia pseudoacacia*, 114, 163, 166, 440, 441, 449, 459
Bubinga—*Guibourtia tessmannii*, 115, 165, 168, 452, 455, 462
Buckeye, Ohio—*Aesculus octandra*, 445
Buckeye, yellow—*Aesculus glabra*, 445
Butternut—*Juglans cinerea*, 442

Carpenter ants—*Camponotus* spp., 225
Canarium (*see* Aiélé)
Carobtree—*Ceratonia siliqua*, 450
Catalpa, northern—*Catalpa speciosa*, 441
Catalpa, southern—*Catalpa bignonioides*, 441
Cedar, Alaska—*Chamaecyparis nootkatensis*, 436, 437
Cedar, Atlantic white—*Chamaecyparis thyoides*, 436, 438
Cedar, eastern red—*Juniperus virginiana*, 436, 437
Cedar, incense—*Libocedrus decurrens*, 436, 437

*Main citations; equivalent common and scientific (Latin) names; extended information on pages **433–465**.

492 SPECIES INDEX

Cedar, northern white—*Thuja occidentalis*, 436, 437
Cedar, pencil (*see* Cedar, incense)
Cedar, port orford—*Chamaecyparis lawsoniana*, 436, 438
Cedar, western red—*Thuja plicata*, 436, 437
Cherry (black)—*Prunus serotina*, 114, 163, 166, 443
Chestnut, American—*Castanea dentata*, 439
Chestnut, European—*Castanea vesca* (*C. sativa*), 23, 114, 448, 459
Common furniture beetle—*Anobium punctatum*, 224
Cottonwood, black,—*Populus trichocarpa*, 445
Cottonwood, eastern—*Populus deltoides*, 114, 163, 166, 445
Cottonwood, swamp—*Populus heterophylla*, 445

Death watch beetle—*Xestobium rufovillosum*, 223, 224
Dibetou—*Lovoa trichilioides*, 115, 165, 168, 218, 454, 457, 465
Difou—*Morus mesozygia*, 115, 165, 168, 453, 456, 463
Douglas-fir—*Pseudotsuga menziesii*, 23, 114, 163, 166, 218, 433, 434, 436, 458

Ehie (*see* Amazakoué)
Elm—*Ulmus*, 440
Elm, American—*Ulmus americana*, 114, 163, 166, 440
Elm, cedar—*Ulmus crassifolia*, 440
Elm, English (*see* Elm, field)
Elm, field—*Ulmus campestris* (*U. procera*), 114, 164, 167, 448
Elm, hard, 440
Elm, mountain—*Ulmus montana* (*U. glabra*), 23, 448
Elm, red—*Ulmus rubra*, 440
Elm, rock—*Ulmus thomasii*, 440
Elm, slippery (*see* Elm, red)
Elm, winged—*Ulmus alata*, 440
Elm, wych (*see* Elm, mountain)

Fir—*Abies*, 435
Fir, balsam—*Abies balsamea*, 23, 435, 437
Fir, California red—*Abies magnifica*, 435, 437
Fir, Fraser—*Abies fraseri*, 435, 437
Fir, grand—*Abies grandis*, 435, 437
Fir, Grecian—*Abies cephalonica*, 23, 446, 447, 458
Fir, hybrid—*Abies alba* × *A. cephalonica*, 458
Fir, noble—*Abies procera*, 435, 437
Fir, Pacific silver—*Abies amabilis*, 435, 437
Fir, silver European (*see* Fir, white European)
Fir, subalpine—*Abies lasiocarpa*, 435, 437
Fir, white (American)—*Abies concolor*, 114, 163, 166, 435, 437
Fir, white (European)—*Abies alba* (*A. pectinata*), 23, 114, 164, 167, 218, 446, 447, 458
Fraké (*see* Afara)
Framire—*Terminalia ivorensis*, 218, 453, 455, 463

Gaboon (*see* Okoumé)
Gedu Nohor (*see* Tiama)
Gum, blue southern—*Eucalyptus globulus*, 115
Gum, manna—*Eucalyptus viminalis*, 115
Gum, red river—*Eucalyptus camaldulensis*, 115

Hackberry—*Celtis australis*, 440, 449
Hazel, European—*Corylus avellana*, 450
Hazel, Mediterranean—*Corylus colurna*, 450
Hemlock—*Tsuga*, 435
Hemlock, Canada (*see* Hemlock, eastern)
Hemlock, eastern—*Tsuga canadensis*, 436, 437
Hemlock, mountain—*Tsuga martensiana*, 436, 437
Hemlock, Pacific (*see* Hemlock, western)
Hemlock, western—*Tsuga heterophylla*, 23, 114, 163, 166, 436, 437
Hickory—*Carya*, 441
Hickory, bitternut—*Carya cordiformis*, 442
Hickory, mockernut—*Carya tomentosa*, 114, 163, 166, 441
Hickory, nutmeg—*Carya myristicaeformis*, 442
Hickory, pecan—*Carya illinoensis*, 23, 441
Hickory, pignut—*Carya glabra*, 441
Hickory, red—*Carya ovalis*, 441
Hickory, shagbark—*Carya ovata*, 441
Hickory, shellbark—*Carya laciniosa*, 441
Hickory, water—*Carya aquatica*, 442
Honeylocust—*Gleditsia triacanthos*, 441
Hophornbeam (American)—*Ostrya virginiana*, 444
Hophornbeam (European)—*Ostrya carpinifolia*, 451, 461
Hornbeam, American—*Carpinus caroliniana*, 443
Hornbeam, eastern—*Carpinus orientalis* (*C. duinensis*), 450
Hornbeam, European—*Carpinus betulus*, 114, 164, 167, 443, 450
Horse chestnut, common—*Aesculus hippocastanum*, 451, 461
House longhorn beetle—*Hylotrupes bajulus*, 225

Idigbo (*see* Framire)
Iroko—*Chlorophora excelsa*, 23, 115, 165, 168, 218, 452, 455, 462

Jarrah—*Eucalyptus marginata*, 115
Juniper, common—*Juniperus communis*, 446, 447
Juniper, Grecian—*Juniperus excelsa*, 446
Juniper, Phoenicean—*Juniperus phoenicea*, 446
Juniper, Syrian—*Juniperus drupacea*, 446, 447

Kambala (*see* Iroko)
Khaya (*see* Acajou)
Korina (*see* Afara)
Kosipo—*Entandrophragma candolei*, 115, 165, 168, 218, 453, 456, 463

Larch—*Larix*, 433
Larch, eastern—*Larix laricina*, 434, 437
Larch, European—*Larix europaea* (*L. decidua*), 23, 114, 446
Larch, western—*Larix occidentalis*, 114, 163, 166, 434, 437
Lauan, dark red (*see* Meranti, dark red)
Lauan, white (*see* Seraya, white)
Limba (*see* Afara)
Lime—*Tilia*, 451, 461 (*see also* Basswood, European; Limetree)
Limetree, common—*Tilia vulgaris*, 451

Limetree, large-leaved—*Tilia grandifolia* (*T. platyphyllos*), 451
Limetree, silver—*Tilia tomentosa* (*T. argentea*), 451
Limetree, small-leaved—*Tilia parvifolia* (*T. cordata*), 451
Limnoria spp., 227, 228
Linden (*see* Lime)
Lingue (*see* Afzelia)

Mahogany, African (*see* Acajou)
Mahogany, American—*Swietenia macrophylla*, 115, 165, 168, 453, 456, 463
Makoré—*Dumoria* (*Tieghmella*) *heckelii*, 23, 115, 165, 168, 218, 453, 456, 463
Mansonia (*see* Beté)
Maple—*Acer*, 444, 460
Maple, Balkans—*Acer heldreichii*, 450
Maple, bigleaf—*Acer macrophyllum*, 444
Maple, black—*Acer nigrum*, 443
Maple, Bosnian (*see* Maple, Norway)
Maple, field—*Acer campestre*, 23, 218, 450, 460
Maple, great—*Acer pseudoplatanus*, 450, 460
Maple, monspessulanian—*Acer monspessulanum*, 450
Maple, Norway—*Acer platanoides*, 450, 460
Maple, Oregon (*see* Maple, bigleaf)
Maple, Pacific (*see* Maple, bigleaf)
Maple, red—*Acer rubrum*, 444
Maple, silver—*Acer saccharinum*, 444
Maple, sugar—*Acer saccharum*, 114, 163, 166, 443
Martesia, 227, 228
Meranti, dark red—*Shorea pauciflora*, 115, 165, 168, 218, 454, 457, 465
Mozambique (*see* Amazakoué)
Mulberry, black—*Morus nigra*, 449, 459
Mulberry, red—*Morus rubra*, 440
Mulberry, white—*Morus alba*, 449, 459

Niangon—*Tarrieta densiflora*, 115, 165, 168, 218, 454, 457, 464

Oak, black—*Quercus velutina*, 439
Oak, broad-leaved—*Quercus conferta*, 448
Oak, bur—*Quercus macrocarpa*, 439
Oak, chestnut—*Quercus prinus*, 439
Oak, cork—*Quercus suber*, 448, 474
Oak, durmast (*see* Oak, sessile)
Oak, English—*Quercus robur* (*Q. pedunculata*), 23, 114, 164, 167, 448
Oak, Grecian—*Quercus aegilops*, 448
Oak, holm—*Quercus ilex*, 442, 449, 459
Oak, kermes—*Quercus coccifera*, 449, 459
Oak, live—*Quercus virginiana*, 442
Oak, mossy (*see* Oak, Turkey)
Oak, northern red (American)—*Quercus borealis*, 114, 163, 166, 439
Oak, overcup—*Quercus lyrata*, 439
Oak, pin—*Quercus palustris*, 439
Oak, post—*Quercus stellata*, 439
Oak, pubescent—*Quercus pubescens*, 448
Oak, red (European—*see* Oak, Turkey)
Oak, scarlet—*Quercus coccinea*, 439
Oak, sessile—*Quercus petraea* (*Q. sessiliflora*), 448
Oak, shumard—*Quercus shumardii*, 439
Oak, swamp chestnut—*Quercus michauxii*, 439
Oak, Turkey—*Quercus cerris*, 23, 114, 164, 167, 218, 448, 459
Oak, valonian (*see* Oak, Grecian)
Oak, white—*Quercus alba*, 439
Oak, white (American *Quercus* spp.), 114, 163, 166, 439
Oak, white (European *Quercus* spp.), 218 (*see also* Oak, English)
Oak, willow—*Quercus phellos*, 439
Oaks, red*—*Quercus* spp., 439, 459
Oaks, white—*Quercus* spp., 439
Obeche—*Triplochiton scleroxylon*, 115, 165, 168, 453, 456, 464
Odoum (*see* Iroko)
Okoumé—*Aucoumea klaineana*, 23, 115, 165, 168, 218, 454, 457, 464
Olive tree—*Olea europaea*, 451, 461
Omu (*see* Kosipo)
Opepe—*Nauclea trillesii* (*N. diderichii*), 115, 165, 168, 454, 457, 465
Ovangkol (*see* Amazakoué)

Padauk—*Pterocarpus soyauxii*, 115, 165, 168, 218, 452, 455, 462
Palissander—*Dalbergia latifolia*, 23, 165, 168, 218, 452, 454, 462
Parana pine—*Araucaria angustifolia*, 23
Persimmon—*Diospyros virginiana*, 442
Pine, Aleppo—*Pinus halepensis*, 23, 447, 448, 458
Pine, Austrian (*see* Pine, black)
Pine, Balkan (*see* Pine, Macedonian)
Pine, black—*Pinus nigra* (*P. laricio*), 23, 114, 164, 167, 218, 447, 448, 458
Pine, Calabrian—*Pinus brutia*, 447, 448, 458
Pine, hard (*see* Pine, Calabrian)
Pine, jack—*Pinus banksiana*, 435, 438
Pine, loblolly—*Pinus taeda*, 23, 114, 163, 166, 434, 438
Pine, lodgepole—*Pinus contorta*, 435, 438
Pine, longleaf—*Pinus palustris*, 433, 434, 438
Pine, Macedonian—*Pinus peuce*, 447, 458
Pine, maritime—*Pinus maritima* (*P. pinaster*), 218, 447, 448, 458
Pine, Monterey—*Pinus radiata*, 458
Pine, parana (*see* Parana pine)
Pine, parasol (*see* Pine, umbrella)
Pine, pitch—*Pinus rigida*, 218, 434, 438
Pine, ponderosa—*Pinus ponderosa*, 23, 114, 163, 166, 435, 438
Pine, red—*Pinus resinosa*, 435, 438
Pine, Scots—*Pinus silvestris*, 23, 433, 434, 438, 458
Pine, shortleaf—*Pinus echinata*, 434, 438

*The distinction of "red" and "white" oaks is made in N. America (*see* p. 439). In Europe, Turkey oak, cork oak, and some other minor species (e.g., *Q. macedonica* = *Q. trojana*) are "red" with regard to wood structures.

SPECIES INDEX

Pine, slash—*Pinus elliottii*, 434, 438
Pine, stone (*see* Pine, umbrella)
Pine, sugar—*Pinus lambertiana*, 434, 438
Pine, umbrella—*Pinus pinea*, 23, 447, 448, 458
Pine, whitebark—*Pinus leucodermis* (*P. heldreichii*), 447, 448, 458
Pine, white eastern—*Pinus strobus*, 23, 114, 163, 166, 433, 434, 438
Pine, white western—*Pinus monticola*, 434, 438
Pines, hard—*Pinus* spp., 434
Pines, soft—*Pinus* spp., 434
Pines, southern—*Pinus* spp., 433, 434, 438
Plane (oriental)—*Platanus orientalis*, 23, 114, 164, 167, 218, 443, 450, 460
Poplar, balsam—*Populus balsamifera*, 445
Poplar, black—*Populus nigra*, 451, 461
Poplar, eastern (*see* Cottonwood, eastern)
Poplar, hybrid—*Populus* × *euramericana*, 23, 114, 164, 167, 218, 461
Poplar, swamp (*see* Cottonwood, swamp)
Poplar, white—*Populus alba*, 451, 461
Powder post beetles—*Lyctus* spp., 224

Ramin—*Gonystylus bancanus*, 115, 165, 168, 218, 452, 455, 462
Redcedar, eastern—*Juniperus virginiana*, 436, 437
Redcedar, western—*Thuja plicata*, 23, 114, 436, 437
Redwood—*Sequoia sempervirens*, 23, 114, 163, 166, 436, 438
Rosewood, Indian (*see* Palissander)
Rowantree—*Sorbus aucuparia*, 450

Samba (*see* Obeche)
Sapele—*Entandrophragma cylindricum*, 23, 115, 165, 168, 218, 453, 456, 463
Sassafras—*Sassafras albidum*, 441
Seraya, white—*Parashorea plicata*, 115, 165, 168, 454, 457, 465
Servive tree—*Sorbus domestica*, 450, 460
Service tree, wild—*Sorbus torminalis*, 450
Shisham (*see* Palissander)
Sipo—*Entandrophrama utile*, 23, 115, 165, 168, 218, 453, 456, 463
Sphaeroma, 227
Spruce—*Picea*, 433
Spruce, black—*Picea mariana*, 434, 437
Spruce, Engelmann—*Picea engelmannii*, 23, 114, 163, 166, 434, 437

Spruce, European—*Picea abies* (*P. excelsa*), 23, 114, 164, 167, 446, 458
Spruce, Norway (*see* Spruce, European)
Spruce, red—*Picea rubens*, 434, 437
Spruce, Sitka—*Picea sitchensis*, 434, 437
Spruce, white—*Picea glauca*, 440
Sugarberry—*Celtis laevigata*, 440
Sweetgum—*Liquidambar styraciflua*, 23, 114, 163, 166, 444
Sycamore (*see* Plane, oriental)
Sycamore, American—*Platanus occidentalis*, 114, 163, 166, 443
Sycamore, plane (*see* Maple, great)

Tamarack (*see* Larch, eastern)
Teak—*Tectona grandis*, 23, 115, 165, 168, 218, 451, 454, 461
Teredo spp., 227-228
Tiama—*Entandrophragma angolense*, 23, 115, 165, 168, 218, 453, 456, 463
Tree-of-heaven—*Ailanthus altissima* (*A. glandulosa*), 441, 449, 460
Tupelo, black—*Nyssa sylvatica*, 445
Tupelo, water—*Nyssa aquatica*, 445

Utile (*see* Sipo)

Walnut, African (*see* Dibetou)
Walnut, black—*Juglans cinerea*, 114, 163, 166, 442
Walnut, European—*Juglans regia*, 114, 164, 167, 449, 459
Whitebeam tree—*Sorbus aria*, 450
Willow, black—*Salix nigra*, 445
Willow, crack—*Salix fragilis*, 451
Willow (European species)—*Salix* spp., 23, 114, 164, 167, 218, 461
Willow, white—*Salix alba*, 451
Wood wasp—*Urocerus* (*Sirex*) *gigas*, 225

Yellow poplar—*Liriodendron tulipifera*, 23, 114, 163, 166, 443
Yew, European—*Taxus baccata*, 23, 446, 459
Yew, Pacific—*Taxus brevifolia*, 437

Zebrano (Zebrawood, Zingana)—*Microberlinia brazzavillensis*, 115, 165, 168, 452, 454, 461